Iron and Steel in the Nineteenth Century

Encyclopedia of American Business History and Biography

Iron and Steel in the Nineteenth Century

Edited by

Paul F. Paskoff
Louisiana State University

A Bruccoli Clark Layman Book

Facts On File
New York • Oxford

Encyclopedia of American Business History
and Biography:
Iron and Steel in the Nineteenth Century

Copyright ©1989 by Bruccoli Clark Layman, Inc., and
Facts on File, Inc.

ISBN 0-8160-1890-1

Full CIP information available on request
British CIP information available on request

Designed by Quentin Fiore

Printed in the United States of America

10 9 8 7 6 5 4 3 2 1

To the memory of Jeanne McHugh, historian

Encyclopedia of American Business History

Railroads in the Age of Regulation, 1900-1980, edited by Keith Bryant (1988)

Railroads in the Nineteenth Century, edited by Robert L. Frey (1988)

Iron and Steel in the Nineteenth Century, edited by Paul F. Paskoff (1989)

Contents

Foreword

The Encyclopedia of American Business History and Biography chronicles America's material civilization through its business figures and businesses. It is a record of American aspirations—of success and of failure. It is a history of the impact of business on American life. The volumes have been planned to serve a cross section of users: students, teachers, scholars, researchers, and government and corporate officials. Individual volumes or groups of volumes cover a particular industry during a defined period; thus each *EABH&B* volume is freestanding, providing a history expressed through biographies and buttressed by a wide range of supporting entries. In many cases a single volume is sufficient to treat an industry, but certain industries require two or more volumes. When completed, the *EABH&B* will provide the fullest available history of American enterprise.

The editorial direction of *EABH&B* is provided by the general editor and the editorial board. The general editor appoints volume editors whose duties are to prepare, in consultation with the editorial board, the list of entries for each volume, to assign the entries to contributors, to vet the submitted entries, and to work in close cooperation with the Bruccoli Clark Layman editorial staff so as to maintain consistency of treatment. All entries are written by specialists in their fields, not by staff writers. Volume editors are experienced scholars.

The publishers and editors of *EABH&B* are convinced that timing is crucial to notable careers. Therefore, the biographical entries in each volume of the series place businesses and their leaders in the social, political, and economic contexts of their times. Supplementary background rubrics on companies, inventions, legal decisions, marketing innovations, and other topics are integrated with the biographical entries in alphabetical order.

The general editor and the volume editors determine the space to be allotted to biographies as major entries, standard entries, and short entries. Major entries, reserved for giants of business and industry (e.g., Henry Ford, J. P. Morgan, Andrew Carnegie, James J. Hill), require approximately 10,000 words. Standard biographical entries are in the range of 3,500-5,000 words. Short entries are reserved for lesser figures who require inclusion and for significant figures about whom little information is available. When appropriate, the biographical entries stress their subjects' roles in shaping the national experience, showing how their activities influenced the way Americans lived. Unattractive or damaging aspects of character and conduct are not suppressed. All biographical entries conform to a basic format.

A significant part of each volume is devoted to concise background entries supporting and elucidating the biographies. These nonbiographical entries provide basic information about the industry or field covered in the volume. Histories of companies are necessarily brief and limited to key events. To establish a context for all entries, each volume includes an overview of the industry treated. These historical introductions are normally written by the volume editors.

We have set for ourselves large tasks and important goals. We aspire to provide a body of work that will help reduce the imbalance in the writing of American history, the study of which too often slights business. Our hope is also to stimulate interest in business leaders, enterprises, and industries that have not been given the scholarly attention they deserve. By setting high standards for accuracy, balanced treatment, original research, and clear writing, we have tried to ensure that these works will commend themselves to those who seek a full account of the development of America.

—William H. Becker
General Editor

Acknowledgments

This book was produced by Bruccoli Clark Layman, Inc. James W. Hipp was the in-house editor.

Production coordinator is Kimberly Casey. Copyediting supervisor is Joan M. Prince. Typesetting supervisor is Kathleen M. Flanagan. The production staff includes Brandy H. Barefoot, Rowena Betts, Charles D. Brower, Amanda Caulley, Teresa Chaney, Patricia Coate, Mary Colborn, Mary S. Dye, Sarah A. Estes, Cynthia Hallman, Judith K. Ingle, Warren McInnis, Kathy S. Merlette, Sheri Beckett Neal, and Virginia Smith. Jean W. Ross is permissions editor. Susan Todd is the photography editor. Joseph Matthew Bruccoli and Penney L. Haughton did photographic copy work and paste-up for the volume.

Walter W. Ross and Jennifer Toth did the library research with the assistance of the reference staff at the Thomas Cooper Library of the University of South Carolina: Daniel Boice, Cathy Eckman, Gary Geer, Cathie Gottlieb, David L. Haggard, Jens Holley, Dennis Isbell, Jackie Kinder, Marcia Martin, Jean Rhyne, Beverly Steele, Ellen Tillett, Carol Tobin, and Virginia Weathers.

Acknowledgments

Introduction

The Nineteenth Century Iron and Steel Industry

The nineteenth-century American iron and steel industry had its origins in the seventeenth century when the Saugus ironworks in newly settled Massachusetts went into operation in 1645. By the beginning of the eighteenth century iron making was under way in almost every other colony from Connecticut and Rhode Island in the north to Maryland and Virginia in the south. At midcentury virtually every colony in British North America supported some measure of iron production, a development which had not gone unnoticed by English ironmasters, who looked with considerable concern upon the growing potential and competitive power of colonial iron producers. Particularly alarming to English manufacturers was the growing American interest in more advanced iron and steel products, especially rolled iron and hardware, and the erection of slitting and nail mills in the colonies.

As a result of the English producers' concern and their not inconsiderable political influence, Parliament passed the Iron Act in 1750 with the explicit aim of inhibiting further development of the colonial iron industry. The act of 1750 was, however, an exercise in futility, as much because of administrative indifference as colonial noncompliance, and the number of iron mills and other specialized facilities increased rather than decreased over the next quarter century.

Although the American Revolution freed American iron producers from the nuisance of parliamentary control, however ineffectual, and offered new business opportunities, it presented grave risks as well. Demand for muskets, cannon, wagon tires, and other war matériel stimulated iron production but probably not enough to offset the loss of output caused by the destruction of ironworks and the disruption of markets. Rising costs of ore, pig iron, fuel, and labor depressed profits and business confidence, forestalling new ventures and bankrupting established firms.

The end of fighting in 1782 and the signing of a peace treaty the next year returned a measure of sta-bility to a shaken domestic economy. The business climate during the so-called "Critical Period," from 1781 to 1789, under the Articles of Confederation was substantially improved over that of the war years, and iron producers in the postwar United States encountered few problems which they had not handled as routine matters of business before the war. Disputes over currency and debts between producers or merchants in one state and those in another were all too familiar to those who had been in business before the Revolution. Iron producers knew, as did other businessmen, that just as profits had been possible before independence, they were attainable with it. Still, the unsettled political environment, especially after 1785, and a postwar recession led many producers to support the movement for a new constitution.

Whether the dramatic improvement in business conditions after 1789 was only coincidental with the ratification of the Constitution or was a consequence of it is largely a matter for conjecture. In either case the improvement was unambiguous and encouraging. In Pennsylvania, for example, the 70 new iron firms established during the 12 years from 1789 through 1800 were five times the number begun during the 6-year period from 1783 through 1788. The increase in the number of iron firms resulted in the growth of the industry's total output but did not result in technological development or in significant geographical expansion of the industry across the Allegheny Mountains.

The American iron industry in 1800 resembled that of 1750 more than that of 1850 or even 1830. Its technology, geographical distribution, and forms of ownership had remained more or less unchanged despite the momentous events of the preceding quarter century. Production of pig iron, castings, and bar or wrought iron still entailed the use of charcoal fuel, ore taken from local bogs or shallow mines, waterpower, fairly small numbers of workers, and modest amounts of capital. With only isolated exceptions such production was confined to the seaboard states, especially Pennsylvania, New

Jersey, New York, and Massachusetts in the north and Maryland and Virginia in the south. Ownership and control of the furnaces and forges and the small number of more specialized facilities—nailworks, wireworks, and rolling mills—rested in 1800, as in 1750, primarily with individuals and small family partnerships.

Capital requirements for new furnaces and forges in the first years of the new century were well within the reach of individual entrepreneurs who, like Joseph McClurg of Pittsburgh, had managed to save the necessary several hundred to few thousand dollars as a wholesale or retail merchant. A number of the early iron producers did, in fact, start their business careers as merchants or, as in the case of David Porter of Pennsylvania, clerks or managers in someone else's ironworks. Such beginnings provided future ironmasters with more than capital; they were also their practical schools of business.

Because furnaces and large forges required a blast of air to raise the temperature of the burning charcoal fuel, almost all ironworks of the early nineteenth century, like those throughout the preceding century, relied on waterpower. Consequently they were located along swift-moving streams or, where these were lacking, close to a mill pond with falls sufficient to provide the motive force necessary to turn a waterwheel which in turn powered the bellows which generated the blast. This dependency on waterpower was not the only consideration regarding the location of iron facilities. The high cost of transportation required that furnaces be located close to iron ore deposits and timber stands and that forges be rather close to furnaces. As a result most early iron installations were located along the eastern seaboard. This geographical constraint continued to influence the nineteenth-century iron industry until about 1840 when three distinct innovations, the canal system, the railroad, and the anthracite iron furnace combined to liberate iron production from the absolute tyranny of geography.

The high relative costs—as much as 5 to 6 percent of the total production cost for a ton of pig iron—incurred in the transportation of the raw materials for iron making had encouraged the concentration of producers near ore and wood sites. As consumption of known ore deposits and wood stocks increased during the latter part of the eighteenth century and, especially, the first third of the nineteenth century, producers and their hinterland economic and political allies began to press for internal improvements, initially in the form of better roads and, later, the construction of canals. Improved roads made little difference to manufacturers concerned with reducing the cost of transporting raw materials since the problem was not really the quality of the roads so much as it was the limited hauling capacity—under two tons—of a freight wagon and its team of horses. Better roads mattered far more in moving goods from producer to market where high hauling costs cut into profits and put producers who were more distant from their markets at a decided disadvantage vis-à-vis their more closely situated competitors.

Iron making generally exhausted supplies of readily available timber more quickly than conveniently located ore deposits. This fact made for the two most pressing problems in the economic geography of iron production by the 1830s. The first of these was to find a cost-effective means by which to bring large quantities of finished iron to the major urban markets from ever more remote production areas. These areas, such as the Juniata Valley in central Pennsylvania, were endowed with previously untapped deposits of high-grade ore on comparatively inexpensive land and promised to relieve the pressure on the more proximate and intensively utilized deposits. The second problem was to provide these new production centers and many of the older sites with sufficient fuel to enable them to continue their iron making without interruption. The beginnings in 1820 of the exploitation of the extensive anthracite coalfields in the Schuylkill and Lehigh valleys of Pennsylvania and the construction of ambitious canals to tap these fields offered the prospect of a practically inexhaustible supply of fuel and a comparatively inexpensive means of transporting it to the production site.

Anthracite coal was not entirely untried by iron makers before its introduction as a primary fuel for pig iron smelting in 1839. Called "stone coal" by colonial producers, various experiments in its use had been conducted, primarily by blacksmiths, on a small scale before and during the Revolution. Joshua Malin of Philadelphia made more ambitious forays into the application of anthracite coal in 1813, during the War of 1812, when he achieved the distinction of making the first large quantity of pig iron castings with the fuel by melting pig iron in a furnace. His achievement spurred other producers to try the mineral fuel and, more significantly, to press for the improvement of southeastern Pennsylvania's Schuylkill River, the main route into the anthracite coalfields. The upshot of these efforts was the incorporation in 1815 of the Schuylkill Navigation Company. Five years later competition for sources of anthracite coal led to the formation of the Lehigh Navigation Company to improve navigation along the Lehigh River. Shortly before or after 1830 (the uncertainty arises from conflicting claims of priority of achievement), Pennsylvania ironmasters made the first wrought iron by the puddling process using anthracite coal. Undoubtedly, the high price of charcoal compared to that of anthracite coal–$10 for the 200 bushels of charcoal necessary to produce one ton of pig iron but only $5 for the corresponding two tons of anthracite–encouraged the interest in anthracite coal.

The significance of these early efforts to use anthracite coal as an industrial fuel in place of charcoal lay mainly in the fact that they resulted in the accumulation of familiarity and experience with the coal and prepared iron producers for the two most important next steps: the use of anthracite in furnaces to make pig iron and in rolling mills to make wrought-iron products. In 1831 and again in 1832 the Franklin Institute in Philadelphia sponsored a competition with the aim of inspiring the first successful production of pig iron in a furnace fueled with anthracite. The first person to achieve this feat and make 20 tons of pig iron would receive an Institute medal. The members of the Franklin Institute had reason to be optimistic that success in this undertaking would not be long in coming. Iron ore smelting with mineral fuels had long been standard practice in Great Britain, and interest in accomplishing the same in the United States was high.

The Franklin Institute's confidence was shared by Pennsylvania's commercial interests and also by influential members of the state government, including the Whig governor, Joseph Ritner. In keeping with the temper of his party Ritner was an advocate of internal improvements, such as canals and railroads, and industrial development, including the promotion of anthracite iron making. His political instincts and his enthusiasm for the application of this new technology prompted him to exult in 1838 over the first successful anthracite smelting, more than a year before that feat would be performed in late 1839. Ritner's jumping the gun was not just a case of the wish being father to the deed. Success in the competition to be first to build a working anthra-

cite blast furnace was expected at any moment by those conversant with the technical progress of the iron industry in the United States. The development in England in 1828 of the hot-blast coke furnace, in which the blast of air injected into the furnace hearth was preheated to raise the furnace's operating temperature and production efficiency, was successfully adapted for use in charcoal blast furnaces in this country in 1834.

With the introduction and rapid spread of the hot-blast method in the United States, the way was open to the construction of an anthracite blast furnace. The underlying principle of the hot blast was simple, and its application quickly yielded abundant evidence of its effectiveness. Although charcoal furnace operators who adopted the hot blast, either by building new furnaces or modifying existing cold-blast installations, never achieved the fuel savings reportedly realized by coke furnace owners in Britain, their own measure of success was enough to encourage the rapid spread of the technology. Significantly, new hot-blast furnaces were built to the same size as the older cold-blast models, a consequence of the use of charcoal fuel which, because it could be easily pulverized by its own weight, limited the size of the furnace stack. Users of the hot blast understood that its full potential as a means of reducing a furnace's fuel consumption and raising its operating efficiency would not be realized in a charcoal blast furnace. The best results would come, they knew, when the hot blast was applied to a furnace smelting with anthracite coal.

Anthracite had long been considered an attractive fuel for blast furnace operation, but its adoption was delayed by the fact that it had to be heated to a higher temperature than did charcoal. The hot blast solved this problem by raising the temperature of the furnace's interior and permitting the ignition of the coal. The use of anthracite solved another problem of blast furnace operation by breaking through the limitation on furnace size imposed by the use of charcoal as a fuel. Of course, the higher temperatures inside the anthracite furnace corroded its brick lining more quickly than the charcoal furnace's lining, and the potentially larger size of the anthracite furnace made its construction an expensive proposition. Still the new fuel promised undeniable advantages and, although the first successful anthracite furnace, built by David Thomas in December 1839, had dimensions comparable to those of the larger hot-blast charcoal furnaces, this prototype was followed within a few years by much larger versions. These had daily output capacities three and four times that of a typical cold-blast or hot-blast charcoal furnace.

David Thomas's achievement was unambiguous proof of the practicality and power of the new anthracite technology; the furnace excited considerable interest among iron makers and those outside the industry who followed its progress. Interest in anthracite furnaces, however, was not accompanied by anything like a headlong rush to build them. Although Thomas built three furnaces in 1840, no one else put another one into blast until 1842, when four came on line. The next year was as barren of new construction as the preceding one, and anthracite furnace building languished until 1844, when three furnaces were completed. The next year was better still, when ten new furnaces began to make pig iron.

The fitfulness with which pig iron producers adopted the new technology may have been due in part to their reluctance to try something new, but a more convincing explanation has to do with the trying economic conditions of the early 1840s. A general economic depression that had begun in late 1839 continued with only all too brief respites until the end of 1843. Not surprisingly, people thinking in those years of building an anthracite furnace, which, on average, entailed a capital cost of more than $56,000, were inclined to hesitate. So, for that matter, were their colleagues who contemplated building charcoal furnaces. Although average capital costs for the latter were considerably lower than for anthracite furnaces–about $36,000 for a cold-blast model and just under $41,000 for a hot-blast furnace–these were nevertheless considerable sums, not to be raised easily at any time and least of all during a depression. As one might expect, construction of new charcoal furnaces fell off almost as sharply as did that of anthracite furnaces, and the building of both types of furnaces did not recover until an improvement in general economic conditions after 1843 began to restore investor and producer confidence.

The anthracite furnace was not the only alternative to the charcoal furnace to be introduced during the 1840s. Almost at the same time that the first success was realized with the anthracite process, encouraging results were achieved in blast furnaces with two other fuels, bituminous coal and coke. Comparable in size but smaller in number

and aggregate output than the anthracite stacks, the raw bituminous and coked bituminous furnaces were introduced in western Pennsylvania to take advantage of that region's extensive bituminous coal deposits. Rolling mill operators in eastern and, especially, western Pennsylvania had extensive experience in the use of bituminous coal as a fuel in their puddling furnaces. There the sulphur content of the coal, which had long precluded its use as a blast furnace fuel, was of little consequence, because the fuel never came into contact with the pig iron in a puddling furnace.

Not surprisingly the primary impetus for the development of bituminous and coke blast furnaces came from the rolling mill owners who sought a cheaper alternative to increasingly expensive charcoal fuel. Unlike, for example, hardware manufacturers for whom bituminous or coke pig iron was too brittle, because of the sulphur imparted to it during smelting, rolling mill operators could be indifferent to the quality of the pig iron from which they were going to make wrought iron to roll into rails and other forms of railroad iron. Had the western rolling mill owners enjoyed easy access to the low-sulphur anthracite coal of eastern Pennsylvania, there can be little doubt but that they would have embraced the anthracite blast furnace as enthusiastically as did their eastern colleagues. In any case the goals of a significant segment of both groups by 1850 were to turn out large amounts of rails and other forms of railroad iron that were competitive in terms of quality and price with those imported from Great Britain and to sell these to the rapidly expanding American railroad industry.

Objectives, of course, are not to be confused with achievements, and American rail makers found the competition from imported British rails difficult to meet. The extent of their problem can be illustrated by the use of some figures on American imports of British railroad iron and domestic American production. In 1847, the penultimate year of the British railroad boom, Great Britain exported just under 25 percent of its total production of railroad iron to the United States. As in previous and subsequent years, the bulk of these exports was in the form of rails. The next year, when the railroad boom in Britain collapsed, British rail makers shipped much of their accumulated inventories to the United States at prices that were substantially below those of the year before and, also, lower than prices of American rail. In 1848 British exports of railroad iron of all types to the United States accounted for 48 percent of worldwide British exports. A year later the proportion going to America jumped to 57 percent. Clearly the collapse of the British railroad mania had no counterpart in the United States, and American railroads soaked up the flood of imported rails, much to the dismay of American producers.

The unfortunate position of American rail mills in their competition with British rails in the domestic market reached its lowest ebb in 1853, when British exports of railroad iron to the United States accounted for 63 percent of Great Britain's total exports of such goods. Happily enough for the American producers of rails and other forms of railroad iron, conditions began to change considerably for the better in 1854. Domestic rail production rose sharply from slightly more than 78,000 gross tons in 1853 to more than 96,000 tons in 1854. The following year the domestic production of just under 124,000 tons came well within 20,000 tons of the total quantity imported. The gap closed to less than 7,000 tons the next year when American mills made almost 161,000 tons of rails. A good part of the credit for this jump in rail production went, and probably rightfully so, to the three-high rail mill developed in 1854 by John Fritz, superintendent of the Cambria Iron Works in Johnstown, Pennsylvania. Fritz's invention revolutionized rail production and made his reputation and Cambria's fortune. At the end of the Civil War about one-third of the nation's rail mills used his three-high mill, a proportion which climbed rapidly.

The newfound success of American rail makers in meeting foreign, especially British, competition head-on was due to a number of factors. By no means the least of these was the superior quality of domestic rails, a claim that American rail makers had long made but that none had heeded because of the price differential which favored the British product in the American market. Quality considerations worked to the Americans' advantage not only because it encouraged increased orders for new rails, but also because as the inferior British rails that had been laid in years past began to deteriorate, demand for rails to replace them increased. Railroads were able to save considerable sums by replacing a significant percentage of worn-out rails, not with new ones, but with rerolled rails from American mills. By 1858, one year into the economic uncertainty associated with the panic of 1857,

American rail production was almost double the volume of imported rails, a state of affairs which became even more extreme during the Civil War.

Such impressive growth was not achieved by a technologically or organizationally static industry. The logic of the rail mill owners' situation dictated their course of action: they vertically integrated their operations by building blast furnaces fueled by anthracite coal and raw and coked bituminous coal. In this way the mills could be assured of supplies of pig iron at low cost and in the quantity and quality required. The move by the larger rail mills into integrated operations began in the 1840s and resulted almost immediately in the concentration of a substantial portion of mineral fuel iron production into their hands. In part this concentration of pig iron output occurred because the rail mill firms had the capital resources which permitted them to build the largest anthracite and coke blast furnaces in the country. But, it was also due to the fact that the mill firms were prolific furnace builders. By the eve of the Civil War integrated rolling mill firms in Pennsylvania accounted for only 9 percent of all iron firms but owned 25 percent of the state's mineral fuel furnaces and produced about the same proportion of its pig iron.

The significant changes in technology of pig and wrought iron production in the United States during the last three decades before the Civil War had done more than simply increase the productive capacity of furnaces and mills. They had also made possible the geographic diffusion and dispersion of the industry by freeing producers from the two traditional locational determinants of pig and bar iron making, the availability of waterpower sites and sufficient supplies of appropriate and accessible wood from which to make charcoal. Moreover the continued development of transportation technology, in the form of new and more ambitious canals and, during the 1840s and 1850s, the construction of a railroad system in the Middle Atlantic and Great Lakes states, extended the producers' draw on new supplies of raw materials, especially iron ore and coal, and their market reach.

By 1850 iron deposits in New York, New Jersey, Pennsylvania, and Ohio supplied the bulk of the iron ore consumed by the charcoal and mineral fuel blast furnaces in the major producing states. And, although these sources of ore, especially the deposits in Pennsylvania, remained dominant until the late 1870s, they were supplemented by the discovery and exploitation of two significant ranges before the Civil War. The earliest of the two discoveries was made in the 1830s in the Iron Mountain district of Missouri. This find was commercially developed during the 1840s and became a significant source for furnaces in Ohio and western Pennsylvania by the late 1850s, when about 35,000 gross tons were produced. In 1870 Missouri produced more than 126,000 tons of ore, a remarkable increase over its prewar output but still only a fraction of the more than 2,337,000 tons extracted from Pennsylvania's mines.

The discovery of rich ore deposits in Michigan's Upper Peninsula during the mid 1840s and their successful exploitation a decade later, upon the opening of the Sault Ste. Marie Canal in 1855, substantially and rapidly altered the ore supply conditions for furnaces in Ohio and western Pennsylvania. The canal was, by itself, a rather modest affair, extending about a mile between Lake Superior and Lake Huron, but its completion obviated the need to unload and then reload Great Lakes ore freighters at the impassible rocks at Sault Ste. Marie. The considerable savings in transportation costs that resulted made Michigan ore competitive in price with supplies from eastern Pennsylvania and New York. Moreover, these western ores had a lower silica content and therefore required less fuel in smelting than did Pennsylvania ores. Michigan and, later, Minnesota ore supplies, along with the rapid urbanization and industrialization of the Great Lakes littoral from Milwaukee and, especially, Chicago on Lake Michigan to Cleveland on Lake Erie, and the intensification of the railroad and canal network that served that area helped to make the region an iron and steel center after the Civil War.

Iron and steel making began in the Midwest decades before the Civil War. The region's leading state, Ohio, had become a center of pig and rolled iron production and a major ore mining state by 1840. Twenty years later its output of iron and iron ore was exceeded only by that of Pennsylvania. With the exception of Kentucky none of the other states of the region supported extensive iron production before the war. Their levels of iron making after 1870, however, were impressive and, in some cases, notably Indiana, Illinois, and Michigan, provided the basis for significant steel production. Much the same may be said of the southern states, which, with the exceptions of Virginia and Tennes-

see, supported little iron making before 1860, a circumstance that contributed to the South's defeat in the Civil War.

In a significant way the Civil War was responsible for the development of extensive steel production in the United States and for the rise of the Midwest as a center of steel production. The war had greatly increased the wear on the North's railroad system, particularly its trackage. The iron rails on which a growing volume of rail traffic was moving began to disintegrate with alarming speed, arousing the interest of railroad corporations in the possibility of using steel rails in their place. The postwar years, especially the 1870s, were the infancy of an American steel industry which quickly matured as it supplied a rapidly growing quantity of steel rails. The formation of an American steel industry was accompanied by a rapid geographical dispersion of production capacity and much of the increased steel rail production came from the Bessemer steel plants and rail mills of Ohio, Indiana, Illinois, and Michigan. Convenient access to high-grade Michigan and Minnesota ores and to Ohio and Kentucky coal and the availability of a sophisticated rail system which provided transportation services contributed greatly to the success of the Midwest steel industry. The railroad, however, was not the only factor in the growth of the iron and steel industry in the United States.

The steel industry, like the iron industry before it, responded to a variety of sources of demand for its products. Although the railroad industry's requirements for railroad iron, including rails, played an important role in stimulating the growth of the iron industry before the Civil War, it probably did not exert the decisive and immediate influence on the growth of iron output that some historians, notably Walt W. Rostow, have attributed to it. As Robert W. Fogel argues in *Railroads and American Economic Growth*, the antebellum growth of the iron industry might be more reasonably attributed to "nails rather than rails." However, Fogel's conclusion is, in its own way, as much an overstatement as that of Rostow, and the truth undoubtedly lies somewhere in between the poles of their interpretations. As levels of residential and commercial construction grew rapidly after 1843, following a sharp depression, demand for heating and cooking stoves and for nails and other forms of rolled iron increased. Industrial consumers of iron also bought greater quantities as production of wagons and their wrought-iron-clad wheels, steamboats, and stationary steam engines expanded.

After the war the variety of iron and steel products demanded by residential and commercial consumers broadened. Rolling mill output of structural iron for buildings and bridges rose as the use of this material became standard practice. Similarly the explosive growth of wire mills was a response to the fencing in of the open cattle ranges. Effective fencing required the production of steel barbed wire, pioneered in 1873 by Joseph F. Glidden and virtually monopolized by the late 1880s by the American Steel and Wire Company of John W. "Bet-A-Million" Gates. The growth of other industries, such as petroleum refining and food packaging and canning, also stimulated the growth of steel production as makers of fabricated steel products turned out containers as varied in size as oil drums and tin cans.

Much of the history of the iron and steel industry during the last quarter of the nineteenth century is the history of the rise of large corporations. From 1870 to 1900 the number of firms in the industry progressively declined as the capitalization of the average firm rose from $150,000 to $880,000. The size and capitalization of the larger firms, most notably the Carnegie Steel Company and other giants such as the Illinois Steel Company and the Bethlehem Steel Company were, of course, much larger. The formation of these and other firms, including the smaller ones, and the lives of the men (and one woman) who founded and ran them comprise much of the subject matter of this volume. This brief general overview of the industry and the discussion, immediately following, of some specific aspects of the iron and steel industry's development are intended to provide the reader with a context within which to evaluate the work of these business leaders and the nineteenth-century iron and steel industry that they built.

Industry and Trade Associations and the Tariff

During the course of the nineteenth century American iron and steel producers organized or participated in a number of industry and trade associations to improve conditions in the domestic market. Although the purpose of individual organizations varied somewhat and changed in emphasis as the industry matured, most were concerned with the collection of accurate information on iron production and market conditions and with the use of

this information to stabilize prices and, especially, to influence national trade policy. The focus of the latter concern was, in almost every instance and throughout most of the century, the tariff.

The tariff as an instrument of national policy was a focus of controversy from almost the first days of the Republic. As a source of revenue to finance the operations of the federal government, tariffs with modest or even nominal rates were widely accepted as necessary evils. Support for such revenue tariffs came even from groups and regions with an aversion to import duties as a means of affording protection to domestic producers from injurious foreign competition. This other purpose of a tariff, that is, protection, was seen by its earliest and most articulate advocate, Alexander Hamilton, as an integral part of a national economic policy.

Called for by Hamilton in 1787, the protective tariff quickly proved to be an unusually divisive issue which all too consistently generated more heat than light during the periodic debates over its efficacy and character. Advocates of a protective tariff argued that its rates had to be sufficiently high so as to discourage importations of foreign, meaning British, goods which would otherwise undercut in price and outsell American products. Their opponents insisted that tariff rates in excess of what was necessary to fund the federal government fostered monopolies and harmed nonindustrial interests, such as hardware makers and agrarian consumers. This initial division of opinion between supporters of a protective tariff and those who advocated only a revenue tariff became the enduring criterion by which positions on the tariff question were to be defined throughout the next century.

The nominal rates of the first tariff, enacted on July 4, 1789, defined it as a revenue tariff in almost every respect. Subsequent increases before 1812 afforded only slight statutory protection for key domestic industrial products such as steel, nails, hemp, and molasses. In 1812 ad valorem rates (those computed as a percentage of declared value) were increased by just less than 50 percent to about 12.5 percent and to about 25 percent during the War of 1812. Despite an often effective British blockade of sections of the American coast, the controversy over tariff rates remained lively as young American industrial interests in New England and the Middle Atlantic states demanded and received a guarantee of tariff protection for a 2-year period immediately following the signing of a peace treaty. This assurance proved to be of comparatively little value because, with the return of peace in early 1815, British exporters began to dump textiles, bar iron, and other primary iron products on the American market. British dumping and a broad decline in prices in Europe and the United States severely pressed American iron producers and intensified their demands for protection by a tariff.

In the late 1820s and early 1830s, as the tariff controversy became increasingly shrill, these efforts initially took the form of rather atomized campaigns waged by groups of pig and bar iron producers in particular cities or states to convince Congress to pass a protective tariff and by opposing groups of manufacturers of iron hardware to defeat any such tariff bill. Although iron manufacturers, as industrial consumers of iron—pig, bar, and, later, rails—lobbied energetically against a protective tariff, most of the industry's members, as producers of these primary iron products, worked strenuously for protection. Their interests coincided with those of producers in other industries, especially textiles, and encouraged their active participation in the welter of protectionist organizations that were active before the Civil War. Although many of these groups were short-lived and had no enduring influence on the development of national trade policy or the iron industry, others, notably the Home League of New York and the Friends of Domestic Industry, also a New York group, were of considerable significance to both.

An association of advocates of a high protective tariff, the Friends of Domestic Industry met in New York City on October 26, 1831, to thwart attempts then under way to slash the rates imposed under the tariff of 1828, the notorious "Tariff of Abominations." This tariff's ad valorem rates, the highest of the pre-Civil War tariffs, amounted to 46 percent on pig iron, 35 percent on hammered bar iron, and 86 percent on rolled bar iron. Because of a substantial drop in the foreign, especially British, prices of these categories of iron, the effective rates were much higher by the time that a serious effort was under way in 1832 to reduce them. The Friends of Domestic Industry realized that it had its work cut out for it and organized itself into specialized committees, one of which was the "Committee on the Product and Manufacture of Iron and Steel." Participants at the convention worked figuratively in the shadow of a convention held earlier that year in Philadelphia that met with the opposing goal of bringing down the rates of 1828 in the interests, ostensibly, of free trade.

Both conventions issued reports intended for public and especially congressional consumption. Part of each convention's report was based on the results of a questionnaire which it had distributed to iron producers and manufacturers to ascertain the industry's size and production for the 3-year period from 1828 to 1830. These were the two questionnaires that had anticipated the one circulated by U.S. Secretary of the Treasury Louis McLane in early 1832 and which sought similar information for the same three years. Many iron producers, especially in Pennsylvania, having just provided detailed information for the Friends of Domestic Industry's circular, belligerently refused to cooperate with the Treasury Department's attempt to collect data, seeing in it only a wasteful duplication of effort. That Secretary McLane decided not to accept at face value the figures compiled by either convention and relied instead on the less complete results garnered from his own questionnaire is not surprising. McLane had circulated his questionnaire in the northern states, including Delaware, in response to a congressional resolution of January 19, 1832, which had urged the Treasury Department to ascertain the degree, if any, to which American manufacturing required protection from foreign competition.

A remarkable aspect to this contest was the intellectual honesty with which both the free trade advocates and the Friends of Domestic Industry conducted their industry surveys. The result of this honesty was a degree of reliability which, for example, prompted the protectionists in October 1831 to accept the earlier findings of the free trade convention's questionnaire, having first subjected them to careful examination and cross-checking against their own results. When the Friends of Domestic Industry subsequently published the report of their committee on the iron industry in 1832, they used the free trade convention's findings, but with their own revised estimate of pig iron output in 1830, as an integral part of their own report.

The fight over the revision of the tariff of 1828 was ultimately lost by the protectionist forces with the passage of the so-called "Compromise Tariff" in March 1833. The construction of this tariff provided for the reduction of all rates in excess of 20 percent over an eight-year period, beginning in

1934 and ending in 1842, by 10 percent every two years. As one might suppose, primary producers of iron, that is, furnace, forge, and rolling mill operators, felt aggrieved, the more so when the bad conditions associated with the panic of 1837 and then the depression of the early 1840s began to press them severely. The widespread perception among iron producers was that salvation lay in a strongly protectionist tariff which would hold their British competitors at bay. To that end they campaigned vigorously through trade and industry associations, especially the New York Home League, which had been founded in 1841 during the depths of the depression as a protectionist lobby with a special interest in the iron industry.

None of these trade associations was, in any strict sense of the term, objective, but the Home League was doubtless better than most. In 1844 the League set for itself the formidable task of generating an estimate of iron output in the United States for 1840, an exercise which should not have been necessary, the U.S. Census of 1840 having just been completed. The census was, however, precisely the problem, in that iron producers considered its figure for iron output to be sharply below the true level. The Home League's review of the census prompted its committee on iron to recommend an estimate of output that was 20 percent higher than the result reported in the census. The League's figure received widespread publicity when a detailed report of its committee's work appeared in the March 1845 issue of the influential *Hunt's Merchants' Magazine*.

The Home League was not alone in its concern over the caliber of the work reporting on the iron industry in the Census of 1840; in fact this concern had been anticipated by a committee of the Pennsylvania Coal and Iron Association which, in 1842, had surveyed that state's iron producers. Tariff reform was also the aim of the Coal and Iron Association, and their efforts and those of the Home League were initially rewarded by the passage of the protectionist Tariff of 1842, which raised duties to what they had been in 1832. Four years after this triumph, however, a Democratic Congress passed the Tariff of 1846, the so-called "Walker Tariff," after Secretary of the Treasury Robert J. Walker, who had formulated it. The Walker Tariff was primarily a revenue, rather than a protective tariff, and resulted in a significant reduction in rates. It was against this background of modest victories and profound defeats on the tariff issue that industry lobbyists and other advocates of protectionism began to muster their forces in 1849 in anticipation of a renewal of the struggle over the tariff. They perceived the stakes to be higher because the iron industry was beset by a severe but narrowly focused business contraction, more or less peculiar to iron industry within the American economy, but which affected British iron producers as well. Domestic producers could only fear the arrival of the inevitable storm surge of British iron in the wake of the collapse of the highly speculative English railroad boom.

In August 1849 ironmasters from the Pittsburgh area met in that city to formulate a course of action by which to achieve greater tariff protection for Pennsylvania iron. Before they went home, however, they laid plans for a more broadly based convention of furnace owners to be held in Pittsburgh in November 1849. Although not every iron-producing state sent delegates to the convention, 166 iron producers from western Pennsylvania, New Jersey, New York, Ohio, Illinois, Kentucky, and Virginia attended the three-day meeting on November 21-23. Little in the way of a consensus emerged from the Pittsburgh meeting, except on the basic point that protection of pig and bar iron required duties of $10 and $20 per ton, respectively.

Within a few weeks of the adjournment of the November Pittsburgh meeting, a "Convention of Iron Masters" from eastern Pennsylvania, New Jersey, and Maryland met in Philadelphia in December 1849 to draw up resolutions and memorials for Congress in which they advocated a return to the tariff rates that had been in force until 1846 under the Tariff of 1842. This convention was more systematic in conducting its business, and its members supported their demands for a resumption of protection by pointing to the results of a comprehensive survey of Pennsylvania's iron industry executed by a committee of the convention. The condensed report of the survey's findings appeared in several influential publications and on its merits should have made a profound impression on Congress. The report reached Congress, however, just at the time that the territorial question and the related but more fundamental issue of slavery began to assume a particularly dangerous cast and to consume the energy and time of congressmen and senators.

The preoccupation of Congress with the debate over territory and slavery did not discourage iron producers from continuing to send petitions and memorials urging greater tariff protection. In 1850 iron producers in New England sent a lengthy statement in support of their request that the rate structure of the Walker Tariff be substantially changed to make the tariff more protectionist. The New Englanders' efforts reinforced those of the Pennsylvania conventions of just a few months earlier, but all of these campaigns failed. The Walker Tariff's rates not only remained in force but in March 1857 Congress passed a new tariff which imposed even lower rates, thereby reducing the level of protection still further.

The success of one Democratic Congress after another in passing antiprotectionist tariff legislation may very well have had much to do with the political fragmentation of forces. But, almost certainly, it was also a result of the return of favorable economic conditions to the industrial states after the recession of the period from 1848 to 1852. Rising iron prices and the reduction of costs through technological change meant increased profits, expanded production, and business optimism which continued until the onset of the panic of 1857 in the late summer and early fall of that year. The panic, which affected Europe as well as the United States, touched off a sharp and broad decline in prices, including those for pig, bar, and rolled iron. A particularly unfortunate aspect to the timing of the price drop was that it had been preceded by the imposition of reduced rates by the new tariff passed about six months earlier. The erosion of iron prices and the increased competition from British iron in the domestic markets stimulated renewed efforts by iron producers to seek passage of a tariff which would afford them greater protection.

Even while business conditions had continued to improve in the years immediately preceding the panic, iron producers had not abandoned their goal of getting a protectionist tariff through Congress. To that end they organized the American Iron Association in 1855. Because Pennsylvania's furnaces turned out about half of the nation's pig iron, the leaders of that state's iron industry, especially Samuel J. Reeves of the Phoenix Iron Company, determined the orientation and political agenda of the new association. Following the practice of earlier Pennsylvania organizations, the association set as one of its first objectives the collection of data on the magnitude and geographical extent of iron production in the United States during 1854. Among the findings of this first survey was the conclusion that the quantity of pig iron made with mineral fuel, that is, anthracite and bituminous coal, was larger than had been thought, amounting to almost half of all pig iron production. This was potentially a useful point to muster in the debate over the tariff, if only to counter antiprotectionist arguments that the adoption by iron producers of advanced, more efficient production methods would allow their industry to be competitive without the artificial and, ultimately, harmful protection of a high tariff.

Such arguments, of course, turned on the very fine points of when and, especially, why producers had jettisoned traditional production processes and had adopted the latest and best techniques. Opinion in this matter arrayed itself along what were by then familiar lines. Antiprotectionists argued that the iron producers' impressive progress along the road to technological advance would never have occurred had they not been compelled to modernize their operations by the force of foreign competition.

Producers maintained that their great progress in introducing more advanced techniques, especially the use of anthracite coal in place of charcoal to make pig iron, had been made during the 1840s and had been possible in large part because of the protection afforded pig iron producers by the tariff of 1842 and the initial rates of the Walker tariff of 1846. These measures had allowed them to undertake the considerable capital investment and accompanying risks without the necessity of contending with British competition, much of which, asserted the iron producers, took the form of misrepresentation of products and dumping of surplus domestic British iron on the American markets to drive down American producers. The subsequent prosperity of the industry in the United States during the mid 1850s, despite the lack of a protective tariff, was very possibly a chance convergence of fortuitous circumstances and was not to be relied upon as a guide to the making of tariff policy. The passage of the antiprotectionist tariff of 1857 was an accurate measure of the producers' influence at that moment.

The collapse of iron prices during the last months of 1857 and throughout the next year following the panic added urgency to the American Iron Association's campaign for tariff protection, from the producers' perspective. Democrats from the northern industrial states began to join in the agitation for higher duties, thereby adding additional stress to the fabric of the party. For the most part, however, Congress remained preoccupied with the succession of events and disputes that marked the trail to the abyss of civil war. The Democrats and their leader, President James Buchanan, found themselves presiding over an impressive erosion of popular support in the congressional and state elections of 1858 throughout much of the North, but especially in Pennsylvania, where the iron and coal interests were hard-pressed by the panic's aftershocks. Hopes among producers that Buchanan and other national Democrats would see the desirability and even the necessity of moving quickly to force a significant upward revision of the rates of the 1857 tariff were dashed when the president exonerated the tariff from any blame for the business woes then besetting the nation.

The rates established by the tariff of 1857 remained in force until the passage of the protectionist Morrill Tariff on March 2, 1861, scarcely a month before the outbreak of the Civil War. The resignation of southern legislators from Congress during the preceding months had removed the main opposition to an effective protectionist tariff. During and immediately after the war the continued absence of traditionally antiprotectionist southern congressional delegations facilitated a series of progressive rate increases which, by early 1869, had reached a level of 47 percent. Sentiment for truly protective tariff rates had been widely held in Congress before southern secession and became intense thereafter. The lead in the campaign for higher duties, however, came from within the industry and was directed by its lobby, the American Iron Association, which was reorganized in 1864 as the American Iron and Steel Association (AISA).

More than cosmetic, the name change reflected the growing importance of steel making as an industry and the increased sophistication of the producers. The first president of the AISA was Eber B. Ward, an innovative and aggressive iron and steel man from Illinois. His second-in-command was the staunch protectionist Samuel J. Reeves of the Pennsylvania-based Phoenix Iron Company. Together they fashioned the AISA into an increasingly active protectionist organization. When Ward resigned in 1869, Reeves assumed the presidency, and under his leadership, which continued until his death at the age of sixty in December 1878, the

AISA became one of the most powerful and influential industry lobbies of the period.

There were, however, limits to what such an organization could accomplish for its members. In the wake of the Civil War a growing popular antipathy to the wartime profiteering of industrial interests helped to erode support for continued tariff protection. For the iron and steel industry this political problem was aggravated by the industry's surplus productive capacity and the rapid collapse of prices in the United States and throughout Europe following the end of the war. As the Republican party's control of Congress, based on a broad consensus which had embraced wartime loyalty to the Union and postwar firmness in the matter of reconstruction, began to unravel, the party's ability and, more to the point, its willingness to press for continued high levels of tariff protection began to dissipate. In 1870 and, again, in 1872, Congress passed tariff acts which, while still protectionist in terms of their overall effects, nevertheless sharply lowered the level of protection afforded industrial producers, including the iron and steel industry.

At the time few producers worried about whether the new tariff rates were sufficiently protective of their interests. Most were more inclined to congratulate themselves on the resumption of prosperity and the renewed strength of prices which was well under way in late 1870. Pig iron prices, for example, which had fallen from a wartime high of $59.25 per gross ton in 1864 to $33.25 by mid 1870 had climbed to almost $49 by 1872. Much of this price recovery resulted from the rapid expansion of existing railroad lines and the construction of new lines in this country and in Europe. In the United States the railroad boom was closely tied to a flood tide of land speculation which it had helped to stimulate. The surge in immigration and the westward migration, briefly interrupted by the war, pushed land prices and railroad stock prices ever higher and lent to the boom an air of seeming inevitability and even immortality. The onset of a financial panic in the fall of 1873 brutally dispelled these illusions.

The panic of 1873 revitalized the cause of protectionism and added weight to the jeremiads of Samuel Reeves, who had remained a fervent advocate of high tariffs even during the short-lived prosperity that had preceded the panic. He and James M. Swank, the AISA's energetic new secretary, began to intensify the campaign for a return to protection.

By early 1875 Congress was disposed to scrap the tariff policy of the preceding four years and reestablish protectionism as a basic component of national political economy. The tariff of that year restored rates to levels which approached those of the immediate postwar years. When Reeves died in December 1878, the tariff then in force was one for which he and the AISA had long fought.

Following Reeves's death, Daniel J. Morrell of the Cambria Iron Company became president of the AISA in March 1879. He held the office until his deteriorating health compelled him to retire in 1884. He could content himself with having held the line, with the able assistance of his old friend, James M. Swank, against any significant deviation from protectionism. Although a tariff passed in Congress only the year before had reduced rates by 5 percent, average rates remained in excess of 40 percent, a level far above that required to insure protection against foreign competition. Critics of the industry, in fact, were quick to point out that, although the tariff on imported iron and steel products might once have been defensible—and even necessary—when American producers were small and vulnerable, that time had long since passed. Their strength was such that now, if anyone needed protection, it was their customers in this country, who, according to the charges leveled against the industry, were being victimized by the concentration of productive capacity into progressively larger but fewer corporate hands.

There was some justice to this indictment. The physical and financial scale of business within the iron and steel industry, as well as within the railroad, petroleum, coal, and many other industries, had increased dramatically during the 1870s and early 1880s. This growth in scale had prompted a drive toward business consolidation which ultimately led to the formation of trusts. One stimulus to the organization of trusts and other, smaller combinations undoubtedly was the broad and sustained decline in prices which had taken hold in the wake of the panic of 1873. As falling prices eroded profit margins, larger firms fought to sustain aggregate revenue and meet their high fixed costs by increasing their shares of the market at the expense of one another or, sometimes, by collaborating to fix prices to limit the growth of supply.

But, as Naomi Lamoreaux has noted in *The Great Merger Movement in American Business*, depressed economic conditions did not necessarily affect all iron and steel firms in the same way or to the same degree. Larger firms apparently fared considerably better than smaller ones and even capitalized on the misfortunes of the latter and seized their market shares when they failed. Still there were sectors of the iron and steel industry, notably the rail and tinplate producers, for whom there was a pressing need to act in concert with one another to stabilize supplies of their products. Although the tariff was a time-tested and venerable means of holding down the level of supply, it was no longer sufficient because of the prodigious capacity of domestic producers.

More conducive to success, at least in the short term, was the formation of pools and price-fixing agreements. These arrangements sometimes suffered, however, from the fatal weakness that their success essentially depended upon the honorable conduct of the parties to them. The high stakes and the curious ethical standards of many of the participants made honor and fidelity to agreements a commodity in short supply. Nevertheless several of these schemes and programs were launched and some achieved a fair measure of success. A particularly outstanding example, because of its success during the perilous decade of the 1890s, was the Rail Association, organized as a pool in 1887 to impose price and production discipline within the rail sector. However, even the Rail Association's formal internal regulatory machinery was unable to handle the likes of John W. "Bet-A-Million" Gates and his Illinois Steel Company, which broke price discipline in February 1897, thereby touching off a vicious price war among all major rail producers.

Although the Rail Association collapsed as a result of the cupidity of Gates, the firms that had been its members managed to reorganize the pool in 1899 and resume their price-fixing. A much different outcome befell the producers of wire nails who had organized a pool, the Wire-Nail Association, in early 1895 for the purpose of supporting and, ultimately, raising prices in the midst of the severe depression of 1893-1896. They achieved this result before 1895 was out, but their success had an interesting and, for them, unfortunate side effect. Rising prices induced other producers who had not joined the Association to increase production, thereby augmenting supplies in the market and putting downward pressure on prices. Among these independent producers was the large and influential firm of Washburn & Moen, with headquarters in Worcester, Massachusetts. The Association's immediate response to these "rogue" wire-nail producers was to pay them not to produce, an answer that was hardly a realistic solution. In any case prices began to fall sharply in summer 1896, and, by the end of the year, the Wire-Nail Association was dead.

Although far more dramatic than the tariff as a weapon against falling prices and rising competition, the accounts of these pools and other combinations should not mislead one into thinking that iron and steel producers no longer embraced the cause of protective tariffs. The AISA remained the industry's chief lobbying agency and pressed the case for protection at every opportunity. Although the AISA had, in fact, enjoyed considerable success in convincing Congress to pursue an essentially protectionist course, tariff rates from 1869 to almost the end of the century behaved something like a roller coaster. This is not to imply that rates assumed anything but protectionist levels by pre-Civil War standards. However, their fluctuations suggest the political passions and pressures which figured in the tariff controversy during a period marked by the rise of organized labor and a powerful agrarian protest movement that had antiprotectionism as one of its cardinal themes.

Protectionist forces had cheered their hard-won victory in 1875 when the tariff of that year restored rates to levels acceptable to an iron and steel industry that was becoming hard-pressed by the depression of the 1870s. Their relief was tempered when Congress reduced rates by 5 percent in 1883 but was intensified upon the passage in October 1890 of the McKinley Tariff, which increased average rates to 49.5 percent, their highest levels up to that time. Four years later, in August 1894, during the depression Congress again slashed rates, this time to an average of 39.9 percent. While still protectionist, these levels reflected the rapidly growing antiprotectionist sentiment among a sizable part of the population. But, what goes down can and often will rise again. The Dingley Tariff, passed in July 1897, well into William McKinley's first year as president and well into the recovery from the depression of 1893, for which he received the credit, increased average rates to their historic high level of 57 percent. Protection on that level or on almost any other level was hardly needed then by the most productive and concentrated iron and steel industry in the world.

Output Levels

The rapid increase in iron and steel production in the United States during the nineteenth century remains one of the most impressive accomplishments in the nation's economic history. But, as is the case with many aspects of the early nineteenth-century economic history of the United States, uncertainty clouds efforts to arrive at some quantitative assessment of the iron industry's levels of production during the first half of the century. In large part this regrettable circumstance is due to the shortcomings of the United States Census reports for 1810 through 1840 and the curious bungling and feuding which characterized the industry's own attempts to gauge its activity. These were not really separable influences but were, instead, intimately bound up with one another and the highly politically charged issue of the protective tariff. The statistical record of the industry's production is presented below in a table of output figures for the major product groupings of the nineteenth-century U.S. iron and steel industry. Almost as important as the figures are the sources from which the statistics were gathered.

The earliest attempt to determine the extent of American iron production was undertaken in the U.S. Census of 1810 through its provision for a survey of manufactures. Unfortunately, its subsequent execution, under the supervision of Tench Coxe, resulted in little more than an enumeration of flour and gristmills, blacksmithies, and ironworks. The chief impetus for the survey, entitled *A Statement of the Arts and Manufactures of the United States of America for the Year 1810*, was the increasing commercial and naval tension between Great Britain and the United States. The American policy of nonintercourse and embargo, though it failed in its object to compel British and French recognition of American neutral rights, had stimulated domestic manufacturing. In an effort to gauge the extent and variety of American manufacturing, Coxe's enumeration more closely resembled the 1810 census's lean decennial enumeration of population than the detailed studies of industry that would characterize the census of 1850 and those which followed. Incomplete and exasperating in its organization, the results of the census do not permit the construction of reliable estimates of aggregate iron production. This is not to say, however, that the report is without considerable value to an examination of the early iron industry.

Despite its limitations, Coxe's report on manufactures contains much of interest to students of the American iron industry. Using its figures one can derive approximations in tons of mean output of forges and furnaces, thereby gaining some understanding of the technical capacity and constraints of the iron industry on the eve of the War of 1812. An equally useful but far bolder exercise is the use of Coxe's figures to arrive at an estimate of the aggregate national output of pig iron in 1810 and, also, some notion of the estimate's reliability. We know that Coxe put the figures at roughly 54,000 long or gross tons, that is, tons of 2,240 pounds each. Of this total output 26,879 tons, or roughly 50 percent, were made in Pennsylvania, by far the leading iron-producing state. Unfortunately, little more of value is to be had using Coxe's report as a guide for a foray into the depths of the "statistical Dark Age" that are the years before 1840, when the first comprehensive census of manufacturing was conducted.

Though it is defective in many respects, the Coxe report is a far better source of data on the iron industry than the three subsequent censuses. Their coverage of industrial activity was so uneven and unsystematic as to appear to have been haphazard. The first of these, the 1820 census enumeration of manufacturing, is notorious for its omissions and errors and was in 1820 considered unreliable. For example its reported figure for the nation's output of pig iron, 20,000 gross tons, is ridiculously low, especially since furnaces in Pennsylvania alone probably turned out at least that much pig iron in 1818. Its shortcomings aside, the census of 1820 does afford the historian some insight into an iron industry still wedded to the traditional charcoal and waterpower technology of production and dependent upon a transportation technology that did not yet include extensive canal building or any railroad lines.

Unlike the census of 1820, with its severely limited tabulation of industry, the census of 1830 did not even include a report on manufacturing. This failure would come back with a vengeance to haunt a Congress that, out of sincere desires both to economize on the census and to avoid further acrimonious debate over the issue of a protective tariff, had decided not to fund that part of the census. Despite its attempt to stifle debate the Congress soon had to confront the issue that it had ducked by underfunding and circumscribing the census of 1830.

The tariff bill that passed Congress during the twilight of John Quincy Adams's presidency in 1828 was a protectionist measure which managed to infuriate its opponents far more than it satisfied the industrial interests who supported it. Damned almost immediately as the "Tariff of Abominations," the 1828 tariff was the issue which Congress sought so assiduously to avoid in 1830 as demands for downward revision of its rates became louder and more insistent. With the inauguration of Andrew Jackson in March 1829, antiprotectionist forces in and out of Congress rejoiced that the time for reform was almost at hand. As matters turned out, however, reform was slow in coming and incomplete in execution.

With a degree of deliberation that admirably suited the dignity of its members, Congress requested in 1832 that the Secretary of the Treasury, Louis McLane, undertake to find out what a proper census of 1830 would have told them: the character and extent of manufacturing and industrial activity in the states of the North, the section from which most of the demands for protection came. McLane proceeded to distribute his questionnaire and, after having compiled the answers, submitted a bound, multivolume report, usually referred to simply as the McLane Report, to Congress in 1833. The report was voluminous and minutely detailed and included the actual answers of the respondents to the questionnaire, as well as an extensive series of tables which summarized the findings. The two industries which received the most attention in the report were the wool industry and the iron industry.

The McLane Report contains a good deal of valuable information about the forms of business organization in the American iron industry just a decade before it began to shift from charcoal to anthracite coal. As such it affords a close look at an industry still dependent for the most part upon its traditional technology, but one that was beginning to adopt steam power to augment and even replace waterpower. As was mentioned earlier, however, the great defect of the McLane Report was its lack of comprehensiveness, a failing dutifully and dolefully acknowledged by those whose job it had been to compile it. Many of those who received questionnaires simply neglected to return them, and some who did, deliberately gave incomplete answers or waxed indignant that their privacy should be so invaded by the federal government. In any case they had already responded to a detailed questionnaire sent out in 1831 by antiprotectionists and resented the further imposition on their time by the treasury.

Although one can gain considerable insight from the report into many technological facets of pig and wrought iron production and steam engine foundry operation, the report does not permit confident estimation of iron production for 1828-1830, the three years with which Secretary McLane and Congress were concerned. Because of the problems associated with the McLane Report, the most reliable estimate of pig iron output in 1830 is one reported the next year by the antiprotectionists and subsequently amended in a report issued by the New York convention of the protectionist Friends of Domestic Industry. The generally accepted figure is slightly more than 180,000 gross tons, a quantity that indicates considerable expansion of the iron industry's size and capacity since 1810.

The tariff question haunted American politics and was considered by neither side, protectionists or free traders, to be a settled issue before or after the Civil War. Throughout the 1830s and 1840s groups and individuals representing the two sides of the dispute increasingly focused their attention on the domestic iron industry and published their estimates of pig iron production. The accuracy of these figures, especially those for the 1830s, was and remains controversial. Questions about their accuracy aside, the figures supplied by the various pro- and antitariff sources offered Congress and the Treasury Department their only reasonable point of departure into the dangerous waters of the tariff controversy.

The next attempt by the federal government to gauge the nature and extent of the nation's industry was made during the census of 1840. Having learned the painful lesson of the 1830 census and still faced with the tariff controversy, Congress appropriated funds to support an enumeration and tabulation of manufacturing. Unfortunately, neither the quality of congressional deliberations nor the work of historians was advanced by the manufacturing schedule of the 1840 census, the results of which were incomplete and unreliable. The schedule's report of output figures for various sectors of the iron industry was particularly faulty and was considered by the iron producers to be so unreliable be-

cause of undercounting that industry leaders in Pennsylvania and New York sponsored their own surveys of iron making in 1841 and 1842. The defects of the 1840 census were all the more serious in terms of making tariff policy because of the profound economic dislocation that had begun in late 1839 and which had severely affected the iron industry.

Once again the federal government found itself faced with the problem of deciding upon the degree of protection to be given to domestic industry without having the necessary information on which to base a decision. The vacuum was at least partially filled by the findings of the surveys of pig iron output conducted by the various associations of commercial and industrial interests, such as the Home League of New York. Of course, while the findings of the various surveys bore some similarity to one another, they were not mutually consistent and varied from a low figure of just under 287,000 gross tons to 347,000 gross tons, a range of about 60,000 tons, or more than 20 percent. More reliable estimates of the nation's pig iron production in 1840 and subsequent years are now available through the recent work of historians, although even these estimates vary, based as they are on somewhat different sets of assumptions.

The schedule of manufactures of the census of 1850 yielded the federal government's first systematically determined estimates of iron output. These were largely due to the schedule's careful design and execution by the Whig superintendent of the census, Joseph Kennedy. And yet, despite Kennedy's care and conscientious oversight and that of his successor, J. D. B. De Bow, the 1850 census figures on aggregate iron production, especially the pig iron sector, leave something to be desired. As had been true of the results of the manufacturing schedule of the 1840 census, those of the new census were impugned by incomplete coverage due to omissions by enumerators, noncooperation by those canvassed, illegibility, and confusing categories of information.

The census put the nation's output of pig iron at 563,755 gross tons, a figure that may or may not have been approximately correct. Because the census used a calendar year which ran from June 1849 through May 1850, the figures most likely understate the slowdown in pig iron production which began in 1849 and became more pronounced in 1850. Because of this economic downturn the figures were at serious odds with contemporary conditions.

Another reason for treating the census estimate of aggregate pig iron output with circumspection is that it is not consistent with a calculated estimate of output based on the observed proportional relationship of the pig iron production of Pennsylvania to that of the nation during the 1840s and 1850s. Pennsylvania's share of national pig iron output during the 1840s rose from under 36 percent in 1840 to about 51 percent in 1850. Calculating 51 percent of 563,755 tons, the result comes to 287,515 tons. This figure is substantially higher than that tabulated in 1849 by a convention of Pennsylvania ironmasters. The convention surveyed the state's furnaces and other ironworks and reported a production of 253,035 gross tons of pig iron during that year from the state's 294 furnaces that were actually in blast. The depression within the iron trade that began in late 1848, following the start of massive dumping of railroad iron on the American market by British producers, makes it likely that production in Pennsylvania and throughout the nation in 1850 was considerably lower than it had been in 1849. A more reasonable estimate of pig iron output in 1850 is the figure of 481,000 gross tons computed by Robert Fogel in his study, *Railroads and American Economic Growth* (1964), though this result, like those it is intended to supplant, is the source of continuing controversy.

Confusion and ambiguity in the matter of pre-Civil War pig iron output extends right up to and, of course, through the war. Even the schedule of manufactures of the census of 1860, patterned after that of the previous census, yielded a figure of 987,559 tons (presumably gross tons, though the 1860 census does not so specify) that is about 20 percent higher than the roughly 821,000 gross tons which was the figure accepted by the American Iron and Steel Association (AISA). Again, there is little reason to suppose that the census figure is the better of the two and considerable evidence to the contrary.

The significance of the Civil War in the development of the steel industry is impossible to fix with precision and all too easy to understate. We know, for example, that total production of pig iron increased from 821,000 gross tons in 1860, according to the most reliable estimate of output for that year, to a wartime high in 1864 of 1,014,000 tons in states outside the Confederacy. This rapid growth represented an ostensible increase of 22 percent but was actually a bit higher because the 1864 figure excluded pig iron output from the southern states. Even more dramatic was the growth in the output of pig iron made with raw or coked bituminous coal, which climbed from about 109,000 gross tons in 1860 to 188,000 tons in 1864, an increase of more than 70 percent.

The sharp increase in aggregate levels of pig iron production during the war stemmed from two distinct sources. One of these was the stimulating influence on all sectors of the iron industry exerted by the war. A combination of government contracts, railroad orders, and hyperinflated prices offered the prospect of high profits and provided a safety net against business failure which only the unluckiest or most incompetent businessman could miss. Thus even charcoal iron producers who, as early as 1850, were a rapidly diminishing part of the pig iron sector of the industry found high levels of demand for their expensive but highly esteemed product. The other source of the increase in pig iron production was, of course, also related to the war and was, ultimately, of far greater significance. The war seems to have accelerated the profound technological change that had been under way within the pig iron sector for more than a decade before 1860. This change was reflected in the substantial increase in the average size and output of all types of furnaces, even charcoal-fueled furnaces.

The Civil War years mark the beginning of an era of careful and reliable iron and steel production records. The American Iron and Steel Association compiled the records which, along with the decennial census reports beginning in 1870, provide detailed information concerning output and prices of a variety of products. The AISA's interest in acquiring and disseminating such information was largely due to the persistence of the tariff as a significant issue of national public policy. It probably also reflected the concern of producers, large and small, for the industry's health during the long deflation following the war; this concern lasted until almost the turn of the century. By 1900 much of the nation's production of iron and, more importantly, steel was controlled by a comparative handful of large corporations, of which the Carnegie Steel Company was by far the most powerful.

The comparatively sudden rise of a steel industry dominated by big business organizations was one of the most significant developments in American history and transformed the iron industry and the general economy, as well. Of relatively little importance prior to 1860, steel making relied on the technique called cementation to make a product called blister steel and, from that, crucible steel from bar or wrought iron. The process, discussed in detail elsewhere in this volume, involved the long-term heating of bar iron at high temperatures—above 1,200 degrees Fahrenheit—in the presence of highly purified and powdered charcoal. The high temperatures permitted the iron to absorb the carbon from the charcoal, thereby transforming low-carbon bar iron into steel. The interaction between the bar iron's slag content and the carbon absorbed through the iron's surface produced gas bubbles which, upon bursting, blistered the steel's surface, giving it its popular name. Once made, the blister steel served as a raw material for the making of finished steels such as shear steel and, early on, a very costly form of crucible steel, also known as cast steel, in which blister steel was reheated in an airtight crucible.

These early steel processes produced steel in small quantities, and the precise amounts are not known. Output in 1850 is thought to have been somewhat more than 6,000 net tons (a net ton is 2,000 pounds) from 13 steelworks in Pennsylvania and another in New Jersey. Steel production was still largely confined to the same area a decade later. In 1860 nine works in Pennsylvania—three in Philadelphia and six in Pittsburgh—and two each in New York and New Jersey produced a reported total of 11,838 net tons, a negligible amount compared to the 800,000 to 1,000,000 tons of pig iron made that year in the nation's furnaces. Output of steel increased modestly during the years immediately following the Civil War from about 22,400 net tons in 1867 to more than 77,000 tons in 1870. Thereafter, the growth of steel production was explosive, reaching almost 1.4 million tons by 1880, about 4 million tons by 1890, and more than 11.4 million tons in 1900.

The steel industry's phenomenal growth rested initially and primarily on the Bessemer process, which accounted for about 86 percent of all steel output in 1880 and 1890 and 66 percent as late as 1900. The decline in the relative importance of the Bessemer process after 1890 was due to the increasingly widespread adoption by steel makers of the competing open-hearth process, which produced less than 575,000 tons in 1890 but 3.8 million tons

a decade later. Together, the two processes accounted for almost all steel made in the United States after the 1870s and imposed unprecedentedly large capital requirements upon the industry. These requirements, perhaps more than any other consideration, stimulated and even impelled the formation of large corporate entities and the drive toward consolidation within the iron and steel industry after the Civil War. The interest of iron producers in the adoption of the Bessemer process and, later, the open-hearth process was, itself, excited by the rapidly increasing demand for steel rails by the nation's railroad industry.

The increased traffic and weight of larger locomotives and tracks of northern railroads during the war had greatly exacerbated one of the railroad industry's most troublesome problems, the deterioration of rolled iron rails. Because the replacement of worn-out rails was expensive and disruptive, the use of longer-lasting rails was imperative. The first use of steel rails by an American railroad was by Pennsylvania Railroad, which had purchased a small quantity of them for testing from their English manufacturers in 1863. Despite the price of the steel rails—twice that of rolled iron rails—the success of the tests of these rails encouraged further purchases by the Pennsylvania and its competitors in the years immediately following the war. The appetite of American railroads for British steel rails, an appetite initially created by brilliant British salesmanship, helped to stimulate the development of steel rail making in this country.

The first steel rails rolled in the United States were made in 1864 at Wyandotte, Michigan, from steel made with the Bessemer process. Production rapidly expanded during the postwar years as American railroad construction accelerated. In 1873 production of rails from Bessemer steel was 115,000 gross tons, a comparatively small amount when compared with the 680,000 gross tons of iron rails rolled that year. Production of steel rails increased during the 1870s, with the entry into the market of the Bethlehem Steel Company in 1873 and the subsequent beginning of operations by Andrew Carnegie's Edgar Thomson Works. By 1877 production of steel rails in the United States had overtaken that of iron rails (386,000 tons of steel rails compared with 297,000 tons of iron rails), and, four years later, output of steel rails exceeded that of iron rails by a margin of almost three to one. The triumph of the steel rail and the demise of the iron rail was signaled in 1884, not a particularly good business year, when the nation's steel rail production was more than 40 times its iron rail output (999,000 tons of steel and 23,000 tons of iron). By 1900 steel rail output was almost 2.4 million gross tons while iron rail output was only 1,000 gross tons.

Labor

The substantial levels of capital investment and the high rate of consumption of raw materials required for iron and steel production during the nineteenth century should not obscure the importance of another factor of production: labor. As the material technology of iron and steel making advanced, leading the companies to become larger and more powerful, the demands upon the industry's workers rose as well. Although the physical scale of production had increased the quantities of iron and steel produced per day and per worker, the production processes remained heavily dependent upon the strong backs and mostly quiet fortitude of the men who did the work.

The types of jobs and degrees of skill required in iron or steel making during the nineteenth century varied from one sector of the industry to another and with each significant stage of the industry's technological development. An eighteenth- or early-nineteenth-century furnace or forge had a relatively simple division of functions and a small number of skilled workers. The latter were the men who actually ran the pigs at the furnace and transformed the pig iron into bar (wrought) iron at the forge's fires and trip-hammer and, of course, the colliers who supplied furnace and forge with the essential charcoal fuel. Most of the other jobs at these installations required far more stamina and strength than skill. This sparely defined arrangement of skills was much less common by the eve of the Civil War and scarcely survived into the postwar decades of the century.

As the technology of iron production and, later, steel production increased the physical scale of production and the productivity of capital and labor, the range of skilled jobs expanded. The two jobs at a rolling mill which demanded the greatest skill and received the highest pay, both before and after the Civil War, were those of the iron boiler and the iron puddler. Comparable levels of skill were required of the founders, keepers, and blast engineers at furnaces.

As one would expect to find, the wages of ironworks and steelworkers during the nineteenth century fluctuated, reflecting changing conditions in the labor market, movements of the general price levels, cycles of business expansion and contraction, and the ebbing and flowing of the workers' organizational strength when bargaining with employers. An unsuccessful strike called in February 1842 by the boilers of Pittsburgh's rolling mills affords an early illustration of the interaction of these factors. Boilers were highly skilled workers whose job was to tend the boiling pig iron in the rolling mills' furnaces and assure the consistency of the iron. In 1837, just before the panic of that year, their wage was $7.00 per ton. After a short-lived recovery in 1838 from the chaos induced by the panic, economic conditions worsened. A general depression began in late 1839 and lasted until mid 1843, with 1842 being the worst year. From 1837 to the end of January 1842, Pittsburgh's rolling mill owners had cut the wages of their boilers by more than 21 percent to $5.50 per ton. No doubt, this substantial but gradual wage reduction figured greatly in the workers' decision to take action. However, the proximate cause of the strike in February 1842 was the further reduction of the boilers' wages by more than 9 percent to $5.00 per ton.

The strike by Pittsburgh's boilers lasted until it collapsed in July, leaving the workers with no choice but to accept the new prevailing wage of $5.00 per ton, which lasted until August 1845 when the boilers won a strike that had begun in May to get an increase in wages of $1.00 per ton. The new wage of $6.00 per ton remained in force until the beginning of 1850 when mill owners again began to cut wages. Thirteen years after their victory in the strike of 1845, boilers in Pittsburgh were compelled to accept wages as low as $3.50 per ton. To some degree, the sharp reductions in the wages of skilled workers—the cuts in the pay of the unskilled were proportionately more severe—reflected the downward movement of rolled iron prices and the substantial pressure that such long-term price changes put on the mill owners. Faced with high fixed costs, including interest on borrowed capital, they turned to the readily available expedient of reducing variable costs, that is, wages.

This interpretation of events is consistent with the collapse of rolled iron prices between 1839 and 1843 and their continued decline during most of the 1840s. However, it fails to account for the actions of the mill owners between 1850, when they again cut the wages of their boilers, and the end of 1857, when the celebrated panic of that year knocked the bottom out of a boom in railroad construction and residential and commercial building. During the latter interval, prices of rolled iron products at first rose considerably and then subsided somewhat before plunging in 1857. As an explanatory vehicle, then, price movements by themselves tell only a part of the story.

American rail makers, most of whom were located in Pittsburgh, were hard-pressed to compete against the wave of competing British rails and did not really begin to hold their own in the contest until 1854 and 1855. By that time they had sustained a fair amount of damage as a consequence of not having been able to do enough business to amortize their physical plants on an efficient schedule. Again the pressure on their fixed costs induced them to put pressure on variable costs and cut their workers' pay. During the dozen years before the Civil War, mill owners in Pittsburgh defeated strike after strike, most of which were called with the aim of winning a partial restoration of lost wages.

The first strikes were somewhat haphazard affairs, but later walkouts were well-coordinated actions led, in some instances, by a necessarily secret organization of puddlers, the Sons of Vulcan. This lodgelike union came into the open during the Civil War when a shortage of skilled labor and booming rolled iron production provided workers with reliable protection against retaliatory dismissals and blacklisting by employers. The membership of the Sons of Vulcan increased rapidly and, on the eve of the panic of 1873, numbered more than 3,300 skilled workers. The size of the membership and its highly skilled character made the union a force to be reckoned with in wage and hour negotiations with employers and enabled it to hold the line against the tide of wage cuts which occurred in every sector of the iron industry following the Civil War, but particularly after the panic.

An early instance of effective unionization, the Sons of Vulcan deserves a prominent place in the labor history of the United States. More impressive, though, was the Amalgamated Association of Iron and Steel Workers, which was formed by the merger of three existing unions, including the Sons of Vulcan, during the depression which followed the panic of 1873. That labor's response to severe and prolonged downward pressure on wages resem-

bled that of its employers who confronted similar pressures on prices was more sensible than ironic. Each group resorted to associationism in an effort to achieve collectively what no individual worker or capitalist could accomplish: stability and a restoration of eroded wages, in the case of labor, and eroded profits, in the case of capital.

Occasionally, as in 1877, 1886, and, especially, in 1892, the contest between capital and labor assumed a violent cast. The clash in July 1892 at Andrew Carnegie's steelworks in Homestead, Pennsylvania, erupted when Henry Clay Frick, whom Carnegie had left in charge of Homestead, deliberately precipitated a strike by the Amalgamated to break the union. His plan included the use of 300 armed Pinkerton agents to intimidate and replace the striking workers. The arrival of the Pinkertons on July 6 was greeted by infuriated and armed strikers who awaited them on shore. After a daylong bloody battle the Pinkerton men surrendered. The strikers' victory was, however, very short-lived, lasting only until July 12 when 8,000 troops of the Pennsylvania state militia entered the mill and broke the back of the strike. By the time the strike had officially come to an end with the union's withdrawal from the mill, Frick had been shot twice by a would-be assassin, seven men had been killed, and many others had fallen wounded. Frick emerged from the strike convinced that he had done the right thing and that his victory had been worth the high cost. His reputation among the workers for brutality made him a hero to other steel mill owners who saw in his harsh and unyielding stand against the Amalgamated the tool which they had sought in their battle with their organized workers.

The Entrepreneurs

Some of the individuals whose lives and business careers are scrutinized in this volume achieved much more than a modicum of success, and a few, most notably Andrew Carnegie and Henry Clay Frick, propelled themselves to a far more rarified level of accomplishment and reputation. Most, however, simply built and maintained going concerns in the iron and steel industry and helped to shape its development during the first three-quarters of the nineteenth century before the rise of Carnegie and Frick and their large corporations. And, because this volume's purview is that entire century, some of the people discussed in it were very small businessmen, indeed. They would appear in any business or economic history of American iron making in an almost fleeting way in the first few decades after 1800 when the iron industry was still partially rooted in the traditional technology of its colonial past.

No claim is made, therefore, that the lives recounted here were typical or representative of all those who were involved in the nineteenth-century iron and steel industry. Such a claim would be difficult if not impossible to substantiate. Moreover, it would be beside the point, which is that the lives of these men, and one woman, illustrate certain characteristics which were common to successful people in many different occupations. Most of them were profoundly determined people, possessed of a quiet stamina and self-confidence and even courage, which permitted them to surmount the considerable uncertainties that capricious markets, sharp competition, and an often haphazardly functioning political economy put in their paths. Some of them were so single-minded—the harsher word is obsessed—in the pursuit of success, measured in money and tons of iron made, social prominence achieved, or control over their surroundings, that they sacrificed friendships, families, and lives to achieve it. A few—again, Henry Clay Frick comes readily to mind—so hated labor unions and so avidly sought monopoly power that they invited the epithet of "robber baron" that was thrown at them.

But most of these people seem to have been fairly decent sorts by the standards of their day or, for that matter, our own. That they were ambitious and that most achieved significant success is obvious; less so is the influence on them of their family and social environment. Fewer than 20 percent came from families with a direct involvement in iron making, which meant that the overwhelming majority fell or were drawn into an industry with which they could have initially had no more than a passing familiarity. Most, however, were not strangers to the ways of business and economic security and came from comfortable backgrounds. Thus, few of them could fairly be described to have been self-made; self-improved, even dramatically self-improved, is a more accurate characterization.

A study of their lives adds weight to the argument that education contributes to success in business: only 20 percent had no formal education at any level. About 30 percent of them had some college education, and more than 10 percent studied at the postgraduate level. More than two-thirds were native-born, and almost all of those who were not came from Great Britain or Germany. Religion does not seem to have been a matter of great moment for these people, most of whom had been born into one of the Protestant denominations. All of this is to say that few of these individuals started out on the bottom of the social ladder looking up; instead, most were well situated and already socialized in the ways of business before embarking on their careers in the iron and steel industry. As a number of studies have shown, this pattern was the common one for American businessmen of the nineteenth century.

Collectively, these people present an interesting demographic profile. Their average age at death was just short of seventy-one, and a quarter of them lived past eighty, a remarkable record when considered in light of nineteenth-century mortality rates and the stress of running a business. Slightly under a third never married, and about half of the total number were either too young or too old to have participated in the Civil War, the pivotal event of the nineteenth century. Of the other half, only about 20 percent saw service during the war (only one of those who did, Joseph Reid Anderson, served in the Confederate army), a proportion that was less than half that of the general northern male population of military age.

Many, perhaps most, of these businessmen (the one woman, Rebecca Lukens, died just before the age of sixty and was preoccupied with her family after retiring from her company's affairs) turned to civic affairs and philanthropy of one form or another upon or shortly before retiring from business. A case in point was Philip Moen of Washburn & Moen, who campaigned vigorously against the saloons frequented by his workers in Worcester, Massachusetts. Peter Cooper and Andrew Carnegie, while the most celebrated for their many good works and passionate interest in education, were hardly alone in seeking a socially constructive outlet for their wealth and energies. Most of those iron and steel men who did indulge in philanthropic work chose colleges and other educational institutions to be recipients of their donations. Many of these businessmen, having long enjoyed social standing and considerable influence in their communities, also became active in civic affairs and local politics. This behavior was a variant of the traditional and still expected form of behavior for the man of affairs:

to serve the interests of the public even as he advanced his own. While actively engaged in doing the latter, these men had not always made the time to do the former. The public interest sometimes had to wait until the close of business.

Table A.1
Estimates of Pig Iron Production, 1810-1859 and Actual Pig Iron Production by Fuel, 1855-1900
(in gross tons; one ton = 2,240 pounds)

PRODUCTION ESTIMATES   PRODUCTION BY FUEL USED (1000 gross tons)

YEAR	Temin[a]	Fogel[b]	Charcoal[c]	Anthracite and coke	Bituminous and coke	Total
1810	53,908					
1815						
1820	20,000					
1825						
1830	165,000					
1835						
1840	286,903	347				
1845		574				
1850	563,755	481				
1855	700,154	700	304	341	56	700
1859	750,560	751	254	421	76	751
1860		821	249	464	109	821
1865			234	428	169	832
1870			326	830	509	1665
1875			367	811	846	2024
1880			480	1614	1741	3835
1885			357	1298	2389	4045
1890			628	2186	6388	9203
1895			225	1271	7950	9446
1900			385	1677	11728	13789

[a] Source: Peter Temin, *Iron and Steel in Nineteenth-Century America: An Economic Inquiry* (Cambridge, Mass.: M.I.T. Press, 1964), Appendix Table C.1, pp. 264-265.

[b] Source: Robert William Fogel, *Railroads and American Economic Growth: Essays in Econometric History* (Baltimore: Johns Hopkins Press, 1964), Table 5.6, p. 166.

[c] Source: Temin, *Iron and Steel*. Appendix Table C.2, pp. 266-267.

Remarks: The course of charcoal iron production is especially interesting because output increased while the number of furnaces decreased markedly. This was possible because of the construction of progressively larger charcoal blast furnaces which operated with increasing efficiency and productivity. See: Richard H. Schallenberg, "Evolution, Adaption and Survival: the Very Slow Death of the American Charcoal Iron Industry." *Annals of Science*, 32 (1975): 341-358; and Schallenberg and David A. Ault, "Raw Materials Supply and Technological Change in the American Charcoal Iron Industry." *Technology and Culture*, 18 (July 1977): 436-466.

Table A.2

U.S. Mineral Fuel Production, 1800-1900

(selected years; 1000s of short tons)

YEAR	Anthracite Coal[a]	Bituminous Coal	Coke
1800		108	
1805		146	
1810	2	176	
1815	2	253	
1820	4	330	
1825	43	437	
1830	235	646	
1835	760	1,059	
1840	1,129	1,345	
1845	2,626	2,097	
1850	4,327	4,029	
1855	8,607	7,543	
1860	10,984	9,057	
1865	12,077	12,349	
1870	19,958	20,471	
1875	23,121	32,657	
1880	28,650	50,757	3,338
1885	38,336	71,773	5,107
1890	46,469	111,302	11,508
1895	57,999	135,118	13,334
1900	57,368	212,316	20,533

a. Pennsylvania anthracite coal.

Source: *Historical Statistics of the United States: Colonial Times to 1970*, 2 volumes (Washington, D.C.: Government Printing Office). See: Series M 123, pp. 592-593 for anthracite coal production; Series M 93, p. 590 for bituminous coal production; and Series M 122, p. 591 for coke production.

Remarks: At first sight, the figures suggest that bituminous coal was far more important as a fuel source than was anthracite coal before the Civil War. In fact, just the opposite was the case. Although bituminous coal has a slightly higher heat value, that is, it emits more heat during combustion than does anthracite coal and is easier to ignite, it also burns at a much faster rate, requiring that a furnace tender feed the furnace far more frequently. The resulting greater inefficiency, as well as the higher sulphur content of the bituminous coal, largely limited its use by antebellum ironmasters to rolling mill operators for whom the coal's sulphur content was of no concern because of their use of puddling furnaces.

Even among rolling mills, however, anthracite coal was preferred because of the greater efficiency with which it could be used. This point can be illustrated by comparing two rolling mills active in 1849 in Pennsylvania—one in the eastern part of the state and one in the western part of the state. Both consumed roughly the same quantity of pig iron—9,000 tons by the western mill and 9,807 tons by the eastern mill—and produced almost identical amounts of rails—7,200 tons by the western mill and 7,357 tons by the eastern mill. The western mill consumed 672,000 bushels of bituminous coal, each bushel having weighed 80 pounds. The eastern mill consumed 19,466 gross tons (2,240 pounds) of anthracite coal. A gross ton was the equivalent in weight of 28 bushels (of 80 pounds each): 28 x 80 lbs. = 2,240 lbs. The eastern mill's consumption of anthracite coal, in terms of bushels, was 545,048 bushels: 28 x 19,466 = 545,048. The eastern mill's consumption of anthracite coal was the equivalent of 74 bushels per ton of rails produced. The western mill's consumption of bituminous coal was 93 bushels per ton of rails produced, more than 25 percent greater than the eastern mill's rate.

Source: Charles E. Smith, "The Manufacture of Iron in Pennsylvania," *Hunt's Merchants' Magazine*, 25 (November 1851): 574-581 and tables following p. 656.

Table A.3
Steel Production in the United States, by Process,
1880-1900
(net tons)

Steel Output, by Process, 1880 to 1900

Year	Bessemer	Open Hearth	Crucible
1880	1,203,173	112,953	80,889
1881	1,539,157	146,946	92,809
1882	1,696,449	160,542	88,104
1883	1,654,626	133,679	86,054
1884	1,540,592	131,617	64,773
1885	1,701,762	149,381	66,207
1886	2,541,493	245,250	83,260
1887	3,288,357	360,717	70,685
1888	2,812,500	352,036	82,837
1889	3,281,829	419,488	90,703
1890	4,131,536	574,820	83,964
1891	3,637,107	649,323	86,318
1892	4,668,647	750,276	99,968
1893	3,601,568	826,437	74,390
1894	3,999,871	879,128	62,477
1895	5,498,223	1,273,644	76,747
1896	4,390,295	1,454,544	70,653
1897	6,132,353	1,801,712	81,727
1898	7,402,099	2,497,927	104,774
1899	8,496,716	3,300,994	118,930
1900	7,486,942	3,805,911	118,075

Source: American Iron and Steel Association, photostat 1880-1886; *Annual Statistical Report*, 1940, p. 15 (1887-1900).

Encyclopedia of American Business History and Biography

Iron and Steel in the Nineteenth Century

Horace Abbott

(July 29, 1806-August 8, 1887)

by John A. Heitmann

University of Dayton

CAREER: Supervisor, (1840?-1847); owner, Canton Iron Works (1847-1865); president, Abbott Iron Company (1865-1879).

Horace Abbott, an iron manufacturer, capitalist, was born in Sudbury, Massachusetts, on July 29, 1806, and died near Baltimore, Maryland, August 8, 1887. Abbott had little opportunity for formal education as a child and was apprenticed to a blacksmith in Westboro, Massachusetts, at age sixteen. After serving his apprenticeship, he became a country blacksmith. In 1836 he and his brother, Edwin Augustus Abbott, moved to Baltimore, where he became interested in the manufacture of iron. New Yorker Peter Cooper had invested heavily in the manufacture and fabrication of iron products in an area of the city known as Canton, and by the early 1840s Abbott had leased Cooper's Canton Iron Works. There Abbott supervised the production of wrought-iron shafts, cranks, axles, and other equipment needed for steamboats and railroads. It was at the Canton Iron Works in 1841 that the first heavy engine forgings (rather than castings) were made.

The business relationship between Cooper and Abbott was often strained during the 1840s–late rent payments, problems with pig iron supplies, and disagreements over equipment were frequent points of contention–until the latter agreed to purchase the ironworks in 1847. However, a cooperative working relationship between Abbott's works and Cooper's Trenton Ironworks later developed; Abbott supplied the New Jersey firm with pig iron, and during the Civil War Cooper's partner Abram Hewitt arranged for government contracts for the Baltimore firm.

In 1850 Abbott constructed the first of a series of roller mills that resulted in great prestige for the company during the decade before the Civil War. The original mill, built for rolling plate and boiler iron, contained four heating and two puddling furnaces, a pair of 8-foot-long roller mills, and a train of muck rolls. A second mill was completed in 1857 and consisted of three heating and two puddling furnaces, a Nasmyth steam hammer, one pair each of 8-foot and 10-foot rollers. Mill number 3 was erected in 1858 and made thin gas pipe and boiler tubes by the use of two heating furnaces and a pair of 5-foot rollers. A fourth mill was completed in the summer of 1861 and consisted of three heating and four double puddling furnaces, a pair of 10-foot rolls, a Nasmyth hammer, and other machinery.

Thus, Abbott's rolling mills were of the most advanced of the day, and this productive capacity played an important role in furnishing critically needed material during the Civil War. It was at the Abbott Iron Works that the armor plates were made for the original *Monitor* and later for other union vessels including the *Roanoke*, *Agamenticus*, and *Monadnock*.

With the conclusion of the Civil War in 1865 Horace Abbott sold his business to a group of capitalists who reorganized the company as the Abbott Iron Company and elected Abbott its first president. As a man of wide business interests, Abbott branched out from manufacturing and became involved in the establishment of several banks in Baltimore, including the First National Bank. He was also a director of the Baltimore Copper Company and the Union Railroad of Baltimore.

Development within Abbott's firm reflected broader changes that were taking place in the nineteenth-century iron industry in terms of organization and scale of production. While the Abbott Iron Works typified the best practices of the day at mid century, the company's production techniques did not change with the times. Soon after Abbott's death in 1887 the firm that he had worked so hard to establish and maintain was eclipsed by corporations using more advanced techniques, employing large sums of capital, and producing larger quantities. In 1891 three blast furnaces owned by the Pennsylvania Steel Corporation were in operation at Sparrows Point, and Bessemer steel was produced in Baltimore. The manufacture of charcoal iron using local ores was displaced by new metallurgical methods and the importation of ores from northern Spain, Algiers, and Cuba. Abbott Iron Company soon disappeared, its place in Baltimore's industrial landscape being replaced by American Can and the Crown Cork & Seal companies.

References:

Baltimore: Past and Present, with Biographical Sketches of its Representative Men (Baltimore: Richardson & Bennett, 1871);

D. Randall Beirne, "Residential Growth and Stability in the Baltimore Industrial Community of Canton During the Late Nineteenth Century," *Maryland Historical Magazine*, 74 (1979): 46;

The Biographical Cyclopedia of Representative Men of Maryland and District of Columbia (Baltimore: National Biographical Publishing, 1879);

J. Leander Bishop, *A History of American Manufacturers from 1608 to 1860*, volume 3 (1868; reprinted, New York: Augustus M. Kelley, 1968);

Eleanor S. Bruchey, "The Development of Baltimore Business, 1880-1914," *Maryland Historical Magazine*, 64 (1969): 146-147;

Industries of Maryland: A Descriptive Review of the Manufacturing and Mercantile Industries of the City of Baltimore (New York: Historical Publishing, 1882);

J. Thomas Scharf, *History of Baltimore City and County* (1881; reprinted, Baltimore: Regional Publishing, 1971).

William Latham Abbott

(April 27, 1852-May 2, 1930)

by Larry Schweikart

University of Dayton

CAREER: Various positions, Carnegie Steel interests (1871-1886); vice-chairman (1886-1889), president, chairman, Carnegie, Phipps & Company (1889-1892); director, Carnegie Brothers (1889-1892).

William Latham Abbott, civil engineer and chairman of Carnegie, Phipps & Company, was born in Columbus, Ohio, on April 27, 1852, and died on May 2, 1930. The son of Timothy Dwight Abbott and Mary Cutler Crosby of New Haven, Connecticut, Abbott entered service with the Carnegie interests on August 14, 1871, as a clerk at the City Mills. He subsequently rose through the ranks of management at the Edgar Thomson Steel Works in Braddock, Pennsylvania, and was promoted to the position of superintendent of the Upper and Lower Union Mills (formerly the Cyclops Iron Works and Kloman-Phipps) in Pittsburgh. After the Carnegie interests became Carnegie, Phipps & Company in 1886, Abbott was elected to the position of vice-chairman. In 1889 he became president of the firm and soon rose to the chairmanship, succeeding

John Walker, who had been named chairman in 1886. In his capacity as chairman, Abbott also served on the board of managers of Carnegie Brothers. At the time of his presidency Abbott was just nearing forty years old and was well known for his handsome looks.

Abbott's tenure as president produced considerable fireworks at the company, and many Carnegie contemporaries regarded Abbott as inexperienced and especially "weak" on labor issues, the last opinion due to his handling of the 1889 Homestead Strike. In the months prior to the strike a sliding pay scale had been implemented at the Edgar Thomson Works over the opposition of the Amalgamated Association of Iron and Steel Workers of North America. The agreement to implement the scale promised to raise wages for most workers but also mandated that the authority to negotiate future contracts be taken away from the union and returned to individual workers. The company next attempted to implement the sliding scale and its conditions at the Homestead plant. Andrew Carnegie, who was in Great Britain, had cabled instructions for Abbott to stand firm against the union and was personally quite willing to have the mills close until Homestead could operate at competitive wage levels.

Homestead shut down on July 1, and although Carnegie had advised Abbott not to use force, the president advertised in local papers for strikebreakers. A small group of black and immigrant workers showed up and attempted to enter the works with an escort of a local sheriff and 125 of his deputies, only to find themselves confronted by an army of nearly 2,000 strikers. The strikebreakers and deputies retreated without a fight. Meanwhile Abbott faced growing resistance from the Edgar Thomson workers, who threatened a sympathy strike. (These workers were, according to Carnegie's plan, to be "de-fused" in a meeting with Capt. W. P. "Bill" Jones, who was under orders from Carnegie to keep the Edgar Thomson group out of the strike.) The workers at the Beaver Falls plant also threatened to walk out, at which point Abbott held a meeting with the strike leaders. The meeting had the effect of recognizing the Amalgamated Association as the bargaining agent for the workers. Despite the fact that the union accepted the sliding scale, which reduced the wages for skilled workers who were paid on a flat tonnage basis, the union to some degree had won a union shop.

Abbott received both praise and blame from Carnegie management. Carnegie himself, while pleased that the contract brought three years of labor peace, implied that Abbott should have been tougher and locked out the strikers, as he assured the president he would have done. Abbott, for his part, was proud that he had not used "the cable" to contact Carnegie about his decision. The role of the Homestead Strike and the resulting agreement with the Amalgamated Association in Abbott's decision to retire when the contract ended in 1892 is unclear. He undoubtedly was aware that the imminent expiration of the contract meant that he might relive the anxious days of 1889. Moreover, he realized that the pending merger of Carnegie Brothers and Carnegie, Phipps put him in an uncomfortable position in regard to the ambitious Henry Clay Frick. While it does not appear that Frick and Carnegie forced Abbott out of the presidency, neither attempted to persuade Abbott to remain in an executive capacity. Abbott's exact title—"president" or "chairman"—remains vague. At least one Carnegie authority maintains that despite references to him as chairman, Abbott never actually attained the position. Ultimately, Abbott lacked the sharp

William Latham Abbott

business edge that Frick and Carnegie still retained, and given his considerable wealth, Abbott retired in June 1892. He maintained holdings of $250,000 in the newly formed Carnegie Steel Company, and he had advised Carnegie to merge the companies. But the company suffered with the loss of Abbott: the merger, the expiration of Abbott's 3-year deal with labor, and Frick's and Carnegie's determination to oust the Amalgamated Association together made conditions ripe for another confrontation. What Abbott had avoided, Frick desired: a second shot at breaking the Amalgamated Association. Frick, relying on the company's surplus, the productive capacity of the firm's nonunion Braddock and Duquesne plants, and new steel-making machinery that was easily operated by unskilled laborers, provoked a strike. He employed the Pinkertons, who engaged in a bloody battle on July 6, 1892, in which the strikebreakers were defeated. Four days later state militia arrived, and the Homestead Works were again in operation. Throughout the affair Frick repeatedly referred to Abbott's weakness in dealing with the union in 1889 as a cause of the violence. Carnegie, however, in his autobiography referred disparagingly to Frick's inexperience as the cause of the trouble in 1892, not Abbott's conciliation with the union.

Abbott most likely retired to enjoy his wealth, which was considerable. According to Carnegie biographer Joseph Wall, Abbott "never enjoyed the fierce competitiveness of business." He nevertheless maintained his holdings in Carnegie Steel as part of the Iron Clad Agreement of 1887 (as did all partners of Carnegie Steel). This agreement, drawn up at the time of Thomas Carnegie's death (and when Carnegie himself was ill) at the suggestion of Henry Phipps, permitted the company to assume the interest of a deceased partner for payment to the estate of book value for the stock, measured over an extended period of time. The agreement also provided that if both three-fourths of the existing associates and three-fourths of the interest (the number of stocks issued) so voted, an individual associate could be forced out, either through resignation, assignment, transfer, or sale to other associates. Carnegie, obviously, was the sole exception to the Iron Clad Agreement, by virtue of his ownership of half of the company. The Iron Clad Agreement survived the 1892 reorganization, but Abbott did not survive the Iron Clad. By 1895 Abbott had voluntarily transferred his stock. Although he received a book value that closely approximated market value, there are suggestions that Abbott transferred his stock to the other associates only after heavy pressure from them to leave the company for what Wall terms "his speculative tendencies."

Abbott, however, never played the role of yes-man to Carnegie, willingly challenging the Scotsman in person or by mail. He, like John Walker, Thomas Carnegie, and Henry Phipps, disagreed with Carnegie's practice of paying small dividends in order to plough the profits back into plant expansion and new or improved equipment. His arguments were so clear and forceful that Carnegie joked that upon his own retirement, the literary mantle of the firm would fall to Abbott.

In his role as president and chairman of Carnegie, Phipps & Company, Abbott presided over the introduction of the "Thomas basic process" to the company's hearths. This process, developed by Sidney Gilchrist-Thomas, an amateur chemist, removed phosphorus from iron so that it was suitable for the Bessemer converter developed by Henry Bessemer. Original Bessemer furnaces had been lined with an acid material, but the basic process used

lime or magnesia to line the converters, thus extracting the phosphorus. This process made vast new iron resources available for steel making, and Carnegie pressed Abbott to expand the open-hearth facilities of the company as rapidly as possible.

Abbott also played a major role in turning the company into a producer of steel armor plate for the United States Navy, which at the time had begun to build a fleet of steel-hulled ships. Clearly, Carnegie himself negotiated the crucial sales and handled the intricate personal dealings for the firm with American and foreign governments. But it fell to Abbott to administer and execute the production of the armor plates in the mills. Abbott once volunteered the company to test its armor plates before signing any agreements with the United States Navy, an unusual display of pre-Pentagon salesmanship. In 1890 Carnegie, Phipps & Company signed a contract to provide 6,000 tons of steel armor plate to the United States government, and that same year obtained a patent from Abel-Ray Corporation for the use of a ferro-nickel process for making compound nickel plates. Abbott had even submitted the bid for steel plates on the battleship *Maine* in 1888–a bid Carnegie thought far too low. When Carnegie communicated his opinions to the president, Abbott responded vigorously by producing data supporting the bid. Carnegie not only retracted his criticism but praised Abbott's eloquent and well-reasoned defense.

In addition to Abbott's investments in Carnegie Steel, he had been invited to invest in other Carnegie interests, which, given Abbott's youth and relatively short tenure in the Carnegie "inner circle," was a remarkable comment on the young engineer. He owned stock and served as a director in the Keystone Bridge Works, a company Carnegie founded in 1865 in response to the rise of the railroads, a major outlet for Carnegie's iron.

Among Abbott's personal activities, he actively supported the Church of the Ascension (Episcopal). He was a member of the American Society of Civil Engineers in 1889 and the Duquesne Club of Pittsburgh. He also belonged to the Pittsburgh Club, the Allegheny Country Club, the Pittsburgh Golf Club, the Oakmont Country Club, the Cobourg Golf Club, (which he served as president), the Union League Club of New York, the Mountain Lake Club of Mountain Lake, Florida, the Carnegie Hero Fund, the Pennsylvania Society of New York, and the Western Pennsylvania Institution for the Blind. He was also president of the Children's Hospital of Pittsburgh and was a guarantor of the Pittsburgh Orchestra. Abbott married Anne Wainright of Pittsburgh on May 17, 1887. They had eight children: Lois, Franklin, William Latham, Jr., Jeanette, Ruth, Wainright, Valerie, and Anne. Abbott and his wife traveled extensively, spending summers in Sidbrook, Cobourg, Ontario, and winters at their residence "Villa Primavera" in Mountain Lake, Florida. They visited or lived at different times in Geneva, Switzerland, and Florence, Italy. Abbott's children received their education in Europe. He was also a member of the Carnegie Veterans Association, which elected him a member of the Carnegie Hero Commission. William Latham Abbott died in New York at the Mayfair House on May 2, 1930, at the age of seventy-eight.

Abbott is notable in that he was the executive responsible for negotiating the first agreement with the union at the Homestead works. His other contributions to the Carnegie companies were also significant and stand as a contradiction to his own view that "most of Andrew Carnegie's partners were most ordinary men." In fact, William L. Abbott was most extraordinary.

References:

Andrew Carnegie, *The Autobiography of Andrew Carnegie* (Boston: Houghton Mifflin, 1920);

William Dickson, ed., *History of the Carnegie Veterans Association* (Montclair, N.J.: Mountain Press, 1938);

Louis Hacker, *The World of Andrew Carnegie* (Philadelphia: J. B. Lippincott, 1968);

Burton Hendrick, *The Life of Andrew Carnegie*, 2 volumes (Garden City, N.Y.: Country Life Press, 1932);

Robert Hessen, *Steel Titan: The Life of Charles M. Schwab* (New York: Oxford University Press, 1975);

William T. Hogan, *The Economic History of the Iron and Steel Industry in the United States* (Lexington, Mass.: Lexington Books, 1971);

Joseph Wall, *Andrew Carnegie* (New York: Oxford University Press, 1970).

Adirondack Iron & Steel Company

by Bruce E. Seely

Michigan Technological University

The Adirondack Iron & Steel Company neither produced the first quality cast crucible steel in the United States nor became a leader of the industry. Rather, the company's performance and fate were typical of those firms that attempted to end the British steel monopoly from 1830 to 1860; the company failed within a few years of its opening in 1848. Despite its failure the Adirondack Iron & Steel Company played a direct role in the development of a domestic steel industry in the United States. The firm's first superintendent, Joseph Dixon, introduced the graphite crucible, which by its ability to withstand the high temperature of molten steel overcame the crucial technical problem in early American steel making. The company also provided a training ground or object lesson for a group of steel men who were highly successful after the Civil War. But to historians the greatest significance of the Adirondack Iron & Steel Company may be the pattern of behavior by the company's investors, who typified an American entrepreneurial spirit that assumed success would follow every venture. This company is a perfect example of the optimistic and enthusiastic thinking that propelled America through its initial industrialization in the first half of the nineteenth century.

The origins of the Adirondack Iron & Steel Company were in the 1826 discovery of the largest iron ore deposit east of the Mississippi near the source of the Hudson River in New York's Adirondack Mountains. An account of the expedition that found the iron is a classic tale of Adirondack lore and has been colorfully retold in Paul Jamieson's *Adirondack Reader* and Alfred Donaldson's *A History of the Adirondacks* (1921). The ore deposit is located about 15 miles from present-day Lake Placid, at the hamlet of Tahawus. From the beginning the discoverers of the deposit were convinced that the site contained ore which possessed the special quality needed to make steel. This seemed highly significant at a time when steel production was controlled by the British, and even they relied on high-quality wrought-iron bars imported from the Dannemora region of Sweden. Thus, from the start an atmosphere of hopeful expectation surrounded this venture.

Two partners dominated the company's affairs until it floundered in 1854. The leading investor was Archibald McIntyre, a man of many interests. He operated an iron forge at Lake Placid during the War of 1812, owned a cotton mill in Broadalbin, New York, served in the New York legislature, and was state comptroller for the second longest term in New York's history. McIntyre guided the efforts to acquire almost 100,000 acres of land and build an ironworks at the remote property. During the summers of 1832 and 1833 a Catalan forge was built to convert the ore into wrought iron directly, but the effort was abandoned because the iron was smelted only with great difficulty. The attempt was renewed in 1839 following enthusiastic reports about the property's mineral resources by Ebenezer Emmons of the New York State Geological Survey. Emmons arranged for Walter Johnson, a Philadelphia consultant who had devised the Franklin Institute's machinery for its steam boiler experiments, to test sample bars. The results, printed in Emmons's reports, confirmed the owner's belief that their deposit was special.

Unfortunately, the ore was still refractory—that is, heat resistant and therefore hard to smelt. The problem was caused by a high concentration of titanium dioxide, a mineral used today as a paint pigment. In spite of five years of indifferent success at the forge, the company decided in 1843 to erect a small blast furnace. At this point another partner, McIntyre's son-in-law David Henderson, became the moving spirit in the company. The proprietor of the first successful commercial pottery in the country, one that trained a generation of successful potters, Henderson was interested in chemistry and determined to identify the cause of the problems. Unfortunately, the furnace also failed to work easily, and Henderson's death in a shooting accident near the works in late 1845 was a major blow to the firm's prospects. No other partner possessed Henderson's combination of technical interest with business skills. Yet even before his death, the ironworks was in real trouble. James T. Hodge, a leading iron expert of the day, visited the works in 1846 and reported that the furnace never yielded a liquid slag, requiring hard work to produce any iron at all.

Yet the owners, led again by McIntyre, persisted, mainly because the iron produced through such struggle was of good quality. Again ignoring the firm's problems, the partners launched an effort to build a steelworks. The inspiration for such an expansion was drawn from contact with a Sheffield steel maker, who was thought to be interested in buying the property. In 1846 the company built a puddling furnace and in 1847 constructed a second dam across the Hudson to supply power for a cast crucible steelworks. This steel plant, however, was not completed, probably because the hoped-for English investment failed to materialize.

But the idea of making steel remained alive, and the company made contact with Joseph Dixon, a multitalented inventor who had developed a graphite crucible. The absence of a suitable melting pot had for years stymied American efforts to make cast crucible steel—the high-quality steel intended for use in tools, saws, and other special purposes. Dixon's invention eliminated this obstacle, and in early 1848 he agreed to build a steelworks for the firm; later he agreed to manage the works for a salary plus a share of the profits. The plant was constructed in Jersey City, the location of Henderson's pottery, after the partners visited the Hawkins & Atwater Company in Derby, Connecticut, a plant which produced blister steel. Hawkins not only opened his works to the visitors but directed Dixon to a Scotsman with 30 years experience in steel making who agreed to build the tilting hammers. The scale of the works was typical of early steel plants: a cementing furnace, 16 melting holes, 3 or 4 tilt hammers, and a boiler and steam engine. The equipment cost approximately $12,000, the building and land another $11,000.

There were problems, however. Late delivery of the steam engine and boiler caused the most severe delay, and the hammers also proved difficult to construct. But the most vexing problem was a dispute between Dixon and McIntyre over the disposition of the original contract when Dixon became the manager of the works. Dixon believed he had money coming; McIntyre disagreed. Quickly each party lost faith in the other. In spite of these problems, the works produced its first blister steel in July 1848. But difficulties with the engine and broken hammer helves delayed production of cast steel, and even in March 1849 the clerk of the works apologized for continued delays.

The Adirondack Iron & Steel Company works, circa 1853

Some things, however, went smoothly. The works secured the help of English immigrants with experience in the steel industry with surprising ease. In addition to the Scots hammer builder, Dixon hired two hammermen and a tilter to pour ingots, all skilled positions. Most importantly, the steel that Dixon finally produced met all of the company's expectations. Samples sent to the Washington Navy Yard, the Springfield Armory, and numerous New York City machine shops produced fine testimonials. One reported, "If the steel manufactured by the American Adirondack Steel Company is equal to the sample you sent me, I shall hereafter give it preference to any European steel we have had in this establishment for years." The Franklin Institute in Philadelphia also awarded Dixon its Elliot Legacy Premium, a prize in recognition of new technical ideas, for his process of making blister steel from pig iron rather than wrought iron. Good notices of the steel appeared in the Franklin Institute *Journal*, the *American Railroad Journal*, and the *North American and United States Gazette*. The ultimate glory would come in London in 1851, when the Adirondack Iron & Steel Company received a prize for quality at the Crystal Palace Exposition. But even prior to this award Dixon's assistant reported that orders were arriving faster than they could be filled, and the partners incorporated in 1849 as the Adirondack Steel Company. They also built an expensive and larger blast furnace, 45 feet high with hot blast, to insure a supply of iron for the steel plant.

The initial promise of the steelworks quickly faded when the company attempted to produce quality steel on a day-to-day basis. By September 1849 the works had produced 140 tons of steel, but only 40 tons had been finished and of that 30 tons had been handed over to agents for sale. The works produced approximately 1 ton of steel per day, but mechanical problems continued to bedevil operations. In January 1850 the works closed to install a larger tilting hammer. Boiler problems forced another suspension of operations in February, and a shortage of crucibles forced a 7-week stoppage in June. Then in late 1849 or early 1850 Dixon left the company. His specific reasons for leaving are unclear, and the partners expressed no regrets. But Dixon left behind several legacies—a working steel plant, knowledge of production processes, graphite crucibles, and a well-trained assistant who became the new superintendent, McIntyre's nephew James R. Thompson.

Unfortunately, the works failed to produce consistently top-quality steel, and Dixon's absence probably worsened the problem. In February 1850 the works reported that while Dixon had been forced to remelt as many as half of the finished steel bars because of quality problems, Thompson remelted almost none. While initially a cause of celebration, this change in procedure likely meant that bad steel was finding its way to customers. A telling blow was the decision of a leading iron house, Jessups of New York, to give Adirondack steel only qualified approval. Not surprisingly, sales had slowed greatly by early 1850. A rolling mill was installed to improve quality, but the initially favorable response of consumers had been eroded by the inconsistent quality. James Swank, in his *History of the Manufacture of Iron in All Ages* (1892), summarized the reputation of the product of the Adirondack Steel Company: "Much of it was good tool steel, but much of it was also irregular in temper." The partners rationalized their difficulties by complaining of prejudice against American steel but in 1853 were forced to close the steel plant. The new blast furnace operated only twice and was closed in 1854.

All told, the partners may have lost $500,000 on their various iron and steel projects. The smelting facilities never produced iron easily, and steel making was plagued by endless technical and quality difficulties. Throughout their ordeal the owners believed that, with a railroad or canal to the works and investment by an English steel firm, their company could rival in size the Merthyr-Tydvil works in Wales. The depth of their optimism, and perhaps a measure of how unrealistic it was, is revealed by the fact that the property continued to be moribund long after the original efforts.

Although several efforts were made by talented iron and steel men in the 1890s, 1900s, and 1910s to revive the works, it was not until 1941 and the advent of war that an open-pit mine was constructed to extract titanium oxide. The U.S. government, through the Defense Plant Corporation, gave $2.5 million to construct a sinter plant and $4.2 million to build a railroad, both operations run by the National Lead Company. The iron sinter, a byproduct of the titanium extraction, was sold to Bethlehem Steel until 1970.

It might seem easy to blame the failure of the Adirondack Iron & Steel Company on the owners'

poor business judgment, for they certainly failed to heed a number of obvious warning signs. They expanded their iron-making operations repeatedly—1833, 1839-1840, 1844, and 1846-1849—even though each addition failed to improve the firm's performance. Nonetheless, it must be remembered that Henderson and McIntyre succeeded spectacularly in their other business endeavors. Henderson's pottery operation was nationally important, while McIntyre was widely respected for his service as comptroller. McIntyre also owned a cotton mill in Auburn, New York, speculated successfully in land in every midwestern state, and operated a coal mine in northeastern Pennsylvania. His only other failure was a gold mine opened in North Carolina in the late 1830s. Nor were the partners the only optimistic thinkers, for they repeatedly retained leading consultants in an effort to alleviate their problems. In every instance the experts endorsed the owner's belief that the iron ore was uniquely suited to steel making. In short, the company's failure illuminates not bad judgment but the optimism and confidence in the future that was a vital component of the drive to industrialization in this country after 1800.

The legacy of the Adirondack Iron & Steel Company is not simply one of repeated, fruitless efforts to bring a promise to reality. After the works were closed, the plant was immediately leased to a Jersey City group called Horner & Company, which retained James R. Thompson as manager until 1857. Horner, however, also met with indifferent success, and at the end of his 10-year lease Dudley Gregory, one of McIntyre's original partners in the steelworks, bought the plant. Gregory installed a new manager and between 1863 and 1866 enlarged the works as well. The remodeled plant had 5 converting furnaces, 40 melting holes, 5 hammers, and 4 trains of rolls, ranging from 9 to 18 inches. Finally, the mill built by the Adirondack Company began to make consistently good steel, joining Hussey, Wells & Company and Park, Brother & Company of Pittsburgh as the first firms to be technically and commercially successful producers of cast crucible steel. By 1876 the plant possessed the capacity to make 2,400 tons of steel per year, although it averaged only 1,750 tons. The addition of two more heating furnaces and two larger hammers in 1880 raised capacity to 3,000 tons. But the sharp cutback in steel demand in the early 1880s finally closed the longest operating steelworks in the country. After Gregory stopped production in 1883, a new manager operated the plant for two more years before another Jersey City steel company–Spaulding Jennings & Company–dismantled the facility in 1885.

Gregory's success was not, however, the only legacy of the original Adirondack Iron & Steel Company. In addition to introducing Dixon's crucibles to the industry, the company also provided an important example for companies with which it later competed. For example, Curtis Hussey's partner, Calvin Wells, visited the Adirondack Company in 1859, specifically to learn how to make crucible steel. It is an intriguing coincidence that Hussey then developed a steel-making process that eliminated a step much as Dixon had. Also learning the steel trade at this facility before moving on to later success was McIntyre's nephew, James R. Thompson. In 1862 he opened the Jersey City Steel Works on a site close to the Adirondack works. Initially Thompson's facilities at the Jersey City Steel Works included 3 puddling furnaces, 7 converting furnaces, 56 melting holes, 2 hammers, 5 steam hammers, and 4 trains of rolls. By 1874 production capacity had increased from 4,000 to 10,000 tons, making J. R. Thompson & Company one of the largest producers in the country. By 1886 the works boasted eight 4-pot melting furnaces, producing 14,000 of steel annually.

Historians rarely examine failure, yet in the case of Adirondack Iron & Steel a failure offers a clear picture of the forces that drove American industrialization and the development of the cast crucible steel industry. Clearly evident is the enthusiasm of American businessmen and their conception of risk. Equally important, the Adirondack Iron & Steel Company played a vital role in the establishment of the cast crucible steel industry in this country.

James Swank offered the final summary when he commented that this company was one of the three or four firms that "dissipated the longstanding belief that this country possessed neither the iron nor the skill required to make good cast steel."

Unpublished Document:

Bruce E. Seely, "Adirondack Iron and Steel Company: The 'New Furnace,' 1849-1854," in U.S. Heritage Conservation and Recreation Service, Historic American Engineering Record (HAER NY-123, 1978).

References:

Alfred L. Donaldson, *A History of the Adirondacks*, 2 volumes (1921; reprinted, Harrison, N.Y.: Harbor Hill Books, 1977);

Henry Dornburgh, *Why the Wilderness is Called Adirondack* (1885; reprinted, Harrison, N.Y.: Harbor Hill Books, 1980);

Harold Hochschild, *The MacIntyre Mine–From Failure to Fortune* (Blue Mountain Lake, N.Y.: Adirondack Museum, 1962);

Arthur Masten, *The Story of Adirondack* (1923; reprinted, with an introduction and notes by William K. Vernor, Syracuse: Adirondack Museum/Syracuse University Press, 1968);

Bruce E. Seely, "Blast Furnace Technology in the Mid-Nineteenth Century: A Case Study of the Adirondack Iron and Steel Company," *Iron Age*, 7 (1981): 27-54;

James M. Swank, *History of the Manufacture of Iron in All Ages* (Philadelphia: American Iron & Steel Institute, 1892).

Archives:

The most important source of information about the Adirondack Iron & Steel Company is the correspondence of Archibald McIntyre with his partners and works managers, which is deposited as the McIntyre Papers in the library of the Adirondack Museum in Blue Mountain Lake, New York. This material spans the entire history of the ore deposit, from 1826 through World War II.

Aetna Iron & Nail Works

by John A. Heitmann

University of Dayton

Organized in 1873 near Wheeling, West Virginia, the Aetna Iron & Nail Works had a historical significance that far exceeded its size and impact upon the local economy. The firm not only introduced innovations that later were adopted widely by the iron and steel industry but also served as a training ground for many leaders who would make their mark after the wave of mergers and consolidations of the late 1890s.

The Aetna Iron & Nail Works, initially capitalized at $200,000 and located between Martins Ferry and Bridgeport, West Virginia, started operations with 20 puddling furnaces and 5 rolling mills. The firm was erected on the former site of a coal-mining and wire-making establishment, and its primary products were iron sheets, bars, bands, and light rails. Led by capitalist J. J. Holloway, who served as the company's president, and William J. Tallman, who was appointed secretary and general manager, the Aetna Iron & Nail Works developed a close working relationship with the Standard Iron Company, a neighboring firm established in 1882 to manufacture iron sheets. The two companies shared many of the same directors and in 1893 were formally merged into the Aetna-Standard Iron & Steel Company.

While both Holloway and Tallman had limited experience in the iron trade before 1873, a lack which caused problems for the company during its early years, the two managers were also disposed to try new approaches that more experienced operators would avoid. As a result of their innovations, which included new techniques in the rolling of gauged steel, the use of improved squaring shears for trimming sheet mill packs, and the introduction of the roller leveler, an essential piece of equipment in the twentieth-century sheet mill, Aetna-Standard was identified as one of the most technologically advanced iron and steel companies of the last quarter of the nineteenth century.

J. J. Holloway, president of Aetna Iron and Nail Works (courtesy of West Virginia Department of Culture and History)

This degree of innovation was not confined merely to improving plant production. Aetna-Standard officers were the first in the Wheeling, West Virginia, area to take advantage of the high McKinley tariff on tinplate by installing in 1893 a tinning department in its sheet and black plate mills. Indeed, Aetna-Standard management had a knack of attracting some of the most talented young men in the iron and steel business during the 1890s. For example, William T. Graham, who rose within the organization to become president during the 1890s, later moved on to become vice-president of the American Tin Plate Company, president of the American Sheet & Tin Plate Company, and president of American Can Company. John A. Topping, appointed secretary of Aetna Iron & Steel at the age of twenty-eight, would later become first vice-president of the American Sheet Steel Company, president of La Belle Iron Works, president of the American Sheet & Tin Plate Company, chairman of the Tennessee Coal, Iron & Railroad Company, and finally president and then chairman of the Republic Iron & Steel Company.

Businessmen like Graham and Topping advanced their careers during the era of combination that characterized late-nineteenth-century American business, and the Aetna-Standard Company played no small role in the merger mania of the day. The firm did some brisk trading, selling its blast fur-

naces and steelworks to the National Steel Company, its tinplate mills to the American Tin Plate Company, its bar mills to the American Steel Hoop Company, and its black sheet mills to the American Sheet Steel Company. When the United States Steel Corporation took control of these horizontal combinations in 1901, not only was Aetna-Standard reunified in a sense, but original shareholders, in the words of historian H. D. Scott, "had a very respectable melon to divide."

References:
Earl Chapin May, *Principio to Wheeling, 1715-1945* (New York: Harper, 1945);

Robert L. Plummer, *Sixty-Five Years of Iron and Steel in Wheeling* (Wheeling, 1938);

Henry Dickerson Scott, *Iron & Steel in Wheeling* (Toledo: Caslon, 1929).

Cyrus Alger

(November 11, 1781-February 4, 1856)

by Alec Kirby

George Washington University

CAREER: Manufacturer, inventor, real estate developer (1798-1856); owner, South Boston Iron Company (1809-1856); alderman, city of Boston (1824-1830).

Cyrus Alger, manufacturer and inventor, was born on November 11, 1781, into a well-established and prosperous Bridgewater, Massachusetts, family. To a striking degree Alger followed the career model of his father, Abiezer. As did his father, Alger became a successful leader in the iron industry and took an active part in the political and social life of Massachusetts. Alger operated an iron-producing furnace in each of three Massachusetts towns—West Bridgewater, Easton, and Titicut.

Early in his life Alger pursued a liberal education; for a time he attended the Tauton Academy in Taunton, Massachusetts. Before graduation, however, he left the academy and went to work for his father, from whom he learned the principles of iron production. Within a few years he was placed in charge of his father's Easton plant. In 1804 he married Lucy Willis, with whom he had seven children.

Alger expanded his business interests in 1809 when he cofounded an iron foundry in South Boston. This foundry received large orders for military supplies during the War of 1812, particularly for cannonballs. In 1814 Alger bought out his partner, Gen. John Winslow, and expanded the company's operations. At about the same time he began to invest in real estate through the South Boston Association, which took an interest in lands recently annexed to the city. Through the construction of a seawall the association had reclaimed low-lying acreage. Alger purchased the reclaimed land, with the stipulation that the deed include land in front of the seawall to the low-water mark. Alger then developed the marginal land, laying out roads and building his own factory and house in this section, and he persuaded others to do the same. Over the next several years wharves and commercial buildings were constructed, and the area became a manufacturing and mercantile section of great importance.

Meanwhile, he was conducting experiments which led to the first of his five patented inventions. The first patent was issued on March 30, 1811, for an improved method of making cast-iron chilled rolls, by which the part of an iron product subject to wear is given added strength. Shortly thereafter Alger introduced anthracite coal as fuel in his Boston furnaces. In 1822 he invented cylinder stoves and reversed the hearths of furnaces for melting iron, so that the molten metal would flow toward the flame. With these innovations Alger rapidly increased the capacity of his iron plants. In 1827 he consolidated his holdings into the South Boston Iron Company and was elected its president, a post he held until his death in 1856. The company—known as "Alger's Foundries"—enjoyed a remarkable reputation as a well-managed and innovative firm. Among the company's employees were several prominent leaders in the iron industry, including Willliam P. Hunt, who was eventually to acquire a reputation in his own right as a manufacturer and inventor.

In 1828 the South Boston Iron Company began to manufacture iron ordnance. Alger invented a method of purifying cast iron which nearly tripled the strength of ordinary iron castings. This innovation gave the firm a great advantage in making iron guns—especially those of large caliber. "Gun iron" was the name by which the strengthened iron came to be known, although the process was used for a multitude of castings. With this special expertise, the company increasingly focused its energies on the manufacture of ordnance and created new technology in weapons production. In 1834 the first rifled cast-iron gun made in the United States was cast and finished by the South Boston Iron Company, and in 1835 the firm began the manufacture of malleable iron guns, for which a patent was granted to Alger on May 30, 1837. One year later he received a patent for the use of malleable iron in the manufacture of plows.

In 1833 the South Boston Iron Company expanded its operations to include the production of bronze cannons, and Alger obtained contracts to supply these weapons to both the federal government and the state of Massachusetts. The cannons were of such high quality that the Mechanics' Association awarded Alger a gold medal. In 1842 his firm constructed the largest gun then cast in the United States–the mortar "Columbiad." He also began to focus his efforts on improving fuses and shells. Over the next two decades Alger's firm became a major supplier of ammunition to the federal government.

Alger distinguished himself in endeavors beyond metallurgical innovations. He was genuinely forward-looking in dealing with his employees, being the first employer in South Boston to introduce as standard a 10-hour workday. He also avoided layoffs during slow times by giving employees reduced hours. Alger's interests went beyond his business. He took an active part in local politics, representing South Boston as an alderman in 1824 and 1827. He received many honors during his lifetime. His memory lived on in Boston; 25 years after his death the city of Boston named a primary school after him.

After Alger's death on February 4, 1856, the South Boston Iron Company continued to prosper under the direction of his son, Francis Alger. Because the firm produced large quantities of weapons for the Union army, the Civil War was a high point for the company. After the war the South Boston Iron slowly declined, finally closing in 1880.

Allentown Rolling Mills

by John W. Malsberger

Muhlenberg College

Allentown Rolling Mills was organized in 1860, the second ironworks to be established in Allentown, Pennsylvania. It expanded rapidly in the 1870s through merger with, and acquisition of, its competitors to become the largest ironworks in Allentown and the only integrated iron company in the Lehigh Valley during the anthracite iron era. It produced a broad diversity of cast-iron and wrought-iron products which contributed to the industrialization of the Lehigh Valley and eastern Pennsylvania in the late nineteenth and early twentieth centuries. It continued operation until 1914 when it was taken over by the Aldrich Pump Company, now part of Ingersoll-Rand Corporation.

Organized by Allentown entrepreneurs Christian Pretz, Samuel A. Bridges, and John D. Stiles, and by Benjamin Haywood of Pottsville, Allentown Rolling Mills was established in 1860 for the manufacture of iron T-rails, which it produced exclusively until 1868. In that year the company, like many other small, local iron mills of the era, embarked on an aggressive course of expansion, primarily through merger and acquisition, that allowed it greatly to diversify its products and brought it significant success in the short term.

The Allentown firm's expansion began in 1868 with the acquisition of the bankrupt Lehigh Rolling Mills, which had been established seven years earlier by local entrepreneur Samuel Lewis to manufacture railroad spikes and boiler rivets. This acquisition tripled Allentown Rolling Mills' original capacity and enabled it to compete in new markets.

The biggest merger in the firm's history occurred three years later when the Allentown company combined with another local firm, the Roberts Iron Works, which had been organized in 1862. With this consolidation the new firm achieved a degree of both horizontal and vertical integration by adding two more blast furnaces to its operations, as well as obtaining leases to iron ore beds in the sur-

rounding area. The new firm conducted business under the name Allentown Rolling-Mill Company until 1882 when it was rechartered as Allentown Rolling Mills.

The company's growth and diversification was furthered by three other actions undertaken in the 1870s. In 1872 the firm constructed a plant for the manufacture of bolts, nuts, and rivets, and it also purchased a local foundry and machine shop, thereby expanding the list of products it manufactured. Finally in 1878, when the founder of Allentown's first foundry retired, Allentown Rolling Mills was able to purchase its assets for $100,000. By the late 1870s the company had become one of the most complete iron manufacturers in Pennsylvania, employing 1,200 men and producing annually 25,000 tons of pig iron and 30,000 tons of finished iron products. Most of the pig iron was used in a wide variety of finished and semifinished products. A partial listing of its products testifies to the great diversification it had achieved and also to the significant contribution it made to the industrialization of Pennsylvania. In addition to pig iron and iron rails, the Allentown firm manufactured merchant bar iron, rolled shafting and car axles, rolled beams and angles, railroad chairs and fish plates, bolts, nuts, rivets, locomotive turntables, steam engines, mill gearing, blast furnaces, rolling mill castings, and mining pumps.

The aggressive policy of horizontal integration pursued by Allentown Rolling Mills in the 1870s followed a pattern common to the American iron industry of the late nineteenth century. By purchasing or merging with established and often thriving competitors, Allentown Rolling Mills was quickly able to achieve the economy of scale and product diversification that enabled it to grow into an important and prosperous regional manufacturer of iron. Ironically, however, it seems likely that the factors accounting for the rapid success it achieved also led to its eventual demise. Believing that its regional market for iron products was secure, and perhaps sobered by the financial panics of the 1870s that had led to numerous bankruptcies, Allentown Rolling Mills chose not to construct a plant to manufacture steel in the decades after the Civil War when many other iron companies were making the transition. As a result, by the 1880s and 1890s, when such technological advances as the Bessemer process and the Siemens-Martin open-hearth furnace had increased both the quantity and quality of steel products and had also substantially reduced their price, Allentown Rolling Mills saw the markets for their more brittle and less durable iron products steadily diminish. The firm's decline was accelerated by the panic of 1893, often regarded as second in severity only to the Great Depression of the 1930s. The furnaces of Allentown Rolling Mills were blown out in 1894 and remained inactive until about 1900, after

which they were operated sporadically until the financial panic in 1907 shut them down permanently. The furnaces were dismantled in 1912.

Although it rose rapidly in the late nineteenth century to become an important and profitable regional manufacturer of iron, Allentown Rolling Mills, like many other iron mills of its era, was rendered obsolete by the technological revolution that rapidly transformed the iron and steel industry from a local and regional to a national and international enterprise.

References:

Craig Bartholomew, "Anthracite Iron Making and Industrial Growth in the Lehigh Valley," *Proceedings*, Lehigh County Historical Society, 32 (1978): 129-183;

The Manufactories and Manufactures of Pennsylvania of the 19th Century (Philadelphia: Galaxy, 1875);

Charles Rhoads Roberts, et al., *The History of the Lehigh Valley, Pennsylvania*, 3 volumes (Allentown, Pa.: Lehigh Valley Publishing, 1914).

Amalgamated Association of Iron and Steel Workers of America

by Northcoate Hamilton

Columbia, S.C.

The Amalgamated Association of Iron and Steel Workers of America was officially formed on August 4, 1876, the product of a merger among four separate labor organizations, the Sons of Vulcan, the Associated Brotherhood of Iron and Steel Rail Heaters of the United States, the Iron and Steel Roll Hands' Union, and the United Nailers. This consolidation created the largest and most powerful labor organization in the iron and steel industry; the union, at its peak in 1891, claimed over 24,000 members. In 1892 a violent and disastrous strike at the Carnegie-owned Homestead Works in Pittsburgh left the Amalgamated Association critically weakened as a force in the Pennsylvania industry. By 1901 the union was almost entirely a western enterprise, the Homestead defeat having left it with little power or credibility in the East. This weakening presaged the almost total defeat suffered by organized labor in the iron and steel industry during the first

decade of the twentieth century. As David Brody writes in *Steelworkers in America: The Nonunion Era* (1960), by 1910 almost the entire industry was "effectively unorganized from the ore to the finished product." Despite the destruction of its power the Amalgamated Association continued to exist, making available a ready-made organization for use when better times for labor returned.

The largest and most important precursor to the Amalgamated Association was the Sons of Vulcan, a union first organized in Pittsburgh on April 12, 1858. Because of the extreme hostility of employers toward any form of organized labor, the founders of the Sons of Vulcan, including Patrick Graham, Matthew Haddock, and James Davies, were forced to keep the organization a secret society with very limited membership. Because of the danger of openly operating and the worsening economic conditions in the wake of the panic of 1857,

Strikebreakers on the grounds of the Homestead plant after the violence had ended in 1892 (courtesy of Stefan Lorant)

the Sons of Vulcan disbanded in fall 1858. These conditions persisted until 1861, when the outbreak of the Civil War and the passage by Congress of the protectionist Morrill tariff rejuvenated the industry and revived interest in the Sons of Vulcan. Miles S. Humphries was elected grand master of the revived union in August 1861, overseeing its transformation into a national organization on September 8, 1862. He was an active leader, organizing union chapters in Pennsylvania, Ohio, New York, New Jersey, West Virginia, Kentucky, Illinois, Maryland, and other states before relinquishing his position in 1866.

In addition to being the principal constituent organization of the Amalgamated Association, the Sons of Vulcan was notable in establishing the concept of the "scale of prices" method of determining wages. Created through an agreement between Humphries and Benjamin Franklin Jones of the Jones & Laughlin Company in February 1865, the first scale agreement set wages of the workman in relation to the price the manufacturer received for his product. Although this method seemed reasonable and predictable, the volatility of prices and therefore of wages during and following the war

doomed it to failure. The conflict between workers and management eventually led to a lockout of the workers at Jones & Laughlin which lasted from December 1866 to May 1867. This pattern–agreement to a scale of wages and then conflict and work stoppage when prices fluctuated–continued during the history of the Sons of Vulcan.

As their efforts to organize and negotiate continued, the leaders of the Sons of Vulcan realized that they were constrained by their small size, numbering some 3,000 in late 1872, and their membership, which was limited mainly to boilers and puddlers. These realizations were further confirmed during the economic upheavals caused by the panic of 1873 and the ensuing depression. Plunging prices and wages, accompanied by plant failures and firings in 1874 and 1875, convinced the Sons of Vulcan that a consolidated labor organization, one that represented a wider range of iron and steel workers, was needed to advance the goals of the employees in the face of hardship and economic turmoil. In August 1875 the union elected Joseph Bishop of Pittsburgh to be its president; his mandate was to prepare the way for the formation of the Amalgamated Association.

TRADES-UNION LEADERS.

Nineteenth-century union leaders; John Jarrett, an official of the Amalgamated Association, is at top left.

Another of the unions consolidated into the Amalgamated Association was the Associated Brotherhood of Iron and Steel Rail Heaters. This organization was formed at Springfield, Illinois, in August 1872 through a merger of several small local lodges, the earliest of which was established in Chicago in 1861. In addition to being a relatively small organization, boasting 412 members in 1876, the Rail Heaters was also in chronically poor financial condition. The union grew until 1874, from which point the organization weakened critically. Regardless of its problems delegates from the Rail Heaters made up the second largest contingent at the organizing convention for the Amalgamated Association.

The largest of the two small organizations included in the Amalgamated Association was the Iron and Steel Roll Hands' Union, which was formed on June 2, 1873, in Springfield, Illinois. At its first convention in 1873 the union membership reached 473 workers. Because of its small size and the prevailing economic conditions the union found it difficult to achieve its goals in the workplace. Especially vulnerable to employer blacklists, the union garnered no significant victories and struggled during its few years of independent existence.

The last union represented in the Amalgamated Association was the United Nailers, which was not really a national organization at all. One delegate from one of the local lodges joined in the loose confederation was present at the organizational convention for the Amalgamated Association and was able to have the United Nailers included. The group exerted no known influence on the larger union.

The constitution of the Amalgamated Association, ratified at the organizing convention in Pittsburgh in 1876, was the product of several years of work by the constituent unions. This extensive preliminary work was necessary because of the history of antipathy among the different groups of skilled workers. The groups' long-standing dislike of each other was the greatest internal threat facing the union and, if not solved, would have doomed the chances for an effective and unified organization. Joseph Bishop, the president of the Sons of Vulcan, was elected to the same post for the Amalgamated Association; it was his mandate to create the unity lacking among the different classifications of workers. John Jarrett, an original trustee of the union and its early historian, writes that this problem "was handled in a masterfully way by President Bishop, and to all appearances, at the end of the first year of the existence of the association, seemed to be almost entirely obliterated."

But disunity was not the only problem facing the fledgling organization. After a relatively quiet first year of operation the union faced a rapidly declining economic situation in 1878. As iron and steel production declined, prices fell, wages stagnated, men were laid off, and the frustration of the workers increased. But faced with slowing membership growth, the number reaching only 4,044 in 1878, and little leverage in times of high unemployment, the Amalgamated Association could only watch as strikes failed and hard times continued. With the return of prosperity to the economy in 1879, both the iron and steel industry and interest in the union revived. By 1880 union membership had more than doubled to over 9,500. During the 1880s the union continued to grow; despite a fallow period during mid decade, the membership stood at over 16,000 in 1889. Despite its success the union at no time was able to completely organize the industry. On the contrary, in 1891 when union membership was at its peak of over 24,000, nearly 100,000 workers were eligible. The union's size, influence, and power were more or less dependent on economic conditions and the success or failure of the most recently undertaken job action. In fact a job action undertaken in 1892 led directly to the decline of the Amalgamated Association as a factor in the iron and steel industry.

The union's dealings with Andrew Carnegie during the period between 1889 and 1892 are legend. The 1892 Homestead strike, about which much has been written, changed the course of labor history in the United States. The violence which occurred during the strike, although it gained the union some public and governmental support, contributed to its tactical defeat and left it with no place in the influential Carnegie plants. In the iron and steel industry the strike led directly to the deunionization of the mills in 1903, a condition that was constant until the mid 1930s. From a peak membership in 1891 the union shrank in size to 10,000 members in 1894. By 1899 the Amalgamated Association had lost its last strongholds in the West, leaving it an organization with members but little power. An abortive strike in 1901 against the new United States Steel Corporation left it weakened further. By 1903 no steel mill in the country was unionized. Carnegie's victory at Homestead in 1892 was realized. It would take organized labor over 30 years to regain its position in the iron and steel industry.

References:

David Brody, *Steelworkers in America: The Nonunion Era* (Cambridge, Mass.: Harvard University Press, 1960);

John Jarrett, "The Story of the Iron Workers," in *The Labor Movement: The Problem of Today,* edited by George E. McNeill (New York: Hazen, 1887);

Jesse S. Robinson, *The Amalgamated Association of Iron, Steel and Tin Workers* (Baltimore: Johns Hopkins University Press, 1920);

Joseph Wall, *Andrew Carnegie* (New York: Oxford University Press, 1970);

Carroll D. Wright, "The Amalgamated Association of Iron and Steel Workers," *Quarterly Journal of Economics,* 7 (July 1893): 400-432.

American Iron and Steel Association

by Northcoate Hamilton

Columbia, S.C.

The American Iron and Steel Association (AISA) was organized in Philadelphia, Pennsylvania, on March 6, 1855, as the American Iron Association (AIA). But the origins of the organization, the first national trade group formed, were in several earlier conventions of iron producers, the earliest of which took place in Pittsburgh, Pennsylvania, on November 21, 1849. The Pittsburgh meeting was called to garner support for a protective tariff for domestic iron producers, a political goal for which the group was to fight throughout the nineteenth century. The convention attracted 160 participants from seven states ranging from Virginia to New York to Illinois, all of them there to protest the Tariff Act of 1846.

The act reduced the duties levied on imported iron and iron products, leaving the domestic industry, from their point of view, easy prey of the British iron makers. The convention could do little but pass resolutions and send petitions to Congress to increase the tariff, but even this had no visible effect. The group met again on December 20, 1849, to continue their efforts to protect the iron industry. Once again resolutions were passed and petitions were sent to Congress requesting that tariffs be returned to the levels under the protectionist tariff of 1846. Once again, the tariffs remained at their 1846 levels. But one noteworthy achievement did arise out of the December 1849 meeting: the agreement to begin the publishing of statistical and polemic material concerning the iron industry and its interests. The first of the group's publications were issued in July 1850 and included a statistical study of the industry by Charles E. Smith, a pamphlet by Abram S. Hewitt, and a reprinted edition of Henry Charles Carey's protectionist classic *The Harmony of Interests*. After these efforts brought about few if any results other than publicizing the plight of the industry, an upturn in the economy brought about by burgeoning railroad construction and the Califor-

nia gold rush turned the group's attention away from the tariff.

Four years of prosperity, however, led to another economic downturn in late 1854; with declining profits the tariff again seemed to be the needed remedy. The Pennsylvania iron interests, who dominated the industry, called another meeting for March 6, 1855. It was at this meeting that the AIA was organized, a constitution adopted, and officers elected. George N. Eckert of Reading, Pennsylvania, was elected president. The office of secretary, which would later become so important, was filled by J. Peter Lesley, a Pennsylvania geologist. The AIA was a poor organization despite the wealth of its individual members. Because of the lack of funds there is no evidence of an organized effort by the group to influence the tariff debate. This lack of effort continued during the next few years; the AIA's most notable achievement was the publication of a statistical study of the industry undertaken by Lesley. Sometime after 1859 the AIA became completely moribund, no records existing for annual meetings after that year.

As the nation approached civil war, events conspired to give the iron and steel producers what they wanted. The export-driven South had always been able to block any protectionist tariff, which would have led to other countries taking retaliatory measures and blocking southern products. When southern legislators had resigned their positions and abandoned Washington in sufficient numbers, the protectionist forces in Congress were able to pass their legislation without conflict. The Morrill Tariff Act, passed on March 2, 1861, raised rates but not in the way preferred by the iron and steel manufacturers.

The rise of the steel-making industry and the introduction of new technology had again given the British the upper hand. Eber Brock Ward, a Detroit capitalist who operated the first successful Bessemer

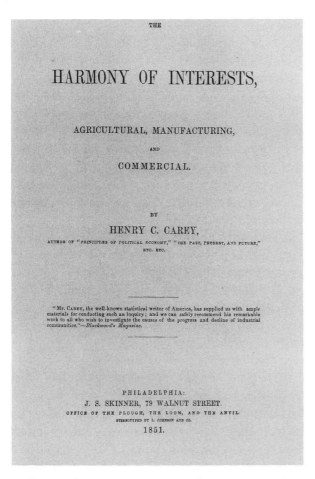

THE

HARMONY OF INTERESTS,

AGRICULTURAL, MANUFACTURING,

AND

COMMERCIAL.

BY

HENRY C. CAREY,

AUTHOR OF "PRINCIPLES OF POLITICAL ECONOMY," "THE PAST, PRESENT, AND FUTURE,"
ETC. ETC.

"Mr. CAREY, the well-known statistical writer of America, has supplied us with ample
materials for conducting such an inquiry; and we can safely recommend his remarkable
work to all who wish to investigate the causes of the progress and decline of industrial
communities."—*Blackwood's Magazine.*

PHILADELPHIA:
J. S. SKINNER, 79 WALNUT STREET.
OFFICE OF THE PLOUGH, THE LOOM, AND THE ANVIL.
STEREOTYPED BY L. JOHNSON AND CO.
1851.

Title page for Henry C. Carey's The Harmony of Interests *(1851 edition, first published 1849-1850), one of the protectionist works reprinted and distributed by the AISA*

steel plant in the United States, was one of the first to see the importance of establishing a strong domestic industry. In 1864 the industry was beginning the search for a method to continue the prosperity they were enjoying because of the war. They met on November 16, 1864, in Philadelphia, Ward being elected the temporary president of the group, which was nameless but soon decided to reconstitute the AIA. On November 17 the AISA was organized with the following officers being elected: Ward, president; Samuel J. Reeves, James M. Cooper, Charles S. Wood, and Joseph H. Scranton, vice-presidents; Charles Wheeler, treasurer; and Robert H. Lamborn, secretary. The wealth of talent in this leadership slate seemed to guarantee that the AISA would be a more effective organization than had been its predecessor. Ward, especially, was already well known as an eloquent and insistent supporter of protectionism.

After the war the AISA worked with other protectionist interests to institutionalize the high tariff policy, successfully doing so. One of the methods used by the AISA to promote its agenda was the publication of a weekly trade journal, the *Bulletin*. The first issue was published on September 12, 1866; the *Bulletin* seems to have been at first a rather dreary affair with little personality or originality. It did, however, provide the AISA with a sounding board for its positions and a framework for gathering statistics. These were especially useful during the tariff debates in 1870. The act, passed on July 14, 1870, for all intents was a subsidy for the industry and did much to create the large companies that began to dominate the industry during the late nineteenth century.

On January 1, 1873, James Moore Swank of Johnstown, Pennsylvania, was appointed secretary of the AISA. Swank for almost 40 years served the AISA and became its leading intellectual and tactical light. During his first year he was faced with the panic of 1873 and the resulting collapse of business; the depression was to last over five years. As prices fell businesses were forced to cut costs or close. As costs were reduced, the iron and steel firms became more efficient and competitive with their foreign counterparts. As the nation emerged from the depression and into a new decade, the domestic market was controlled by domestic producers.

Swank was a talented statistician, historian, writer, and was also a staunch protectionist. He more than anyone in the AISA, by the sheer force of his continuity, talent, and industry, put forth the case for protectionism and a strong, sheltered domestic industry. This was made much easier by the consolidation of several industry trade groups, the AISA, the National Association of Iron Manufacturers, and the American Pig Iron Manufacturers Association, into a new AISA in 1874. As the AISA began to speak for more and more of the industry, its influence, and Swank's, began to increase. He was a visible and important figure in Washington, and an appearance or favorable mention in the *Bulletin* was regarded as an imprimatur by protectionists. This importance was further enhanced with the de facto merger of the AISA with the Industrial League of Pennsylvania, one of the most powerful voices of protection in the country, in 1878.

But propaganda was not the only function of the AISA; the organization, under Swank's initia-

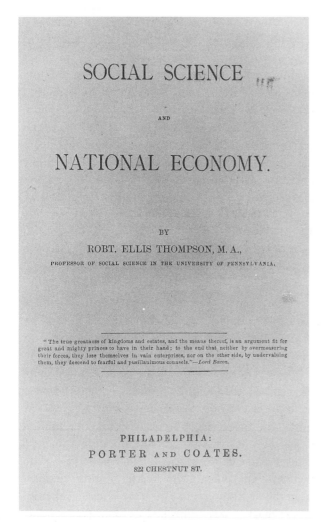

SOCIAL SCIENCE

AND

NATIONAL ECONOMY.

BY

ROBT. ELLIS THOMPSON, M. A.,

PROFESSOR OF SOCIAL SCIENCE IN THE UNIVERSITY OF PENNSYLVANIA.

"The true greatness of kingdoms and estates, and the means thereof, is an argument fit for great and mighty princes to have in their hand; to the end that, neither by overmeasuring their forces, they lose themselves in vain enterprises, nor on the other side, by undervaluing them, they descend to fearful and pusillanimous counsels."—*Lord Bacon.*

PHILADELPHIA:

PORTER AND COATES.

822 CHESTNUT ST.

Title page for Robert Ellis Thompson's Social Science and National Economy (1875), *a work placed in libraries by the AISA*

tive, produced extensive and thorough statistical reports on the iron and steel industry, reports that still provide some of the best information available on the nineteenth-century iron and steel business. But agitating for protectionism was the AISA's raison d'être, and even the statistical studies were used to further that effort.

In July 1876 the AISA began the distribution of free copies of tariff tracts, a method and medium which was to become a powerful weapon in the tariff wars of the 1880s. The AISA also instituted a program of placing protectionist literature in college libraries, one of the early examples being Robert Ellis Thompson's *Social Science and National Economy* (1875). It is clear that Swank saw himself as a tireless prophet spreading the word of a messianic protectionist philosophy. Yet, despite his efforts, the academic community remained hostile to protection-

ism, and in prosperous times the business interests were at best apathetic. The return of good times in 1879 brought a resumption of the arguments for free trade. The 1882 Tariff Commission, a concept that the AISA had supported, brought forth the 1883 tariff bill, an act which raised the tariff on most steel and steel products but lowered it slightly on pig iron and steel rails. Overall, the bill offered more protection for the industry than had the legislation it replaced. Swank, having distributed over one million tracts in 1882 alone, was disappointed at the slight gains in the 1883 act.

He was to be more upset at the 1884 election, when Grover Cleveland won the presidency and began an attempt to dismantle the tariff. Not surprisingly, Swank began an intensive campaign for the Republicans and protectionism in the 1888 presidential race. Much invective was published in the *Bulletin*, and hundreds of thousands of pamphlets were distributed in order to paint the 1888 race as a referendum on the tariff, a referendum that Swank was sure the protectionists would win. The Republican candidate Benjamin Harrison was victorious in the closest election in United States history; it was seen as a victory for protection. Swank's efforts paid off in the passage of the 1890 McKinley tariff bill, the most protectionist tariff until then passed in the Congress.

After the panic of 1893 the AISA and Swank seemed to lose energy, retreating to the sidelines as new controversies such as the money question began to dominate the political landscape. The change in the political winds in 1892 led to the passage, despite Swank's objections, of the 1894 tariff, a bill which reduced the rates significantly if not radically. As the century closed, American commerce had become more complex, and Swank's somewhat simplistic position in favor of protection was not enthusiastically supported by the AISA's members. Still Swank had formed an organization which was poised when needed to argue forcefully the interests of the iron and steel industry.

References:

Victor S. Clark, *History of Manufacturers in the United States*, 3 volumes (New York: McGraw-Hill, 1929);

Joseph Dorfman, *The Economic Mind in American Civilization*, 5 volumes (New York: Viking, 1946-1959);

Frank W. Taussig, *The Tariff History of the United States*, eighth edition (New York: Putnam's, 1931);

Paul H. Tedesco, *Patriotism, Protection and Prosperity: James Moore Swank, the American Iron and Steel As-*

sociation, and the Tariff: 1873-1913 (New York: Garland, 1985);

Peter Temin, *Iron and Steel in Nineteenth Century America: An Economic Inquiry* (Cambridge, Mass.: M.I.T. Press, 1964).

Archives:

The AISA papers are located at the offices of American Iron and Steel Institute in New York. The papers of James M. Swank are located at the Cambria Public Library, Johnstown, Pennsylvania.

Joseph Reid Anderson

(February 6, 1813-September 7, 1892)

by Brady Banta

Louisiana State University

CAREER: Assistant engineer, state of Virginia (1837-1840); commercial agent (1840-1843), manager (1843-1848), owner, Tredegar Iron Works (1848-1867); president, Joseph R. Anderson & Company (1867-1892).

Born on February 6, 1813, the youngest of nine children reared by William and Anna Anderson, Joseph Reid Anderson grew up in rural Botetourt County, Virginia. Although his father was a man of only moderate means, Anderson received an appointment to West Point, partly as a result of connections and favors garnered through his father's long career in politics and public service. Anderson's stay at West Point appeared to presage a promising military career; he achieved the distinction of being appointed cadet captain and upon graduation in 1836 ranked fourth academically in a class of forty-nine. But while enrolled at the military academy his interest turned toward civil engineering. When graduation presented no immediate opportunities to pursue his preferred vocation, Anderson accepted a commission as a second lieutenant.

Initially assigned as an artillery officer, he secured a transfer to the Corps of Engineers before the year ended. Although this assignment paralleled his professional interest, Anderson resigned his commission in 1837 to become the assistant engineer for the state of Virginia. An additional inducement to resign his commission was the fact that he had recently married Sally Archer. Assigned to supervise the construction of the Valley Turnpike from Staunton to Winchester, Anderson emerged as an outspoken advocate of state-sponsored internal improve-

Joseph Reid Anderson (courtesy of Meserve Collection)

ment projects. Having observed New York's economic and commercial development while at West Point, he had concluded that public encouragement and sponsorship of internal improvements underlay northern prosperity. His advocacy of similar programs in Virginia brought him into contact with two organizations through which he would influence nineteenth-century Virginia's iron industry, the

Tredegar Iron Works and the Whig party.

Joseph Anderson's introduction to executives from the Tredegar Iron Works of Richmond, Virginia, came in November 1838 when he attended a commercial convention in Norfolk as a delegate from Staunton. Mutual interests in Whig politics and internal improvements blossomed over the next three years into an offer to become the Tredegar's commercial agent, a position of considerable financial and business authority in the company. Anderson attacked the job with characteristic zeal, motivated both by the new challenge and the prospects for increased compensation, a factor that had become increasingly important following the birth in 1838 of the first of his six children.

The Tredegar Iron Works, having been chartered by the Virginia House of Delegates in 1837, received an inauspicious baptism into the world of free enterprise. The company's directors had predicated their decision to launch this venture upon the expectation that lucrative sales of iron to Virginia's rapidly expanding railroads would return an immediate profit. Unfortunately, the opening of the Tredegar Iron Works coincided with the panic of 1837, and the ensuing depression decimated the market for finished iron products, especially iron rails. As commercial agent, Anderson's foremost responsibility was the identification and solicitation of customers, and he focused attention on the cultivation of government markets.

By negotiating contracts for the sale of chain cable, shot, and shells to the navy, Anderson demonstrated his business acumen and brought sorely needed financial stability to Tredegar. These transactions also established lines of communication that led to a contract to provide the United States Navy with 100 cannons. Unfortunately, deficiencies in quality caused the navy to reject the last 40 pieces produced.

While Anderson could not persuade the navy to accept the artillery in question, its rejection precipitated a change in his relationship with the Tredegar Iron Works. Irked by what he considered undue interference from the company's directors, Anderson proposed to lease the ironworks for an annual rental of $8,000; the Tredegar's directors accepted his offer. Assuming both managerial and operational control in November 1843, Anderson imposed rigorous quality standards and predicted renewed success for the company. His optimism was soon justified, as the army and navy once again or-

dered Tredegar-produced artillery. Indeed, prior to secession the Anderson-managed Tredegar Iron Works produced almost 900 artillery pieces for the federal government.

Anderson's lease was for five years, and during its life the company prospered. Confident that this success would continue under his leadership, Anderson purchased the Tredegar Iron Works in 1848 for $125,000, to be paid over six years. Unfortunately, this transaction coincided with a general slump in profits for the American iron industry. The railroad construction boom in Great Britain having run its course, British rail makers invaded the American market and consistently undersold domestic manufacturers. Having lost a lucrative market to foreign competition, the Tredegar's profits dropped sharply, and only by bringing in partners did Anderson avoid losing his investment in the ironworks.

Prudent management and Anderson's attention to details enabled the company to survive this recession, and throughout the 1850s the financial and competitive condition of the ironworks improved. Anderson's drive and determination underlay this progress, and, as one might expect, the profitability of the ironworks frequently turned on the new owner's actions and decisions. Confronted with the influx of inexpensive British rails, Anderson realized that he must reduce operating and production costs in order for Tredegar to become competitive once again. In an attempt to reach this goal, Anderson took steps to reduce the company's labor costs.

Wishing to limit the need to pay premium wages to lure northern and foreign laborers to Richmond, in 1848 Anderson initiated the use of slaves in skilled ironworking positions. Slaves had been used at the Tredegar Iron Works for several years, but they had been limited to subordinate jobs. While white laborers had generally tolerated this situation, they vigorously opposed skilled positions, such as that of puddler, being given to slaves at Tredegar's armory mill. Puddlers conducted the crucial task of transforming molten pig iron into wrought iron by removing impurities, thus preparing the substance for manufacture into finished iron products. Steadfastly determined to guard the economic and social status associated with puddling, skilled white employees mobilized a strike against the Tredegar Iron Works. Anderson reacted with characteristic determination. He announced that the striking workers, through their own actions, had fired themselves and insisted that those living in company-owned housing vacate their accommodations. Anderson's victory was total. Strikers were not rehired, and slaves were thereafter employed whenever and wherever needed.

The determination with which Anderson responded when challenged did not, however, always contribute to the company's success. The prime example of this was his unwillingness to adopt improvements in artillery-casting technology. During Anderson's tenure the Tredegar Iron Works conducted a lucrative ordnance business with the army and navy. As intersectional tensions worsened in the late 1850s Anderson anticipated that the government would accelerate these purchases. Acting upon this belief, he expanded the Tredegar's ordnance-producing capabilities and laid in a large supply of gun iron.

In November 1859 Anderson believed his planning had been rewarded as the company received a $20,000 cannon production order. But the Army Ordnance Bureau insisted that these weapons be produced according to the Rodman Plan, a technological innovation that involved casting a cannon around a hollow core. The circulation of a cooling medium through the core allowed the molten iron being cast in the form of a cannon to cool from the inside out. This eliminated a great deal of the stress placed on the metal during the standard practice of casting a solid gun tube, allowing it to cool from the surface inward, and then laboriously boring out the cannon. Army officials maintained that the new technology facilitated the casting of larger weapons, accelerated the pace of production, and produced safer ordnance. To reinforce this commitment, Secretary of War John B. Floyd made government funds and advisers available to assist foundry owners in implementing the new technology.

Talk of technical advancements and promises of government cooperation, however, had little impact on Joseph Anderson. Characterizing the new method as a "Yankee catch penny," he dismissed the claims of technical superiority and campaigned for the withdrawal of the order mandating use of the Rodman Plan. Despite direct appeals to Secretary Floyd, the government's decision remained unchanged. Anderson's stubborn loyalty to the solid cast method cost the Tredegar its share of the lucrative government ordnance business. Moreover, according to Charles B. Dew, Anderson's biographer, when the Civil War erupted, Anderson's rejection of the Rodman Plan denied the South the capability of employing the most modern artillery production technology.

As this resistance to the Rodman Plan demonstrated, Joseph Anderson was a man of strongly held convictions, a characteristic that he revealed in politics as well as business. Even before launching his business career, Anderson displayed an affinity for Whig politics, finding the party's positions favoring government support for internal improvements and a protective tariff particularly attractive. Lobbying for these positions stimulated his interest in partisan politics, and during the late 1840s Anderson entered the political fray by running for, and being elected to, the city council in Richmond. Anderson advanced from the local arena into state politics in 1852 by winning election to the Virginia House of Delegates, a theater where he remained active throughout most of that decade.

But by the mid 1850s Anderson found it increasingly difficult to reconcile his probusiness whiggery with the antislavery sentiments of Conscience Whigs. The emergence of the Republican party and its nomination of John C. Fremont in 1856 destroyed Anderson's wavering Whig loyalty and prompted his jump to the Democrats. The switch in party affiliations underscored Anderson's transformation from a nationalist Whig into a sectionalist Democrat. Thereafter he distrusted Northern intentions and advocated that Southern consumers patronize local manufacturers whenever possible.

John Brown's raid and the election of Abraham Lincoln intensified Anderson's awareness that a critical juncture in history had arrived. Throughout the winter of 1861 the Tredegar Iron Works received ordnance and munitions orders from states throughout the South. The pace of this business accelerated as secession swept through the Deep South. The magnitude of these orders more than compensated for the Tredegar's loss of business stemming from Anderson's stubborn refusal to embrace the Rodman Plan. Moreover, the company's prospects improved so dramatically during this period that management made plans to expand their ordnance production facility.

These developments not only proved beneficial for the company's balance sheet, they also reinforced Anderson's sectional leanings. Yet Virginia had not seceded, largely because Gov. John Letcher was determined that the state should stick with the Union. As much as this inaction frustrated Anderson, he took no overt steps to encourage Virginia's se-

Portrait of Joseph Reid Anderson by H. Bebie (courtesy of Cole of Baltimore)

cession until the news of the attack on Fort Sumter reached Richmond. Immediately Anderson orchestrated mass public demonstrations designed to sway Governor Letcher and the Virginia Convention, then in session, to adopt an ordinance of secession. When this public pressure failed to produce the desired result, Anderson stood ready to participate in an extralegal convention to force Virginia's secession. Such drastic action was not necessary, however, as President Lincoln's call for 75,000 volunteers to put down the insurrection prompted Governor Letcher to drop his opposition, and the Virginia Convention rapidly passed a secession ordinance in April 1861.

His ardor for Southern independence knowing no bounds, Anderson offered the Tredegar Iron Works, through sale or lease, to the Confederate cause. The government, however, saw no reason to end private ownership of the works. Anderson also organized his Southern-born employees into a volunteer militia known as the Tredegar Battalion. The principal function of this unit, other than ceremo-

nial occasions, was to guard the ironworks against arson and sabotage.

Still wishing to do more for the Southern cause, Anderson requested a field command in the Confederate army. No less a figure than Robert E. Lee responded that Anderson's greatest service would be efficiently managing the Tredegar Iron Works. Not to be put off, Anderson insisted that the Tredegar management team, led by longtime associate John Tanner and father-in-law and business partner Dr. Robert Archer, was capable of running the firm in his absence. However, should the need arise, he would return to the Tredegar without protest. This assurance having been provided, President Jefferson Davis appointed Anderson a brigadier general on September 3, 1861. While troops under Anderson's command participated in the defense of Richmond in the spring of 1862, the exigencies of wartime operation, specifically the Tredegar's entry into pig iron production, demanded his return to the ironworks. Therefore, on July 19, 1862, Anderson resigned his commission.

For the remainder of the war Anderson wrestled with financial, labor, supply, and transportation problems confronting the Tredegar Iron Works. His skill and innovative leadership notwithstanding, the Tredegar's performance deteriorated. Regardless, Anderson's contribution to the Confederate war effort should not be minimized. Only one Northern competitor produced more ordnance during the war. And despite quality deficiencies that multiplied as the company's supply and provision problems worsened, artillery fashioned at the Tredegar Iron Works helped sustain Confederate armies in the field. Indeed, Anderson's biographer maintains that the Tredegar Works, given the environment in which it operated, could have done little more for the Confederacy.

Anderson dedicated himself and his company to the cause of Southern independence, but when the Confederacy collapsed in 1865 he sought an accommodation with the victorious Union. The Confederates evacuated Richmond during the night of April 2, 1865, and on the following day federal troops occupied the city. Rather than flee, Anderson emerged as a leader in the movement to sever the state's association with the Confederacy and to bring an immediate end to the hostilities. Clearly among his principal motivations was the Confiscation Act of 1862, which authorized the seizure and sale of property held by insurrectionists. Since Anderson had served in the Confederate army and owned and directed a business so vital to the Southern war effort, the army of occupation had seized his property, and federal officials had demonstrated a willingness to sell it.

In an attempt to defuse the crisis, Anderson sought a presidential pardon, his military participation in and commercial support of the insurrection having excluded him from President Andrew Johnson's May 29, 1865, amnesty proclamation. In this campaign Anderson stressed the brevity of his military service and grossly understated his business dealings with the Confederate government. Moreover, Anderson prevailed upon Virginia's Unionist Gov. Francis H. Pierpont, to intervene on his behalf. Armed with a letter of introduction from Pierpont, Anderson traveled to Washington and on four occasions met with President Johnson. Representing himself as thoroughly reconstructed, Anderson obtained both pardon and amnesty, thus saving the Tredegar Iron Works from confiscation.

The presidential pardon cleared the way for Anderson to resume business at the Tredegar. The facility emerged from the war in remarkably good condition. Supplies, raw materials, and labor were widely available, and local and southern markets eagerly awaited Tredegar products. But the task confronting Anderson was enormous. While he had almost $200,000 in foreign bank accounts, money which he had accumulated during the war by selling cotton shipped to England despite the federal blockade, it had to be used to satisfy the claims of prewar creditors. Therefore, to put the ironworks back into production Anderson had to arrange financing, and the ravages of the war dictated that it must come from the North.

Anderson obtained the initial infusion of northern capital by selling coal pits acquired during the war. When the sale of assets failed to produce sufficient revenue, Anderson solicited investments by northern industrialists and financiers. Several capitalists expressed genuine interest, but only if the company was able to secure government contracts. Anderson's diligent efforts notwithstanding, such contracts were not forthcoming, and the northern-

ers lost interest. To make the Tredegar more attractive to northern capitalists, Anderson dissolved the southern-dominated partnership and reorganized the firm as a corporation. Anderson emerged as the majority stockholder and retained control of the business, but northern investors purchased enough stock to enable the refinanced Tredegar to resume production.

Having survived the transition from war to peace under Anderson's leadership, the Tredegar experienced a period of unparalleled success. Much of its business came from rebuilding and expanding southern railroads, but in the wake of the panic of 1873 many of them were unable to pay their bills. Now financially vulnerable, in 1876 Anderson took the company into receivership to avoid bankruptcy. While Anderson brought the ironworks back from the brink of collapse before his death in 1892, these difficulties occurred as technological advances were accelerating the transition from iron to steel in the industrial markets traditionally served by Tredegar products. Its inability, or unwillingness, to stay current with technological innovations caused changes in market preference to transform the Tredegar Iron Works from a leading southern industry into a profitable, but parochial, iron foundry.

References:

Kathleen Bruce, *Virginia Iron Manufacture in the Slave Era* (New York: Century, 1930);
Charles B. Dew, *Ironmaker to the Confederacy: Joseph R. Anderson and Tredegar Iron Works* (New Haven & London: Yale University Press, 1966).

Archives:

There are several collections of material concerning Joseph Reid Anderson. The most important is the collection of Tredegar Company Records of the Virginia State Library, Richmond, Virginia. Three deposits in the National Archives, Washington, D.C., record Anderson's government dealings: Record Group 156, Records of the Office, Chief of Ordnance; Record Group 74, Records of the Bureau of Ordnance (United States Navy); and Record Group 109, War Department Collection of Confederate Records. Anderson's personal papers are located in the Anderson Family Papers of the University of Virginia, Charlottesville, and the Francis T. Anderson Papers of the Duke University Library, Durham, North Carolina.

Bay State Iron Company

by David B. Sicilia

The Winthrop Group, Inc.

The Bay State Iron Company became the first commercially successful open-hearth steelworks in the United States in 1870. The company was founded in South Boston during the 1840s, its first president being Samuel Hooper, a leading Boston merchant who also held an interest in furnaces at Port Henry, New York. Ralph Crooker, an ironmaker with decades of experience, was superintendent, John H. Reed served as treasurer, and J. Avery Richard was the company's assistant treasurer and clerk.

By 1860 Bay State had grown into the largest enterprise of its kind in New England and was one of the only leading ironworks of its day not located near the rich anthracite coalfields of Pennsylvania. By that year total capitalization had reached $300,000, and the works located at East First and "I" streets employed some 300 men and ran continuously to meet a thriving demand for rolled railroad and plate iron.

After the Civil War Bay State sought to increase the durability and flexibility of its wrought-iron rails by melding steel heads onto them. Crooker first tried puddled steel heads but this method proved unsatisfactory. He then attempted to roll Bessemer ingots shipped in from Troy, New York, where Alexander L. Holley had produced the first Bessemer steel in the United States in 1864. At this time Bessemer steel was still irregular and difficult to roll, and this second experiment failed as well.

In 1868 Crooker learned of another attempt to produce steel-headed iron rails by Cooper, Hewitt & Company (a rolling mill in Trenton, New Jersey), an attempt made after Abram S. Hewitt had secured the rights and plans for a 5-ton Siemens regenerative furnace to manufacture open-hearth steel. Crooker visited the facility and decided to adopt a similar approach.

Although Cooper, Hewitt abandoned the project after many trials, Crooker suspected that suc-

cess could come with important modifications to the French design and the use of ferromanganese in the final stages of the process to control the carbon content. This view was shared by a young Siemens engineer named Samuel T. Wellman, who had assisted in starting up the Cooper, Hewitt plant. When Siemens's American agents refused to design a special furnace for Bay State, Cooper hired Wellman, now working independently, to build the works.

During summer 1868 the capitalization of Bay State was increased to $500,000. Construction of the new plant began in late 1869 and was completed a few months later in 1870. The heart of the facility was an efficient, 5-ton melting furnace complete with gas producers and a coal-fired preheating furnace. Wellman designed a furnace with a bath deeper than European models so that pig iron and scrap could be added to adjust the molten mixture and also replaced ingot railcars with a more reliable turntable arrangement. Most notably, he added several anthracite crucible steel furnaces to produce the first ferromanganese manufactured in the United States.

It was soon apparent that Wellman's modifications enabled the Siemens furnace to operate profitably, making it the first commercially successful open-hearth facility in the United States. However, the market for steel-headed rails proved weak, so the company dedicated its new furnace to the manufacture of flange plates from soft steel. As Wellman recalled, "The first were made from their own bars, puddled from charcoal iron; but this was not satisfactory, as the iron contained too much phosphorus. . . ." After blooms from Lake Champlain were tried, however, "a beautifully soft, ductile steel was produced that was perfectly malleable, hot or cold; plates rolled from it never showed a blister, and seldom defects of any kind, and most satisfactory of all was the fact that it could be produced cheaper than the best iron flange plates."

Wellman moved on to other installations, where he continued to improve and disseminate the process and equipment; more than any other American, he became responsible for the triumph of the open-hearth over Bessemer as the leading method of steel making in America by the early twentieth century. Bay State's success proved less enduring. Judging from reports filed by the R. G. Dun credit agency, Bay State began the 1870s as "rich and good as ever" but started "losing money" after the onset of the depression of 1873. Still, the works continued to expand; capitalization reached $1.8 million (and debts $646,400) by summer 1876, when the firm was reported to be roughly breaking even.

However, Bay State closed before the end of the following decade.

Unpublished Documents:

R. G. Dun & Co. Collection, Baker Library, Harvard University Graduate School of Business Administration, Massachusetts, volume 72, pp. 320, 331.

References:

Thomas C. Simonds, *History of South Boston* (Boston: David Clapp, 1857);

John J. Toomey and Edward P. B. Rankin, *History of South Boston* (Boston: Privately printed, 1901);

Samuel T. Wellman, "The Early History of Open Hearth Steel Manufacture in the United States," in *The Open Hearth: Its Relation to the Steel Industry; Its Design and Operation*, edited by Victor Windett (New York: U. P. C. Book Company, 1920).

Bellaire Nail Works

by John A. Heitmann

University of Dayton

Organized in 1867 in Bellaire, Belmont County, Ohio, the history of the Bellaire Nail Works illustrates the importance of local elites in financing manufacturing enterprises and the influence of a dynamic market in forcing changes in the technological basis of production.

During Reconstruction the Wheeling, West Virginia, area emerged as the center of iron-cut nail production in the United States. Several small nail-making firms were established in this region during the late 1860s, and one was launched on the Ohio side of the river across from Wheeling in 1867. Initially capitalized at $155,000, the Bellaire Nail Works started operations with 25 nail machines and had just opened its doors when a fire destroyed almost the entire factory.

Fortunately, however, Bellaire had many resourceful and financially well-endowed backers, including the Oglebays, the "first family" in Wheeling in terms of social status. Earl W. Oglebay, educated at Bethany College, was born in Wheeling and in 1877 succeeded his father as president of the National Bank of West Virginia. With the support of the Oglebays, Bellaire was rebuilt, and in 1872 its capital increased to $375,000. Im-

provements were soon made in manufacturing capacity and production facilities, and the rebuilt plant featured a 65-by-16-foot blast furnace, 21 boiling furnaces, and 90 nail machines. Historians of manufacturing often fail to emphasize properly the significance of finance and the role of local, non-technically trained elites to the success of an industry, and Bellaire's story provides an excellent example of these groups' contributions to the economic growth of a region.

Bellaire survived the panic of 1873 and prospered during the next decade. In 1875 J. R. McCortney was elected the firm's president, and McCortney, like so many of his contemporaries, viewed organization as power. To that end McCortney played an active role in the Western Nail Association, but more importantly in 1882 proposed a novel plan to combine the Wheeling Nail Manufacturers in an association of producers that would cosponsor the erection of a large capacity Bessemer steelworks in Wheeling. According to McCortney's scheme, steel blooms made at this plant would serve as the feed material for local nail manufacturers, including Bellaire, Belmont Nail Company, Benwood Iron Works, La Belle Iron Works,

Laughlin Nail Company, Riverside Iron Works, and the Wheeling Iron & Nail Company. Preliminary tests indicated that steel-cut nails were superior to iron varieties, and McCortney's efforts were approved by stockholders in the participating companies. For some unknown reason, however, the joint venture was never implemented, and this failure to consolidate most certainly had long-term consequences for the Wheeling economy. One historian of Wheeling's iron and steel industry, H. D. Scott, interpreted this inability of local firms to join together as the pivotal event in Pittsburgh's ascendancy over Wheeling as a steel center during the 1880s.

Undaunted by this organizational setback, McCortney pushed for Bellaire's own Bessemer facility, which was started in April 1884 and marked the first blow of steel in the Wheeling area. The switch from iron to steel in the manufacture of cut nails was only a temporary expedient to an industry experiencing increased competition from the rapidly growing wire-nail industry. By the early 1890s the cut-nail industry was in rapid decline, and in 1893 Bellaire discontinued the manufacture of nails. Instead of developing other finished goods such as tin plate or steel tubing, Bellaire management decided to specialize in making steel sheet bars for local tin works, a market stimulated by the high McKinley tariff of 1890. Bellaire developed a national reputation for producing the highest quality Bessemer sheet and tin bar in the country, and it was considered an attractive plum by financiers and businessmen involved in the wave of consolidations of the late 1890s. In 1899 National Steel Company, led by Elbert H. Gary, purchased Bellaire, and it was soon merged as a subsidiary of National Steel into the Carnegie Steel Corporation. On the eve of World War I, Bellaire, now an operating unit of United States Steel, continued to be known within the trade for its specialty of rolling sheet bar.

References:

John N. Ingham, *The Iron Barons: A Social Analysis of an American Urban Elite, 1874-1965* (Westport, Conn.: Greenwood Press, 1978);

Henry Dickerson Scott, *Iron & Steel in Wheeling* (Toledo: Caslon, 1929).

Philip Benner

(May 19, 1762-July 27, 1832)

by John W. Malsberger

Muhlenberg College

CAREER: Merchant and ironmaster (1781-1832); owner, Logan's Branch Woolen Factory (1825-1832); owner and publisher, *Centre Democrat* (1827-1831).

Philip Benner was a pioneer ironmaster and merchant of central Pennsylvania in the late eighteenth and early nineteenth centuries. The owner of the first iron forge in Centre County, Pennsylvania, Benner also operated a slitting mill and a nail mill. He originated the famous "Juniata iron" that was given renown for its high quality by such early manufacturers as Eli Whitney.

Born on May 19, 1762, in Chester County, Pennsylvania, Benner was the son of Henry and Dinah Thomas Benner. During the American Revolution Benner, although still a youth, enlisted as a private in the Continental army after his father, a vocal Whig, was captured and imprisoned by the British. He fought under the command of his relative and neighbor, Gen. Anthony Wayne. The military experience he gained during the Revolution enabled Benner in later life to receive a commission as major general in the Pennsylvania state militia, thereafter being known as Gen. Philip Benner.

After the Revolution Benner embarked on the two careers which he would follow for the rest of his life: he took charge of a store in Vincent Township and, more importantly, learned the iron trade at Coventry in northern Chester County, the site of some of the earliest iron plantations in Pennsylvania. Sometime around 1785 Benner married Ruth Roberts, and together they had eight children, four sons and four daughters. In May 1792, displaying the industry and willingness to take risks that were the hallmarks of his career, Benner took his iron-making skills to central Pennsylvania where he purchased the land of Rock Forge on Spring Creek in Centre County. Benner took with him experienced ironworkers from Chester County and in May 1793 began improving his site by building first a house and a sawmill and then in 1794 erecting his first iron forge, likely the first forge in Centre County. Later a gristmill was added to the works, followed in 1799 by the construction of a slitting mill and in 1800 a second forge and nail mill.

In order to establish a successful iron business in the wilds of central Pennsylvania in the early nineteenth century, Benner was forced to overcome a number of substantial obstacles. To construct his iron mill Benner had to pack provisions by horse through the woods from the eastern counties to sustain 93 people. Once his mills had been built, Benner faced a series of legal challenges to his ownership of the land at Spring Creek. In 1802 he was served with a notice of eviction because of the existence of prior claims to the land. Benner fought these claims through the courts of Pennsylvania, finally losing in the Supreme Court in 1811, a loss that forced him to repurchase his land.

When his legal difficulties were finally resolved, Benner faced what was certainly his most substantial problem, that of finding profitable markets for his iron products. When he began to manufacture iron in the early 1800s, the only markets for his goods were on the Atlantic coast. Ever the risk-taker, Benner carved out a new trade route, sending his iron goods to Pittsburgh and the West. Originally Benner shipped his iron to Pittsburgh on horseback over nearly impassable roads, bringing salt back in return. It cost him $75 per ton to ship his iron overland to the West, a cost that would have been prohibitive in most markets. But Benner realized that the selling price for a ton of iron in Pittsburgh at the time was about $250, allowing him a comfortable profit margin even with the high cost of transportation. After the War of 1812, when turnpikes began to improve the efficiency of transportation, Benner turned to six-horse teams to ship his iron westward. Benner's gamble paid off handsomely, and for many years he was the sole supplier of "Juniata iron" to Pittsburgh and the West.

Like many other entrepreneurs of the early nineteenth century, Benner's success in business rested on a variety of enterprises in addition to the iron trade. A major real estate speculator in his region, Benner was one of the founders of Bellefonte, Pennsylvania. Additionally, Benner continued in the mercantile trade by operating general stores in Bellefonte and in rural Ferguson Township. In 1824 he opened a factory to manufacture woolen goods. Because his iron business depended heavily on transportation, Benner also became an early advocate of internal improvements, serving in 1821 as the first president of the Centre & Kishacoquillas Turnpike Company. Benner was also active in politics. An ardent Democrat, he established in 1827 a Jacksonian newspaper, the *Centre Democrat*, and twice served as a presidential elector, most notably for the Jackson-Calhoun ticket in 1824.

In many ways Benner's business practices combined the pioneering qualities of the American frontiersman with the shrewd business sense and attention to detail that marked the successes of business leaders in the late nineteenth century. A strong believer in the work ethic, Benner always began his workday between 4:00 A.M. and 5:00 A.M. and expected his employees to adhere to the same stringent standards. On one occasion he refused to hire a man whose pants were worn in the seat because in his judgment "a man . . . who sat down so much as to wear out the seat of his breeches was too lazy a man to be tolerated at Rock [Forge]." At the same time the risks Benner took and the obstacles he surmounted in establishing a thriving iron business in the wilderness of central Pennsylvania reflect the same pioneering spirit and determination that was responsible for the steady westward movement of the American frontier throughout the nineteenth century. Finally, in establishing his businesses, Benner exhibited a constant attention to detail similar to that which helped to account for the success of such business leaders as Andrew Carnegie or John D. Rockefeller. Benner, for instance, regularly examined the feet of the horses used to transport his iron to make certain that they were well shod and insisted that all his drivers be equipped with a hammer, nails, and extra horseshoes to make emergency repairs. Thus, while Benner was not responsible for any important innovations in the manufacture of iron, his entrepreneurial abilities helped to spur the geographical spread of an industry which before his time had been restricted largely to the Atlantic seaboard, and to point the way to the integrated iron companies of the mid nineteenth century.

References:
John Blair Linn, *History of Centre and Clinton Counties, Pennsylvania* (Philadelphia: L. H. Everts, 1883);
John B. Pearse, *A Concise History of the Iron Manufacture of the American Colonies Up to the Revolution and of Pennsylvania Until the Present Time* (Philadelphia: Allen, Lane & Scott, 1876);

Sylvester K. Stevens, *Pennsylvania: Titan of Industry*, 3 vols. (New York: Lewis, 1948);
James Moore Swank, *History of the Manufacture of Iron in All Ages*, 2d ed. (Philadelphia: American Iron and Steel Association, 1892).

Bessemer Process

by William H. Becker

George Washington University

Large-scale production of steel (a strong but malleable alloy consisting basically of iron with the impurities removed and between 1 and 2 percent carbon) became possible during the 1850s. Before that time steel was difficult to make; only small quantities were produced for use primarily in cutlery and special tools. A British inventor, Henry Bessemer, is credited with developing the process that resulted in marked increases in steel production. In August 1856 Bessemer read a paper to the British Association for the Advancement of Science, describing his discovery that carbon and silica could be removed from hot pig iron by a blast of cold air. The infusion of oxygen easily cleaned silicon and other impurities from the iron.

Bessemer's ideas were improved upon by others. An English metallurgist, Robert Mushet, was able to control precisely the carbon content of steel by the addition of spiegeleisen, a variety of pig iron containing between 15 and 30 percent manganese. Important to steel making in the United States was research into the quality of iron ores at a laboratory in the Wyandotte ironworks near Detroit, the site of the first commercial production of steel in the United States. Others happened onto the same discovery as Bessemer. Before the Englishman, a Kentucky ironmaster, William Kelly, had discovered the value of injecting air into hot pig iron. He patented the process, and eventually there was a litigation between him and Bessemer. Ultimately, their patents were consolidated, along with those that Mushet held.

Alexander L. Holley of Connecticut was instrumental in introducing the Bessemer process into the United States. He studied firsthand Bessemer's

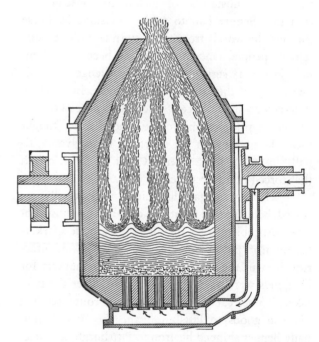

Diagram of a Bessemer converter during the cold air blast

plants in Sheffield, England, and he was impressed by the egg-shaped, open-topped furnace or converter used there to make steel. In 1864 he experimented with the process in Troy, New York, and organized a group that gained rights to both the Bessemer and Kelly patents. Holley licensed new producers and helped them design plants, including Andrew Carnegie's massive J. Edgar Thomson Works in Pittsburgh.

Steel makers altered the Bessemer process during the years that it dominated steel production in the United States and England. Initially the converters served well when ores from the Lake Superior area were used. They could not make good use, however, of ores rich in phosphorus or sulphur. In the 1870s lime or magnesia was used to line the converters, which removed phosphorus from the ore. The new process was known as the "basic" or the "Thomson-Gilchrist," after the two Englishmen who developed it.

Bessemer steel production dominated the U.S. industry until the 1880s. Rails accounted for the bulk of Bessemer steel output. But there were problems with Bessemer steel, especially a tendency to break under sudden strain. By the time Bessemer steel came to monopolize production, the alternative open-hearth process had been developed. Early in the 1860s Emile and Pierre Martin began to use shallow-hearth furnaces in France. They depended on a regenerative gas furnace invented in 1856 by William and Werner Siemens in Germany. Siemens-Martin steel making used a shallow bowl to hold the pig iron, scrap iron, old steel, and ferro-

manganese; heat was applied from the outside. More time was required by the open-hearth process, but it allowed for greater sampling and adjustment of the molten mass. Open-hearth steel could be made to more exact specifications than Bessemer. As a result, those who made such things as axles and structural steel stipulated the open-hearth product because of its greater reliability. Nevertheless, it was not until the 1890s, when rail construction slackened, that open-hearth steel began to compete seriously with Bessemer. In 1908 open-hearth steel production surpassed Bessemer for the first time in the United States. By the 1950s only 1 out of every 20 tons of steel was produced by the Bessemer process.

Reference:

Douglas Alan Fisher, *The Epic of Steel* (New York: Harper & Row, 1963).

Bethlehem Iron Company

by John W. Malsberger

Muhlenberg College

The forerunner of the Bethlehem Steel Corporation, the Bethlehem Iron Company was in the late nineteenth century the largest manufacturer of iron and steel products in the Lehigh Valley region of Pennsylvania and one of the most important corporations within the iron and steel industry. Organized on the eve of the Civil War simply to produce pig iron for the local market, Bethlehem Iron grew rapidly under the leadership of its general superintendent, John Fritz, by introducing the latest technological advances and by constantly expanding its line of products. By the 1890s Bethlehem Iron was one of the most modern and diversified iron and steel corporations in America.

The origins of the Bethlehem Iron Company are rooted directly in the use of anthracite coal in the manufacture of iron in America. Anthracite coal was first used as a fuel in nearby Catasauqua, Pennsylvania, in the early 1840s. Bethlehem Iron was the idea of Augustus Wolle, a Bethlehem businessman and one of the chief stockholders in the

Thomas Iron Company of Catasauqua, the firm organized by the Welsh ironmaster David Thomas, noted for demonstrating the commercial viability of anthracite fuel. Wolle and Bethlehem businessmen Charles Brodhead and Charles W. Rauch were granted a charter on April 3, 1857, for the Saucon Iron Company, which proposed to build a blast furnace for the manufacture of pig iron. Wolle and his partners believed that their iron company would spur the economic development of their city and were convinced that the close proximity of ore deposits and anthracite coalfields, together with the development of railroads in their region, would assure the commercial success of their venture.

Throughout the summer and fall of 1857 the partnership attempted to raise sufficient capital to construct their iron company through the sale of stock. Their efforts were hindered by the panic of 1857, however, and the sale of stock was halted for two years. In late 1859, in an effort to rekindle the venture, the Saucon Iron Company was reorganized

The Bethlehem Iron Company works in 1893 (courtesy of Bethlehem Steel Company)

through the addition of two new principals, Charles B. Daniel, a Bethlehem merchant, and Robert H. Sayre, the head of the Lehigh Valley Railroad. With the reorganization the firm also made its first effort to diversify, a process that ultimately helped to insure its success. Influenced undoubtedly by the business failures associated with the panic of 1857, the directors attempted to insulate their firm from the vagaries of the business cycle by adding a rolling mill to their proposed blast furnace. To reflect the changes achieved in the reorganization, the firm was renamed the Bethlehem Rolling Mills & Iron Company.

In the spring of 1860 the new company took perhaps the most important step to assure its future growth and success when it hired John Fritz to oversee the construction of its facilities and to serve as the firm's general superintendent and chief engineer. By 1860 Fritz had already gained a strong reputation in the iron industry because of the numerous innovations, most notably the three-high roll train, he had introduced at the Cambria Iron Works of Johnstown, Pennsylvania. For the next 30 years Fritz's imaginative and innovative leadership was one of the chief factors behind Bethlehem Iron's growth from a small iron company serving a local market to a major diversified iron and steel corporation serving a national and international market.

Under Fritz's supervision ground was broken for the first blast furnace on July 16, 1860, and in early 1861 work was begun on the rolling mill. The outbreak of the Civil War in April 1861, however, seriously interfered with the company's plans. In May 1861, for instance, the financial uncertainty spawned by the war forced the iron company into yet another reorganization, the new firm being

named the Bethlehem Iron Company. With the infusion of new capital, construction was resumed, and on January 6, 1863, the first batch of pig iron was produced. The rolling mill, which consisted of 4 engines, 14 puddling furnaces, 9 heating furnaces, and 3 trains of rolls, was completed in July 1863. Bethlehem Iron rolled its first rails on September 26, 1863.

For several months after the completion of its rolling mill Bethlehem Iron subsisted on a variety of small contracts with local concerns. The first major contract landed by the company came in early 1864 when the New Jersey Central Railroad ordered 2,000 tons of rails at $62.50 per ton. Buoyed by this sizable contract, Bethlehem Iron embarked on a program of expansion in May 1864 with the construction of a second blast furnace which was blown in three years later. Similarly, to enable the company to build or repair any of the equipment needed for the production of iron, it constructed a machine shop in 1865 and a foundry in 1868. Finally, in September 1868 Bethlehem Iron acquired the assets of the Northhampton Iron Company, a local concern then under construction. This acquisition brought to the Bethlehem firm a third blast furnace, blown in at the end of the year, and expanded its annual output of pig iron to 30,000 tons.

Up to this point the path of expansion followed by Bethlehem Iron was common to many other small iron manufacturers throughout the United States. But in 1868, under the constant prodding of John Fritz, the directors of Bethlehem Iron approved several actions that caused its path to diverge substantially from that followed by most of its competitors. Fritz had been interested in what became known as the Bessemer process for manufacturing steel since Henry Bessemer and William Kelly had begun experimenting separately with it in the 1850s. In July 1868, sensing that the more durable steel rails would almost certainly be the choice of America's burgeoning railroad industry, and convinced that the low phosphorous content of the iron ore located near Bethlehem would permit the use of the Bessemer process, Fritz persuaded the directors of Bethlehem Iron to approve the construction of a Bessemer plant. Work was begun in fall 1868, only three years after the first successful manufacture of Bessemer steel in America. When Bethlehem Iron rolled its first steel rail on October 18, 1873, it took a step that expanded the market for its products to the national level.

In the 1870s and 1880s, while many small iron companies in America were disappearing because of economic panics or the monopolization movement then rampant in the industrial community, Bethlehem Iron continued to improve, diversify, and expand. In 1871, again at John Fritz's urging, Bethlehem Iron increased its efficiency by adopting another technological advance in the manufacture of steel, the Siemens-Martin open-hearth furnace, only three years after the method had been introduced successfully in the United States by Cooper, Hewitt & Company. The year 1875 brought the addition of two new larger blast furnaces, so that by the end of the year Bethlehem Iron had become the largest producer of steel rails and pig iron in the Lehigh Valley, employing 1,400 men to manufacture more than 80,000 tons of iron and steel per year.

The expansion of Bethlehem Iron into a major national producer of iron and steel was completed in the 1880s when the company became the chief supplier of armor plate and large caliber guns to the United States Navy. When in 1882 Secretary of the Navy William Chandler began the process of rebuilding the nation's fleet (which had been allowed to deteriorate badly after the Civil War), Bethlehem Iron seized the opportunity. The principal stockholder of the company, Joseph Wharton of Philadelphia, together with John Fritz, succeeded in persuading the directors of the firm to authorize construction of an armor plate and gun forging mill. When the government asked for bids on the navy projects, Bethlehem Iron was the only company in the nation to bid on all the contracts. As a result it was awarded $4.5 million in government contracts, which pro-

vided the means to modernize the plant still further. To produce steel in sufficient quantity to fulfill the government contracts, Bethlehem Iron spent $3 million over the next two years to construct a series of open-hearth furnaces which were put into operation in August 1888.

The decision to diversify into the manufacture of armor plate and large caliber guns was clearly of seminal importance to the fortunes of Bethlehem Iron. The steady expansion of the U.S. Navy in the 1890s continued to provide lucrative government contracts to Bethlehem Iron and allowed the company to prosper despite the general economic collapse that began with the panic of 1893. Indeed, Bethlehem Iron came to rely so heavily on government contracts that after 1887 it was able largely to abandon the less efficient Bessemer furnaces and rollings mills which had brought it earlier success.

By the late 1890s, therefore, Bethlehem Iron, by following a steady course of innovation and modernization, had been transformed from a small, local producer of pig iron into one of the most modern steel corporations in the nation. It was reorganized as the Bethlehem Steel Company in 1899, reflecting its near total concentration on steel. Two years later Charles Schwab of U.S. Steel gained controlling interest in Bethlehem Steel in an effort to merge it into his United States Shipbuilding Company. When this company declared bankruptcy in 1903, Schwab resigned from U.S. Steel, reorganized his personal finances, and in December 1904 created the Bethlehem Steel Corporation.

In a very real sense Bethlehem Iron Company was one of the great success stories in the nineteenth-century iron and steel industry. The rapid growth it achieved in the 30 years after its founding underscored the immense importance of innovation and modernization in the iron and steel industry. In this manner Bethlehem Iron played an integral role in the maturation of the iron and steel industry from one composed of hundreds of decentralized producers to one that was highly integrated and dominated by a few, large national corporations. In a larger sense, however, Bethlehem Iron also contributed to the transformation of America from a rural, agrarian nation to an urban, industrial nation. Its innovations helped change steel from a relatively scarce and expensive product to an abundant commodity that fueled the industrialization of America in the last half of the nineteenth century.

References:

Craig Bartholomew, "Anthracite Iron Making and Industrial Growth in the Lehigh Valley," *Proceedings,* Lehigh County Historical Society, 32 (1978): 129-183;

John W. Fritz, *The Autobiography of John W. Fritz* (New York: John Wiley & Sons, 1912);

The Manufactories and Manufactures of Pennsylvania of the 19th Century (Philadelphia: Galaxy, 1875);

James Moore Swank, *History of the Manufacture of Iron in All Ages,* second edition (Philadelphia: American Iron and Steel Institute, 1892);

W. Ross Yates, *History of the Lehigh Valley Region* (Allentown, Penn., 1963).

Brady's Bend Iron Works

by Paul F. Paskoff

Louisiana State University

Brady's Bend Iron Works, located just to the northwest of Pittsburgh in Armstrong County, Pennsylvania, was the first vertically integrated rolling mill in the United States. As such, it established the pattern for subsequent business organization throughout much of the American iron and steel industry.

The company was the successor to the innovative but failed Great Western Iron Works, which had been established at Brady's Bend in 1839 by Philander Raymond. Raymond erected his first furnace in 1840 and built a rolling mill in 1841. He built two more furnaces in 1841 and 1843, and these, like the first one, were fired with coke made from the local bituminous coal. The decision to make iron with this coke rather than charcoal was a radical departure from the accepted practice and conventional wisdom within Pennsylvania's iron industry.

Built on a large scale during difficult economic times and using a new fuel, the Great Western Iron Works was a daring initiative. His first year of operation, 1840, was the first full year of a severe economic contraction, the depression of 1839-1843, and was therefore hardly an auspicious time to embark on so ambitious a capital venture. Perhaps even more audacious was Raymond's decision to build mineral fuel furnaces using coked bituminous coal, only a year after the first anthracite coal furnaces had come on line in the eastern part of the state. Although Raymond's plan was innovative and, in principle, sound, the decision to use coke made from bituminous coal proved to be the source of unforeseen and grievous difficulties. The problems began almost as soon as the ironworks went into operation.

Raymond had assumed that his furnace's coke pig iron would find a ready market in nearby Pittsburgh, as coke pig iron sold there at a lower price than pig iron made with charcoal. He quickly learned otherwise as buyers there spurned Great Western pig iron specifically because it had been made with coke. Buyers of pig iron in much of western Pennsylvania preferred charcoal pig iron to that made with coke because the former could be more readily worked up into bar iron for agricultural and mechanical purposes. This resistance to his pig iron must have been particularly frustrating to Raymond because his furnaces operated with considerably more efficiency than did his competitors' charcoal furnaces. Faced with this dilemma, Raymond responded with a certain entrepreneurial flair.

The nub of the problem facing Raymond was the necessity of finding a profitable way of disposing of the tons of pig iron which his furnaces were producing. With external sources of demand closed to him, he turned his attention to the rolling mill of his own Great Western Iron Works. Raymond had built the mill to roll bar iron, using bituminous coal as fuel, and now had to rely on it to absorb his furnaces' pig iron. His coke-fired furnaces fed their pig iron directly to his bituminous coal-fired rolling mill. This innovation, the creation of a vertically integrated rail mill, had largely been born of necessity, even desperation, and had not been anticipated when Raymond began his business. What mattered, however, was that the thing had been done and that it set an example for others to imitate.

The crowning irony for the company's managers was that just as the works might have begun to make money, with the end of the economic crisis in

The Brady's Bend Iron Works, date unknown (courtesy of the Kittanning Free Library)

1843, the venture failed. Precisely why the firm died is not known, but its difficulties almost certainly had to do with the enormous capital expenditures associated with its construction program and its inability to find a market for pig iron. In 1844 the Great Western Iron Works was sold at a sheriff's auction to M. P. Sawyer and his associates, capitalists from Boston, who reorganized the defunct firm under the name of Brady's Bend Iron Works.

Sawyer and his colleagues quickly began to expand the works and to reorient their operation from one geared to the production of bar iron for manufacturers of agricultural implements to the making of rails for the burgeoning Pennsylvania railroad industry. In 1846 they completed and brought into production a fourth coke-fired furnace to produce pig iron and retooled the rolling mill to produce rails from the pig iron. The next year the total output of the four furnaces approached 10,000 tons while their operation consumed more than 1 million bushels of bituminous coal and the labor of about 900 workers.

By 1849 the rolling mill at Brady's Bend was producing more than 7,000 tons of rails annually, and with considerably greater efficiency than its competitors in the eastern part of the state. Each year the mill employed 200 men and boys and consumed more than 600,000 bushels of bituminous

coal and about 9,000 tons of pig iron. Although somewhat smaller than the rail mills in the East, Brady's Bend Iron Works, as the only significant rail operation west of the Allegheny Mountains, enjoyed a near monopoly of the railroad business. This privileged situation did not last.

The success of Brady's Bend as a vertically integrated rail producer encouraged the founding in 1851, along similar lines, of the larger and ultimately more successful Cambria Iron Works at Johnstown, just to the west of the Alleghenys. Cambria's owners realized that the rapid growth of the Pennsylvania Railroad and particularly its connection to Pittsburgh would mean large orders for rails, orders which, in the absence of effective competition from Cambria, would of necessity go to Brady's Bend Iron Works. Having studied the vertically integrated structure of the Brady's Bend Works and its resulting efficient operation, Cambria's owners resolved to imitate the structure and count on the railroad's voracious demand for rails to sustain their new enterprise.

Even before plans for the Cambria Works were formulated, the owners of Brady's Bend Iron Works had planned to expand the scale of their operation and to increase its efficiency. Although all four coke-fired furnaces were out of blast in 1849, the owners proceeded in 1850 to make H. A. S. D.

Dudley general manager, a position which he was to hold until 1864. Because of his astute management and the strong stimulus provided, first, by the railroad industry and, subsequently, by the Civil War, the mills at Brady's Bend operated at near capacity and the company became a leader in the railroad supply industry.

References:

Fritz Redlich, *History of American Business Leaders: A Series of Studies,* 2 volumes (Ann Arbor, Mich.: Edwards Brothers, 1940);

Charles E. Smith, "The Manufacture of Iron in Pennsylvania," *Hunt's Merchants' Magazine,* 25 (November 1851): 574-581 and tables following p. 656;

James M. Swank, *History of the Manufacture of Iron in All Ages* (Philadelphia: American Iron and Steel Association, 1884, 1892);

Peter Temin, *Iron and Steel in Nineteenth-Century America: An Economic Inquiry* (Cambridge, Mass.: M.I.T. Press, 1964).

Cambria Iron Company

by Stephen H. Cutcliffe

Lehigh University

Cambria Iron Company, located on the Conemaugh River in Johnstown, Pennsylvania, and later reorganized as the Cambria Steel Company, was arguably one of the two or three most important iron and steel producing facilities in America during the second half of the nineteenth century. It was the location of numerous improvements and developments in iron and steel production and, much like the role the Erie Canal played in training civil engineers, Cambria served as a practical school and source of on-the-job training for a whole generation of iron and steel men.

Cambria first came into being as a unified entity in 1852 and 1853. In the early 1840s George S. King, a Johnstown businessman, in conjunction with several partners, including Dr. Peter Shoenberger, constructed several charcoal iron furnaces. He subsequently recognized that the geographic advantage of Johnstown's iron ores, combined with the arrival of the Pennsylvania Railroad in 1850 and its completion to Pittsburgh two years later, held out great financial promise, if iron suitable for rails could be produced. King convinced a group of Boston, New York, and Philadelphia businessmen and investors to buy and convert these charcoal furnaces to coke furnaces. However, the money ran out in 1854 before the work could be completed, and the Philadelphia creditors sent one of their men, Daniel J. Morrell, to investigate the situation and determine how best to preserve their investment. Although Morrell, a dry-goods merchant, knew little of the iron operation, he perceived its potential and recommended further investment, to which the creditors agreed. In 1855, because of a shortage of funds, the company was again forced to suspend operations before the works could be completed.

In that year the Philadelphia investors organized a new firm, Wood, Morrell & Company, to lease the works for a period of five years. Six investors, several of whom were already involved—Charles Wood, David Reeves, Mathew Newkirk, Edward Y. Townsend, George Trotter, and Morrell—each contributed $30,000 to the reorganized firm in a final attempt to put into practical operation the heretofore only promised potential. Charles Wood became head of the firm with Edward Townsend as his assistant, and Morrell was sent to Johnstown as the general manager.

John Fritz, an engineer of the practical school who had gained his experience in the iron mills at Norristown and Safe Harbor, Pennsylvania, had arrived at Cambria in June 1854, at the suggestion of David Reeves, to take over as general superintendent of the plant. It was clearly Fritz's knowledge and day-to-day supervision that brought the mill to completion and into operation as a fully integrated system. However, low-grade iron ore from the Johnstown area made rolling good rails without the flanges tearing off very difficult, and even the importation of a better grade iron from outside the area did not provide the solution.

The real problem lay with the standard technique mills used to roll rails at the time, which was

The Cambria Iron Company on the Conemaugh River in Johnstown, Pennsylvania

by passing them through a set of two rollers with a sequence of appropriately sized notches. As the rails passed through the rollers, workers picked them up with heavy tongs and brought them over the top or around to the front side of the rollers, whereupon they were passed through successfully smaller notches until finally reaching the correct size. At each pass the rail cooled, becoming successively more brittle and hence liable to tear and break up.

Fritz's now well-known solution was the building of a three-high rolling mill, thereby eliminating the hand-carrying stage and making the process more nearly continuous. Although the solution was obvious to Fritz, it was not so to the financially hard-pressed management of Cambria. It was thus with great difficulty, and not without several reversals of decision, that Fritz was finally able to convince Edward Townsend, and subsequently the rest of management, to back his plan. In addition, the workers at Cambria treated the new project with suspicion and opposed it; tradition often died hard in the craft environment of the mid-nineteenth century.

The idea of the three-high mill was not entirely new, having been broached a century earlier by Christopher Polhem, the Swedish engineer. Also, several "reversing" mills were then in use in England, and on a smaller scale three-high mills were being utilized for iron shaping in the United States. Fritz's creation, however, was conceptually different, due to several key elements. First was the scale. Fritz proposed three rolls of 20 inches diameter each, although as built, they were only 18—a compromise Fritz later judged to have been "a great mistake." Secondly, Fritz eliminated the gear train for powering the mill, driving it instead directly off the

flywheel of a steam engine. Finally, he replaced the so-called "breaking pieces," which, in a standard two-roll design, were intended to fail under excessive stress in order to save the rolls themselves. However, this meant the machinery was constantly breaking, resulting in both a loss of time and of money. Despite the safety objections of the plant manager, Daniel Morrell, against the warning of his former employer, James Hooven, owner of the Norristown Iron Works, that he would ruin his reputation, Fritz forged ahead with his plans. The new mill was finished on July 29, 1857, but because of the fear of failure and hostility of the workforce, Fritz conducted the first trial in secret. To almost everyone's surprise, except Fritz's, the experiment was a complete success. The plant operated smoothly for the next two days, but during the night of July 31 the wooden building caught fire and burned. Luckily the walls fell inward, saving the adjoining wooden buildings from a similar fate. Within a month Fritz had the mill reerected and running again, albeit without walls. These were rebuilt with brick and roofed over with slate, the job being finished by late December 1857.

Fritz remained at Cambria for another two and a half years until July 1860 when he severed his ties and moved to the Bethlehem Iron Company. During his last years at Cambria Fritz was responsible for several improvements to the Cambria facility, including increasing the output of the puddling furnaces, for the mill could now roll four times the amount of rails previously produced. He also improved the flow and handling of materials and redesigned numerous pieces of equipment, including the blowing engine. Fritz's contribution to

Cambria and to the iron industry in general were crucial. Upon his departure Cambria was a smoothly running operation and, with 1,948 workers, the second largest firm in the United States in terms of employees, placing it behind only the Montour Iron Works, which employed 3,000. In 1860 the original lease was extended for an additional two years. In 1862 the company was reorganized, absorbing Wood, Morrell & Company. The new firm had capital stock of $1.5 million and was also the owner of 30,000 acres of land. Charles S. Wood became president and Edward Y. Townsend, vice-president, with Morrell serving as general manager.

Fritz's career and especially his time at Cambria were indicative of the process of the transfer of technological knowledge and its importance to the growth of industry. Frank Jones of Jones and Laughlin was later to say that "Cambria was the cradle in which the great improvements in rolling-mill practice were rocked" and that "the three-high mill was the commencement of the great improvement that took place in the iron works after 1857 paving the way for the introduction of the phenomenal Bessemer process." By 1865 one-third of all mills had adopted the three-high mill, and by the 1880s the figure was 100 percent. Fritz had brought with him a knowledge base gained in his early years at Norristown and Safe Harbor and would again transfer that knowledge, this time expanded by experiences at Cambria, to the great Bethlehem Iron Company. But Fritz was also important as a trainer of other iron and steel "pupils" who went on to prominent positions in other companies as well, a tradition that continued on at Cambria long after Fritz's departure.

At the same time Fritz was contemplating his three-high mill, another experiment that would eventually lead to equally important changes at Cambria was taking place. In 1856 William Kelly, the Kentucky kettle maker who had developed a pneumatic process for making steel similar to that of Sir Henry Bessemer in England, came to Cambria and asked Morrell for an opportunity to demonstrate his process. If, as Fritz suggested, Morrell "knew nothing about the iron business" and was hesitant about the three-high mill, he must certainly have been receptive to some new ideas, for he not only gave Kelly space to conduct his experiment, but provided an assistant, James H. Geer, to help build the converter. Although the first blast resulted in a disastrous explosion that emptied the converter of its con-

tents and became known, much to local amusement, as Kelly's "fireworks," a second attempt was somewhat more successful. By some accounts Kelly continued his experiments sporadically up to the time of the Civil War, but no further serious attempts to produce steel were conducted at Cambria until after the war.

Although apparently impressed with the potential of the steel-making process, Morrell was unable to pursue its development due to the panic of 1857 and the disruptions of the early war years. In 1862, however, Morrell joined forces with a group of men that included among others E. B. Ward, a midwest iron producer and shipping magnate, and Zoheth Durfee, an expert iron maker, in an attempt to purchase the Kelly and Bessemer steel-making patents in order to build a Bessemer plant at Wyandotte, Michigan. Although failing to acquire the Bessemer patents, the group was successful in obtaining Kelly's and also the American rights to the Mushet patent, which controlled the recarburization process. On the strength of these two patents, they formed the Kelly Pneumatic Process Company in 1863, later consolidated in 1866 into the Pneumatic Steel Association, and then in 1877 the Bessemer Steel Association, in which Ward and Morrell were the driving forces. The Cambria Iron Company was thereby very loosely connected by virtue of Morrell's direct involvement. Under the supervision of William F. Durfee, Zoheth's cousin, the plant was built, and on September 6, 1864, the first large-scale batch of Bessemer steel was produced. And on May 24, 1865, several experimental steel rails were rolled at the North Chicago Rolling Mill Company using steel from Wyandotte.

Despite the success of this experiment, it would take Morrell eight years to convince the Cambria management to install a Bessemer converter to make steel. In the interim, however, the honor of rolling the first commercially produced steel rails would fall to Cambria. In 1866 the Pennsylvania Steel Company, established by a group of railroad men including J. Edgar Thomson of the Pennsylvania Railroad, began construction under the direction of Alexander Holley of a Bessemer plant near Harrisburg (now Steelton) and conducted their first successful blow in June 1867. Because their rail mill would not be completed until 1868, the blooms were sent to Cambria to be rolled into rails for the Pennsylvania Railroad. Robert Hunt,

Cambria's metallurgist, who in 1865-1866 had been sent to Wyandotte to observe the process in anticipation of the company building a converter, was placed in charge of the rolling. This brought him into close contact with Holley and cemented a close working relationship between the two men that lasted throughout their careers.

Following some initial difficulties with the machinery, which was designed for rolling iron rails and that occasioned Holley to say "there is an inherent cussedness about rolls, which so far, no man has been able to find," the first batch of commercially produced steel rails was completed in August 1867. Despite this success, however, Morrell was still unable to convince the Cambria management to avail themselves of the Bessemer opportunity. This is perhaps not surprising in that Cambria had only just weathered the economic instabilities of the 1857 panic and five years of war to become one of the largest iron companies, with 48 double puddling furnaces producing 1,000 tons of iron per week. To risk this success on a still largely experimental process with an unclear market was as yet too uncertain. However, after three years of Morrell's prodding, and the example of five operating Bessemer plants, management finally gave George Fritz, John's brother and successor as chief engineer and general superintendent, the go-ahead, and so Cambria became the sixth company to install a Bessemer converter.

Alexander Holley was called in to design the new Bessemer plant, and Robert Hunt had the job of hiring the crew. While Holley was building the new converter, George Fritz contributed several new features to the blooming mill. Two of these were roller-driven tables and a hydraulic pusher for moving and turning the ingots, improvements which made the new plant more efficient than Holley's plant at Troy, New York, by requiring only three men and a boy in contrast to eight men. This design now became the standard for Bessemer works in the United States. Fritz's blooming mill utilized 21-inch-diameter rolls to reduce an 8 1/2-inch-square ingot to a 6 1/2-inch bloom in preparation for the rail mill. Cambria's two 6-ton converters were finished in summer 1871, and the first blow took place on July 10. The rails subsequently rolled from the steel were sold for $104 per ton. Upon the Bessemer plant's completion Robert Hunt was placed in charge until he left Cambria for the Troy plant in

1873; Holley was retained as a consultant for an annual fee of $1,000.

The experience gained in the design and operation of this plant, as with the three-high mill, again placed Cambria in the forefront of technical knowledge in the industry. Holley drew on this experience when he designed subsequent plants, while men like Hunt and Capt. Bill Jones, who worked as an assistant to Fritz and later went to the J. Edgar Thomson works, transferred and disseminated their knowledge of steel production throughout the industry.

In 1873, following the death of Charles Wood, Edward Y. Townsend acceded to the presidency of Cambria. At that time the company's capital stock was $2 million with a large floating debt. Under Townsend's direction the company was able to extinguish the debt and gradually transform itself from an iron to a steel plant with about $15 million worth of capital property. In 1878 Cambria absorbed the Gautier Steel Company when the latter changed its location from Jersey City to Johnstown. Cambria's annual output was generally about 10 percent of the total production of American rails, and in 1876 the company rolled the largest total tonnage of rails ever produced by a single mill—103,743 tons, of which 47,643 tons were iron rails and 56,100 tons, steel rails.

During this period the new open-hearth process, which utilized a slow regenerative furnace allowing greater control over the steel product, began to emerge, not so much as an immediate threat to the Bessemer process, but as more of a signal of the future. By July 1877 16 companies had adopted the open-hearth process and were turning out 25,000 tons of steel per year which, although smaller than the 100,000 ton production of the Bessemer plants, was still significant. Holley tried to convince Morrell and Cambria of the advantages of the open hearth but initially failed in part due to the high royalties on the required patents. However, Cambria finally agreed to the conversion in 1878 and became the first American plant to adopt the Pernot furnace and its revolving hearth.

Cambria continued to grow during the 1880s and 1890s, establishing its position as one of the leading independent, integrated steel-making firms in the country. In 1880 Cambria had 4,200 employees, a number that had grown to more than 8,000 by 1900, placing it within the ranks of the nation's top three steel producers measured by number of workers. Cambria's solid economic position enabled it to recover from the great Johnstown flood of 1889 despite the great loss of property suffered. In 1898 Cambria was among the six largest U.S. beam manufacturers, having expanded greatly beyond the production of steel rails. During this period Cambria also operated and established a "Scientific Institute" where its employees could take courses in science, engineering, mechanical drawing, and language.

In 1898 the Cambria Steel Company was formed, to which Cambria Iron Company leased its properties at a fixed rental equal to 4 percent of its $8.5 million capital, in effect absorbing the parent company. The original shareholders had the option of subscribing to all $24 million of the stock of the new company. Its real assets at this time were about $20 million.

Under the leadership of Powell Stackhouse, who served as president of Cambria Iron from 1892 until his death in 1927 and of Cambria Steel from 1898 to 1910, the company subsequently embarked on a period of rapid expansion, including the acquisition of the Conemaugh Steel Company. Other companies affiliated with Cambria Steel for which Stackhouse also served as president included the Mahoning Ore and Steel Company, the Penn Iron Mining Company, the Republic Iron Company, and the Manufacturers' Water Company. Cambria's plants covered almost 400 acres and were equipped with 8 blast furnaces, 4 Bessemer converters, 28 open-hearth furnaces, 4 blooming and slabbing mills, 10 rail mills, 13 merchant bar mills, 1 wire rod mill with seven shops for finished wire products, 372 by-product Coke ovens, six testing laboratories, four electric power and light plants, four coal mines, a brickyard, and a lumberyard. By 1901, only three years after the transformation, the company reorganized again under the same name,

this time with a capital of $44 million.

In the early twentieth century Cambria increased its production of finished products, from 467,000 tons in 1901 to 1,193,000 tons in 1913, a growth of 155 percent. In 1915 J. Leonard Reploge bought $15 million of Cambria's stock, which he then sold to Midvale Steel Company in February 1916. Midvale in turn was acquired by Bethlehem Steel in 1923, so ending Cambria's history as an independent iron and steel producer.

References:

John Fritz, *The Autobiography of John Fritz* (New York: John Wiley & Sons, 1912);

William Hogan, *Economic History of the Iron and Steel Industry in the United States* (Lexington, Mass.: Lexington Books, 1971);

Robert W. Hunt, "Evolution of the American Rolling Mill," *Transactions of the American Society of Mechanical Engineers,* 13 (1892): 45-69;

Hunt, "History of the Bessemer Manufacture in America," *Transactions of the American Institute of Mining Engineers,* 5 (1877): 201-215;

Jeanne McHugh, *Alexander Holley and the Makers of Steel* (Baltimore: Johns Hopkins University Press, 1980);

Elting E. Morison, *From Know How to Nowhere* (New York: Basic Books, 1974);

Morison, *Men, Machines, and Modern Times* (Cambridge, Mass.: MIT Press, 1966);

Fritz Redlich, *History of American Business Leaders: A Series of Studies,* 2 volumes (Ann Arbor, Mich.: Edwards Brothers, 1940);

James M. Swank, *History of the Manufacture of Iron in All Ages* (Philadelphia: American Iron and Steel Association, 1892);

Swank, *Introduction to a History of Ironmaking and Coal Mining in Pennsylvania* (Philadelphia: J. M. Swank, 1878);

Peter Temin, *Iron and Steel in Nineteenth-century America: An Economic Inquiry* (Cambridge, Mass.: MIT Press, 1964).

Henry Charles Carey

(December 15, 1793-October 13, 1879)

by John W. Malsberger

Muhlenberg College

CAREER: Partner, Carey & Lea Publishing Company (1814-1835); political economist (1835-1879).

A partner in the largest American publishing house of its time, Henry Charles Carey was most well known as the leading American political economist in the nineteenth century. Beginning in 1835 Carey published numerous books and pamphlets in which he developed an optimistic set of theories that challenged directly the prevailing economic ideas espoused by David Ricardo, Thomas Malthus, John Stuart Mill, and others. In his writings Carey was also a vigorous champion of high protective tariffs, believing that the free trade policy promoted by Great Britain was designed largely to increase her economic hegemony over the world. Thus, Carey represents an important transitional figure between eighteenth-century mercantilism and nineteenth-century economic nationalism. His ideas, which influenced the economic thinking of Stephen Colwell, Condy Raguet, Henry C. Baird, William Elder, and, later, Simon Nelson Patten, also provided the theoretical underpinnings used by the iron and steel industry, as well as other businesses, to justify the protective tariffs adopted by the United States in the last half of the nineteenth century.

Henry Charles Carey was born in Philadelphia on December 15, 1793, the eldest son of Mathew Carey. His father, the son of a Dublin baker, had immigrated to Philadelphia in 1784 where he became noted as a bookseller, publisher, economist, and philanthropist. Instructed mainly by his father, Henry displayed at an early age a great capacity for learning that remained with him all his life. At age eleven, for instance, Carey was placed in charge of the Baltimore branch of his father's bookstore and for six weeks ran it successfully. In 1814, at age twenty-one, Carey became a partner in his father's publishing company, Carey & Lea, and when his fa-

ther retired seven years later, Henry assumed control of what was then the largest publishing house in America. Included among the authors published by Carey's firm were Washington Irving, Thomas Carlyle, and Sir Walter Scott. Apparently absorbing a great deal of knowledge by reading the wide variety of manuscripts submitted to Carey & Lea for publication, Carey retired from the firm with a fortune in 1835 and devoted the remainder of his life to the study of political economy.

At the time Carey began to write on economic matters, the discipline was dominated by the classical theories of Adam Smith and his nineteenth-century descendants, including David Ricardo, Thomas Malthus, and, later, John Stuart Mill. Although the nineteenth-century classical economists embraced Smith's idea of the beneficence of a competitive economy, they looked to the future with great foreboding. Malthus's theories, for instance, saw war, famine, and pestilence as the inexorable result of an expanding population whose growth rate exceeded the ability of the land to sustain it. Ricardo's theories on land and rent came to similarly dire conclusions. Because, in his mind, the best lands were settled first, continued population growth would bring into cultivation more and more marginal land until at some point the land would be unable to feed all of the people. In addition, because the supply of fertile land was inelastic, Ricardo predicted that as a society grew, wealth would become increasingly concentrated in the hands of the landlords who were able to charge rent for their productive lands. To the classical economists of the early nineteenth century, then, their discipline was truly a dismal science. Arguing that there were inherent limits to a society's growth, they foresaw a future which was dominated by hunger, starvation, and an inequitable distribution of wealth, leading to inevitable warfare between labor and capital.

Henry Charles Carey

Beginning with the publication, in 1835, of *Essay on the Rate of Wages* and expanding on these ideas in *Principles of Political Economy* (three volumes, 1837-1840), Carey took issue with the predictions of the British classical economists, suggesting that the future prospects of mankind were much more promising. He challenged Malthus's ideas, for instance, arguing instead that as industrialization enhanced the technical abilities of society, the output of goods and services would increase more rapidly than population. He also correctly foresaw that the birthrate of industrial societies would decline rather than continue its geometric expansion. In these early works Carey also questioned Ricardo's theories of land and rent. Basing his conclusions on a survey of the settlement and patterns of cultivation of the United States, Mexico, Great Britain, France, Italy, Greece, India, South America, and the Pacific Islands, Carey asserted that historically settlement began on the poorest lands and progressed to richer lands only as increases in

population and wealth made it possible. Therefore, he concluded, the future did not hold war, pestilence, and famine, but rather peace, prosperity, and happiness. Similarly, he contended, land had little intrinsic value but was merely one of the factors of production. Land could yield little without the application of labor and capital. Therefore, Carey reasoned, the return to landowners (rent) was no different than the return to capital (interest) or the return to labor (wages). Thus, Carey contended that society was not composed of competing interests, but of a harmony of interests which he believed would insure a steady improvement in the well-being of all. Finally, he maintained that the increasing power of accumulated capital would not impoverish the workers, as classical theory believed, but would actually improve their standard of living. Reasoning that no commodity can command more than the value of labor that is required to reproduce it, Carey posited that every increase in the power of accumulated capital correspondingly reduced the relative value of labor in those commodities already in existence and thereby expanded the purchasing power of the worker.

Although Carey's writings on political economy aimed to refute many of the gloomy predictions of the nineteenth-century British classical economists, his ideas continued to reflect many of the basic principles espoused by Adam Smith. Throughout his writings, for instance, he tended to define human affairs in purely economic terms. Man, to Carey, was economic man, whose well-being was defined in terms of wealth. Thus, Carey did not reject classical economic theory completely but rather refined it and adapted it to the conditions existing in nineteenth-century America. His optimistic theories, though they were ridiculed by contemporary economists, were perfectly attuned to the buoyant capitalism of the Jacksonian era. America in the 1830s seemed to be a nation of unlimited land and resources and thus had no use for the gloomy predictions of a Malthus or a Ricardo. Henry Carey's ideas captured the hearts and minds of American businessmen, politicians, and newspaper editors because they also seemed to separate the United States still further from the depravity of the Old World, suggesting that the potential for American economic growth was boundless. By providing a hopeful set of assumptions with which to view the future, then, Carey's ideas played a major role in expanding the climate of confidence essential for American industrialization in the nineteenth century.

Perhaps contributing even more importantly to America's industrialization were the arguments Carey began to develop in 1844 on protective tariffs. Until 1844 Carey remained sympathetic to the principles of free trade espoused by Adam Smith and Jean Baptiste Say. In late 1842, however, his conversion to economic nationalism began when he observed the economy, depressed since the panic of 1837, rebound under the effects of the protective tariff of 1842. This experience forced him to the conclusion that throughout America's history it had prospered under protection and had faced bankruptcy under free trade. Carey's change of mind was also apparently influenced by two of his business ventures that failed during this period. In the early 1840s, for instance, a New Jersey paper mill in which Carey had invested heavily went bankrupt when the manufacture of paper ceased to be profitable, because of the gradual reduction in import du-

ties initiated by the compromise tariff of 1833. Similarly, in 1839 Carey became one of the major stockholders in the Clearfield Coal & Iron Company, which was organized in Karthaus, Clearfield County, Pennsylvania, to manufacture iron using coke. The company succeeded in producing pig iron, but the furnaces were shut down in late 1839 because of inadequate inland transportation.

All of these influences are apparent in the protectionist views Carey trumpeted in an 1845 pamphlet, "Commercial Associations in France and England," and expanded in two books, *The Past, the Present, and the Future* (1848) and *The Harmony of Interests, Agricultural, Manufacturing, and Commercial* (1849-1850). In these works Carey contended that Britain's policy of free trade was not designed to better the interests of all commercial nations, but rather was an effort to increase England's economic strength at the expense of other countries. To counter this plan, Carey advocated that the United States pursue a strict course of economic nationalism, using high protective tariffs both to protect its infant industries and to stimulate internal improvements, much in the way Henry Clay's "American System" had earlier proposed. Should America fail to protect its home industries, Carey warned, it would quickly become an economic colony of Great Britain. Following a protectionist course, he promised furthermore, would not come at the expense of anyone in America. Quite the contrary, Carey asserted that because economic interest groups in America were complementary rather than antagonistic, a system of high tariffs would benefit agriculture and commerce, as well as manufacturing, by increasing the wealth and efficiency of society.

Carey's ideas on economic nationalism were disseminated to a wide audience between 1849 and 1857 when he was a regular contributor of editorials to Horace Greeley's *New York Tribune*, then one of the most avowedly protectionist newspapers in America. Although his arguments made him appear to some of his contemporaries as nothing more than an apologist for big business, Carey's theories had a broad appeal to many Americans of the mid nineteenth century, partly because of their implicitly chauvinistic tone, but also because they seemed to provide a comprehensive, scientific theory in support of high tariffs. Carey's ardent advocacy of protectionism, then, helped to provide the intellectual rationale for American tariffs just as the

United States was beginning to industrialize in the second half of the nineteenth century. The rapid development of big business in this era, most notably the iron and steel industry, thus owed much to the ideas developed by Carey.

Although Carey is best known for his uncompromising advocacy of protective tariffs, his influence was also felt in other areas. Moved by the publication of Harriet Beecher Stowe's *Uncle Tom's Cabin*, Carey in 1853 added his views to the abolitionist crusade by producing an antislavery tract, *The Slave Trade, Domestic and Foreign: Why it Exists, and How It May Be Extinguished*. In this work Carey accepted the view that slavery was morally wrong, but his main concern seemed to be a desire to twist the British lion's tail once again. Although Britain had already abolished slavery, Carey contended in this work that she could claim no moral superiority to America. British rule of Ireland, he maintained, had produced pauperism and social degradation and was therefore no less reprehensible (and in some ways was worse) than the South's peculiar institution. Additionally, Carey's influence was also evident in Pennsylvania politics. Elected as a delegate to the 1872 convention charged with revising the Pennsylvania state constitution, Carey, despite his advanced age, took an active part in the debates dealing with banking, usury, corporations, and railroads.

Selected Publications:

The Geography, History, and Statistics of America and the West Indies (London: Sherwood, Jones, 1823);

Essay on the Rate of Wages (Philadelphia: Carey, Lea & Blanchard, 1835);

The Harmony of Nature (Philadelphia: Carey, Lea & Blanchard, 1836);

Principles of Political Economy, 3 volumes (Philadelphia: Carey, Lea & Blanchard, 1837-1840);

The Credit System in France, Great Britain, and the United States (London: J. Miller, 1838; Philadelphia: Carey, Lea & Blanchard, 1838);

Answers to the Questions: What Constitutes Currency? What are the Causes of Unsteadiness of the Currency? and What is the Remedy? (Philadelphia: Lea & Blanchard, 1840);

Beauties of the Monopoly System of New Jersey (Philadelphia: C. Sherman, 1848);

The Frauds, Falsifications, and Impostures of Railroad Monopolists (N.p., 1848);

Letters to the People of New Jersey, on the Frauds, Extortions, and Oppressions of the Railroad Monopoly (Philadelphia: Carey & Hart, 1848);

The Past, the Present, and the Future (Philadelphia: Carey & Hart, 1848; London: Longman, Brown, Green & Longmans, 1848);

The Harmony of Interests, Agricultural, Manufacturing, and Commercial, 2 volumes (Philadelphia: Skinner, 1849-1850);

The Railroad Monopoly (Philadelphia: L. R. Bailey, 1849);

Proceedings of the Late Railroad Commission (Philadelphia: L. R. Bailey, 1850);

The Prospect: Agricultural, Manufacturing, Commercial, and Financial (Philadelphia: J. S. Skinner, 1851);

How To Increase Competition for the Purchase of Labor, and How to Raise the Wages of the Laborour (New York: Finch, 1852);

How To Have Cheap Iron (New York, 1852);

Letter Addressed to a Cotton Planter of Tennessee (New York: Finch, 1852);

The Working of British Free Trade (New York: Finch, 1852);

Letters on the International Copyright (Philadelphia: A. Hart, 1853);

The Slave Trade, Domestic and Foreign: Why It Exists, and How It May Be Extinguished (Philadelphia: A. Hart, 1853; London: Low, 1853);

The North and the South (New York: Office of the Tribune, 1854);

Letters to the President, on the Foreign and Domestic Policies of the Union (Philadelphia: Lippincott, 1858; London: Trubner, 1858);

Principles of Social Science, 3 volumes (Philadelphia: Lippincott, 1858-1859; London: Trubner, 1858-1859);

The French and American Tariffs Compared (Detroit: J. Warren, 1861);

Financial Crises: Their Causes and Effects (Philadelphia: H. C. Carey, 1863);

The Paper Question (Philadelphia: Collins, 1864);

The Farmer's Question (Philadelphia, 1865);

The Iron Question (Philadelphia: Collins, 1865);

Letters to the Hon. Schuyler Colfax (Philadelphia: Collins, 1865);

The Way to Outdo England Without Fighting Her (Philadelphia: H. C. Baird, 1865);

British Free Trade, How It Affects the Agriculture and the Foreign Commerce of the Union (New York: The Iron Age, 1866; Chicago: J. A. Norton, 1866);

Contraction or Expansion? Repudiation or Resumption? (Philadelphia: H. C. Baird, 1866);

The Public Debt, Local and National: How to Provide For Its Discharge While Lessening the Burden of Taxation (Philadelphia: H. C. Baird, 1866);

The Resources of the Union (Philadelphia: H. C. Baird, 1866);

Reconstruction: Industrial, Financial & Political (Philadelphia: Collins, 1867);

Review of the Decade 1857-67 (Philadelphia: Collins, 1867);

The Finance Minister, the Currency, and the Public Debt (Philadelphia: Collins, 1868);

How Protection, Increase of Private and Public Revenues, and National Independence March Hand in Hand Together (Philadelphia: Collins, 1869);

Resumption: How It May Profitably Be Brought About (Philadelphia: Collins, 1869);

Shall We Have Peace? (Philadelphia: Collins, 1869);

Currency Inflation: How It Has Been Produced, and How It May Profitably Be Reduced (Philadelphia: Collins, 1870);

Wealth: Of What Does It Consist? (Philadelphia: H. C. Baird, 1870);

A Memoir of Stephen Colwell (Philadelphia: H. C. Baird, 1871);

The International Copyright Question Considered (Philadelphia: H. C. Baird, 1872);

The Unity of Law (Philadelphia: H. C. Baird, 1872);

Capital and Labor (Philadelphia: Collins, 1873);

Of the Rate of Interest; and of its Influence on the Relations of Capital and Labor (Philadelphia: Collins, 1873);

Miscellaneous Papers on the National Finances, the Currency, and Other Economic Subjects (Philadelphia: H. C. Baird, 1875);

Appreciation in the Price of Gold (Philadelphia: Collins, 1876);

Commerce, Christianity, and Civilization, versus British Free Trade (Philadelphia: Collins, 1876);

Resumption: When, and How Will It End? (Philadelphia: Collins, 1877);

How To Perpetuate the Union (N.p., n.d.);

Ireland's Miseries, Their Cause (New York: Tribune, n.d.).

References:

Biographical Encyclopedia of Pennsylvania of the 19th Century (Philadelphia: Galaxy, 1874);

William Elder, *A Memoir of Henry C. Carey* (Philadelphia: Henry Carey Baird, 1880);

Charles Levermore, "Henry C. Carey and his Social System," *Political Science Quarterly*, 5 (December 1890): 553-582;

James Moore Swank, *History of the Manufacture of Iron in All Ages*, second edition (Philadelphia: American Iron and Steel Institute, 1892).

Andrew Carnegie

(November 25, 1835-August 11, 1919)

by Stuart W. Leslie

Johns Hopkins University

CAREER: Superintendent, Pennsylvania Railroad (1859-1865); partner, Keystone Bridge Company (1865-1892); partner, Carnegie, Kloman & Company (1870-1881); partner, Carnegie, McCandless & Company (1872-1876); partner, Carnegie Brothers & Company (1881-1892); partner, Frick Coke Company (1882-1899); partner, Carnegie, Phipps & Company (1886-1892); partner, Carnegie Steel Company (1892-1899); partner, Carnegie Company (1899-1901).

Andrew Carnegie, steel master, was born on November 25, 1835, in Dunfermline, Scotland, north of Edinburgh, and died August 11, 1919, in Lenox, Massachusetts. In between was a lifetime of truly remarkable achievement, a career in investment, railroads, bridge building, and iron and steel making that shaped and measured America's Age of Steel. Cynics may have mocked Carnegie's uncritical paeans to progress, or dismissed his *Triumphant Democracy* (1886) as "sunshine, sunshine, sun-

shine," but Carnegie resolutely believed in the American dream because he knew he had lived it. His rise from immigrant poverty to the title of "the richest man in the world," remains its classic testimonial.

The Carnegies had for generations made their living in Scotland as linen weavers until in Andrew's boyhood one of the last campaigns of the British textile revolution drove them out of business. Carnegie's father, William, was a hand-loom weaver prosperous enough at one point to be managing several looms and apprentices. But when British textile manufacturers adapted cotton power looms for linens, the days of the hand-loom weaver were numbered. The family's fortunes steadily declined in the 1840s, and Andrew's mother, Margaret, took on cobbling and opened a small grocery in the family home to make ends meet. "Shortly after this I began to learn what poverty meant," Carnegie wrote in *Autobiography of Andrew Carnegie* (1920). "Dreadful days came when my father took the last of his webs to the great manufacturer, and

Andrew Carnegie

I saw my mother anxiously awaiting his return to know whether a new web was to be obtained or that a period of idleness was upon us." Those periods came more and more frequently thanks to increasing mechanization in the industry, and when a steam-powered weaving factory arrived in Dunfermline itself in 1847, William found himself out of work completely. With no real prospects in the Old World, the Carnegies reluctantly joined thousands of their countrymen on an exodus to the New. To pay the passage they auctioned off their furniture and borrowed the rest of the money from relatives and neighbors.

Andrew, then thirteen, left his homeland with little formal education. Although Scotland boasted the best public schools of the day, Carnegie's parents had kept him home until he was eight. Nonetheless, he did well enough once he got to school to earn a reputation as the teacher's pet. Except for a few night courses later on, Carnegie's five years at Mr. Martins's Rolland Street School in Dunfermline were all the schooling he would ever have. This did not, however, keep him from acquiring a first-rate ed-

ucation, not only in business but also in literature, history, music, art, and philosophy. Somehow he had mastered the most important lesson of all—how to learn—and devoted much of his life to a rigorous course of self-education.

It was Carnegie's informal education from the Dunfermline years that proved perhaps the more invaluable and enduring. One favorite uncle, George Lauder, filled his head with Scottish history, literature, and lore, and instilled a lifelong love for Robert Burns. Another uncle, a notorious local Chartist who once landed in jail for his cause, introduced the boy to radical politics, and passed on to him a fanatical (if rather naive) faith in democracy. Lessons in the workings of the world were offered too for those who could grasp them. Recalling how he once talked his friends into feeding his growing family of pet rabbits in return for the privilege of naming them, Carnegie wrote in his autobiography, "I treasure the remembrance of this plan as the earliest evidence of organizing power upon the development of which my material success in life has hung, a success not to be attributed to what I have known or done myself, but to the faculty of knowing and choosing others who did know better than myself. . . . I did not understand steam machinery, but I tried to understand that much more complicated piece of mechanism—man." Whether self-taught or gifted in this regard, the many successes of Carnegie's career came back to this simple truth—to manage business you must manage men.

Like most immigrants, the Carnegies followed the route of friends and relatives who had already made the trip to America. Margaret had family in Pittsburgh, and they took the Carnegies in, got them settled, and found them work. In those days Pittsburgh was all business. It was rebuilding after a major fire a few years before and had already earned a well-deserved reputation as the dirtiest and ugliest city in America. "Significantly, there were no parks in Pittsburgh," one of Carnegie's biographers pointed out. "Apparently the townspeople had no time for leisure and no confidence that in this atmosphere green grass and flowers could survive." It was in this new world that the Carnegies, crowded into a two-room flat on a muddy alley in what was called Slabtown (more properly, Allegheny), Pennsylvania, embarked upon their new life.

Ironically enough, Andrew got his first job in a textile mill. For a time his father tried hand-loom weaving again, with no more financial success than in the old country. Margaret went back to cobbling. William swallowed his pride and accepted a job in a cotton mill owned by a Scot who liked to lend a hand to his fellow countrymen. Teenage Andrew accompanied William and worked as a bobbin boy (changing the bobbins on the power looms), 12 hours a day, for $1.20 a week. William simply could not adjust to the factory, so he went back to weaving tablecloths and selling them door-to-door. Andrew, on the other hand, persisted.

Like his political contemporaries at Tammany Hall, Carnegie, even in his lowly position, saw opportunities and took them. He found a better job dipping bobbins and tending a steam boiler for $2.00 a week and convinced the owner with his excellent penmanship (something useful from his days at Mr. Martin's school) to give him some part-time bookkeeping assignments. Carnegie then walked to night classes several times a week all winter to learn how to do them. His next job (procured in 1849 thanks to more Scottish connections) was as messenger boy in the O'Reilly Telegraph Company. At $2.50 a week the money was a step up from the bobbin factory. More important, though, was that for the first time this job thrust him into the world of finance and commerce. Looking back, he saw it as a turning point: "From the dark cellar running a steam engine at two dollars a week, begrimed with coal dirt, without a trace of the elevating influences of life, I was lifted into paradise, yes, heaven, as it seemed to me, with newspapers, pens, pencils, and sunshine about me. . . . I felt that my foot was on the ladder and that I was bound to climb."

And climb he did, by working hard, learning fast, and impressing everyone with his efforts. He made a point of remembering important clients and delivering their messages to them on the street. He taught himself code, and how to take messages by ear, and earned promotions to part-time and then to full-time telegraph operator, at $20 a month. The prices, orders, and shipping schedules he sent out and received made the young Carnegie feel that he was at the center of Pittsburgh commerce, and the experience proved to be an excellent education in business. By age sixteen, he was making more money than his father. No wonder his long letters to relatives back in Scotland spoke so glowingly of his adopted land, of its miles of new telegraph and railroad lines, and its economic prosperity. "America was promises," he said.

In December 1852 the Pennsylvania Railroad opened service between Pittsburgh and Philadelphia. Because the railroad depended so heavily on the telegraph, district superintendent Thomas Scott felt he needed his own operator. Impressed with what he saw on his frequent visits to the O'Reilly office, he hired Carnegie for $35 a month. Scott, although not yet thirty, had already established a reputation as someone to watch. He had worked his way up from station agent to head of the railroad's western (Altoona to Pittsburgh) division and was clearly destined for even bigger prospects. He recognized and admired Carnegie as a slightly younger version of himself, and took him under his wing. For his part, Carnegie sized up Scott as someone with talent and ambition, and just as important, long coattails.

As Scott's personal secretary and protégé, Carnegie learned the railroad business from the inside. American railroads laid down record miles of track in the 1850s, and America's first "big business" was of such unprecedented scale and complexity that unprecedented means for controlling the business were required. Thus, new management techniques were laid out along with the track as the railroad companies struggled to systematize their operations.

For its day the Pennsylvania Railroad was considered a model of progressive management, particularly in accounting. Its methods were so advanced it was said that "A charge or entry of a day's labor, of the purchase of a keg of nails, or the largest order goes through such a system of checks and audits as to make fraud almost an impossibility." The Pennsylvania became the business school for a generation of young executives. Many did not make it through its harshly competitive, if informal, curriculum. Others went off and applied what they had learned to other railroads and other industries, including iron and steel. Carnegie learned the fundamentals of business during the years he spent in this pressured environment—hiring and firing, motivating and organizing men, determining and cutting costs, lowering the margins and making the profit on volume. The most significant lesson, perhaps, was that whatever paid for itself—a better machine, an improved process, a new source of raw materials, even a more skilled executive—was a bargain at any price.

The Pennsylvania days also gave Carnegie some invaluable business contacts for the years

Carnegie's Homestead plant where the violent 1892 strike occurred (courtesy of E. E. Moore, USX Corporation)

ahead. Putting himself, as always, at center stage, he made it a habit to take charge, occasionally coming close to insubordination. During one crisis, for instance, he sent out orders under Scott's name. Rather than being distressed by this sort of brashness, Scott bragged about it to other executives. Soon most of the railroad's key officers, including president J. Edgar Thomson, knew of "his Andy," as Carnegie was always proud to be called. When Carnegie went into business for himself, such contacts at one of the nation's largest railroads, which was also one of the largest consumers of iron and steel, paid off handsomely. As Scott rose through the corporate ranks Carnegie advanced with him, and by 1859, when Scott was named as a vice-president, Carnegie moved into his mentor's old job of superintendent of the western division. He was twenty-four years old.

Scott's education of his protégé extended to investing money as well as earning it. He gave Carnegie his first stock tip (on Adams Express) and then loaned him the money to take advantage of it, since as Carnegie remembered, he had closer to 50¢ than the $500 Scott's investment required. But he recalled, "I was not going to miss the chance of becoming financially connected with my leader and great man." Nor did he forget the thrill of that first $10 dividend check. "I shall remember that check as long as I live, and that John Hancock signature of 'J. C. Babcock, Cashier.' It gave me the first penny

of revenue from capital—something that I had not worked for with the sweat of my brow."

Following Scott's investment advice, most of it inside tips from the Pennsylvania, Carnegie began to ensure a future that would not entail much sweat. A tip from Scott on the newly formed Woodruff Sleeping Car Company turned Carnegie's $200 investment into three times his annual salary in just two years. Carnegie later parlayed his sleeping car investment into an extremely lucrative arrangement with George Pullman to supply sleeping cars for the Union Pacific. In a similar inside deal with the Keystone Telegraph Company, Scott, Thomson, and Carnegie netted $150,000, essentially for selling the Pennsylvania telegraph right-of-way concession to a legitimate telegraph company.

During these years Carnegie somehow found time to continue his cultural self-improvement program. He bought a home in the upscale suburb of Homewood and went back to night school to polish his French and elocution. Sensing the advantages of good manners, the weaver's son set out to make himself a gentleman: "I began to pay strict attention to my language, and to the English classics, which I now read with great avidity. I began also to notice how much better it was to be gentle in tone and manner, polite and courteous to all—in short, better behaved." Climbing the social ladder as quickly as its business counterpart, he was soon at ease in the New York literary salons, as an intimate to presidents and prime ministers, and as a best-selling author.

Although Carnegie cast his first presidential vote for Lincoln, denounced slavery and the South at every opportunity, and bragged in his memoirs about his experiences at Bull Run, he actually spent most of the Civil War fighting financial battles. He passed the first blistering summer in Washington working under Thomas Scott in the War Department, before deciding he could perhaps better serve his country running the Pennsylvania Railroad from Pittsburgh. Offered another opportunity to serve when he was drafted a couple of years later, he felt himself too busy to go to the front and hired an Irish immigrant to go for him.

Naturally, the war presented plenty of potentially profitable investments. Carnegie put $11,000 into the Columbia Oil Company, established in 1861 to tap into the "black gold rush" of the day, and received twice that much within the year, a return eventually totaling $1 million. He moved next into the iron business. To cash in on record iron prices and to take advantage of new demand for iron railroad bridges, he became a partner in both the Superior Rail Mill and Blast Furnaces and the Keystone Bridge Company. Carnegie's income tax return for 1864 (the personal income tax was passed as an emergency revenue measure during the Civil War) revealed that of a total income of $42,000, all but his $2,400 salary derived from investments.

Looking ahead to the postwar world, Carnegie was only certain of the fact that he did not want to run a railroad. Realizing that the managerial ladder could only take him so far, he resigned from his superintendent's post in 1865, commenting later that "A man must necessarily occupy a narrow field who is at the beck and call of others. Even if he becomes president of a great corporation he is hardly his own master unless he holds control of the stock." It was from this time on that Carnegie resolved to be the dominant stockholder in all of his enterprises and to leave the day-to-day managing to others. It was quite natural, given his investments, that he should turn from the railroad to the industries which allowed it to run–bridges, iron, and steel.

As with the telegraph and the railroad, Carnegie chose a new technology at exactly the right time. The construction of railroad bridges presented to mid-nineteenth-century civil engineers the same kind of unprecedented challenges running the railroad offered their managerial counterparts. Wooden trestles simply could not stand up to the ever increasing loads of the new locomotives and freight trains, now larger and longer as the railroads grew to meet increasing demand for service. By 1860 a brilliant generation of British engineers, led by Robert Stephenson and Isambard Kingdom Brunel, had met the challenges with bridges of imaginative new design in wrought iron. Stephenson's tubular (or girder) Britannia Bridge (1850) had main spans of 450 feet, while Brunel's lenticular Royal Albert (1859) had main spans of 460 feet. American designers, in contrast, lagged behind their European counterparts. John Roebling's stunning iron wire suspension bridges at Niagara Falls (1855) and Cincinnati (1867), with spans of 821 and 1,057 feet respectively, were the exceptions. Most American engineers thought suspension bridges were not sufficiently rigid for railroads. Ignoring his own famous business pronouncement–that pioneering did not pay–Carnegie's decision to found a company to build iron bridges put him into the middle of one of the most rapidly changing technologies of the day and at the same time opened spectacular opportunities for achievement.

Carnegie once called Keystone Bridge "the parent of all the other works." It had been, he felt, a template for future business organization and a model for the shaping of a technology. Carnegie, when he started, knew nothing about building bridges, though he did know a great deal about building companies. As might be expected, he looked first to the Pennsylvania Railroad for expertise, capital, and connections. Thus the original venture included among its partners his old friends from Pennsylvania days, Scott and Thomson, plus the railroad's chief mechanic J. L. Piper, chief bridge engineer J. H. Linville, and general bridge superintendent A. Schiffler.

Carnegie left the design and construction details to Piper and Linville–though occasionally he could not resist lecturing bridge company directors on such matters as the relative merits of wrought and cast iron–and concentrated on what he knew best, salesmanship and finance. Carnegie himself developed the inside connections with the railroads, invested judiciously in the bridge companies in order to ensure a favorable hearing for Keystone's construction bids, and nailed down the contracts. "If you want a contract," he advised, "be on the hand when it is let. . . . And if possible, stay on hand until you can take the written contract home in your pocket." When he could, he also arranged for the sale of the construction bonds in the financial markets of New York and London, where even more money could be made.

Keystone's greatest success was the Eads Bridge across the Mississippi. Carnegie had long had his eye on the "big bridges at St. Louis and Omaha." When Congress finally gave its approval for the construction of a bridge at St. Louis "for the ages, of a material that shall defy time and of a style that will be equally a triumph of art and contribution to industrial development," Carnegie was ready. He was not as prepared, perhaps, for the many challenges of working with the idiosyncratic James Eads, who had been chosen to design the bridge.

The self-taught engineer and the self-made businessman never did see the project in quite the same way. As Carnegie so perceptively, if ruefully, noted, while "this Bridge is one of a hundred to the Keystone Company–to Eads it is the grand work of a distinguished life. With all the pride of a mother for her firstborn, he would bedeck the darling without much regard to his own or other's cost." Eads had made his fortune as a river salvager, and picked up his engineering knowledge as a builder of the Union fleet of ironclads during the Civil War. Although he probably knew as much about the strength of materials as anyone in America at the time, he had never built a bridge and would never build another. What he lacked in bridge-building experience he made up for by hiring well-qualified assistants and by paying meticulous attention to the details of what he did know, the quality of his materials. He established rigid standards that Carnegie's company could not always meet and in place of the usual cast iron in the arches, insisted on exotic chrome steels that Carnegie's company could not always make, leaving Carnegie to complain that "Nothing that would have pleased, and does please, other engineers is good enough for this work." In the end Carnegie had to subcontract most of the steelwork to an outside firm.

Somehow Carnegie and Eads managed to deliver a bridge worthy of the lofty congressional mandate. Carnegie kept the contract out of the hands of a rival Chicago syndicate, sold the construction bonds in London, arranged the subcontracts, and for all his carping, got the materials delivered more or less on schedule. One student of Carnegie's career has suggested that "The resulting business structure was as complicated and as carefully engineered as the bridge itself. Stress and support had to be determined with the same exactness and the same insistence upon perfection that Eads had shown in his engineering specifications. Had Carnegie been as willing as Eads to submit his blueprints to public scrutiny, the student of commerce might have found the St. Louis bridge enterprise as instructive as did the mechanical engineer." Although the project came in three years late and well over budget, most of the extra time and money went into the pneumatic caissons which Eads had to develop for anchoring the piers in the turbulent Mississippi. When it opened in 1874, the Eads Bridge was the world's first really large steel structure. With three spans over 500 feet, it was also the longest metal-arch bridge in the world. Then and now, it was acknowledged a classic achievement of nineteenth-century American civil engineering. "In the end," noted one biographer, "Carnegie would regard the St. Louis bridge as one of his proudest achievements." Over the next decade Carnegie was instrumental in the building of a number of other important American bridges, including the Brooklyn Bridge, for which Keystone won the superstructure contract. More important, perhaps, his bridges also inspired some of the best structural engineers of the next generation, like Louis Sullivan, whose pioneering steel girder architecture was adapted for skyscrapers from Keystone's designs.

At the close of 1868 Carnegie totaled his personal and business accounts and set himself a private agenda for the years ahead. Proudly, he recorded that at thirty-three he had an annual income of $50,000. In a couple of years he would retire, go to Oxford and get a real education, publish newspapers, and give away money for worthy causes: "Man must have an idol–The amassing of wealth is one of the worst species of idolitary [*sic*]. No idol more debasing than the worship of money. . . . To continue much longer overwhelmed by business cares and with most of my thoughts wholly upon the way to make more money in the shortest time, must degrade me beyond hope of permanent recovery."

Sooner or later Carnegie actually got around to doing nearly everything on his list, excepting the Oxford degree. He did not retire to philanthropy at age thirty-five as he had planned, but he did shift the focus of his career from finance to production,

Carnegie at the reins with a coaching party in Pittsburgh (courtesy of Stefan Lorant)

from merely making money to making iron and steel. Years later he boasted with some conviction that he "manufactured steel, not securities," as if the one were somehow more virtuous than the other.

Before he made any steel, Carnegie first had to learn how to make iron. Typically, he found himself in the right place at the right time—at the end of a technological revolution which had replaced a world of wood with one of iron, and at the beginning of another which would replace a world of iron with one of steel. Even as late as Carnegie's day, only a small fraction (maybe 5 percent) of blast furnace output was used directly. In cast form, as it comes from the furnace, iron contains a relatively high amount of carbon (2.5 percent to 3 percent) which makes it quite brittle, suitable perhaps for water pipes, cooking pans, and machine frames but not much else. In order to be used for such things as bridge members or iron rails, it has to be reworked into a malleable form known as wrought iron. Traditionally, ironmasters had done this by reheating the cast-iron pigs and then pounding them with a hammer until the remaining carbon combined with atmosphere oxygen. This slow,

labor-intensive process kept wrought iron scarce and expensive.

Starting about 1750 ironmasters learned how to make iron cheaper and in far greater quantity by substituting coal (in the form of coke) for charcoal as fuel, by adopting Henry Cort's famous puddling and rolling processes (patented in 1783 and 1784) for converting cast into wrought iron, and by incrementally improving virtually every other aspect of iron production. By about 1850 a parallel series of innovations and improvements in machine tool design and application had made iron the predominant material of industrial civilization.

Carnegie had invested in local iron forges as early as 1861, and as the market for iron grew and the price rose, he expended his investments accordingly. He organized the Cyclops Iron Company in 1864, along with his associates Piper and Linville, primarily to supply iron for Keystone Bridge. The next year he merged Cyclops with a nearby competitor company owned in part by his brother, Thomas, creating the Union Iron Mills.

From his years on the Pennsylvania, Carnegie had learned that profit depended upon getting costs down and volume up. That deceptively simple idea, which Carnegie proceeded to apply–through cost accounting and scientific management, vertical integration, "driving" his works, and technological innovation–did nothing less than transform the business of making iron.

Cutting costs, Carnegie discovered, would not be an easy matter because astonishingly enough, few experienced iron men of the day really knew what they were. "As I became acquainted with the manufacture of iron I was greatly surprised to find that the cost of each of the various processes was unknown. Inquiries made of the leading manufacturers of Pittsburgh proved this. It was a lump business, and until stock was taken and the books balanced at the end of the year, the manufacturers were in total ignorance of results ... Owners who, in the office, would not trust a clerk with five dollars without having a check upon him, were supplying tons of material daily to men in the mills without exacting an account of their stewardship by weighing what each returned in the finished form." Carnegie's railroad experiences had taught him better. He immediately put into place formal procedures for weighing and accounting, so he could keep track of every pound of material as it moved through the works. Carnegie eventually got the cost sheets good enough so that he could tell at a glance how each department in the business was doing. The same sort of accounting system could work just as well on men, Carnegie thought, documenting "what each man was doing, who saved material, who wasted it, and who produced the best results."

Knowing the costs precisely, Carnegie could systematically begin to reduce them. One of the most effective ways he found of doing so was by controlling the production process from beginning to end and thus reducing overhead and profits needlessly diverted to middlemen. Carnegie first integrated backwards, from finished products (beginning with bridges) toward raw materials. Creating Cyclops and Union Mills to supply Keystone Bridge with plates and beams had been one important step in this direction. In 1870 Carnegie took the next step by building the Lucy blast furnace to supply pig iron to Union Mills. Completed in 1872 and named for Carnegie's sister-in-law, it was one of the largest furnaces of its day and the most productive. Within a couple of years it was setting world tonnage records, 642 tons in a week (when the average was about 350) and 100 tons in a day, and con-

tinued setting new records into the 1880s. The blast furnace ensured Carnegie of a dependable source of raw material at a dependable price for his iron mills and his bridge company, but integration may have been easier on the production than the management side. Carnegie at one point had to remind a manager that "You [and Union Mills] are not competitors; on the contrary you are necessary to each other–the true policy is to work together." Even though Lucy Furnace, Union Mills, and Keystone Bridge shared many of the same partners, besides Carnegie, breaking down traditional barriers between different parts of the industry and convincing the various managers that they had to work as a team would take some doing.

Foreign visitors found the pace in Carnegie's mills "reckless." They wondered, for instance, how Carnegie could afford to push his blast furnaces so hard that he had to reline their interiors every three years or so. Carnegie wondered how his British competitors could afford not to. By driving his works as he did, Carnegie dramatically lowered his unit cost. When a British rival pointed out that he had been using some of his manufacturing equipment for decades, Carnegie retorted, "And that is what is the matter with the British steel trade. Most British equipment is in use twenty years after it should have been scrapped. It is because you keep this used up machinery that the United States is making you a back number." Indeed, when Lucy went into operation in 1872, America produced less than 2 million tons of pig iron while Great Britain produced nearly three times as much. By the turn of the century, thanks in large measure to Lucy and her sisters, America was producing nearly 18 million tons a year, and Great Britain less than 6 million.

Through scientific analysis and technological innovation, Carnegie learned how to make iron both cheaper and better. Distressed that his blast furnace was being run by men with no training in the science of metallurgy, Carnegie hired a German-trained chemist to analyze the furnace's raw materials and processes. Carnegie's chemist soon discovered that some of the ores for which the company had been paying premium prices were not nearly as good as their reputations and that some supposedly inferior ores were actually superior. "Nine tenths of all the uncertainties of pig-iron making were dispelled under the burning sun of chemical knowledge," Carnegie later bragged. Such scientific sophistication helped for a time to make

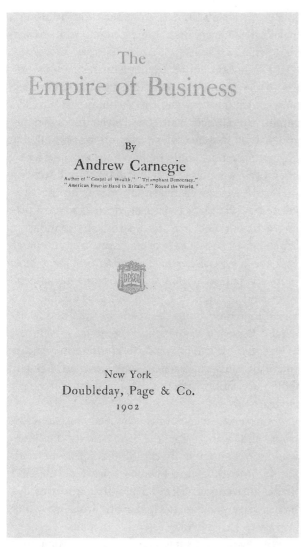

The
Empire of Business

By
Andrew Carnegie
Author of "Gospel of Wealth," "Triumphant Democracy,"
"American Four-in-Hand in Britain," "Round the World."

New York
Doubleday, Page & Co.
1902

*Title page for one of Carnegie's books espousing
his business and social theories*

the Lucy Furnaces the most profitable of Carnegie's iron operations. Competitors might question the extra expense of a professional chemist, but they certainly could not dispute the results.

Carnegie, with some experience from his railroad- and bridge-building days, became quite adept at sizing up technical genius and then managing it effectively. In Andrew Kloman, an ingenious German ironworker who had already earned a reputation around Pittsburgh as the best fabricator of axles in the railroad business, Carnegie saw someone who could improve the manufacturing side of his operation. Union Mills was renamed Carnegie, Kloman & Company when Carnegie made Kloman a partner in the venture. "How much this German created!" Carnegie recalled. "He was the first man to introduce the cold saw that cut cold iron the exact

lengths. He invented upsetting machines to make bridge links, and also built the first 'universal' mill [for rolling plates in any width with finished edges] in America." Kloman picked up good ideas from talking with newly hired foreign workmen (he learned of the universal mill that way), from visits to competitors' plants (personally arranged by Carnegie), and from his own tinkering. In one decade, Kloman's "genius," as Carnegie always called it, transformed virtually every aspect of iron making, from the blast to the rolling mill.

Carnegie himself was forever in search of good ideas as well as good men. Even during his annual summer vacation abroad–a tradition he began during the Civil War–he set aside time for investigating leads on promising innovations. In 1865 he heard about a new process for improving the durability of iron rails by facing them with steel. Tracking down its inventor in London, he persuaded him to sell the American rights for the invention. As it turned out, not even Carnegie's best promotional efforts–including published endorsements from J. Edgar Thomson, who had privately urged him to forget the whole thing–could make these so-called Doddized rails worth much. Carnegie took out a license on a similar British scheme called the Webb process, with similarly disappointing results. Exasperated with the difficulties of forecasting the technological future, he told Scott, who had sought an opinion on a new technique for making chrome steel: "My advice (which don't cost anything if of no value) would be to have nothing to do with this or any other great change in the manufacture of Steel or Iron. I know at least six inventors who have the secret, all are anxiously awaiting . . . That there is to be a great change in the manufacturing of iron and steel some of these years is probable, but exactly what form it is to take no one knows. I would advise you to steer clear of the whole thing. One will win, but many lose and you and I not being practical men would very likely be among the more numerous class." In spite of the odds, Carnegie remained willing, when instinct and experience encouraged him, to gamble on new processes. Taking another chance, on Bessemer steel, turned out to be one of the shrewdest business decisions of his career.

No one had to tell Carnegie, or any other railroad man, what steel rails would mean for the business, and the man who could figure out how to make them cheap and plentiful enough to compete with iron. "Upon certain curves at Pittsburgh," Carnegie recalled, "on the road connecting the Pennsylvania with the Fort Wayne, I had seen new iron rails placed every six weeks or two months." The advantages of steel were obvious, and as early as 1862 J. Edgar Thomson had started experimenting with steel rails on the Pennsylvania's lines. Even with imported steel rails costing more than twice as much as domestic iron rails, he ordered several thousand tons worth for the most critical sections on the road. Because steel rails could wear as much as seven times longer than iron, they easily paid for themselves.

Until the mid nineteenth century the term "steel" was used to mean a particular type of expensive, high-carbon, high-quality iron produced through a week-long process of heating, reworking, and tempering wrought iron into a product of unmatched toughness and hardness. While a number of ways of making such steel had been developed over the years, none of them could be applied to making anything larger than razors or sword blades. Not until the early nineteenth century did iron manufacturers master ways of producing steel by the ton instead of the pound. Even then steel remained too expensive for all but the most crucial instruments of war.

All this changed dramatically, in theory at least, in 1856 when Henry Bessemer told the members of the British Association for the Advancement of Science (BAAS) that he had discovered a way of converting molten pig iron into low-carbon steel in 15 or 20 minutes, simply by sending a blast of ordinary air through it. Bessemer did not call his product steel, and for some time older steel makers contemptuously referred to Bessemer's product as "homogenous iron." When put to the test, if Bessemer's process did not produce high-quality steel of the traditional sort, the material was an excellent compromise between the strongest, most durable steel and ordinary low-cost iron. There was considerable feeling that Bessemer had not as yet perfected his idea. The blast, it was found, burned out not only the carbon, but some essential minor ingredients as well. Further, Bessemer had not had time to work out the details for scaling up the operation. All the same, Bessemer's idea intrigued nearly everyone who learned of it, including such influential British engineers as George Rennie, who talked Bessemer into giving the paper to the BAAS in the first place, and James Nasmyth, who afterwards proclaimed the small sample Bessemer brought along a "true British nugget; its commercial importance is beyond belief."

Despite the initial excitement, Bessemer's process at first produced very little steel, and even less money. Attempts to use the technique in Sweden, Britain, the United States, and even India demonstrated that Bessemer's process did work well, but only with certain kinds of low-phosphorus ores. If other ores were used, the iron produced was far too brittle. Moreover, the situation was further complicated by what was to become one of the most prolonged patent fights in the history of the industry. The so-called Bessemer process, it turned out, had already been used by a backwoods ironmaster from Kentucky named William Kelly. Someone else already held a patent for the process of adding back crucial ingredients into the steel after the blast. Ultimately, Bessemer could only patent the machinery he invented for the process. By the time the various American interests reached agreement on a patent pool, the whole affair had dragged on for nearly ten years.

Carnegie, meanwhile, watched the developments and waited for the opportunity he knew would sooner or later appear. "I had not failed to notice the growth of the Bessemer process," he said. "If this proved successful I knew that iron was destined to give place to steel; that the Iron Age would pass away and the Steel Age take its place."

He would let someone else take the first steps to bring the Bessemer process to America. Like Carnegie, Alexander Holley knew about running railroads. During the Civil War he got some other railroad men to provide him with financial backing to build the first American Bessemer plant. Holley then went to Great Britain to learn the process firsthand and to secure the necessary license for making Bessemer steel. Upon his return in 1864 he supervised first the construction of one Bessemer plant for the Albany and Rensselaer Iron Works in Troy, New York, another for the Pennsylvania Steel Company in Harrisburg, and a third for the Cambria Iron Works in Johnstown. He made so many improvements on Bessemer's basic idea—detachable bottoms (so the converters could be kept in constant blow), elevated converters, better handling methods—that it was said that "every ingot has 'Holley fecit.'" Holley literally built the American Bessemer steel industry and instructed most of the best engineers of the next generation in the process, including Carnegie's future superintendent William (Bill) Jones.

Carnegie entered the Bessemer steel business in 1866, when he reorganized the Freedom Iron Company into the Freedom Iron and Steel Company to make steel for rail headings. Freedom imported British equipment and went into blow in 1868, three years behind Holley's plant in Troy and a year behind the Pennsylvania Steel Company, and only the third American firm to turn out Bessemer steel. A more dubious distinction, it was also the first American Bessemer company to go out of business, just one year later, having exhausted its capital working out the bugs in its production system.

By 1872 the success of the Pennsylvania Steel and the Cambria Iron Companies and his own observations in England convinced Carnegie that it was time to try Bessemer again. Back in Pittsburgh, he began forming a partnership to finance a Bessemer rail mill. When his former partners, familiar with the recent troubles at Freedom Iron and Steel, declined the offer, Carnegie found new ones, including his friend William Coleman from his oil-investing days (now his brother's father-in-law) and David McCandless, an old family friend with important connections in local banking. Carnegie eventually persuaded most of his old iron partners to invest something in the new enterprise, but his $250,000 (out of a total capitalization of $700,000) retained for him the controlling interest in Carnegie, McCandless & Company. The partners bought some property south of town on the Monongahela River (on the very site of Braddock's defeat in the French and Indian War, the skeptics enjoyed pointing out) convenient to raw materials by water and markets by rail, via the Pennsylvania and the Baltimore & Ohio. Having seen the railroad business from the other side, Carnegie knew that the only way to ensure competitive freight rates was to force the railroads to compete. However, to keep the goodwill of his former employer, as well as a huge potential customer, Carnegie astutely invited J. Edgar Thomson to join the company and then named the new works after him. The company was later reorganized as the Edgar Thomson Steel Company.

As he had done in building his iron companies, Carnegie started by hiring the most experienced people in the business he could find. By this time too, some of the best men came looking for Carnegie, drawn by his considerable reputation. One of them was Alexander Holley, who, hearing rumors of Carnegie's steel interest, wrote to one of Carnegie's partners in the summer of 1872 and was immediately hired on as a consulting engineer. Holley brought with him Bill Jones, his chief lieutenant from Cambria, who had recently quit after being passed over for promotion to mill superintendent. Jones became, over the years, one of the most knowledgeable production men in the business, not for any one major innovation, but rather for a series of incremental improvements in Bessemer steel making that perhaps doubled the efficiency of Holley's original scheme. Jones, in turn, hired some 200 former Cambria employees for key positions in Carnegie's mill, including the head of the rail department, head furnace builder, superintendent of machinery, and superintendent of transportation.

No sooner had construction gotten under way than the panic of 1873 brought everything to a halt. Carnegie, forced to buy off worried partners and reassure his creditors, called it "the most anxious period of my business life." He could raise money only by selling bonds in England and selling some of his own stock at home. And when Carnegie refused to bail Scott out of a bad railroad investment, the panic even cost him his oldest and closest business associate. "This was to me the severest blow of all," he said. On the positive side, as Carnegie built the Edgar Thomson works the panic taught him some of the advantages of buying new equipment when business was slack and prices low. For the rest of his career one of Carnegie's business strategies was to expand during recessions, and then use his updated facilities to undersell rivals when business returned to normal.

Holley's cost estimates projected that the mill would pay for itself in a year, so Carnegie was confident he could turn a profit even in the middle of a depression. Cost, Carnegie believed, not profit, was the key to business success. While profits might reflect "special conditions in the trade," costs were the direct result of a company's own efforts and could thus be compared and controlled from month to month and year to year. Watch your costs, he often said, and the profits will take care of themselves.

William P. Shinn, a former railroad auditor, was a man who knew how to watch costs, and Carnegie hired him as general manager for Edgar Thomson. Shinn's concise financial statements, usually on one sheet so Carnegie could get right to the bottom line, told Carnegie just what he needed to know, and he liked what he saw. The costs (which Shinn always listed first) for Thomson's second month of operation showed that it was making rails for $56.64 a ton and selling them for $66.32 a ton for a total profit of $18,000.

"You [were] expected always to get it ten cents cheaper the next year or the next month," one old-timer recalled. Shaving 15¢ on a mold here, a couple of pennies of materials handling there, Carnegie drove down his costs to the lowest in the industry, proving that small consistent efforts could add up to big savings in a high volume operation. He encouraged and rewarded employees like Henry Phipps, a former bookkeeper who figured out how to recycle the scale, or shavings, from the rolling mills. Carnegie then purchased the refuse at low cost from other mills, which were only dumping it into the river. Phipps ended up as a major partner in the enterprise. Even the company's fire insurance premiums were scrutinized for cost-saving opportunities. When Carnegie learned that it would cost less than the premiums he was paying to replace his wooden buildings with iron ones, he rebuilt the works and cancelled his policy.

Long after nearly every other steel company had been incorporated, Carnegie maintained his company as a partnership, believing that people who owned part of the company would be more likely to give their best effort. He also preferred that top employees join the partnership ranks. "I want somehow or other to get you root and branch—compensation can be arranged—I don't care about money so much as about success," he once told Shinn. Carnegie got Shinn by doubling his number of shares, thereby making him the firm's second largest stockholder. Carnegie continued to reserve blocks of stock for talented young managers. While the percentages may have seemed tiny (less than 1 percent in most cases), Carnegie's "young geniuses" knew that these shares could one day make them millionaires if they held on to them. Carnegie liked to say that "Mr. [J. P.] Morgan buys his partners, I grow my own." The only catch was Carnegie's so-called "Iron Clad Agreement," which required any departing partner to sell his holdings back to the other partners at book value. Because Carnegie was notoriously stingy about raising the book value of the company, the only way the junior partners could realize the true value of their shares was by staying with the company.

Like many other self-made men of his day, Carnegie feared that if his company went public, he would lose control to outsiders and speculators. So he kept the company (and the books) closed, and financed expansion by retaining earnings and holding back dividends. Able to invest in plant and equipment while his rivals were paying off stockholders, Carnegie gained a significant advantage over publicly held companies, which were under pressure to pay dividends to their shareholders in good and bad times. Some of the partners (including Carnegie's own brother) did not share Carnegie's obsession with the long term over the short run and would have preferred a more immediate payoff, but with Carnegie the dominant partner, they found they had no choice. As Carnegie liked to point out, they could always sell out—at book value, of course.

Dissatisfaction in the ranks under this type of rule was, of course, not uncommon, and Carnegie's response to it was often harsh. He demanded top performance and uncritical loyalty from his partners—anything less was grounds for immediate dismissal. "If he can win the race he is our race-horse," Carnegie often said, "if not he goes to the cart." Even senior partners could find themselves pushed out, especially if Carnegie thought they had betrayed him in some way. Kloman lost his partnership for recklessly investing in outside interests. "Mr. Kloman's ambition had been to be in the office, where he was worse than useless, rather than in the mill devising and running new machinery, where he was without a peer," Carnegie commented. Shinn was cut off after he formed his own limestone company and then tried to sell his product to Carnegie at a fat profit.

Men who measured up to Carnegie's exacting standards, however, usually found themselves going places in a hurry. In his memoirs Carnegie recalled how William Borntraeger, one of Kloman's distant relatives, earned himself a promotion from clerk to partner by pointing out a way of saving energy costs on reheating iron. "All the needed labor in preparing this statement he had performed at night unasked and unknown to us. The form adapted was uniquely original. Needless to say, William soon became superintendent of the works and later a partner, and the poor German lad died a millionaire. He well deserved his fortune."

Among Carnegie's lieutenants, only Jones never succumbed to the lure of capitalism. After years in front of the blast furnace ("This is my home," he liked to tell visitors to the mill, "a good preparation for the next world"), he felt he should be able to enjoy a little more success in the here and now. Jones insisted on "a hell of a salary if you think I'm worth it," which Carnegie obviously did. He began paying Jones $25,000 a year, the same salary the president of the United States was making at the time.

It was money well spent, for Jones was probably the best steel maker of the day. Jones, like his mentor Alexander Holley, appreciated that materials handling was as critical as metallurgy in the making of iron and steel. Over the years he devised dozens of improvements for moving materials through the works, with significant savings in time, resources, and energy. The Jones mixer dramatically speeded up production by eliminating traditional sand casting and providing a virtually uninterrupted flow of pig iron from the blast furnaces to the Bessemer converters. Writing to a company vice-president, Carnegie insisted that the importance of this invention for the company and the steel industry could not be denied: "Yours received in regard to our Mixer Patent. I feel very strongly about the action of our friends the Illinois and the Pennsylvania Steel Companies. No one pretends that a mixer was ever used successfully except ours. These companies saw the results and had already realized that a mixer was necessary for the direct process. They deliberately appropriated our invention without arranging with us for a liscence [sic]. We can prove that the invention was worth to the Illinois Steel Co. at least $150,000 per year probably $200,000, and it will be worth the same to the Pennsylvania and the Sparrows Point works."

The men had to move as fast as the materials. Like other steel mills of the day, the Edgar Thomson Works ran 12-hour shifts seven days a week, with a 24-hour swing shift on Sunday. Rebuked by the clergy for running the mill on Sunday, Jones threatened that if they persisted in agitating his workers, "I will retaliate by promptly discharging any workman who belongs to their Churches and thereby get rid of the poorest and most worthless portion of our employees. If they don't want to work when I want them, I shall take good care that they don't work when they want to." Jones was killed in an industrial accident in 1889, when one of his converters exploded. Carnegie paid $35,000 to his widow in return for rights to all of the patents which Jones had previously licensed to the com-

pany at a nominal fee. Of all the young men who came up through the system both before and later, Jones remained Carnegie's favorite. Until the end of his life, it was said, Carnegie kept a portrait of Jones on his bedroom wall.

Carnegie knew he could not depend solely on technical experts like Jones, so while he had neither formal training in steel making nor shop floor experience, he studied the product and the process carefully. His letters to subordinates revealed a respectable command of technical details. He watched constantly for news of innovations, which he felt might shift the balance of power within the industry. Always he urged his managers to buy new equipment and try out new ideas: "Jones can't do it successfully without new rolls and a shear," he pointed out, but "with them he can beat Cambria badly." Even before the first Bessemer blow at the Edgar Thomson Works, Carnegie installed two open-hearth furnaces, part of a different steel-making process in use for only a few years and which Carnegie considered promising.

The so-called Siemens-Martin process had been developed by the Siemens brothers in Germany and Pierre Martin in France in the 1860s, and although it was slower and more expensive than Bessemer (about ten hours instead of ten minutes), it did offer some advantages in control and accuracy and could also use scrap. "I well remember the criticism made by older heads among the Pittsburgh manufacturers about the extravagant expenditure we were making upon these new-fangled furnaces," Carnegie recalled. It would be 5 years before anyone produced steel rails in America with open-hearth furnaces, and 20 before total open-hearth production surpassed Bessemer, but this early experience certainly helped ease the company's transition to the new technology and once again gave Carnegie a jump on the competition.

Carnegie was also among the first American steel makers to appreciate the significance of the "basic process" invented by British cousins Sidney Thomas and Percy Gilchrist. In the late 1870s, after years of tinkering, they figured out a way to draw phosphorus out of iron ore at high temperatures with a lime-rich slag, and so made it possible for steel makers to use cheaper, high-phosphorus ores in both the Bessemer and open-hearth processes. They also found that a magnesite rock lining would prevent the acid slag from corroding the furnace inte-

The first class of the Carnegie Technical Schools, 1905 (courtesy of Carnegie Library of Pittsburgh)

rior. Carnegie heard about the process during one of his European tours, and arranged a personal meeting with Thomas. He succeeded in getting Thomas to sell the American rights to the invention to him for $300,000. Carnegie later sold the rights to the American Bessemer Steel Association, earning himself a $50,000 commission and free use of the patent in his own plants. Carnegie considered the process something of a breakthrough and spread the word quickly. "I see it is done," he told one of his managers, "by a lime lining in the converter which is found to absorb the phosphorous. Dr. Siemens is out in a letter saying he had tried that for his open hearth furnaces, and found it could not be made to stand; this thing appears to have been overcome . . . it is too important not to invite our earnest attention. . . ."

Social Darwinist Herbert Spencer once called Carnegie his best American friend, and for Carnegie, business truly was a struggle for the survival of the fittest. Carnegie loved a good fight, and respected the same instinct in his competitors no less than in his own subordinates. His business axiom became—"Cut the prices; scoop the market; run the mills full." As he once told Shinn, "Two courses are open to a new concern like ours—1st Stand timidly back, afraid to 'break the market' following others and coming out without orders to keep our works going—that's where we are going to land if we keep

on. 2nd To make up our minds to offer certain large consumers lots at figures which will command orders—For my part I would rather run the works full next year even if we made but $2 a ton."

Efficient production methods may have put him in a position to undersell his rivals consistently, but it was not his style to sit back and wait for the business to come to him. Always a master salesman, he went after it with a variety of creative marketing schemes—credit sales, testimonials, and advertising. He called in favors from his railroad days, and he tried to anticipate what future markets would be. Just as he had earlier won the first contract for bridge steel, he would later win the contract for the steel in America's first skyscraper, Louis Sullivan's Home Insurance Building (1885) in Chicago.

Carnegie presented his men with one competitive challenge after another—to beat Cambria's output for next month, or their own for the previous one. Seeking, perhaps, to encourage competitive zeal equal to his own, Carnegie's instructions to his subordinates very nearly sounded like campaign orders. "Joliet [Steel] is in a death struggle," he told Shinn. "Having faith in our ability to manufacture cheaper than others I do not fear the results of a sharp fight." So fierce was the pace in the Carnegie mills that it was said that when a distinguished British visitor asked to sit down and simply watch, he was told that he would only find ingots cool enough for sitting back in England.

Though Carnegie was among the most visible American spokesmen for laissez-faire capitalism, he remained enough of a pragmatist to exploit restrictions on competition when it served his purposes. From the very beginning of American Bessemer steel production, domestic manufacturers had sought legislative protection from what they insisted was unfair British competition. In 1870 they got it, when Congress placed a $28-per-ton tariff on imported Bessemer steel. Still, for several years after, the British continued to undersell the Americans despite the tariff and higher transportation costs because of their more advanced steel making. As the Americans closed the manufacturing cost differential, however, the tariff became a formidable barrier, and after about 1875 the British were no longer a factor in the American rail market. The tariff proved to be of enormous benefit to Carnegie as he was putting the Edgar Thomson Works into production and working out initial problems in its manufacturing operations. With increasing sophistication in production and without significant competition from some of the most experienced Bessemer plants in the world, Carnegie could brag to a London banker in 1876 that "Even if the tariff were off entirely, you couldn't sent [sic] steel rails west of us." He preferred not to test the claim, though, and continued to support tariff protection for domestic steel.

Carnegie also joined the Bessemer Steel Association, which was established to control the Bessemer patents, divide the American market, and, in concert with the railroads, set prices. Invited to join even before he had put his mill into operation, Carnegie's first meeting with the "Fathers-in-Israel," as he insisted on referring to the leaders of the Association, made quite an impression on the other members. Informed that as the newest member he would be given only 9 percent of the market, the smallest share of the group, Carnegie refused to accept the allotment and instead demanded as large a share as anyone else. As he later told the story: "I informed each of the other representatives, all Presidents of their companies, that I was a stockholder in their concerns and as such had access to their financial reports. I singled out each President and said, 'I find that you receive a salary of $20,000 a year and expenses of $80,000,' etc.—instancing each one, telling him just what his salary

was, and how much he spent in expenses, etc. Then I told him that the President of Edgar Thomson received a salary of $5,000 a year and no expense allowance. Moreover, I said Mr. Holley, the engineer who built the Edgar Thomson works had informed me that it was the most complete and perfect in the world and would turn out steel rails at cost far lower than its competition. 'So gentlemen,' I concluded, 'you may be interested to know that I can roll steel rails at $9 a ton. If Edgar Thomson Co. isn't given as high a percentage of this pool as the highest, I shall withdraw from it and undersell you all in the market–and make good money doing it.' The committee at once got off its high horse, stopped snickering at me and met my demands." While the cost figure was a bluff ($50 a ton was closer to his actual cost), the threat, as those who came to know his methods could attest, was real enough.

Carnegie recognized that a pool like the Bessemer Association could provide significant advantages for a new firm easing into a market. He initiated a number of pooling agreements himself and was quick to cry foul whenever he discovered violations. Throughout the 1870s he even went so far as to arrange secret deals with archrival Cambria in order to shape and control the market. Carnegie of course had too much experience in the railroad business to expect such "gentlemen's agreements" to last for very long, and when a pool did collapse there were other steps that could be taken. With the demise of one deal for dividing the beam market looming he told his partners: "Let us fill our Beam Mill full if we can even at $5 per ton profit, run it double time & see just what we can do. . . . In conclusion having abandoned the idea of small tonnage & exorbitant profits there is but one sound idea to embrace & that heartily—immense tonnage, small profit—let us take the beam orders of the country."

Carnegie pursued this strategy of taking the orders of the country from the opening of the Edgar Thomson works in 1875 until the mid 1880s when, it was clear, as one observer put it, that "He became himself the pool." He reorganized the company in 1881 as the Carnegie Brothers & Company, Ltd. but kept on, as he liked to say, putting all his eggs in one basket and then watching that basket. When other companies started experimenting with full vertical integration, he continued doing what he knew best, making steel more cheaply than anyone else in the business. Through boom and bust he improved his production methods, drove down his costs, and made more steel and more money than any of his competitors. Carnegie's company consistently returned a 60-percent profit on investment. Even in a protected industry where annual dividends on the order of 50 percent were not unheard of, this was exceptional performance. As before, Carnegie plowed most of the profit back into the business, despite objections from his cash-hungry partners. Jones was allocated much of the surplus for his unceasing campaign to improve production methods and cut costs even further. According to Carnegie's personal secretary, "The famous scrap-heap for outgrown, not outworn, machinery was instituted by Jones, who never hesitated to throw away a tool that had cost half a million if a better one became available. And as his own inventions saved the company a fortune every year, he was given a free hand."

Carnegie once recalled that "The one vital lesson in iron and steel that I learned in Britain was the necessity for owning raw materials and finishing the completed article ready for its purpose." Yet aside from the Lucy Furnaces he had been slow to apply that lesson to his own operations. With the tremendous expansion of his mills in the 1880s, Carnegie began to pay closer attention to the availability and cost of materials. Ferromanganese, for example, a minor but crucial additive in the steel-making process, had to be purchased from foreign companies at a cost of about $80 a ton. This, Carnegie decided, was both expensive and too risky. Locating a suitable supplier in Virginia, Carnegie bought the mine, and went into the ferromanganese business himself. So successful was this venture, that he was soon supplying the entire American market with ferromanganese at $50 a ton.

Carnegie saw even greater opportunities in becoming independent of the market in all the basic ingredients of steel making—iron ore, limestone, and coal. He had long viewed steel making not as a series of independent processes, but as an integrated manufacturing operation. Gradually Carnegie also began to see steel making as only one part of a developing technological system, which extended directly from iron mines and coke furnaces to finished rails, boiler plates, and nails. On the scale at which he was operating, an imbalance in any part of that system could quickly upset the entire enterprise. Carnegie's rolling mills, for instance, needed steel in-

gots from the Bessemer converters. The converters required sufficient supplies of cast iron, which could in turn be produced only from enormous quantities of ore and coke. Running out of any of the crucial raw materials brought the whole system to a halt. "We found," Carnegie said after studying the situation carefully, "that we could not get on without a supply of fuel essential to the smelting of pig iron; and a very thorough investigation of the question led us to the conclusion that the Frick Coke Company had not only the best coal and coke property, but that it had in Mr. Frick himself a man with a positive genius for its management."

As the biggest coke customer in steel business, Carnegie knew exactly what Frick could do. Starting as a $12 a week department store clerk, Frick had worked his way up to bookkeeper at $50 a week, and was the head of his own company by the time he was twenty-two. When he founded the Henry C. Frick Coke Company in 1871, he owned only 300 acres of coal land and 50 coking ovens in his native Connelsville, Pennsylvania, 50 miles southeast of Pittsburgh. But like Carnegie, he found himself in the right place at the right time. Pittsburgh's iron and steel men were discovering that with the insatiable appetites of the new Bessemer converters they had to turn from charcoal and anthracite fuels to coke. To meet the demand Frick borrowed to expand, then borrowed and expanded again. Like Carnegie he recognized that it was good business to expand when nobody else was—when prices were low and people were nervous. Frick bought out his partners, and his competitors as well, when the panic of 1873 sent coke prices plunging to less than a dollar a ton. By the end of the decade he controlled 3,000 acres, 1,000 ovens, and 80 percent of the region's coke industry.

Carnegie knew that to control his source of coke, he must somehow control Henry Frick. He proposed partnership to Frick, who immediately saw the advantages of going into business with his largest customer. Partnership meant a dependable and virtually unlimited market for his coke, an inexhaustible supply of capital for future expansion, and an opening into the steel business if he chose to move in that direction. Frick was also aware that if he refused the partnership offer, Carnegie might just go into the coke business himself. In January 1882 Frick accepted the offer and sold Carnegie 10 percent of his company. Within two years, as Frick sold more and more stock to Carnegie to raise capi-

Carnegie and his wife wave farewell on his last visit to Pittsburgh, October 30, 1914 (courtesy of Carnegie Library of Pittsburgh)

tal for expansion, Carnegie became the majority partner in the Frick Coke Company. Now Carnegie could buy his own coke at cost, and at the same time share in the profits from Frick's sales to his competitors.

Beyond guaranteeing his coke supply, Carnegie also sought to acquire Frick's impressive managerial skills to help fill the gap left when his brother Tom died in autumn 1886. Over the years Tom had taken on much of the day-to-day decision making within Carnegie's growing enterprises. Although he had been overshadowed by his older brother, Tom had genuine business talent of his own and handled his responsibilities with rare skill. After Shinn's "resignation" in 1879 Tom had served as chairman and general manager for Carnegie Brothers. But if Tom had a head for business, he did not have the stomach, which Carnegie sensed when he once wrote to Tom while in the midst of one of his European tours: "I'm sure you have had a trying time of it and often you must have felt disposed to throw up the game. [But] the more I find myself drinking in enjoyment, the deeper is my appreciation of your devoted self-denial. . . . It is a heavy load for a youngster to carry." Heavier, perhaps, than either of the brothers fully realized. Tom eventually turned to the bottle and died from alcoholism.

Meanwhile, Carnegie's expansion program never slackened. He continued to enlarge his own plants and buy up competitors at a furious pace. In 1883 he bought the two-year-old Homestead Works, which had been designed for the Pittsburgh Bessemer Steel Company by former partner Andrew Kloman. Explaining how the acquisition came about, Carnegie wrote, "These works had been built originally by a syndicate of manufacturers, with the view of obtaining the necessary supplies of steel which they required in the various concerns, but the steel-rail business, being then in one of its booms, they had been tempted to change plans and construct a steel-rail mill. They had been able to make rails as long as prices remained high, but, as the mills had not been specifically designed for this purpose, they were without the indispensable blast furnaces for the supply of pig iron, and had no coke lands for the supply of fuel. They were in no condition to compete with us." In the face of so formidable an adversary and recurring labor disputes, Homestead's owners sold out to Carnegie for little more than the original cost of the plant.

Accumulating huge sums of cash, Carnegie found that he could always overcome a late start by simply buying his way into new markets and technological processes. The same year that he acquired Homestead, Carnegie took his first step toward full forward integration by purchasing the Hartman Steel Company, a nail and steel wire plant northwest of Pittsburgh. "So now we made everything in steel from a wire nail up to a twenty-inch steel girder," Carnegie wrote. A new challenge to Carnegie's supremacy occurred in 1888 when the Allegheny Bessemer Steel Company opened its Duquesne works, with a new assembly line system for casting ingots and then sending them directly to the rolling mill without a further reheating. Attempting first to discredit Duquesne with rumor and innuendo, Carnegie told friends in the railroad industry that direct rolling, because it did not reheat the ingots, would make steel without sufficient "homogeneity," a "defect" he invented to serve his own purposes. When that failed, he just went ahead and bought the plant, and promptly introduced the very same production system into his other plants. Nothing further was heard about "homogeneity."

By the late 1880s Carnegie had controlling interest in two integrated steel companies (Carnegie Brothers and Carnegie, Phipps, formed by the merger of Homestead with the Lucy Furnaces in 1886), a finishing facility (Hartman), and a coke company (Frick), and was working on additional acquisitions (Duquesne). At the same time Carnegie was spending more and more of his own time on outside interests. There was a real danger, he knew, that without an expert manager at the top, this sprawling empire would lose its competitive edge. Carnegie had watched Frick run the coke company for three years and was convinced that Frick was his man. He offered him at first a small (2 percent) share in Carnegie Brothers in 1887, then two years later made him chairman with an 11-percent interest. "Take supreme care of that head of yours," he told Frick at the time. "It is wanted. Again, expressing my thankfulness that I have found THE MAN, I am always yours, A. C."

Carnegie's faith in Frick's managerial talents proved to be well founded. During Frick's first year as chairman, Carnegie Brothers profits increased from $1.9 million to $3.5 million, an astonishing gain even by Carnegie's standards. Frick also played a key role in the Duquesne acquisition. And of even greater significance in the long run, he consolidated Carnegie's scattered investments into the one gigantic enterprise that was to become known as the Carnegie Steel Company. Officially created on July 1, 1892, it was capitalized at $25 million and merged the Edgar Thomson, Homestead, Hartman,

and Duquesne mills, Keystone Bridge, the Lucy Furnaces, and several minor mining and coke companies (though significantly not the Frick Coke Company) into the biggest steel company in the world. "F[rick] is a marvel," Carnegie proclaimed. "Let's get all F[rick]s."

With the day-to-day management of the company in Frick's capable hands, Carnegie allowed himself to spend more time on his long-deferred program of self-improvement. He toured the Far East and there found reassuring evidence for his theories of cultural evolution. Japan especially impressed him as an example of what western technology could accomplish, though he worried about its infatuation with western military technology. Ever the businessman, Carnegie could not help speculating about the potential Asian market for rails and steel. He took his mother for a coaching tour of Britain, and gave her the honor of laying the cornerstone for the first foreign Carnegie library, to be built in their old hometown of Dunfermline. He even bought the newspaper chain he had once promised himself he would (seven dailies and ten weeklies scattered across Great Britain) and became a crusading publisher. After several frustrating years of losing both elections and money, he sold the papers, deciding to concentrate on the businesses he understood.

In 1886, with the publication of *Triumphant Democracy*, Carnegie became a best-selling author. One biographer called it "A Fourth of July oration with statistical tables," but to Carnegie's satisfaction, at least, the work distilled the great lesson of the democratic experience: "The old nations of the earth creep on at a snail's pace; the Republic thunders past with the rush of the express." The book's popularity established Carnegie as an influential political and social commentator, something all his business achievements had not done. He began writing articles for leading journals on both sides of the Atlantic, attacking protective tariffs in Britain and defending them in the United States, championing Gladstone Liberalism in England and McKinley Republicanism in America. Such inconsistencies puzzled his friends and exasperated his enemies, who denounced him for playing the radical abroad, where it won him friends and cost him nothing, and the conservative back home, where it protected his economic interests.

Finally, too, Carnegie found time for romance. In 1887 he married Louise Whitfield, whom he had known for years but had delayed marrying until after the death of his possessive mother. (Louise once said that Margaret Carnegie was one of the least likable people she had ever met.) They made an odd couple—she 20 years younger and several inches taller—but a happy one. A decade later a daughter, Margaret, completed the family, and she made her father as proud as any of his business successes.

In 1886 Carnegie published two influential works under the titles, *An Employer's View of the Labor Question* and *Results of the Labor Struggle*. In the midst of some of the bitterest confrontations in American labor history, Carnegie asserted, astonishingly enough, that workers had as much right to organize as employers, that scab labor was un-American, and that management was as much to blame for strikes as the unions. He even proposed a profit-sharing scheme. Labor hailed Carnegie as a new (and unexpected) champion of its cause.

Carnegie gloried in his new role as a workingman's hero and filled a drawer labeled "Gratitude and Sweet Words" with laudatory telegrams and letters. Meanwhile, his own business strategies made a confrontation with labor virtually inevitable. Like other American steel makers, Carnegie had, over the years, achieved his greatest productivity gains by replacing men with machines. For instance, when the Edgar Thomson works was "modernized" in 1885, 12 men at the heating furnaces could do the work previously done by 69. Twelve men could roll more rails than 63. Jones fought for wage increases to go along with increased productivity while he ran Edgar Thomson, despite Carnegie's constant calls for wage savings. "Our labor is the cheapest in the country," Jones told Carnegie. "Low wages does not always imply cheap labor. Good wages and good workmen I know to be cheap labor." He even convinced Carnegie to run Edgar Thomson on 8-hour shifts, arguing that "it was entirely out of the question to expect human flesh and blood to labor incessantly for twelve hours." Over the long run, however, Carnegie's preoccupation with efficiency induced a growing inclination to treat labor like any other factor of production, something to be analyzed, rationalized, and reorganized in the name of lower costs. One labor historian wrote of this process, "That impulse for economy shaped American steel manufacture. It inspired the inventiveness that mechanized the productive operations. It formed the calculating and objective mentality of the industry. It selected and hardened the managerial ranks. Its technological and psychological consequences, finally, defined the treatment of steelworkers. Long hours, low wages, bleak conditions, antiunionism, flowed alike from the economizing drive that made the American steel industry the wonder of the manufacturing world."

With the formation of Carnegie Steel, Carnegie planned some major new technological initiatives. The experience with Duquesne worried him, demonstrating that rivals could mechanize too, sometimes just as effectively. Carnegie knew that to stay ahead he would have to push even harder and cut costs even further. He was convinced that to accomplish this he would have to eliminate what he considered to be the last roadblock to innovation, the Amalgamated Association of Iron and Steel Workers. Echoing Carnegie's thoughts, one of his partners later said, "The Amalgamated placed a tax on improvements, therefore the Amalgamated had to go."

The Amalgamated Association had been formed in 1876 from a consolidation of three older puddler and ironworker unions. It was a traditional craft union, open only to skilled workers who learned their trade by apprenticeship and practiced it within time-honored constraints. By 1891 the Amalgamated had about 24,000 members. Most of them worked in older iron mills where technological change had not yet completely eroded craft skill. In modern steel facilities like Carnegie's, where mechanization was much more advanced, the union had not fared as well. Recognizing new conditions, the Amalgamated made numerous concessions. Its members gave up their traditional right to hire and fire their own crews, agreed to let management set hours, and dropped their objections to output restrictions. "The Association never objects to improvements.... They believe in the American idea that the genius of the country should not be retarded," said union president William Weihe. What the union wanted, as most American unions have always wanted, was simply a larger slice of the productivity pie. At Homestead, however, it ran headlong into a management equally determined to keep that slice for itself.

When Carnegie purchased Homestead in 1883, he acquired the Amalgamated along with a state-of-the-art steel mill. Although only 800 of the plant's 3,800 workers belonged to the Amalgamated, the union ran Homestead. Its privileges were protected by a 60-page agreement, and at one point the national union had to censor its own local for wildcat strikes "for little frivolous purposes." In 1889 it won a strike at Homestead, and forced the company to recognize the Amalgamated Association as the mill's sole bargaining agent.

Carnegie, meanwhile, had been constantly improving and updating the plant. "We had put in new improvements in some departments which increased the output and reduced the work, and we thought we were entitled to some of the benefits. . . . We were paying more money than our competitors in the same class of work and we had also invested more money in machinery to do that work than our competitors." Carnegie thought there were "far too many men required by Amalgamated rules." He proposed to eliminate more than 300 skilled positions and to hold wage rates at the prevailing industry standard.

Carnegie expected a fight when the Amalgamated's contract, which had been signed after the successful strike of 1889, ran out in the summer of 1892. He gave Frick a bulletin to post at the plant saying that because of the impending consolidation with nonunion plants at Edgar Thomson and Duquesne, the entire company would be open shop in the future. Since Homestead had a Navy order that could not be delayed, Carnegie attached a handwritten note to the bulletin urging Frick to have the plant superintendent "roll a large lot of plate ahead, which can be finished, should the works be stopped for a time." He told Frick further that when the strike came, he should simply wait it out. "Shut down and suffer," he said. "Let them decide by vote when they decide to go to work. Say kindly, 'all right, gentlemen, let's hear from you; no quarrel, not the least in the world. Until a majority vote (secret ballot) to go to work, have a good time; when a majority vote to start, start it is.' " On that note he left for his annual six months in Scotland.

Frick had his own ideas. Like Carnegie, he chafed under the constraints of the Amalgamated, complaining, "The mills have never been able to turn out the product they should, owing to being held back by the Amalgamated men." Frick's reputation as one of the toughest antiunion men in the business, which he earned while serving as president of the coke company, was well deserved. As chairman of Carnegie Steel he had no intention of backing down. He thought Carnegie's waiting games "soft," and favored direct confrontation. By June he had turned the mill into an armed camp, complete with turrets, searchlights, and barbed wire. Frick then gave the Amalgamated a list of demands he knew it could not accept and awaited the fight he had been seeking.

The violent response of the workers caught even Frick off guard. The entire mill, union and nonunion alike, went out on strike on July 1, and then together seized control of the mill. Frick ordered a hired army of Pinkertons to sneak into the mill under cover of darkness and reopen it to strikebreakers. But a union man noticed the Pinkerton barges moving upriver in Pittsburgh, and alerted the strikers. By the time the Pinkertons reached Homestead at four o'clock in the morning on July 6, the whole community was waiting along the riverbank. In perhaps the most famous battle in American labor history the strikers fought it out with the Pinkertons with rifles, rocks, dynamite, a flatcar piled with burning rags, and even the old courthouse cannon (which blew up in the attempt). By late afternoon the Pinkertons had had enough and surrendered with the understanding that they would be given safe passage out of town. As they were being escorted from the river to the village jail, though, the angry mob exploded. Several Pinkertons were killed, with many more severely beaten. The governor of Pennsylvania ordered in the state militia to restore order.

The press and public blamed Carnegie for the tragedy. "Nothing I have ever had to meet in all my life, before or since, wounded me so deeply," Carnegie recalled in his memoirs. "No pangs remain of any wound received in my business career save that of Homestead. . . . I was the controlling owner. That was sufficient to make my name a by-word for years." However sincere these feelings of self-recrimination, public opinion, so memorably captured in a widely reprinted editorial in the *St. Louis Post-Dispatch*, was decidedly negative:

> Count no man happy until he is dead. Three months ago Andrew Carnegie was a man to be envied. Today he is an object of mingled pity and contempt. In the estimation of nine-tenths of the thinking people on both sides of the ocean he had not only given the lie to all his antecedents, but confessed himself a moral coward. One would naturally suppose that if he had a grain of consistency, not to say decency, in his composition, he would

favor rather than oppose the organization of tradesunions among his own working people at Homestead. One would naturally suppose that if he had a grain of manhood, not to say courage, in his composition, he would at least have been willing to face the consequences of his inconsistency. But what does Carnegie do? Runs off to Scotland out of harm's way to await the issue of the battle he was too pusillanimous to share. A single word from him might have saved the bloodshed–but the word was never spoken. Nor has he, from that bloody day until this, said anything except that he had "implicit confidence in the managers of the mills." The correspondent who finally obtained this valuable information, expresses the opinion that "Mr. Carnegie has no intention of returning to America at present." He might have added that America can well spare Mr. Carnegie. Ten thousand 'Carnegie Public Libraries' could not compensate the country for the direct and indirect evils resulting from the Homestead lockout. Say what you will of Frick, he is a brave man. Say what you will of Carnegie, he is a coward. And gods and men hate cowards.

The Amalgamated never had a chance to recover from Homestead. Its leaders blamed an unrelated assassination attempt on Henry Frick for turning public sympathy against the union. "The bullet from Berkman's pistol went straight through the heart of the Homestead strike," said one. It was, however, neither the assault of an anarchist (who had no connection with the Amalgamated), nor the 8,000 state militia men that really defeated the union. The union was beaten on the shop floor, by production methods which transformed skilled craftsmen into machine tenders and could transform farmhands (and strikebreakers) into steelworkers in a matter of weeks. "We had to teach our employees a lesson, and we taught them one they will never forget," crowed Frick, who reopened Homestead with 700 immigrant strikebreakers and broke the union. The lesson that both management and labor learned was that mechanization had made craft unions every bit as obsolete in the industry as the cementation steel they had once made.

Ironically enough, Homestead also brought down Henry Frick. Publicly, Carnegie supported Frick throughout the strike. Privately, though, he could never forgive Frick for so badly tarnishing his public image. Carnegie began pushing Frick to the periphery of the company, removing him in 1894

as chief executive officer of Carnegie Steel, and cutting his partnership share in half. A few years later, after a bitter fight over the price the Frick Coke Company (of which Frick was still president) charged Carnegie Steel, he took away his remaining titles and threatened to invoke the "Iron Clad," and thus take over Frick's holdings at book value. "For years I have been convinced that there is not an honest bone in your body," Frick exploded. "Now I know that you are a god-damned thief." Frick sued to get a more equitable settlement, and forced Carnegie to recapitalize the company and pay him $31 million for his shares, which had been worth only $5 million at book. The two never saw each other after this final confrontation, and it was said that when Carnegie, as an old man, offered to reconcile with his old associate, Frick's only reply was "Tell Mr. Carnegie I'll meet him in hell."

With Frick on the way out, Carnegie turned for managerial leadership to the last, and most precocious, of his "young geniuses," Charles Schwab. Like most of them, Schwab started out on the bottom, as a stake driver at Edgar Thomson at a dollar a day, but his performance soon attracted attention. He became Jones's protégé, studied his methods, and put them into practice with ferocious energy. "You are a hustler," Carnegie told him. It took Schwab less than six months to work his way up to head of blast furnace construction at Edgar Thomson. At twenty-five he was superintendent at Homestead, and at twenty-seven (after Jones's death) superintendent at Edgar Thomson. Before reaching the age of thirty-five, he became president of Carnegie Steel, a partner in the company, and a very wealthy man.

He seemed to thrive under Carnegie's approach to motivating his managers with equal doses of shameful bullying and shameless flattering. Taking charge after the Homestead fiasco (though he was not officially appointed president until 1897), Schwab pushed for renovation and innovation at a pace and on a scale that occasionally intimidated even Carnegie. "The task, therefore, which I have set myself for the year 1895," Schwab said, "is to save one-half million dollars in the cost of manufacture over the cost of the year 1894. In other words, [I] expect to make a clear savings by practice, labor, running expenses, etc. of one-half million dollars for this year." Like Carnegie, he insisted on knowing his precise costs–"We made a careful . . . statement of each manufacture, with the cost as compared with each department . . . [and] had the manager of each department make such explanations as were necessary." And like Carnegie, he manipulated his managers into competing with one another—"[I] rivalled one against the other, and in that way got better results." Keeping close track of the relationship between cost and investment, he could report, for example, that while Edgar Thomson was producing pig iron for $1,584 in investment for each ton of iron per day, at the Lucy Furnaces the figure was $1,900. When Carnegie hesitated about switching Edgar Thomson from Bessemer to open-hearth ("The basic rail involves grave consequences, perhaps, and we should take no risks," he said), Schwab barraged him with facts and figures until he relented. At Homestead, Schwab added ten new open-hearth furnaces and new blooming and plate mills. He remodeled Duquesne so that it could manufacture sheets, pipes, and other non-rail items, purchased additional blast furnaces, and bought land for further plant expansion. "Believe me," Carnegie told him, "I am rejoicing equally with yourself at your brilliant success and at the improvements which I am sure you are going to make. . . . I do not have a single word of adverse criticism. There has never been a time on my return that I can recall when everything seemed moving so smoothly." Schwab's modernization program cut labor costs alone some 15 percent and eliminated any last chance for unionism at Carnegie Steel. "We have completely knocked out any attempt to organize Homestead workmen," he reported to Carnegie. "I now feel satisfied no further attempt will be made for sometime at least."

Sensing that the company was reaching the limits of what it could do with mechanization alone, Schwab urged Carnegie toward fuller horizontal integration to achieve further economies of scale. Although the company already controlled its own sources of coke and limestone, iron ore was still purchased on the open market, mostly from mines in Michigan's upper peninsula. As these deposits ran out, the center of ore production moved west, to the incomparably rich Mesabi range of Minnesota. Uncharacteristically, Carnegie failed at first to anticipate the consequences of this shift and let John D. Rockefeller, of all people, get the jump on him. During the depression of 1893, while Carnegie was recovering from the trauma of Homestead and contending with the worst rail market in years,

Rockefeller bought up control over most of the Mesabi and built a railroad line and a steamboat fleet to transport his ore. Rumors spread that he planned to construct a modern mill in Cleveland and go into the steel business himself. By then, of course, Carnegie was well aware of the seriousness of his predicament. "Remember," he told an associate, "that Reckafellows [as he aptly dubbed the oil baron] will own the R. R. and that's like owning the pipeline. Producers will not have much of a show." Facing up to the situation, Carnegie worked out an arrangement to lease Rockefeller's ore lands and agreed to pay Rockefeller a 25-cent royalty on every ton mined and to ship the ore on Rockefeller's rail and steamboat lines. Although he said he found Rockefeller drove a "hard bargain," Carnegie was convinced that he had gotten a good deal in the end.

Carnegie next turned his attention to railroad shipping rates. For years he had known that he was paying higher rates than competitors which were serviced by several railroads. After more than a decade of pleading with the Pennsylvania Railroad's management and lobbying state legislators to lower rates, he took matters into his own hands and built a railroad—the Pittsburgh, Bessemer & Lake Erie—which ran from the ore docks on Lake Erie to his steel mills in Pittsburgh. That, at last, brought results. The Pennsylvania promptly cut his freight schedules by half, saving Carnegie $1.5 million a year, in return for his promise not to build more roads into coke country. In 1898 Carnegie also bought his own fleet of ore boats and modernized his port facility. These investments probably saved him an additional $2 million a year.

In the decade from 1888 to 1898, Carnegie Steel doubled its investment capital from $20 million to $45 million and tripled its production of coke from 6,000 tons to 18,000 tons a year, pig iron from 600,000 tons to 2 million tons a year, and steel from 2,000 tons to 6,000 tons a day. Profits increased steadily from $1.9 million in 1888 to $3 million in 1893 and to $11.5 million in 1898.

Social critics, appalled by the size of those figures, accused him of being a 'robber baron,' but Carnegie understandably held a somewhat different view. " 'The robber baron' has ceased to rob and is now being robbed," he told a reporter. "The eighth wonder of the world is this—two pounds of iron-stone purchased on the shores of Lake Superior and transported to Pittsburg [*sic*]; two pounds of coal mined in Connellsville and manufactured into one and one-fourth pounds of coke and brought to Pittsburg; one-half pound of limestone mined east of the Alleghenies and brought to Pittsburg; a little manganese ore, mined in Virginia and brought to Pittsburg, and these four and one-half pounds of material manufactured into one pound of solid steel and sold for one cent. That's all that need be said about the steel business." Some years before he had bluffed his way into the Bessemer Association by claiming that he could manufacture rails for $9 a ton; now he was actually selling them for $15 a ton.

As the great merger wave of the 1890s hit the steel industry, as it had many other American businesses from cottonseed oil and sugar to whiskey and petroleum, the competition Carnegie faced became more intense than ever before. "Trusts," backed by impressive Wall Street resources, were seeking to accomplish through merger and acquisition what pools and informal agreements had not been able to do–stabilize prices and profits by limiting competition. The largest of these was the Federal Steel Company, created in 1898 by financier J. P. Morgan. It had manufacturing plants in Chicago, Milwaukee, and Joliet; its own coke, limestone, and iron ore supplies; and a total capital of $200 million. On paper Federal may have looked like a dangerous adversary, but Carnegie recognized its weaknesses. "I think Federal the greatest concern the world ever saw for manufacturing stock certificates," he said, "But they will fail sadly in steel." Except where he saw a temporary advantage, he declined invitations to join with it in dividing the market. "Put your trust in the policy of attending to your own business in your own way and running your mills full regardless of prices and very little trust in the efficacy of artificial arrangements with your competitors, which have the serious result of strengthening them as they strengthen you," he said. Sure, as always, that it would be a "question of the survival of the fittest," he instructed his lieutenants to issue no dividends and to spend the surplus on new hoop, nail and wire, and tube mills. "Believe a continuance of war much better for us than any peace," he told Schwab, "a good start for Federal being the last thing desired." In 1899 Carnegie Steel doubled its profits over the year before to $21 million, and the next year doubled them again. Elbert H. Gary, Federal's president, later admitted, "It is not at all certain that if the management that was in force at the time had continued, the Carnegie Company would not have driven entirely out of business every steel company in the United States."

If Carnegie could not be beaten, perhaps he could be bought. For years he had talked of selling and devoting himself to philanthropic and literary interests. In 1899 his partners thought they had a deal, until Carnegie learned that the prospective buyers included such infamous speculators as the Moore brothers (creators of the match trust) and John "Bet a Million" Gates. Then in December 1900 J. P. Morgan heard Schwab speak about the potential advantages of streamlining the industry by eliminating excess capacity and duplication. That sort of talk intrigued him, for he had been doing exactly the same thing in the railroad industry for years. Afterwards he arranged a private meeting with Schwab. Again, he liked what he heard. "Well, if Andy wants to sell," he told Schwab, "I'll buy. Go and find his price." Schwab (at Mrs. Carnegie's suggestion) tried out the proposal on his boss the next afternoon on the golf course. After a few rough calculations, they came up with an asking price of $480 million for Carnegie Steel. Morgan accepted Carnegie's terms immediately, and a couple of days later a simple handshake completed the largest single business deal in history.

The United States Steel Corporation was formed on April 1, 1901. Besides Carnegie Company, it included Federal Steel, National Tube, American Steel and Wire, American Steel Hoop, American Tin Plate, American Sheet Steel, American Bridge, Shelby Steel, enormous holdings of iron, coal, and limestone properties, several railroads, and a fleet of ships. With a total capital of $1.4 billion, it was the world's largest corporation. Carnegie took his share (nearly $300 million) in first-mortgage U.S. Steel bonds, making him, some said, the richest man in the world.

Back in 1889 Carnegie had written an extremely popular essay entitled *The Gospel of Wealth* (published in 1900), in which he stated that "the man who dies . . . rich, dies disgraced." Now, finally, he had the opportunity to put into practice what he had preached. He often said that it was ten times harder to give away a fortune than to make it in the first place, and, with $300 million in bonds earning 5 percent a year, he may have had a point. He became best known as a giver of libraries and church organs, and paid for 2,811 free libraries worth more than $50 million, and 7,689 organs worth some $6 million, over the years. He took a special interest in education, particularly practical education. He created the Carnegie Trust for the Universities of Scotland with $10 million (with money for every discipline except the classics), the Carnegie Institute of Technology with $27 million, and the Carnegie Institution of Washington with $10 million. He snubbed the Ivy League (though he did give Princeton a lake, costing $400,000, as an inducement to crew over football) in favor of endowing smaller, vocationally oriented schools, including black colleges such as Hampton Institute and Tuskegee Institute. To provide a secure retirement for professors he started the Carnegie Teachers Pension Fund with another $10 million.

For a while Carnegie found in philanthropy the same competitive exhilaration he had once found in business. He especially enjoyed seeing the newspaper box scores which showed him out in front of Rockefeller in the charity business. But by 1904 Carnegie had given away nearly $180 million and still had an equal amount left. He told a friend that "The final dispensation of one's wealth preparing for the final exit is I found a heavy task . . . You have no idea the strain I have been under." So in 1911 he took most of what was left and created the Carnegie Corporation. With an endowment of $125 million it was the first great private foundation and the largest until the Ford Foundation was established in 1947.

Bored with philanthropy, Carnegie now threw himself into the international peace movement with all the old energy and enthusiasm. All of his life Carnegie had been a pacifist, though he was not above taking military orders in his steel mills and had once gotten involved in a scandal over defective iron plate intended for the Navy. In *Triumphant Democracy* he had remarked that "It is one of the chief glories of the Republic that she spends her money for better ends and has nothing worthy to rank as a ship of war." With most nations of the world, including the United States, embarked upon a massive military build up, Carnegie now saw in the peace movement a final, noble purpose for his millions. He endowed four new foundations, including the Carnegie Endowment for International Peace at a cost of $25 million, and built a Temple of Peace at the Hague. He became an early supporter of the League of Nations idea, and after arranging a personal meeting between Theodore

Roosevelt and Kaiser Wilhelm, concluded *Autobiography of Andrew Carnegie* (1920) with a chapter entitled "The Kaiser and World Peace." Just a couple of months later, in August 1914, he had to revise that chapter, noting sadly, "As I read this today, what a change! The world convulsed by war as never before! Men slaying each other like wild beasts." His wife later revealed that "Optimist as he always was and tried to be, even in the face of the failure of his hopes, the world disaster was too much. His heart was broken." He left his beloved Scotland at the outbreak of World War I and never returned. He died at a summer retreat in Massachusetts on August 11, 1919, at the age of eighty-three.

The *New York Sun* obituary called Carnegie "the personification of 'Triumphant Democracy,'" embodying for his generation the American success story. He accumulated millions and gave away millions, but the legacy he left was greater still, for he built an industry as well as a fortune. As one observer noted, "He invented nothing. But he made the inventions of others enormously effective, and the growth of the American steel industry was largely his masterpiece." Almost single-handedly, Carnegie transformed steel making into big business, the first in American manufacturing. The result was steel so cheap that it became the distinguishing material of the late nineteenth and early twentieth centuries.

Paradoxically, one effect of the transformation to the modern steel industry was that Carnegie, and others like him, became virtually obsolete in the business world they had done so much to create. Carnegie remained to the end an entrepreneurial capitalist in an era increasingly dominated by managerial enterprise and by the professional managers and financiers who ran the great businesses they inherited. Carnegie would never make that transition himself, but he had trained many younger men who did, including Charles Schwab, the first president of U.S. Steel. Carnegie was the last of his breed, as the later history of the American steel industry would demonstrate. Certainly he would have been surprised to see how quickly the American steel industry lost its world leadership to aggressive Asian competitors, even though he had forseen such a possibility as early as 1908. In an article on protective tariffs he quoted John Stuart Mill: "The superiority of one country over another in a branch of production often arises only from having begun it sooner. . . . A country which has this skill and expertise yet to acquire may in other respects be better adapted to the production than those which were earlier in the field." Still, it is evidence of Carnegie's enduring legacy that companies which now dominate the industry owe much of their success to methods he pioneered a century ago.

Selected Publications:

An American Four-In-Hand in Britain (London: Lowe, 1864);

Notes of a Trip Around the World (New York, 1879);

The Employer's View of the Labor Question (New York: J. J. Little, 1886);

Results of the Labor Struggle (New York: J. J. Little, 1886);

Round the World (New York: Scribners, 1886);

Triumphant Democracy; or, Fifty Years' March of the Republic (New York: Scribners, 1886);

Andrew Carnegie's Response To A Radical Address (Dunfermline: W. Clark, 1887);

Home Rule in America (Glasgow: Glasgow Junior Liberal Association, 1887);

America and the Land Question (Glasgow: Glasgow Junior Liberal Association, 1888);

The A B C of Money (New York: North American Review, 1891);

Wealth and Its Uses (New York, 1895);

The Reunion of Britain and America: A Look Ahead (New York, 1898);

The Gospel of Wealth (New York: Century, 1900);

The Opportunity of the United States (New York: Anti-Imperialist League of New York, 1901);

The Empire of Business (New York: Doubleday, Page, 1902);

Peace By Arbitration (New York: R. G. Cooke, 1902);

The Industrial Future of Nations (Manchester: Hinuhliffe, 1903);

Britain's Appeal To The Gods (New York: World's Work, 1904);

Drifting Together: Will the United States and Canada Unite? (New York: World's Work, 1904);

Industrial Peace (New York, 1904);

James Watt (New York: Doubleday, Page, 1905);

Education of the Negro: A National Interest (Tuskegee Institute: Tuskegee Institute, 1906);

Edward M. Stanton (New York: Doubleday, Page, 1906);

A League of Peace (Boston: Ginn, 1906);

My Experience With Railway Rates and Rebates (New York: M. J. Roth, 1906);

Ezra Cornell (New York: M. J. Roth, 1907);

The Negro In America (Philadelphia: E. A. Wright, 1907);

Problems of Today: Wealth–Labor–Socialism (New York: Doubleday, Page, 1908);

Armaments and Their Results (New York: Peace Society of the City of New York, 1909);

The Path to Peace (New York: Peace Society of the City of New York, 1909);

Peace Versus War: The President's Solution (New York: American Association for International Conciliation, 1910);

War As the Mother of Valor and Civilization (New York: Peace Society, 1910);

Arbitration (London: Peace Society, 1911);

Britain and Her Offspring (London: Wortheimer, Lea, 1911);

Business (Boston: Hall & Locke, 1911);

The Industrial Problem (New York: North American Review, 1911);

Autobiography of Andrew Carnegie (Boston & New York: Houghton Mifflin, 1920);

Miscellaneous Writings of Andrew Carnegie, edited by Burton J. Hendrick (Garden City, N.Y.: Doubleday, Doran, 1933).

References:

David Brody, *Steelworkers in America: The Nonunion Era* (Cambridge, Mass.: Harvard University Press, 1960);

Louis M. Hacker, *The World of Andrew Carnegie, 1865-1901*;

Burton J. Hendrick, *The Life of Andrew Carnegie* (Garden City, N.Y.: Doubleday, Doran, 1932);

Jonathan R. T. Hughes, *The Vital Few: American Economic Progress and Its Protagonists* (Boston: Houghton Mifflin, 1966);

Harold Livesay, *American Made: Men Who Shaped the American Economy* (Boston: Houghton Mifflin, 1979);

Livesay, *Andrew Carnegie and the Rise of Big Business* (Boston: Houghton Mifflin, 1975);

Elting Morrison, *Men, Machines and Modern Times* (Cambridge, Mass.: Harvard University Press, 1966);

George Swetnam, *Andrew Carnegie* (Boston: Twayne, 1980);

Joseph F. Wall, *Andrew Carnegie* (New York: Oxford University Press, 1970).

Archives:

Carnegie's personal papers are deposited in the Library of Congress. Most of his business correspondence, however, is still in the archives of the USX. Joseph Wall was given permission to use that material in preparing his biography and quotes extensively from it, but the papers remain company property and are not yet open to scholars.

Thomas Morrison Carnegie

(1843-October 19, 1886)

by Larry Schweikart

University of Dayton

CAREER: Partner, assorted Carnegie companies (1865-1886).

Thomas Morrison Carnegie, a partner in the Edgar Thomson Steel Company, Ltd., and Andrew Carnegie's brother, was born in 1843 in Dunfermline, Scotland, the son of William and Margaret Carnegie. Thomas, or Tom, as Andrew usually referred to him, came to the United States with his brother in 1848. While Andrew worked in the telegraph office, Carnegie went to school. He finished two years of high school at Altoona and learned telegraphy skills from his brother. After this instruction Andrew gave his younger brother a job as his personal secretary in the Western Division of the Pennsylvania Railroad, a capacity in which he proved to be totally discreet and absolutely loyal. Carnegie joined his brother in a number of ventures, investing in the Storey farm oil fields, the Central Transportation Company, and the Pacific & Atlantic Telegraph Company. He also invested in one of Andrew's earliest businesses, the Keystone Bridge Works, a company in which Andrew took advantage of the expansion of the railroads.

Much like his cousin George "Dod" Lauder, Carnegie had conservative and cautious qualities, and he, more than anyone else, understood Andrew's goals for his companies. Even so, he disagreed with certain of Andrew's practices, such as his penchant for retaining earnings for reinvestment in capital equipment. As a result of Carnegie's loyalty, when Andrew took a 9-month tour of Europe in 1865 and 1866 at a delicate time in the company's history, he left his interests entirely in Tom's hands. At age twenty-two Carnegie assumed all responsibilities for the Carnegie companies, including the Keystone Bridge Company and investments in 15 other businesses, mostly railroads and iron mills. He immediately received a torrent of letters bearing contradictory advice from Andrew. On

Thomas Morrison Carnegie

the one hand he was told to expand; on the other Andrew urged him to be careful. But Carnegie had the trust of many of Andrew's business associates, partly because he occasionally opposed his brother's profit retention policies. Just as Walt Disney needed Roy to reassure and soothe his movie company's creditors, Andrew needed his brother's quiet, secure demeanor with partners, bankers, and customers.

When Andrew was stretched too thinly, Tom agreed to take a $20,000 interest in Kloman & Phipps, a partnership between a Prussian immigrant iron maker, Andrew Kloman, and Henry ("Harry") L. Phipps, Jr. That investment gave Tom a 50-percent share in that company. For Carnegie it offered an opportunity to get out from under Andrew's shadow in a venture that his older brother had seen as a worthy investment. Tom essentially provided Kloman with the capital he needed to sever a relationship with another associate, Tom Miller, who then joined with Carnegie to form the Cyclops Iron Company.

In March 1865 Carnegie entered into several businesses that proved lucrative. First he worked from his position inside Kloman & Phipps to merge Carnegie's Cyclops Iron Company with Kloman & Phipps. Tom and Andrew handled the negotiations for each company because, fortuitously, a feud between Miller and Kloman prevented the two men from speaking to each other. Andrew saw an opportunity to bring into his firm the business talents of Phipps and Tom Carnegie, as well as the considerable mechanical talents of Andrew Kloman. Tom's vote swung the merger deal for Andrew. The new partnership was known as the Union Iron Mills. Moreover, through Kloman & Phipps, Carnegie had developed a strong friendship with Henry Phipps, and when a seat on the Union Iron Mills board opened, the Carnegies asked Phipps to fill it. Carnegie and Phipps became so close that Andrew, in *The Autobiography of Andrew Carnegie* (1920), referred to their relationship as "a partnership within a partnership." Miller's dissatisfaction with Phipps's position on the board forced another reorganization in 1869 in which Tom sold his shares in the newly named Carnegie, Kloman & Company to Andrew, giving Andrew controlling interest.

A second source of business opportunities for Carnegie developed through his marriage to Lucy Coleman in summer 1867. Her father, William Coleman, had been a baker and bricklayer, but eventually he began producing iron springs and had achieved a reputation as a fine ironmaster. In the 1870s Carnegie joined Coleman in erecting a Bessemer steelworks called Carnegie, McCandless & Company (1873) that in 1876 was dissolved when Coleman sold his interest and was reformed as the Edgar Thomson Steel Works. Along with Carnegie and Coleman, the other original partners included Pittsburgh merchant David McCandless; two railroad men, John Scott, president of the Allegheny Valley Railroad; and David Stewart, who was also president of the Columbia Oil Company; and William Shinn, the vice-president of the Allegheny Valley Railroad. Carnegie held 6 percent of the Edgar Thomson Company.

Carnegie also parlayed his contacts with other businessmen into avenues for growth. As manager of the Union Iron Mills in Pittsburgh, he became acquainted with Henry Clay Frick, who sold coke to the mills. Frick and Carnegie cooperated in the H. C. Frick Coke Company, laying the groundwork for vertical combination in the steel industry. Indeed, it seems that Tom recognized the benefits of vertical integration long before Andrew, and he began to sing the praises of such combinations to his brother at every opportunity.

In 1870 the Union Mills partners decided to build their own blast furnace, and on December 1 of that year the Carnegie brothers and Kloman & Phipps formed a new company, Carnegie, Kloman & Company, to construct a furnace. This furnace, completed in 1872, was named the Lucy Furnace after Tom's wife. Kloman later left to pursue his own business interests, and on April 1, 1881, the name of the company was again changed, this time simply to Carnegie Brothers. Tom was named chairman of the new company. As partners had left, been bought out, or retired, Carnegie had attained an increasing share of the companies. In 1886 Carnegie, Phipps & Company was organized—and operated separately from Carnegie Brothers—and by the end of the year the combined net profits of the Carnegie companies totaled over $2.9 million, which represented 60 percent of the capitalization of the companies. Virtually no business in America had witnessed such a remarkable return on assets, and Tom led the partners' charge to convince Andrew that some of the profits should be dispersed in dividends. The elder Carnegie, however, would not hear of it, and his majority shares carried every vote.

Thomas Carnegie contracted a fatal case of pneumonia in 1886. At the time of his death, he held 17.5-percent interest in Carnegie Brothers and 16-percent interest in Carnegie, Phipps & Company, an interest second only to Andrew's. His shares were purchased with some difficulty by his brother through a friendly arrangement with Lucy so that Andrew could acquire them over a period of time. But the threat posed by the death of one of the major partners immediately rippled through the ranks of the partners, and they realized that Andrew Carnegie's death would bankrupt the company. After Tom's death the notorious Iron Clad Agreement was instigated by Henry Phipps and allowed company associates the right to purchase the stock of a deceased member, with up to 15 years allowed to purchase Andrew's interest. The Iron Clad Agreement also included provisions that if three-fourths of the outstanding stock interests or three-fourths of the associates so voted, an associate could be forced to sell his interest to the remaining partners, a provision that was invoked on several occasions. Andrew, because of his majority interest, was clearly excluded from this last provision.

Carnegie was survived by his wife and their son, Thomas Morrison. Several sources suggest that Tom and Andrew had an agreement by which Tom was free to marry while Andrew would remain unmarried as long as his mother was still alive. Tom and Lucy built a mansion shortly before his death, which Andrew disparaged at every opportunity. Nevertheless, the two brothers remained close, with Tom serving as a restraining influence on some of Andrew's more impetuous tendencies.

Living in Andrew Carnegie's shadow, Thomas Carnegie anticipated the most important trend in nineteenth-century industrialization, vertical integration. A talented and competent businessman in his own right, Tom complemented Andrew, and it is unlikely that the frequent reorganizations of Andrew's companies could have occurred as smoothly as they did without Tom. Thomas Carnegie was the forerunner of the new breed of "top level management" that was beginning to appear in the American steel industry.

References:

Andrew Carnegie, *The Autobiography of Andrew Carnegie* (Boston: Houghton Mifflin, 1920);

Burton Hendrick, *The Life of Andrew Carnegie*, 2 volumes (Garden City, N.Y.: Country Life Press, 1932);

William T. Hogan, *The Economic History of the Iron and Steel Industry in the United States*, volumes 1 and 2 (Lexington, Mass.: Lexington Books, 1971);

Joseph Wall, *Andrew Carnegie* (New York: Oxford University Press, 1970).

Archives:

Much of Thomas Carnegie's personal papers are collected with Andrew Carnegie's personal papers, which are located at the Library of Congress, Washington, D.C.

Carnegie Steel

by Larry Schweikart

University of Dayton

Andrew Carnegie's empire effectively dated from 1861 and the founding of the Iron City Forge company of Pittsburgh, Pennsylvania, by Andrew Kloman, a Prussian immigrant, and his brother Anthony. The Klomans joined Thomas Miller, whose name did not appear on any official documents, and Miller's front man Henry Phipps, Jr., as partners in the venture. Iron City Forge produced finished iron and steel products for the Union army during the Civil War but also established a reputation for its manufacture of fine axles for railroad cars. A restructuring of the company occurred in 1862 when Miller backed the lease of land for a new mill and the construction of the works. The new firm, Kloman & Company, was capitalized at $80,000 and consisted of the Kloman brothers,

Miller, and Phipps, although Andrew Kloman soon purchased his brother's interest. Internal dissension between Kloman and Miller over the direction of the firm soon resulted in a struggle between Miller and Phipps.

On September 1, 1863, Andrew Carnegie, invited by Miller to negotiate a settlement between the partners, helped create Kloman & Phipps, with Kloman, Miller, and Phipps joined by a new partner, Thomas Carnegie (Andrew's brother), who invested $10,000. Due to a buy-out clause in the partnership, Miller was forced out by the other partners. He sought revenge and in October 1864 opened with Andrew Carnegie a rival firm, the Cyclops Iron Company in Pittsburgh, to drive Kloman & Phipps out of business. Of course, Carnegie had no intention of destroying his brother's interest; instead, he planned all along to merge the two companies. With the natural production advantages of a heavy demand from his own bridge company (Keystone Bridge Company, organized in 1865), and with Tom Carnegie working inside Kloman & Phipps, Andrew Carnegie effected the consolidation of the two ironworks in May 1865. Capitalized at $500,000, the newly formed Union Iron Mill Company produced railroad specialties and structural beams and plates. This represented Carnegie's first real venture in the iron and steel business, and it proved profitable, generating profits of over $250,000 in 1867. The mills were known as the Upper Union (ex-Iron City and Kloman & Phipps) and the Lower Union (ex-Cyclops) and were boosted by the revival of the railroads from 1868 to 1872.

In 1865 Carnegie had successfully coordinated the reorganization of the Piper & Shiffler Company (organized in 1862) into the Keystone Bridge Company, a firm in which he held a 20-percent interest ($1,250). Keystone provided a steady source of demand for iron from all of the Carnegie companies. Carnegie also maintained an interest in the Superior Rail Mill Company and the Pittsburgh Locomotive works but was not actively involved in iron production before the Cyclops venture.

Having established a successful milling operation with the Union mills, Carnegie, Kloman & Company (as it was reorganized in 1867 upon Miller's resignation) searched for inroads into other areas of iron and steel production, especially pig iron and rolled steel. This expansion required a blast furnace. Although Carnegie was approached about investing in the construction of the Isabella Furnace in 1870, he and his partners began their own furnace which was completed and blown in by 1872. Designed with a larger diameter by Andrew Kloman, Lucy #1, as it was called (after Tom's wife), was soon followed by Lucy #2. The two furnaces set records for pig-iron production for the rest of the decade.

The company undertook expansion into rolled Bessemer steel products and on November 5, 1872, reorganized as Carnegie, McCandless & Company. Investors in the new firm included David McCandless, a Pittsburgh merchant; John Scott, president of the Allegheny Valley Railroad; David Stewart, president of the Columbia Oil Company; William Shinn, vice-president of the Allegheny Val-

ley Railroad; J. Edgar Thomson; Andrew and Tom Carnegie; Tom's father-in-law, William Coleman; Kloman; and Phipps. The company had a capitalization of $700,000, of which Andrew Carnegie held $250,000.

Carnegie focused on supplying rails to Edgar Thomson's railroad interest, as well as the Allegheny Valley Railroad, the executives of which were investors in Carnegie, McCandless & Company. To handle the increased business a new plant was required, which Carnegie called the Edgar Thomson Steel Works, or simply E.T. The works featured the most modern machinery, and its construction was supervised by Alexander Holley, the leading authority on Bessemer steel mills in the United States. Built in the panic years of 1873 to 1875, this mill only cost a fraction of what it would have cost in less financially stringent times. Carnegie and Holley also opened negotiations with agents of the Siemans Company, a German producer of open-hearth furnaces, to have the firm's gas furnaces installed at E.T.

Kloman, meanwhile, had engaged in questionable investments in the Cascade Iron Company and the Escanaba Furnace Company, both of which he had secured with his holdings in the Union Iron Mills, the Lucy Furnace, and E.T. The two companies failed in the panic of 1873, leaving Kloman deeply in debt. By 1876 Carnegie and his partners had forced Kloman out, and Carnegie, Kloman & Company became Carnegie Brothers, the reorganized firm having a capitalization of $507,629. Carnegie held just under 50 percent but kept his eye on further reorganizations that would leave him with majority control. When William Coleman withdrew from Carnegie, McCandless & Company in 1876, Carnegie purchased Coleman's shares and reorganized the company as the Edgar Thomson Steel Company, with $1 million in capital (which was increased to $1.25 million in 1878). Carnegie thus owned majority interest in the Edgar Thomson Steel Company and slightly less than majority in Carnegie Brothers (although with his brother he had a majority). As production rose to unprecedented levels, Carnegie had yet other reorganizations in mind.

The furnaces and rail mills at E.T. began production in 1875 under the remarkable supervision of Capt. William R. "Bill" Jones. More than 45,000 net tons of ingots and 32,000 net tons of rails were produced that year. The company earned $181,000 profit on an investment of $731,000 and by 1880 would clear $1.6 million on $1.25 million

Carnegie's Lucy Furnace #1, first opened in 1872 (courtesy of USX Corporation)

of investment. From 1879 to 1883 Julian Kennedy supervised the mills, which attained record outputs. As he had hoped, Carnegie succeeded in attracting large investments and orders from J. Edgar Thomson's Pennsylvania Railroad. On April 1, 1881, the Carnegies, Phipps, McCandless, Stewart, Scott, and John Vandervort organized Carnegie Brothers & Company, Ltd., a steel consolidation that included the Upper and Lower Union Iron Mills (an annual capacity of 45,000 net tons of beams, angles, and bars); the Edgar Thomson Steel Works (an annual capacity of 100,000 net tons each of Bessemer ingots and rails); the two Lucy furnaces (an annual capacity of 95,000 net tons of pig iron); the Unity coal mines and coke ovens; and an 80 percent interest in the Lorimer Coke Works. Total capitalization was $5 million, and Thomas Carnegie was named as chairman of the board. Andrew, though, owned 55 percent of the capital, or roughly $2.7 million. In 1881 this $5 million partnership made profits of $2 million.

In 1879, out of a desire for revenge against Carnegie for what he perceived as wrongdoing, Andrew Kloman joined William Singer and several other entrepreneurs to purchase land and begin construction of a rail mill at Homestead, Pennsylvania, under the name of the Pittsburgh Bessemer Steel Company, Ltd. This company combined Kloman's Bessemer rail mill, universals, and other equipment, which Kloman had begun erecting in 1880, with the Pittsburgh Bessemer mill. Kloman's mill, how-

ever, had been entirely dependent on the E.T. works for its ingots, and while collectively the group could have posed a threat to E.T.'s rail market, this dependence in part led to its absorption by Carnegie in 1882. Kloman died before the mill actually began production. The Pittsburgh Bessemer Steel Company entered production in 1881, and the price of steel rails fell substantially. Carnegie was chagrined to see that a new competitor could wedge itself so easily into the market that he had dominated; it quickly acquired almost 20 percent of E.T.'s business. A labor dispute in 1882 with the Amalgamated Association of Iron and Steel Workers of North America set back production and made Pittsburgh Bessemer vulnerable. In October 1883 Carnegie and his associates acquired the company from its desperate owners. Carnegie offered face value for the stock and offered to exchange Carnegie company stock. In return they received the Homestead works, the most modern rail mill in the United States.

Acquisition of Homestead led to yet another reorganization. Carnegie's holdings were too extensive to fit under the aegis of Carnegie Brothers & Company, Ltd. without increasing that company's capitalization. The Lucy furnaces had been held by Wilson, Walker & Company, which had purchased the Lower Union Iron Mill in 1873. On January 1, 1886, John Wilson agreed to sell his shares to Carnegie, and, as part of the merger, John Walker was named chairman of the new company, called Carne-

Carnegie's Edgar Thomson Steel Works in 1890 (courtesy of USX Corporation)

gie, Phipps & Company. There were now two companies–Carnegie, Phipps & Company and Carnegie Brothers & Company, Ltd.–heading for a final consolidation. It was inevitable, given Andrew's concern over maintaining personal control and Tom's farsighted interest in vertical integration, that another consolidation would occur. One company not under this corporate umbrella was Hartman Steel Company, a finishing mill and merchant steel manufacturer built in 1882. The owners had sold their firm to Carnegie in 1883, but the plant consistently lost money.

Carnegie's fear that competitors could continue to cut costs was well founded. Cheaper and easier methods of producing steel appeared at a number of upstart plants, such as the Duquesne Steel Company. Organized in June 1886 with a capital of $350,000, the company located its works southeast of Pittsburgh. After production began in February 1888, the company reorganized as the Allegheny Bessemer Steel Company. Carnegie realized that this company used radical methods and machinery that could dramatically shave production costs. To acquire control of it, Carnegie spread stories that the rails were unsafe, which undercut Allegheny's sales. But Carnegie needed more capital, and a new Carnegie associate, Henry Clay Frick, entered the scene.

Frick founded the Henry C. Frick Coke Company in 1871 to build coke ovens, and he soon found there was no limit to the market. A millionaire by age thirty, Frick seemed to be a natural ally for Carnegie. Together, they would achieve unprecedented levels of vertical integration. Frick's coke company was reorganized in 1881 with a capital of $2 million. By 1883 Carnegie had become the largest shareholder, and that same year Carnegie obtained majority interest in the company. Frick, of course, remained as the manager. On January 31, 1887, Carnegie sold Frick 2 percent of Carnegie Brothers & Company, Ltd. in order to bring him into the corporate fold. In January 1889 Frick replaced Phipps as chairman of Carnegie Brothers & Company, Ltd. In 1890 he sealed the acquisition of the Allegheny Bessemer Steel Company for $1 million. The new plant was incorporated into Carnegie, Phipps & Company.

Frick also presided over the organization of the largest steel company in the world, when on July 1, 1892, Carnegie, Phipps & Company and Carnegie Brothers & Company, Ltd. sold their assets to a new corporation, Carnegie Steel Company, Ltd. The new firm was capitalized at $25 million, of which Andrew Carnegie held 55.33 percent ($13.8 million), Frick and Phipps each held 11 percent ($2.75 million), and 19 other partners shared

1 percent each. Still another 3.66 percent was reserved for the purpose of rewarding outstanding individuals within the company at a later date. This company had four steel plants—Homestead, Hartman, E.T., and the newly acquired Duquesne—plus the Upper and Lower Union Iron Mills, the Lucy Furnaces, the Keystone Bridge Works, mines, and coke works. The Henry C. Frick Coke Company remained independent. Only Carnegie and Phipps had survived from the nine original partners that had formed Carnegie, McCandless & Company. Frick was the new chairman, and the company continued to make profits, averaging over $4.5 million a year from 1892 to 1894.

Frick led the company to acquire several coal, ore, and transportation companies as well as valuable lands. Unfortunately, Frick also dealt with the 1892 Homestead strike, in which Pinkertons violently battled with striking workers. Following a 5-month lockout and government military intervention, the Amalgamated Association of Iron and Steel Workers of North America was defeated. But so was Carnegie's reputation.

Relations between Frick and Carnegie, strained because of the strike, became even more difficult as the chairman's ambitions overreached the bounds that Carnegie and his associates would permit. In 1899 Frick and a group of lesser partners attempted a buyout of Carnegie's interest, obtaining the backing of John W. "Bet-a-Million" Gates and William and James Hobart Moore of Chicago, but failed to raise sufficient capital. Charles Schwab, the president of Carnegie Steel since 1897, played the role of intermediary between Frick and Carnegie. In 1900 an arrangement was negotiated in which both Carnegie Steel and the Frick Coke Company were superseded by a new company, the Carnegie Company (capitalized at $320 million). Frick was forced out as an active partner for the sum of over $30 million in stocks and bonds. His demise had been the result of his insistence on an increased price per ton of coke, which Carnegie refused to pay on grounds of principle, not price. A year later, on February 26, 1901, Carnegie sold Carnegie Steel Company, Ltd. for $480 million to J. P. Morgan's United States Steel Corporation—the world's first billion-dollar corporation. Thus, Carnegie's legacy was absorbed into an even larger corporate legend, U.S. Steel.

References:

Andrew Carnegie, *The Autobiography of Andrew Carnegie* (Boston: Houghton Mifflin, 1920);

Louis Hacker, *The World of Andrew Carnegie: 1865-1901* (Philadelphia: Lippincott, 1968);

William T. Hogan, S.J., *Economic History of the Iron and Steel Industry in the United States* (Lexington, Mass.: Lexington Books, 1971);

Paul Louis, *Success Unlimited: Great Industrialists of the World* (Centerville, Ohio: Orient House, 1973);

Joseph Wall, *Andrew Carnegie* (New York: Oxford University Press, 1970).

Archives:

Andrew Carnegie's personal papers are located in the Library of Congress, Washington, D.C.

Cast Crucible Steel Making

by Bruce E. Seely

Michigan Technological University

Most accounts of the American steel industry begin with the development of the Kelly/Bessemer process after 1860, but the initial effort to develop a domestic steel industry began at least three decades earlier when Americans began to adopt the cast crucible process. This technique was developed in England and was, in the period before Bessemer, the only method of producing a homogeneous steel. Americans only slowly mastered it, and the efforts of American entrepreneurs to learn this process provide a case study in the transatlantic transfer of technology that characterized so many aspects of American industrial development during the nineteenth century. Yet almost as soon as the process of producing cast crucible steel was finally adopted, it was overshadowed in importance by the new steel making technologies that appeared after 1870. Nonetheless, the crucible process remained the leading source of high-quality steels for tools and specialty purposes to the end of the century.

Cast crucible steel was developed in England by Benjamin Huntsman in the 1740s, and his process became the basis of Sheffield's preeminence as a steel-making center. Until Huntsman the only method of producing steel required converting (cementing) high-quality wrought iron bars into "blister" or cementation steel by packing the bars in chests surrounded by charcoal and heating the chests in a furnace for about ten days. This process permitted the iron to absorb the carbon that converted it into steel, but the absorption was uneven through the bar, as evidenced by swellings on the bar's surface which gave it the name "blister" steel. Huntsman succeeded in melting the blister steel in clay crucibles holding about 70 pounds of steel, then pouring the molten metal into ingots. Because the process yielded small batches of melted steel, with close attention to conditions Huntsman produced a uniform material of high quality. Yet Huntsman's process proved difficult to emulate out-

side of England, for it was demanding of both labor and materials. His success depended on the use of the best Swedish charcoal wrought iron, the discovery of special clay formulas for the crucibles, closely guarded formulas for mixing blister steel, scrap, and other materials in the crucibles, and British experience with coal-fuel technology—in this case, the combination of coke and special melting furnaces that permitted temperatures approaching 2,500 degrees Centigrade. Finally, Huntsman relied on the enormous skill of Sheffield's melters, who not only hefted crucibles weighing more than 150 pounds while enduring enormous heat at close range but also monitored the progress of steel making without the aid of measuring tools.

Americans made no attempt to copy Huntsman's process until the late 1820s, limiting their efforts to producing blister steel of indifferent quality. The requisite combination of skill and materials was elusive, and Sheffield cast steels enjoyed a reputation for quality that provided a tremendous competitive advantage. In 1831, 14 companies produced blister steel in the United States, but it was used exclusively for agricultural implements. British blister steel, which sold for 14¢ to 16¢ per pound as compared to 6¢ to 7¢ per pound for American steel, dominated other uses, while British cast steel captured all special uses, in spite of a premium price of 19¢ to 20¢ per pound. By the 1850s Americans annually imported an average of 13,000 tons of Sheffield cast steel, as Sheffield excellence strangled every American venture before 1860. Only with intense efforts did Americans erase this quality deficit after 1865, although an image of inferiority long shadowed domestic crucible steels.

This dependence on Sheffield was, of course, galling to American producers. The first challenge to England's monopoly in cast crucible steel was directed by William and John H. Garrard in Cincinnati from 1832 to 1837. Their steel proved satisfactory, but the depression of the late 1830s, the credit terms of English producers, and consumer preference for English steel halted the operation. Others soon repeated the Garrards' efforts, with Pittsburgh being the site of most of the early attempts to make cast crucible steel. Usually these attempts were launched by producers of blister steel eager to enter the high-quality market. Altogether, about a dozen different American companies made cast steel before 1859, with pioneering efforts made by G. & J. H. Shoenberger ironworks in 1840, Jones & Quigg in 1845, Coleman, Hailman & Company in 1845, Singer, Nimick & Company in 1853, and Isaac Jones, and later Anderson & Woods, in 1855. A few toolmakers, such as Tingle & Sugden and McKelvy & Blair in Pittsburgh, and Henry Disston in Philadelphia made steel for their files and saws and eventually for the general market. But in every case, the quality was inconsistent and failed to overcome the doubts of American users. The same problem bedeviled the Adirondack Iron & Steel Company of Jersey City, founded in 1848. Despite producing some excellent steel, the works foundered through two managers until the early 1860s, struggling against British firms with superior financial resources and products. A good indication of the problem was the decision of Singer, Nimick & Company of Pittsburgh to name their plant the Sheffield Works, although it is not clear whether from respect for the competition or as a subterfuge. The latter explanation is entirely plausible, for many American producers routinely doused their ingots with saltwater to give them the appearance of rusty Sheffield ingots that had crossed the Atlantic. British competition was not unique to the steel industry, but these difficulties affected cast steel makers longer than in many other areas of the economy.

After 1860 the American cast crucible steel industry finally began to match Sheffield quality. Three firms made this breakthrough, led by Hussey, Wells & Company of Pittsburgh, founded in 1859 and offering tool-quality steel in 1860. In 1862 Park, Brothers & Company founded the Black Diamond Steel Works in the same city and built a reputation for a superior quality product. Then in 1863 the moribund Adirondack Iron & Steel Company was revived in Jersey City. One obvious reason for these successes was the opportunity afforded by the Civil War, which disrupted imports as demand increased. Some credit also belongs to emigrant steelworkers from Sheffield. More important,

however, was Joseph Dixon's creation of a durable graphite crucible, first used by the Adirondack Iron & Steel Company. This melting pot eliminated one of the most obvious disadvantages faced by American producers, and all of the successful companies used Dixon's crucible or ones like it. Finally, the American industry located a source of high-quality wrought iron equal to the Swedish iron favored in Sheffield–charcoal iron made in traditional Catalan forges in the Lake Champlain region. It is worth noting, however, that both Hussey and Park entered the steel trade after successfully smelting and manufacturing copper, thereby gaining experience in working with metals and securing the capital necessary for supporting the steelworks during their initial money-losing experiments.

From this point the American industry grew quickly. Many plants were built on the East Coast, but firms in Pittsburgh dominated production; even in 1866 seven firms were in business there: Hussey, Wells & Company, Park Brothers & Company's Black Diamond Works; Singer, Nimick & Company's Sheffield Works; Anderson & Woods; Miller, Metcalf & Parkin's Crescent Works; Brown & Company's Wayne Iron and Steel Company; and Smith Brothers & Company's LaBelle Works. Moreover, perhaps the most important innovation in the industry, the Siemens gas-fired regenerative furnace for melting steel was further refined in Pittsburgh. By 1876, 10 of the 38 cast crucible steel makers in the country were located in Pittsburgh. Other large firms in the industry included James R. Thompson & Company; Gregory & Company; and D. G. Gautier & Company; all of Jersey City, Benjamin Atha & Company of Newark, New Jersey, and in Philadelphia, the Midvale Works; Disston's Keystone Saw, Tool, Steel, and File Works, and the Philadelphia Steel Works of William Baldwin. But Pittsburgh manufacturers accounted for about 27,000 tons of the 40,000 tons of crucible steel made in the United States in 1876, and nearly 52,000 tons of the 72,000 tons produced in 1880.

During the 1870s American output grew steadily, but from 1880 through 1897, demand was erratic and fluctuated between 60,000 tons and 90,000 tons. While some firms expanded, others found the 1880s a difficult time. After 1880 the cast crucible process was slowly overwhelmed by newer, large-scale production technologies. Even as annual tonnages increased, crucible steel's share of total steel production fell from about 20 percent in

1872 to about 5 percent in 1880 and to 1 percent in 1900. Especially threatening was the open-hearth process, which, because of its slower, more controllable conversion process, yielded uniformly high-quality steel in quantity. As open-hearth plants slowly appeared after 1880, markets previously held by crucible steel, such as locomotive boiler plate, car springs, and agricultural implements, were lost. This market shift coincided with the roller coaster business cycle of the Gilded Age to drive a large number of crucible steel makers, including the Adirondack Iron & Steel Company in Jersey City, to close their doors in the early 1880s. Those crucible steel firms that survived did so by adopting new technology; indeed, the first open-hearth furnace in the country was built in 1873 by Anderson & Woods of Pittsburgh. Within a decade Park, Brothers & Company, Hussey, Howe & Company, Singer, Nimick & Company, and several other large firms had built open-hearth facilities. To retain their markets, they substituted open-hearth steel for products that had always before required cast crucible steel. Also, they melted open-hearth steel instead of blister steel in their crucibles.

In an even more surprising move, crucible steel producers adopted Bessemer steel as a raw material. By 1880 both English and American producers had found that for certain uses, Bessemer steel could replace blister steel. One outgrowth of this shift was the 1879 establishment by several crucible steel producers of the Pittsburgh Bessemer Steel Company to supply their needs. In 15 months they erected the most modern Bessemer steel facility in the world at Homestead, Pennsylvania. Labor and market difficulties led the owners to sell the plant to Andrew Carnegie in 1883.

Obviously, the American crucible steel industry was changing radically. In an age that stressed quantity production, crucible steel was replaced outright by open-hearth or Bessemer steels, or the cheaper, mass-produced steels replaced blister steel. Yet even with these substitutions, in no year during the last quarter of the century did American crucible steel producers work at full capacity—about 110,000 tons by 1890. Not surprisingly, the number of producers declined while surviving firms grew larger.

To make matters worse, as American steel began to equal the British in quality, several Sheffield firms—William Butcher, Sanderson Brothers, and Thomas Firth—opened or purchased American plants to retain a hold on the American market. Moreover, English manufacturers' could still claim they produced superior steel, as Sheffield firms introduced alloy steels after 1870, a development that culminated in high-speed tool steels at the turn of the century. American steel makers responded to these British initiatives more quickly than in the past; indeed, American engineer Frederick W. Taylor pioneered high-speed tool steel. But Sheffield firms were the technical leaders in this field.

At the end of the century, then, the American cast crucible steel industry had developed impressively, but it faced a limited future. On the surface, the outlook was rosy, for crucible steel output peaked between 1898 and 1918, and annual production fell below 100,000 tons only twice. The best year was 1907, when 146,892 tons were produced. Yet these figures were deceiving. The market for cast crucible steel was shrinking; it could compete only when quality demands made price no object. Moreover, American steel had not fully emerged from the shadow of British leadership. The formation of the Crucible Steel Company of America in 1900 was thus the high point of this branch of the American steel industry as well as an act of desperation for a dying trade.

A trust organized largely by William and David Park of the Black Diamond Steel Works in Pittsburgh and capitalized at $50 million, the Crucible Steel Corporation brought together the 13 leading crucible steel makers. The corporation soon closed the oldest facilities, taking advantage of technological innovations that produced high-quality specialty steels without such back-breaking, highly skilled labor. Electric arc furnaces—Crucible Steel installed the first in Syracuse in 1906—completed the trend started by open-hearth furnaces and eventually displaced the crucible process. Although World War I held demand for crucible steel temporarily high, and production continued through 1945, the end was in sight by 1920. The cast crucible process has never attracted the attention of the more spectacular steel production processes, but it was an important method nonetheless. It was, above all, a product of the early Industrial Revolution, when the traditional linkage between human skill and quality still remained.

References:

K. C. Barraclough, *Steelmaking Before Bessemer*, volume 2: *Crucible Steel: The Growth of Technology* (London: Metals Society, 1984);

Victor S. Clark, *History of Manufactures in the United States, Volume II, 1860-1893* (New York: McGraw-Hill, 1929);

Harrison Gilmer, "Birth of the American Crucible Steel Industry," *The Western Pennsylvania Historical Magazine*, 36 (March 1953): 17-36;

J. S. Jeans, *Steel: Its History, Manufacture, Properties, and Uses* (New York: E. & F. N. Spon, 1880);

William P. Shinn, "Pittsburgh and Vicinity—A Brief Record of Seven Years Progress," *Transactions of the American Institute of Mining Engineers*, 14 (1885-1886): 662-663;

Shinn, "Pittsburgh—Its Resources and Surroundings," *Transactions of the American Institute of Mining Engineers*, 8 (1879-1880): 19;

James M. Swank, *History of the Manufacture of Iron in All Ages* (Philadelphia: American Iron and Steel Institute, 1892);

G. H. Thurston, *Pittsburgh and Allegheny in the Centennial Year* (Pittsburgh, 1876);

Geoffrey Tweedale, *Sheffield Steel and America: A Century of Commercial and Technological Interdependence, 1830-1930* (New York: Cambridge University Press, 1987);

Erasmus Wilson, ed., *Standard History of Pittsburgh, Pennsylvania* (Chicago: H. R. Cornell, 1898).

Catasauqua Manufacturing Company

by John W. Malsberger

Muhlenberg College

Organized in 1863 by the noted ironmaster David Thomas, who earlier had perfected the method for producing commercial quantities of anthracite iron, the Catasauqua Manufacturing Company was in many respects typical of the mid-nineteenth-century single-purpose iron manufacturers in America. The Catasauqua firm, which operated several rolling mills in the Lehigh Valley region of Pennsylvania, was established to fill a need created by the boom in anthracite iron production, in which its founder, David Thomas, had played a prominent role. By manufacturing a wide variety of finished cast iron products from the locally produced pig iron, the Catasauqua Manufacturing Company contributed significantly to the economic development of its region. Despite its initial success, however, the Catasauqua firm, like many other regional iron mills of its day, chose not to diversify into the manufacture of steel, concentrating instead exclusively on cast iron. This decision inevitably destroyed its ability to remain competitive in an industry dominated by the large, integrated iron and steel corporations of the late nineteenth century. Thus, by linking the era of anthracite iron production with that of the large, integrated steel corporations, the Catasauqua Manufacturing Company played only a supporting role in the tranformation of the nineteenth-century American iron and steel industry.

The origins of the Catasauqua Manufacturing Company date to February 20, 1863, when a group of local investors headed by David Thomas established the Northern Iron Company to construct a rolling mill in Catasauqua, Pennsylvania, on a site next to the works of the Lehigh Crane Iron Company. Capitalized initially at $100,000, the Northern Iron Company was organized to manufacture armor plate and iron rails for the booming market of the Civil War. Chosen as officers of the new concern were David Thomas, president; Charles G. Earp, secretary and treasurer; and David Eynon, general superintendent. Before the company could complete its manufacturing facilities, however, the Civil War ended and forced Northern Iron to retool in order to produce tank-, flue-, and boiler-plate to meet the demands of the civilian economy.

The expense of retooling, together with the generally depressed economic conditions following the Civil War, limited the success of the new firm's production. In an effort to remedy these problems, Northern Iron made a number of changes in 1866. The original general superintendent was replaced by a new man, William P. Hopkins, and more importantly, the firm increased its capitalization by $75,000 to enable it to diversify its products and improve its efficiency. Added to the production facilities in 1866 were an 18-inch bar iron train and a

10-inch guide-mill train. Northern Iron's expanded line of products was well received by the local market, and by the end of 1867 the firm's immediate economic success had been secured. Thus, only four years after its founding, David Thomas's rolling mill had, through adaptation and diversification, achieved an annual output of 6,000 tons of finished iron products, making it one of the region's leading iron mills.

Northern Iron continued its pattern of diversification and expansion throughout the late 1860s and early 1870s. In 1868, for instance, it leased a rolling mill at Ferndale (now Fullerton), Pennsylvania, a plant which had been constructed four years earlier by a competitor, the East Penn Iron Company (later the Lehigh Manufacturing Company). The Ferndale mill enabled David Thomas's company to expand its product line to include bar and skelp iron. To reflect its expanded facilities, Northern Iron was renamed in 1868 the Catasauqua Manufacturing Company. David Thomas continued to serve as the president of the firm until his 1879 retirement, when he was replaced by Oliver Williams, one of the principal stockholders. In 1872 the Catasauqua firm purchased the Ferndale mill outright, and three years later it assumed control of the Hope Rolling Mill in Allentown, Pennsylvania, permitting still further diversification of the company's product line to include rod, plate, and sheet iron. Finally, in 1878 the Catasauqua firm added a 6-ton steam hammer to the Ferndale mill.

As a result of the steady expansion and modernization of its physical plant, the Catasauqua Manufacturing Company had developed by the early 1880s into a major producer of finished cast iron products. Its annual output was more than 36,000 tons, valued in excess of $2 million. By this time, moreover, its labor force had grown to 600 with an average monthly payroll of $28,000. Indeed, the Catasauqua firm's efficiency and economic success had been so great that even during the generally depressed years of the 1870s it was not forced to shut down except for several minor labor strikes. Thus,

in the early 1880s the future of the Catasauqua Manufacturing Company must have seemed quite bright. Unfortunately, however, the rapidly changing nature of the American iron and steel industry quickly reversed the firm's fortunes.

Like many other regional iron companies of its era, the Catasauqua Manufacturing Company was apparently convinced by its early success that its markets were secure. Thus, in the 1870s and 1880s, when many other American iron manufacturers were adopting such technological advances as the Bessemer process and the Siemens-Martin open-hearth furnace to produce ever larger quantities of steel, the Catasauqua firm continued to concentrate exclusively on cast iron. Similarly, although many iron and steel producers in this era achieved significant economies of scale through vertical and horizontal integration, Catasauqua Manufacturing remained a relatively small manufacturer, serving only a regional market. Throughout the 1880s, as the demand for steel surged, firms such as the Catasauqua Manufacturing Company found themselves increasingly unable to compete against the integrated, national iron and steel corporations that had come to dominate the industry. With the market for its cast iron products eroded and its capacity to compete diminished, Catasauqua Manufacturing was forced to declare bankruptcy in 1892 as a result of both a protracted labor strike and the depressed economic conditions that culminated in the panic of 1893. Despite the firm's failure, its linkage with David Thomas assures it a place in the history of the iron and steel industry in the United States.

References:

Craig Bartholomew, "Anthracite Iron Making and Industrial Growth in the Lehigh Valley," *Proceedings*, Lehigh County Historical Society, 32 (1978): 129-183;

The Manufactories and Manufactures of Pennsylvania of the Nineteenth Century (Philadelphia: Galaxy Publishing, 1875);

Alfred Matthews and Austin N. Hungerford, *History of the Counties of Lehigh and Carbon* (Philadelphia: Everts & Richards, 1884).

Charcoal Fuel

by Paul F. Paskoff

Louisiana State University

Until well into the first half of the nineteenth century, charcoal was the primary fuel used in America to smelt iron ore into pig iron, to work pig iron into bar or wrought iron, and to produce nails, rails, and other finished iron goods. By the early eighteenth century, coal and, later, coke had replaced charcoal as the fuel used by the British iron industry because of diminishing wood supplies. The substitution of coal and coke for charcoal in the United States, however, occurred at a much later date because of the ready availability of comparatively inexpensive and abundant hardwood stocks and the preferences of iron producers and consumers for iron made from charcoal.

Charcoal was the product of a time-consuming process which required considerable skill and stamina from the men, called colliers, who made it. Typically, wood that had been cut into rough cordwood size and inspected for defects was hauled to circular coaling pits of 10 to 15 yards in diameter. There it was carefully arranged in a dome-shaped mound, 6 to 8 feet high, which was layered with sod into which vent holes were made. The mound of wood was then fired, or ignited, and was allowed to char for anywhere from three to ten days under the alert eyes of the colliers who watched the color and volume of the smoke coming from the vent holes and adjusted the burn accordingly. Because charcoal making also produced noxious fumes, colliers did their job at a considerable distance from the furnaces and forges which consumed the charcoal. Once the coaling process had been completed, the colliers dismantled the remnants of the mound and broke and raked the often still-glowing coals into several small, shallow mounds to allow them to cool and to minimize loss due to fire. The coals were then transported by wagon to a stone-walled coal house at the ironworks where the furnacemen, forgemen, and other

production workers could conveniently draw upon the stored fuel.

Charcoal making and woodcutting were the two jobs at any ironworks that required the largest number of workers, although their employment was decidedly seasonal. Because most forges and furnaces purchased at least a part of the wood for coaling from independent farmers, the main period for woodcutting was the late fall, winter, and early spring. The peak period for charcoal making, however, was the middle six months of the year, from May through October. The particular seasonality of the colliers' work was dictated by the fact that charcoal was best made under dry, calm conditions. Weather was therefore a factor of considerable importance in determining the quality and quantity of the charcoal.

Although the quantity of charcoal made from a cord of wood depended in large part on the weather and the colliers' skill, the basic determinant of the yield was the wood itself. The harder woods, such as hickory, oak, beech, ash, and black walnut, because of their tight grain, low water content, and resulting high specific gravity, made the best charcoal per cord but not, it should be noted, the most. The softer woods yielded about the same number of bushels per cord as did the harder woods, but the weight of the bushel of coal made from a soft wood was as little as half that of one made from a harder wood. However, most wood used to make charcoal was likely not to have been of optimal quality, particularly in those areas of the mid-Atlantic states which had long supported iron making and had therefore lost much of their hardwood stands. A dry cord of wood of average quality, chestnut, for example, made 30 bushels of charcoal, each of which weighed about 20 pounds. By way of contrast, a cord of hickory could yield 36 bushels of charcoal with a weight of 33 pounds

A completely covered charcoal burn (courtesy of Hagley Museum and Library)

A charcoal-burning pile in the process of being covered (courtesy of Hagley Museum and Library)

An unidentified, though typical, coal-blast charcoal furnace

each. At the other extreme was a wood such as white pine, from which 30 bushels of charcoal could be made, each of these weighing less than 16 pounds. Ultimately, the weight of a bushel of charcoal was the more important consideration to the ironmaster because it, rather than the volume of charcoal made from a cord, indicated the coal's quality, frangibility (that is, its susceptibility to being broken and pulverized in the furnace stack), and durability.

The use of charcoal in the iron industry began a relative decline with the introduction of the anthracite coal blast furnace in the 1840s and declined absolutely after 1850 as use of anthracite coal and, later, coke became widespread throughout the iron industry. Interest in the fuel among some ironmasters revived during the last quarter of the nineteenth century when changes in the architecture of the charcoal-fueled blast furnace permitted the construction of larger furnace stacks and, therefore, the production of large quantities of pig iron.

Unpublished Documents:

Brinton Collection, Hagley Museum and Library, Wilmington, Del. Photographs of charcoal making by N. R. Evans, Whiting, N.Y., 1939.

References:

Arthur C. Bining, *Pennsylvania Iron Manufacture in the Eighteenth Century* (1938; reprinted, Harrisburg: Pennsylvania Historical and Museum Commission, 1973);

Frederick Overman, *The Manufacture of Iron*, third edition (Philadelphia: H. C. Baird, 1854);

Paul F. Paskoff, *Industrial Evolution: Organization, Structure and Growth of the Pennsylvania Iron Industry, 1750-1860* (Baltimore: Johns Hopkins University Press, 1983).

Albert Huntington Chester

November 22, 1843-April 13, 1903)

by Terry S. Reynolds

Michigan Technological University

CAREER: Professor, Hamilton College (1870-1891); mining and geological consultant, Charlemagne Tower & Samuel Munson (1875, 1879-1880); chemist, New York Board of Health (1881-1890?); Professor, Rutgers University (1891-1907).

Albert Huntington Chester, the mining geologist whose explorations and reports convinced Charlemagne Tower to invest in opening the first of Minnesota's iron ranges, was born November 22, 1843, in Saratoga Springs, New York, the son of Albert Tracy and Elizabeth Stanley Chester. Chester was reared in New York and received his higher education at New York institutions, spending two years at Union College before transferring to Columbia, where he attended the School of Mines and graduated in 1868. The next year he married Alethea S. Rudd of New York City. In 1870 he was appointed professor of chemistry, mineralogy, and metallurgy at Hamilton College in Clinton, New York. He worked there from 1870 to 1891. In this same period he completed work on his Ph.D. from Columbia (1878) and served as a chemist with the New York State Board of Health from 1881 to about 1890.

In 1875 Chester was hired by Charlemagne Tower, at the recommendation of Tower's partner Samuel Munson, to lead an expedition to northeastern Minnesota, where he was to investigate reported iron ore deposits. Guided by Duluth, Minnesota, prospector and surveyor George Stuntz, Chester spent the summer of 1875 in the Minnesota wilderness collecting and analyzing samples. He collected samples both from the eastern end of the Mesabi Range and from the Vermilion Range slightly further north. The ore samples from the Mesabi Range were lean and did not seem to support the vague reports of mountains of iron in the area which had stimulated the explorations. On the

Albert Huntington Chester (courtesy of Hamilton College)

other hand Chester and Stuntz found very rich deposits of Bessemer grade ores near Vermilion Lake. Based on these results, Chester recommended that Tower ignore the Mesabi and concentrate additional exploratory efforts on only the Vermilion. Chester was unaware that he had investigated only the far eastern end of the Mesabi Range and that this range extended to the southwest for a considerable distance beyond the point of his explorations. Thus the much richer iron ores of the western Mesabi escaped his notice. Chester's negative judgment of the area of the Mesabi which he examined was accurate and supported by later explorations.

His ignorance, however, of the extent of the Mesabi Range and the richer deposits to the west delayed exploitation of its enormous ore deposits for a decade.

In 1879 Tower and Munson employed Chester to lead a second expedition. In the summer of 1880 Chester spent considerable time around Vermilion Lake examining the area he had recommended for further exploration. The glowing reports he submitted on iron deposits in the area led Tower to finance the sinking of Minnesota's first iron mine, the Soudan, in 1883 and 1884 and the construction of a railroad from Lake Superior to the mine.

In 1891 Chester resigned his position at Hamilton College to accept employment at Rutgers as professor of chemistry and mineralogy. Chester taught and conducted research at Rutgers until his death at New Brunswick, New Jersey, on April 13, 1903.

Although Chester's explorations in Minnesota missed the enormous iron ore deposits of the Mesabi Range, they did reveal significant, high-grade deposits on the Vermilion Range and provided the stimulus for the opening of the first Minnesota iron mine in 1883 and 1884.

Selected Publications:

The Iron Region of Central New York (Utica: Utica Mercantile and Manufacturing Association, 1881);

"The Iron Region of Northern Minnesota," *Annual Report of Minnesota Geological Survey*, 11 (1884): 154-167;

A Dictionary of the Names of Minerals, Including Their History and Etymology (New York: John Wiley & Sons; London: Chapman & Hall, 1896);

A Catalogue of Minerals, Alphabetically Arranged, With Their Chemical Compositions and Synonyms (New York: John Wiley & Sons, third edition, 1897).

References:

Hal Bridges, *Iron Millionaire: Life of Charlemagne Tower* (Philadelphia: University of Pennsylvania Press, 1952);

David A. Walker, *Iron Frontier: The Discovery and Early Development of Minnesota's Three Ranges* (St. Paul: Minnesota Historical Society Press, 1979).

Henry Chisholm

(April 22, 1822-May 9, 1881)

by Marc Harris

Ohio State University

CAREER: Carpenter (1839-1844); building contractor (1844-1854); partner, Chisholm & Jones (1857-1860); partner, Stone, Chisholm & Jones (1860-1864); partner and manager, Cleveland Rolling Mill Company (1864-1881).

Henry Chisholm, like so many prominent figures in the nineteenth-century American iron industry, was an aggressive and capable Scotsman with the ability to adapt to changing opportunities and new places. Like others, he was also fortunate in his business and family connections as well as in his personal gifts. He began his career in the United States as a building contractor, but within a few years moved into ironworking, and at the end of his life had built the Cleveland Rolling Mill Company into one of the industry's leaders.

Chisholm was born on April 22, 1822, in Lochgelly, Fifeshire, across the Firth of Forth from Edinburgh and not far from Andrew Carnegie's birthplace of Dunfermline. His brother William, a partner in the iron business, was born three years later. Stewart Chisholm, their father, worked as a mining contractor, but his career was cut short by his untimely death in 1832, throwing the family on diminished resources. Nevertheless, Henry and his brother both managed to complete their educations through the age of twelve, in the fine Scottish school system, at which time they were bound over as apprentices. Henry learned carpentry, and at seventeen went to Glasgow to practice his trade. The city was then a thriving commercial center and just beginning to develop the industries that would later make it the second most important city in the British Isles.

At the end of three years, however, Chisholm decided to pull up stakes, and on the eve of his twentieth birthday arrived in Montreal with his new wife and little else beyond his training. Montreal then served as the commercial capital of Britain's North American dominions, deeply involved in the growing lumber, grain, and staple trade between

Henry Chisholm

northern North America and Europe. When Chisholm arrived, the city's merchants were locked in a desperate battle with New York City's merchants for the trade of the Great Lakes states, a battle which by midcentury Montreal had obviously lost. But Chisholm managed to ply his trade as a journeyman carpenter for two years and then for five additional years contracted on his own account for various projects in and around the city.

In 1850 he and a Montreal associate with contacts in Cleveland won a contract to build a breakwater for a railroad terminus there, and Chisholm moved to Cleveland to supervise construction. The Cleveland & Pittsburgh Railroad's breakwall, intended to protect its coal dock, was an ambitious project of apparently unusual design and took three years to complete. For another four years Chisholm remained in the construction business, building piers, depots, and other commercial and railroad-related structures. It was probably through his work as a railroad contractor that Chisholm met Amasa Stone, Jr., who had built the Cleveland, Columbus & Cincinnati Railroad and acted as one of

its officers; he had begun to build his industrial empire.

In the meantime, Chisholm's brother William had been following a different path to Cleveland. William, not liking the dry-goods trade to which he had been apprenticed, took to the sea in 1840 and worked his way up to captaincy of a merchant ship in the Atlantic trade. In 1847, however, he settled in Montreal as a builder and contractor. He remained there until 1852, when he relocated in Cleveland to captain a new steamer partly owned by Henry. Following the wreck of that vessel he moved once again, this time to Pittsburgh. There he may have become acquainted with the iron industry; he took an interest in mechanical things, and his experience with the steamer may have piqued that interest.

But in 1857 he returned to Cleveland once again to join forces with his brother and with John and David Jones, experienced Welsh-born ironworkers, to build a rolling mill. The firm, Chisholm & Jones, was capitalized at $30,000, and the Chisholms were responsible for $25,000 of the total. It was one of the first such mills in the Cleveland area. It is possible that Henry's railroad connection encouraged him to invest in the mill, or perhaps he simply saw that Cleveland needed rolling-mill capacity to serve the several important new railroads in the city.

In 1860 Stone's brother Andros joined the firm and helped reorganize it into the partnership of Stone, Chisholm & Jones, which involved the three pairs of brothers. The new firm controlled the mills at Newburgh along with a blast furnace which Chisholm & Jones had installed in 1859 to ensure a steady supply of iron for their mills. Another new blast furnace was added when the firm reorganized. In 1864 the partners recapitalized and incorporated the business and brought all its properties under the aegis of the Cleveland Rolling Mill Company. William Chisholm had withdrawn from active management of the firm in 1860 in favor of further experimentation with iron and iron products, although he did retain his investment.

The leading share of management in the Cleveland Rolling Mill Company fell upon Henry Chisholm, and his management was in some ways adventurous as well as shrewd. The company's two blast furnaces, for example, were intended to be fueled by imported ores; the Cleveland area, unlike the Mahoning Valley and western Pennsylvania, lacked native ore supplies and had to import coal or other suitable fuels and limestone as well. The city's lakeside location favored such a scheme because it minimized handling and reshipping of ores and fuels. The company depended primarily on Lake Superior ores, newly available since the Soo canal opened at St. Mary's Rapids in 1855. For coal and limestone the company probably relied on Mahoning Valley supplies at first; later it would look to Connellsville coke. Chisholm's firm rapidly became a major integrated producer of iron and iron products.

Chisholm was also astute enough to realize that steel rails would dominate the rail business in short order, and accordingly he directed the firm's investment in a pair of Bessemer converters in 1868. It was the first continuously operating installation west of the Alleghenies and only the third successful one in the nation. Chisholm soon expanded the firm's steel capacity with more Bessemer converters and open-hearth furnaces, and it became a major producer of steel and steel products.

It did so because Chisholm insisted, as a principle of business, that he would never sell raw metal out of the furnaces and mills if he could help it. A pithy axiom compressed his point of view in typical Scottish manner: "Make up as much of your steel as possible. Do not sell it as raw material." In accordance with his basic policy Chisholm oversaw the firm's development of new products. Originally the steel capacity was intended to meet the business's main demand for rails, but in the years after the converters blew in, the inventive brothers developed several new lines for railroad, agricultural, and commercial use, as well as the early mainstay, rails. In this policy he resembled his fellow Scot and steel man Andrew Carnegie.

Many developments came about because of Chisholm's near obsession with efficiency. He quickly became dissatisfied with the amount of scrap left over from rolling rails and ran the stock through the company's rod mills. In so doing he solved a major problem in the steel industry—waste could run as high as 20 percent—and upset traditional notions about how steel could be worked. With refinements this process of producing rod and wire stock from Bessemer steel became the mainstay of Cleveland Rolling Mill's business later in the century. Chisholm's drive for further efficiency induced his superintendent, William Garrett, to developed a continuous mill that produced wire from

ingots with no reheating stages. Although Chisholm at first hesitated to adopt the new mill, the innovation allowed the firm to weather a huge reduction in the tariff on imported rod and wire stock; the firm was the only American producer able to do so. The mills also produced commercial and railroad shapes, plates, and spring steel, among many products.

Cleveland Rolling Mill served the Chisholms and the Stones as the center of a collaborative empire, the industrial side of which was based on Chisholm's planning. As early as 1863 Chisholm had decided that the railroad industry's growth would make Chicago an attractive base, and he established a rolling mill there under his son William's direction. In 1871 Chisholm, Stone, and other partners reorganized the works and installed a pair of Bessemer converters, the first in Chicago. Union Rolling Mill thereafter became a major steel and rail producer. As did the parent company, it produced its own iron from upper Great Lakes ores shipped by associates of the investors, converted it to steel, and rolled it in its own plants.

Henry's brother William, continuing his experiments, demonstrated in 1871 that Bessemer steel stock could be used to manufacture machine screws. To capitalize on this new outlet for rail scrap, the brothers established the Union Steel Screw Company. Other stockholders included the Stone brothers, Jephtha Wade of Western Union, and a Lake Superior mine operator named Fayette Brown.

The Chisholm enterprises achieved an important degree of integration in their businesses through partnerships and family connections. In this respect they represent a halfway point in the development of modern industry. Integrated operations increased the size of businesses and allowed industrialists to achieve large economies of scale, but partnership and family ownership were more typical of earlier and smaller businesses. The bureaucratic management structure which Carnegie adapted from railroads may have given his company some advantages over Chisholm's plan in coordinating operations. However, bureaucratic integration was not easily adopted universally, as it demanded enormous capital resources, and many parts of the iron industry remained partnerships for some time. This was particularly true of the Cleveland-based mining and shipping companies. Differences in form, however, should not obscure the fact that with his partners, Chisholm controlled ore and shipping, integrated iron- and steel-producing facilities, and access to rail transportation. In the business conditions prevailing after the Civil War, such control proved vital to competitive strength.

In all, Chisholm's own industrial investments amounted to some $10 million by 1875. To that must be added his brother's and his son's investments and some of the Stone brothers' holdings in order to arrive at an understanding of the value of his properties. Through these properties the partners exerted a major influence on the pattern of employment and of industrial growth in Cleveland.

Chisholm remained an entrepreneurial type and reputedly never developed any extravagant tastes or habits. Perhaps his religious commitment was significant; a devoted Baptist who belonged to the same congregation as John D. Rockefeller at the time Rockefeller began his consolidation of the oil industry, Chisholm gave a tenth of his possessions in 1871 as a contribution to a new church building. He also acted on the boards of several charity organizations as well as local businesses. Having made a success of himself in his adopted country, Chisholm was also a Scottish-American patriot. Burns's verse never failed to bring tears to his eyes, and for Cleveland's celebration of the national centennial in 1876, he donated on behalf of his companies a 168-foot flagpole of Bessemer steel rolled by the Cleveland Rolling Mill Company. Although a very robust man, he is thought to have scattered his energies and overworked himself, leading to his early death at age fifty-nine.

References:

William Garrett, "Landmarks in the Rolling Mill History of the United States," *Iron Trade Review* (November 14, 1901): 1-7;

Stephen L. Goodale, *Chronology of Iron and Steel* (Pittsburgh: Pittsburgh Iron & Steel Foundries, 1920);

"A Group of Cleveland Manufacturers," *Magazine of Western History*, 3 (1885-1886): 174-177;

William T. Hogan, *Economic History of the Iron and Steel Industry in the United States*, volume 1 (Lexington, Mass.: Lexington Books, 1971);

William Howe, *Historical Collections of Ohio*, centennial edition (Cincinnati, 1902);

Maurice Joblin, *Cleveland, Past and Present* (Cleveland: World, 1869);

James Harrison Kennedy, *A History of the City of Cleveland: Its Settlement, Rise, and Progress, 1796-1896* (Cleveland: World, 1896);

Emilius Randall and Daniel Ryan, *History of Ohio: The Rise and Progress of an American State* (New York: Century, 1912);

William Ganson Rose, *Cleveland: The Making of a City* (Cleveland & New York: World, 1950);

Harriet Taylor Upton, *History of the Western Reserve*, volume 1 (Chicago & New York: Lewis, 1910);

"William Chisholm," *Magazine of Western History*, 4 (1886): 247-250.

Pierre Chouteau, Jr.

(January 19, 1789-September 6, 1865)

by Terry S. Reynolds

Michigan Technological University

CAREER: Partner, Berthold & Chouteau (1813-1831); partner, Berthold, Chouteau & Company, (c. 1820-1831); partner, Bernard Pratte & Company, agents for Western Division, American Fur Company (1831-1834); manager, Western Division, American Fur Company (1834-1838); president, Pierre Chouteau, Jr. & Company (1838-1865); partner, Chouteau, Harrison & Vallé (1845-1865).

Pierre Chouteau, Jr., whose community status and financial resources were critical to the exploitation of Missouri iron ores and the establishment of a Missouri iron industry, was born in St. Louis, Missouri, on January 19, 1789. His father was Jean Pierre Chouteau, one of the founders of St. Louis and an early merchant and trader in the area.

Pierre, Jr. (also called Cadet [second-born]), received his early education from the village schoolmaster but before age sixteen had begun serving as his father's clerk. Shortly after, he worked as a clerk to Julien Dubuque at the lead mines of the upper Mississippi River valley. In 1809 he traveled with his father on an expedition for his St. Louis Missouri Fur Trading Company, and soon after he went into business on his own.

Between 1813 and 1831 he engaged in fur trading in partnership with his brother-in-law, Bartholomew Berthold. During this period he became a central figure in the fur trade in the upper Mississippi and Missouri river valley regions, transforming the hitherto haphazard business into a methodical and efficient system with fixed trading posts extending throughout the length and breadth of the region. This organization enabled him to fight John Jacob Astor's American Fur Company to a standoff. In 1831 Astor convinced Chouteau to join his American Fur Company, the Western Division, which was then placed under his control. Under Chouteau's direction the Western Division ruthlessly drove all other major fur companies out

Pierre Chouteau, Jr. (engraving by A. H. Ritchie, courtesy of Missouri Historical Society)

of the Upper Missouri. In 1834 Chouteau purchased Astor's interest in the Western Division, and in 1838 he organized the firm of Pierre Chouteau, Jr. & Company to run his enterprises. Chouteau's fur trading firms for some years had a virtual monopoly on trade in the upper Missouri area.

Tall and strong, Pierre, Jr., had a taste for adventure and, in his younger days, often accompanied the fur-trading expeditions which he sponsored. Although he had a somewhat sober countenance, he became animated and cheerful in conversation. He was able to move in polished society as well as hold the respect of unlettered hunters and trappers. Throughout his life he gave generously of his services to scientific expeditions and government surveys of the American West.

As Chouteau, Jr.'s business enterprises grew, he was drawn into other investments. In 1841 he established a branch of his fur business in New York and soon moved there. In the 1840s and early 1850s he established a commission house in New York, as well as a firm for selling iron. For a period he was a major financier.

Several of Chouteau's investments in the 1840s and early 1850s were directed toward the exploitation of Missouri iron resources. In 1843 he worked with James Harrison on the organization of the American Iron Mountain Company, which began to mine the iron deposits of the then famous Iron Mountain region southeast of St. Louis. In 1850, as part of the firm of Chouteau, Harrison & Vallé, he participated in the erection of a large rolling mill in north St. Louis. In 1865 the same firm purchased membership in the Kelly Process Company with the intention of entering steel production.

Although Chouteau directed his business empire from New York for a time, he returned to St. Louis in his old age and died there, a multimillionaire, on September 6, 1865. Although Pierre Chouteau never assumed close management of his iron enterprises as he did with his fur-trading empire, his community standing and financial power were crucial to the creation of Missouri's iron industry.

References:
Howard L. Conard, *Encyclopedia of the History of Missouri*, volume 1 (New York, Louisville-St. Louis: Southern Historical Co., 1901), pp. 590-591;

Bernard DeVoto, *Across the Wide Missouri* (Boston: Houghton Mifflin, 1947);

Thomas Scharf, *History of Saint Louis City and County*, volume 1 (Philadelphia: Louis H. Everts, 1883);

James M. Swank, *History of the Manufacture of Iron in All Ages* (Philadelphia: Privately printed, 1884).

Archives:
Chouteau's papers are located in the Missouri Historical Society, St. Louis, Missouri.

Cleveland-Cliffs Iron Company

by Terry S. Reynolds

Michigan Technological University

The Cleveland-Cliffs Iron (CCI) Company, one of the premier mining companies of the Michigan iron ranges, was incorporated under West Virginia laws for tax purposes on May 7, 1891. It merged with The Cleveland Iron Mining Company and the Iron Cliffs Company, two of the largest producers on the Marquette iron range in the Lake Superior district. The merger had been promoted for several years by Jeptha H. Wade, Sr., a director in both firms, and negotiated by Samuel Livingston Mather, the head of The Cleveland Iron Mining Company. The new firm was, from its creation, the leading iron mining company on the Marquette Range, a position it has held to the present. CCI was, moreover, one of the largest of the independent ore producers after the creation of United States Steel in 1901.

By the mid 1880s iron mining had become an expensive proposition on the Michigan ore ranges. The easy-to-reach ores had been depleted and capital-intensive underground mining had become the rule rather than the exception. Moreover, partially as a result of the opening of new iron ranges in Michigan, Wisconsin, and Minnesota, ore prices in the mid 1880s declined. High costs and low prices favored large, efficiently operated companies able to reduce costs through ownership of their own ore carriers and railroads or those able, through their bulk shipping power, to negotiate lower rates with ore carriers. The deep financial depression which began in 1893 exacerbated problems for small companies and prompted a series of consolidations culminating with the formation of United States Steel in 1901. The creation of CCI in 1891 through the merger of two of the biggest producers on the Marquette Range can be considered as the first of these major consolidations. The new organization controlled nearly a dozen iron mines, several charcoal blast furnaces, almost 60,000 acres of land, and two modern lake ore freighters.

Under the leadership of William Gwinn Mather, son of Samuel Livingston Mather, CCI was able to maintain its prominent position in the Lake Superior mining district. But this position was not held without initial problems. Shortly after the consolidation the company was hit by the depression of 1893. CCI stock dropped from 100 to 25, and the company was forced to close many of its mines and substantially reduce its work force. Moreover, in August 1895 the company was hit by a major strike, which ended only after Michigan's governor, at Mather's urging, ordered five companies of Michigan militia into the Marquette district to protect the company's properties. CCI's size and strength, however, enabled it to weather the crises of the 1890s when many smaller companies collapsed. By 1900 its annual ore production was above one million tons.

To forestall further labor difficulties CCI, under Mather's direction, embarked in 1898 on a far-reaching policy of corporate paternalism. In 1905 the company created a formal welfare department to administer sick, injury, and death benefits. In 1909 the company initiated one of the earliest pension plans in the iron mining industry. In 1910 it created one of the first safety programs in the industry, and the following year created a corporate safety department and organized the first mine rescue team in the Lake Superior iron mining district. These services were supplemented by an extensive visiting nurse program and evening English literacy programs for its largely foreign-born labor force. Moreover, by 1915 CCI either owned or was affiliated with hospitals in Ishpeming, Negaunee, Gwinn, and Republic, Michigan, and Nashwauk, Minnesota.

The apex of CCI's corporate paternalism was the creation of the model town of Gwinn, Michigan. Following mineral explorations which had begun in 1902, CCI decided to open several mines

The land office of the Cleveland-Cliffs Iron Company (courtesy of Marquette County Historical Society)

about 20 miles south of Marquette. To house the work force which the company needed, CCI hired a professional landscape architect, Warren H. Manning of Boston, to lay out a new town. When completed in 1908 the company town of Gwinn had broad, landscaped avenues, a sewer system, a first-class company-built school, three company-built churches, an outdoor swimming area, a company-built hospital, and a company-built clubhouse for recreation which included reading rooms, pool tables, baths, lockers, a gymnasium, and a bowling alley. The houses CCI constructed in the new town were substantial structures and avoided the sameness of design typical of most mining towns. These dwellings were sold to employees at cost. The company also erected the town's commercial buildings.

At the same time CCI moved to solve its labor problems, it sought to strengthen its transportation system. To cut overland shipping costs CCI and the Pittsburgh & Lake Angeline Iron Company jointly built the 16-mile Lake Superior & Ishpeming (LS&I) Railroad from the CCI mines to a new dock erected at Presque Isle, on the north side of Marquette. Completed in 1896, the LS&I quickly became a major ore carrier. In 1897, for example, it carried more than 1 million tons of iron ore. LS&I

long remained one of the major Lake Superior district ore railroads, steadily expanding its lines and services. CCI also expanded the ore fleet it had inherited from The Cleveland Iron Mining Company. In both 1892, the year after CCI was created, and 1893 it contracted for the construction of ore freighters. Later in the decade CCI officials organized the Grand Island Steamship Company and the Presque Isle Transportation Company to operate ore-carrying vessels for the company. By 1900 CCI controlled ten steamers and barges and expansion continued. In 1905, for example, it added two more new vessels, and in 1906 three.

As it attempted to mollify its labor force and cut costs through control of its own transportation system, CCI also attempted to reduce the costs and increase the profits of its mining and smelting operations. In 1896 it erected a by-products chemical plant adjacent to its charcoal blast furnace at Gladstone, Michigan, and in 1903 erected another by-products plant at the Pioneer Furnace in Marquette. By linking the manufacture of charcoal iron to the production of chemical by-products, CCI was able to sell charcoal iron at a smaller premium over ordinary coke iron and thus remain competitive in charcoal iron production longer than most other companies. In 1892 CCI pioneered the application of electricity to iron mining operations, introducing underground electric haulage equipment at its Lake Mine to reduce labor costs. Between 1910 and 1916 it constructed five hydroelectric generating facilities, mostly along the Carp and Dead rivers. The subsidiary company which operated the stations, Cliffs Power and Light, provided over 90 percent of the electric power needs of CCI and until 1953, when it was sold, was one of the major electric power producers on Michigan's Upper Peninsula.

As economic conditions improved, CCI steadily expanded its mineral and landholdings. In 1900 the company organized a geology department for exploration purposes and shortly after hired a professional geologist to guide these activities. Initially, CCI did not join the rush to claim lands on the Mesabi Range because of the faith and the mineral reserves it had in the Marquette Range. Around 1900, however, the company began acquiring new properties in several districts, including the Mesabi Range. In 1900 it purchased the Munising Railroad Company, also acquiring nearly 85,000 acres of hardwood timber, most of it just east of the Marquette Range. In 1902 it leased its first property on the Mesabi, the Crosby Mine. That same year it purchased the Maas Mine on the Marquette Range. In 1905 CCI acquired the Jackson Iron Company, one of the original companies operating on the Upper Peninsula, and at about the same time began operating the Negaunee Mine under contract. As previously noted, CCI also opened several mines in the Gwinn district, south of Marquette, in the first decade of the twentieth century. Between 1891 and 1913 CCI landholdings jumped from under 60,000 acres to more than 714,000. And by 1920 the company owned or operated 28 iron mines and 24 ore carriers.

Because CCI primary holdings were on the Marquette Range, where mining by the 1890s was already largely underground, CCI's production costs were significantly higher than the production costs at the new, open-pit mines, which opened on the Mesabi Range in the 1890s. CCI was able to survive and even flourish, not only in the 1890s, but in the twentieth century, in the face of this competition for several reasons. First, the company made systematic efforts to keep its ore production costs to a minimum by pioneering in such areas as mine electrification and labor welfare practices. Second, as outlined above, the company kept its transportation costs low by acquiring its own railroad and ore fleet. Third, transportation costs from CCI mines to the lower Great Lakes' ports were lower than from the Mesabi Range because of the shorter rail haul from mine to port and the shorter water route to Cleveland. Fourth, Marquette Range ores were hard and high quality. They thus commanded a price premium over the softer ores of the Mesabi, which often could be used only when mixed with harder ores. Finally, Marquette Range mines were not burdened with royalty rates and state taxes as high as those of the Mesabi Range mines. These factors and good corporate management enabled CCI to mine underground yet still remain competitive both in the 1890s and well into the twentieth century.

Unpublished Document:

Burton Boyum, "Cliffs Illustrated History," unpublished manuscript, c. 1986, in the Michigan Iron Industry Museum, Negaunee, Mich.

References:

"A Bond of Interest," *Harlow's Wooden Man* Marquette County Historical Society, volume 13, no. 5 (Fall 1978);

S. R. Elliott, "The Cleveland-Cliffs Iron Company: History and Organization," in *The Cleveland-Cliffs Iron Company and Its Extensive Operations in the Lake Superior District* (Cleveland: Cleveland-Cliffs Iron Company, 1929), pp. 5-8;

H. Stuart Harrison, "The Cleveland-Cliffs Iron Company," *Transactions of the Newcomen Society in North America*, 44, no. 6 (1974): 8-12;

Harlan Hatcher, *A Century of Iron and Men* (Indianapolis: Bobbs-Merrill, 1950).

The Cleveland Iron Mining Company

by Terry S. Reynolds

Michigan Technological University

The Cleveland Iron Mining Company was one of the two "parents" of the Cleveland-Cliffs Iron Company and one of the most important early mining companies on the Marquette Range.

Attracted by reports of mineral wealth, especially copper and silver, on Michigan's Upper Peninsula, a group of Cleveland merchants and professional men formed the Dead River Silver and Copper Mining Company of Cleveland in the winter of 1845-1846. The group hired J. Lang Cassels, dean of the new Western Reserve Medical School and an expert mineralogist, to travel to the Upper Peninsula and explore for minerals. At Sault Ste. Marie, while en route to explore for silver and copper, Cassels met Abraham Berry, president of the Jackson Iron Company. Berry persuaded Cassels to consider iron properties near those of his company, apparently in the hope that the two companies could jointly build a road from their mines through thick forests and rough terrain to Lake Superior. After looking at the territory, Cassels in 1846 took "squatter's possession" of a square mile of land for his backers, and on his return to Cleveland persuaded them to transfer the focus of their activities to iron.

On November 9, 1847, 11 men, including the backers of Cassels' expedition, formed the Cleveland Iron Company, and shortly after began to develop their property. Some work was done on the site in 1848, but a conflict with the Marquette Iron Company over landownership clouded affairs. After the courts in late 1850 awarded title to the Cleveland company, more extensive work began. At first the company supplied ore for the forge of its old rival, the Marquette Iron Company. When the Marquette Iron Company made the first commercial shipment of iron ore from the Upper Peninsula in July 1852, the ore shipped had been originally mined by the Cleveland company.

In 1850 the Cleveland Iron Company was reincorporated as the Cleveland Iron Mining Company of Michigan to meet Michigan laws but soon had to reorganize again. By 1852 it was clear to the owners of the Marquette Iron Company that they could not smelt, forge, and transport bar iron to markets in the East and sell it at a profit. Thus the principal owner approached the Cleveland Iron Mining Company in 1852 to determine if they wished to purchase his company. In March 1853 the Cleveland Iron Mining Company reincorporated as The Cleveland Iron Mining Company, increasing its capitalization from $300,000 to $500,000 in order to purchase the Marquette Iron Company.

Although the company struggled in the 1850s, it recognized early that the sale of iron ore rather than the production of bar iron was the path to profit. Thus when the forge the company had purchased from the Marquette Iron Company burned in late 1853, it was not rebuilt. Between 1853 and 1855 the company cooperated with the Jackson Iron Company in constructing a plank road linking their lines with Marquette harbor, providing the first reliable overland transportation from mine to port. Production at the company's Cleveland Mine first passed the 10,000-ton mark in 1857. The coming of the Civil War and higher iron prices finally brought prosperity. The company paid its first dividend in 1862 and paid large dividends for several years after. By the early 1860s The Cleveland Iron Mining Company was producing over 40,000 tons annually, and it surpassed the 100,000 ton mark in 1868.

The driving spirit behind the success of The Cleveland Iron Mining Company was Samuel Livingston Mather, the company's first secretary and treasurer, and its president from 1869 to 1890. Largely due to his influence, The Cleveland Iron Mining Company was a leading producer and innovator in the Great Lakes iron ore district through most of

The Cleveland mine on the Marquette range, date unknown (courtesy of Marquette County Historical Society)

the late nineteenth century. In 1855, when the opening of the Sault Canal provided an all-water route from Marquette to markets to the south and east, the first ore-carrying ship to move through the canal carried The Cleveland Iron Mining Company ore. That year the company shipped 1,447 tons, the only company to ship ore through the new locks. In 1859 the company completed one of the first substantial ore-docking facilities at Marquette. Earlier than other companies, the firm sought control of its own shipping facilities, purchasing in 1867 half interest in a sailing vessel. In 1872 the leading stockholders of The Cleveland Iron Mining Company played a major role in forming the Cleveland Transportation Company. This company controlled a fleet of four steamers and four schooner barges and worked under long-term contract to The Cleveland Iron Mining Company until 1886 or 1887. In 1888 the company elected to build its own fleet, the first iron mining company to do so, contracting for the construction of two of the largest and fastest lake steamers of the period. In addition, The Cleveland Iron Mining Company was the first iron mining company to use dynamite in its mines, instead of the traditional black powder, and it was usually among the leaders in introducing new mining machinery to reduce dependence on labor. In the 1870s the company was an early user of the diamond drill for exploration.

Depressed ore prices and the growing costs of underground mining had convinced Mather by the mid 1880s that only large and efficient companies could survive in the iron mining business. Thus when the Iron Cliffs Company, one of his most powerful rivals on the Marquette Range, faltered and when its owners showed an interest in selling, Mather initiated the negotiations and the series of corporate maneuvers which culminated in the formation of the Cleveland-Cliffs Iron Company on May 7, 1891. Although the formal corporate existence of the Cleveland Iron Mining Company did not come to an end until 1914, the creation of Cleveland-Cliffs as owner of both The Cleveland Iron Mining Company and the Iron Cliffs Company essentially terminated the company's existence.

References:

Burton Boyum, "Cliffs Illustrated History," unpublished manuscript, c. 1986, copy in Michigan Iron Industry Museum, Negaunee, Mich.;

H. Stuart Harrison, "The Cleveland-Cliffs Iron Company," *Transactions of the Newcomen Society in North America*, 44, no. 6 (1974): 8-12;

Harlan Hatcher, *A Century of Iron and Men* (Indianapolis: Bobbs-Merrill, 1950);

Raphael Pumpelly, T. B. Brooks, and Adolf Schmidt, *Iron Ores of Missouri and Michigan* (New York: Putnam's, 1876).

Cleveland Rolling Mill Company

by Marc Harris

Ohio State University

The Cleveland Rolling Mill Company, an integrated producer of pig iron and Bessemer steel and steel products, was an early innovator in the production and use of steel and the focal point of a Great Lakes industrial empire controlled by the Chisholm and Stone families. One of its forerunner firms, the Railroad Rolling Mill, was established in 1856 by A. J. Smith in partnership with others. In 1857 the company's major precursor took shape in Newburgh (an early competitor to Cleveland, south on the Cuyahoga River) as a partnership between Henry Chisholm, an immigrant Scottish construction contractor, and entrepreneurs John Jones and David Jones. Of the firm's original $30,000 of invested capital, the Welsh-born Jones brothers put up $5,000; their contribution to the business came primarily in their knowledge of iron making.

Both the Railroad Mill and the Chisholm & Jones mill were established to reroll worn rails, a necessary and prospectively lucrative procedure in the early days of the railroads. Iron rails were soft and wore very quickly—rails might last anywhere from six months to four years, according to some estimates—and railroads were consequently forced to replace them fairly often. The old rails were then recycled by simply hot-rolling them into oblong bars and then re-forming the bars into rails. This procedure did not require particularly large capital investment. Henry Chisholm abhorred waste and reliance on a single product, so the Chisholm & Jones mill also produced merchant iron and bar products. In 1859 the partners augmented their independence from outside suppliers when they invested in a blast furnace.

The Cuyahoga Valley was then an unusual place to build a furnace as it had no access to local ore or charcoal. Blast furnaces operated on the Western Reserve as early as 1803, but their sites had been determined by availability of nearby ore deposits and charcoal supplies. Cleveland Rolling Mill's

partners pioneered in taking advantage of new transportation routes, including the Sault Sainte Marie canal and the Cleveland & Pittsburgh Railroad, to feed its furnace with newly proven ores from the Lake Superior region along with Mahoning Valley block coal and later Connellsville coke.

In 1860 a new partner entered the firm and helped bring about major changes. Andros B. Stone brought into the business his own capacities and capital and, as well, the capital and business connections of his brother and partner Amasa, Jr., a prominent railroad man and industrial investor. The Stone interests involved several important railroads, including the Cleveland, Columbus & Cincinnati and what would later become the Lake Shore & Michigan Southern. Both roads served as strategic links in the New York Central system's midwest-to-east transportation network, and the new partnership of Stone, Chisholm & Jones was in a good position to serve their demands for iron. With Stone's infusion of capital, the firm invested in another blast furnace and enlarged its mills. It also organized a Chicago branch plant in 1863. In 1864 the partnership reorganized as the Cleveland Rolling Mill Company, capitalized at $500,000, and purchased the Railroad Rolling Mill's facilities. Incorporators included Chisholm, A. B. Stone, Amasa Stone, Stillman Witt (a contractor partner of Amasa Stone), Jephtha Wade of Western Union, and H. B. Payne, a Rockefeller associate whose selection as a senator in 1884 caused great controversy. Amasa Stone and Henry Chisholm between themselves held nearly half the shares outstanding.

Chisholm's hand guided the company's investments in new furnaces and, later, in Bessemer converters. He believed strongly that Cleveland Rolling Mill should control as much of the manufacturing process as possible; he was also convinced very early on that steel rails would prevail over iron. Iron from the new furnaces was first used on the riding surfaces of T-rails milled from old stock, a practice that improved their wear characteristics, and was also rolled into bar stock and made into several salable end products. In late 1868, at Chisholm's instigation and under Alexander Holley's direction, the company installed two Bessemer converters. This was only the second Bessemer installation west of Pittsburgh and the first to operate continuously. Two more converters were added within three years, and the company also replaced its original converters in 1875 with three open-hearth furnaces. With these installations the company had the capacity to produce 110,000 tons per annum of Bessemer steel and 40,000 tons of open-hearth steel, and by 1880 was the largest producer in the Cleveland area. By 1888 the firm employed some 5,000 people locally.

At Chisholm's insistence, and with the inventive aid of his brother William, the company followed an aggressive policy of producing a wide range of finished products for sale. Steel rails naturally formed a large proportion of the mill's output, and iron rails continued to be produced for some time, although in decreasing quantities. In addition, the mill also rolled plate steel, steel wire, spring steel, and agricultural and merchant stock.

The company was the first to produce rod and wire stock from Bessemer steel, an 1870 innovation attributable to Henry Chisholm. In the process of rolling steel rails, as much as 20 percent of the ingot weight was normally discarded as scrap. Chisholm, like others in the industry, was anxious to find a use for this material and as a newcomer to the field was not bound by earlier notions of how steel could be processed. Perhaps fortuitously, he was faced with a problem in rolling iron rods and hit on the solution of using the iron-rolling mill to roll steel rods from scrap stock. The firm thereafter became an industry leader in the production of wire and rod products of all kinds, both in volume and in technique.

In 1881 the wire rod industry was rocked by a 75 percent decrease in the tariff on imported rods, from 1.2 cents per pound to 0.3 cents. The Cleveland Rolling Mill Company was the only American producer to keep its plant open, and it was able to do so by rigorously eliminating intermediate processing involved in producing rods from ingots. In 1882 the company hurriedly introduced the key device in this drive, the Garrett rod mill. Designed by superintendent William Garrett, this type of con-

tinuous mill allowed rod stock to be rolled directly from ingots, thus bypassing intermediate blooming stages. It was also long enough to allow several pieces to be processed at the same time. Originally Garrett granted Cleveland Rolling Mill exclusive rights to his process, but he cancelled the agreement over a monetary dispute and instead installed the mills in several competing plants that were later to buy out his original employer.

In addition to its forward integration to the production of commodities and its early backward integration into iron and steel production, the company benefited by access to a number of holdings in coal and ore which were controlled by the Chisholm interests or their partners, primarily through Amasa Stone. This was part of a larger pattern of very close cooperation and some overlap in ownership among iron and steel producers, operators of the Lake Superior ore beds, and Great Lakes ore shipping firms. Such integration through family ownership allowed the Chisholms and the Stones to control many aspects of production in very much the same way Carnegie eventually did.

Bringing in $25 million annually in the late 1880s, Cleveland Rolling Mill was the center of an important industrial empire that might, it is said, have rivaled Carnegie Steel if Henry Chisholm had lived beyond his fifty-ninth year. At the height of its independent existence near the turn of the twentieth century, the firm employed more than 8,000 men. It provided capital and experience which the Chisholm and Stone brothers drew upon to form several other major enterprises in the iron and steel industry. One of these, the Union Rolling Mill Company of Chicago, had begun as a branch plant and was reorganized and expanded in 1871. Managed by Henry Chisholm's son William, it became an important producer in the Chicago region; it too relied on Lake Superior ore controlled by related interests. Later it was absorbed as an integral part of the Illinois Steel combine. Other related companies included the Union Steel Screw Company of Cleveland and William Chisholm & Sons, which manufactured steel shovels and other products.

In 1899, during the first wave of mergers and reorganizations, the Cleveland Rolling Mill Company was absorbed into the American Steel & Wire trust of New Jersey. Its iron- and steel-making facilities furnished the trust's other Cleveland-area mills with raw material. American Steel & Wire was itself absorbed into J. P. Morgan's giant United

States Steel combine when it was organized in 1901.

References:

William Garrett, "Landmarks in the Rolling Mill History of the United States," *Iron Trade Review* (November 14, 1901): 1-7;

Stephen L. Goodale, compiler, *Chronology of Iron and Steel* (Pittsburgh: Pittsburgh Iron & Steel Foundries, 1920);

Harlan Hatcher, *The Western Reserve: The Story of New Connecticut in Ohio* (Indianapolis & New York: Bobbs-Merrill, 1949);

William T. Hogan, *Economic History of the Iron and Steel Industry in the United States* (Lexington, Mass.: Lexington Books, 1971);

Emilius O. Randall and Daniel J. Ryan, *History of Ohio* (New York: Century History, 1912);

William Ganson Rose, *Cleveland: The Making of a City* (Cleveland & New York: World Publishing, 1950).

Colorado Coal & Iron Company

by Paul F. Paskoff

Louisiana State University

Colorado Coal & Iron Company, located near Pueblo, Colorado, was founded by A. H. Danforth in 1880. A railroad man, Danforth had been drawn in the latter part of the 1870s to the Pueblo district by its substantial coal deposits and large tracts of available land. He retained this interest even after the Colorado Coal & Iron Company proved to be profitable as an iron- and steelworks. In regarding the firm in this light, Danforth was not simply holding to a railroad man's suspicion of manufacturing enterprises. The fact was that the making of iron and steel yielded a smaller return on investment than did the company's land and mining ventures. Still, Danforth persevered for a time.

The company erected its first coke blast furnace in 1881 and started to make steel on September 7 of that year. In 1882, after almost two years of work, the construction of a Bessemer steel plant was completed. These undertakings were soon followed by additional works so that by 1885 the company's physical plant embraced two blast furnaces, a Bessemer steelworks, a nail works, and a rolling mill for making steel rails for the railroad industry.

Danforth's intimate knowledge of the railroad industry and his realization that, upon completion, his rail mill would be the only such facility west of the Missouri River—and therefore insulated from the competition of the eastern rail mills—undoubtedly encouraged him to bring the mill to a swift completion. In any event, he was right to push its construction. The company soon found a market for its rails and began to sell them to the Denver & Rio Grande Railroad. Before too long, however, other firms began to compete for this lucrative market, and, with their appearance, the initial advantage enjoyed by any innovator began to erode.

In 1892 the Colorado Coal & Iron Company merged with the Colorado Fuel Company, an anthracite coal mining firm founded by J. C. Osgood, to form the Colorado Fuel & Iron Company. Led by Osgood, the combined enterprise significantly enlarged its physical plant and expanded its operations. Its enlarged capacity to produce rails was sufficiently great to have induced the Rail Association, an industry pool formed in 1887 to fix the supply and price of rails, to pay the company not to make any. The disruption of the association in 1895 put a temporary end to price stability in the rail market and, for a while, pitted Colorado Fuel & Iron Company against such giants as Carnegie Steel Company and John W. "Bet-a-Million" Gates's Illinois Steel Company.

The reestablishment of the rail pool in 1899 brought renewed stability to the market. By 1900 Colorado Fuel & Iron Company had $46 million in capitalization, a modernized physical plant, and about 15,000 workers, all of which made it an attractive target for a corporate takeover. An attempt to wrest ownership of the company from Osgood by E. H. Harriman and John W. Gates was not

The Pueblo, Colorado, Bessemer works of the Colorado Coal & Iron Company in 1892 (courtesy of Colorado Historical Society)

long in coming. Osgood was able to turn back this assault in 1900, thanks largely to the backing which he received from John D. Rockefeller, Jr., and George J. Gould. Ironically, it was to these two men, to whom Osgood had turned in desperation to defeat the earlier takeover of his company, that he lost the company not long after they had come to his assistance.

References:

Naomi R. Lamoreaux, *The Great Merger Movement in American Business, 1895-1904* (New York: Cambridge University Press, 1985);

Fritz Redlich, *History of American Business Leaders: A Series of Studies,* 2 volumes (Ann Arbor, Mich.: Edwards Brothers, 1940);

James M. Swank, *History of the Manufacture of Iron in all Ages* (Philadelphia: American Iron and Steel Association, 1892);

Peter Temin, *Iron and Steel in Nineteenth-Century America: An Economic Inquiry* (Cambridge, Mass.: M.I.T. Press, 1964).

Peter Cooper

(February 12, 1791-April 4, 1883)

by John A. Heitmann

University of Dayton

CAREER: Apprentice, John Woodward (1808-1810); manufacturer, Hempstead, New York (1810-1815); grocer, Manhattan (1815-1821); owner, glue factory (1821-1828); partner, Canton Iron Works (1828-1836); owner, Cooper Iron Works (1838-1845); owner, Trenton Iron Works (1845-1883).

Peter Cooper, an inventor, manufacturer, and philanthropist, was born in New York City on February 12, 1791, and died there April 4, 1883. As an old man Cooper looked back upon his life as being divided into three periods: 30 years to get started, 30 years to gain a fortune, and 30 years to wisely give away what he had worked so hard to earn. From the pages of his autobiography it almost seems as if Cooper had planned his career that way. To what degree Cooper controlled his fate is a matter of debate, but his success was no accident. His wealth and achievements in business can be ultimately traced to his perseverance, thrift, mechanical ingenuity, and above all an uncanny sense of knowing and responding to the market.

Cooper's father was a hatter who had served in the Continental army and who had resumed his trade in New York City after the conclusion of the Revolutionary War. One of Cooper's earliest memories was that of helping his father make hats by pulling hair out of rabbit skins. The elder Cooper, unlike his son, never became successful in the various business ventures that he attempted. As a result of these false hopes and trying failures, the Cooper family moved from Peekskill to Catskill to Brooklyn and finally to Newburgh, as the father vainly tried to manufacture profitably various products from brewing ale to making bricks and hats. As a result of economic circumstance and the unsettled nature of the Cooper family during these early years, Peter had little time to acquire a formal education. It would be this perceived disadvantage that later in

Peter Cooper

life dominated Cooper's thoughts concerning philanthropy and learning.

In 1808 Cooper was apprenticed to John Woodward, a carriage maker, and it was during this time that he began to display a talent for mechanical improvisation and invention. Cooper made a machine for mortising the hubs of carriages, and he became so valuable to his employer that he was offered a partnership. However, Cooper decided to steer an independent course. He settled in Hempstead, Long Island, where he began making machines for shearing cloth, an enterprise that prospered during the War of 1812. The peace that followed resulted in a decline in the textile business and Cooper's lucrative operation, but the profits that were made during this time were soon invested in land and houses in lower Manhattan where Cooper now started a grocery business.

For the first 30 years of his life Cooper had drifted rather aimlessly, from apprenticeship to machine maker to grocer, shifting from one pursuit to another. However, in 1821 Cooper made a momentous personal decision, one that would put him on a new course and eventually lead him to wealth and success; he became a manufacturer. His accumulated money–from the cloth-shearing machine venture and the grocery–were invested in a glue factory located between Thirty-first and Thirty-fourth streets in New York City, where he directed a business of manufacturing glue, oil, whiting, prepared chalk, and isinglass. The high quality of these products soon resulted in a near monopoly of the American market. Later this business would be moved to Brooklyn, but despite Cooper's extensive activities in the iron and steel trade, his glue-making operations formed the cornerstone of business interests that became increasingly diversified during the 1830s, 1840s, and 1850s.

Cooper's entrepreneurial vision soon extended beyond New York City. In 1828 he and two partners purchased 3,000 acres of land in Baltimore in an area called Lazaretto Point. It was a risk, but at the time a good one, for all of Baltimore was caught up in a speculative fever brought on by the chartering of the Baltimore & Ohio (B&O) Railroad. Faced with increased economic competition from the larger port cities of Boston, New York, and Philadelphia, the citizens of Baltimore had boldly decided to tap the resources of the West with a yet unproven transportation technology, the railroad. Cooper and his associates, possessing land with excellent harbor facilities, large quantities of iron ore, and a strategic location at the terminus of this proposed rail line, stood to gain much if the gamble succeeded.

Cooper's Baltimore venture soon ran into problems, however. To begin with, Cooper discovered that his two partners, residing in Baltimore while he remained in New York, were bilking him of his investments. He dealt with them by buying them out and then personally taking charge of the property and its development. He began by digging up nearby iron ore deposits, devising a conveyor system to transport the ore, erecting charcoal kilns,

and constructing smelting furnaces. Cooper's Canton Iron Works quickly became profitable.

However, a much more serious question soon surfaced: that of the technological viability of the B&O track with its numerous tight curves. Basing his opinion on second-hand reports that had reached Britain, the pioneering railway engineer George Stephenson had concluded that locomotives could not be constructed to negotiate the numerous tight curves that had been laid on the B&O line. News of Stephenson's evaluation soon led to widespread discouragement in Baltimore, and by early 1830 the railroad was facing bankruptcy. Cooper realized that if the B&O failed, he would fail as well. In this moment of desperation he harnessed his mechanical ingenuity to avert what promised to be a calamity. He visited the president of the railroad company and proclaimed that he would "put a small locomotive on" that would negotiate the small curves. As a result of his determined effort he constructed the legendary Tom Thumb in 1830 at the Mount Clare Railroad shops in Baltimore.

The Tom Thumb was a reflection of Cooper's ability to improvise from existing equipment and materials. For the power plant he used a rotary steam engine that he had built in New York. The boiler for this engine was so small that one commentator had derisively stated that it was not powerful enough to drive a coffee mill. To connect the engine to the boiler Cooper needed pipe, but none was sold in the United States at that time. Yet, he was only temporarily stumped, for he took two muskets, broke off the wood stocks, and used the barrels for tubing. To maintain steam pressure, Cooper conceived of using a blowing apparatus driven by a drum connected to one of the car wheels. Even obtaining the wheels proved to be a challenge, since during the initial trials of July 1830 wheels kept breaking. To complicate matters a thief stole copper from the engine, which further delayed a public trial of the device.

These obstacles were overcome, however, and in August of 1830, with six men on the platform of the Tom Thumb and another dozen packed in a car behind, Cooper's locomotive went on a 13-mile trip from Baltimore to nearby Ellicott Mills in an hour and 12 minutes and returned in 57 minutes. Confidence in the B&O venture was thus restored, and Cooper's investment in Baltimore was more secure than ever.

Soon afterward Cooper expressed an interest in selling out his Baltimore holdings, which he eventually did in 1836. Profits from the sale of the Canton Iron Works were invested in other activities during the later 1830s and 1840s. But Cooper gained much from his Baltimore years. For the remainder of his life he distrusted partnerships, undoubtedly the result of his unfortunate and unpleasant experiences with the Canton Iron Works. On a positive note, however, Cooper now knew the iron business and its unwritten methods. He soon established ironworks in New York, and emerged during the 1840s as one of the nation's leaders in the antebellum iron trade.

In 1838 Cooper decided to reenter the iron trade, this time setting up a factory in New York City. It was a good time to enter the business, since the demand for iron was growing rapidly as a result of pioneer railroad expansion. Also, the basic process of iron making was being transformed during the late 1830s as eastern manufacturers were beginning to switch from charcoal to anthracite coal for smelting ore. In 1833 Fredrick A. Geisenhainer had taken out a patent for making iron and steel with anthracite, and three years later he had succeeded in making pig iron exclusively from anthracite at the Valley furnace in Schuylkill County, Pennsylvania. Cooper had kept abreast of these technical developments, and he and his brother Thomas patented various devices that took advantage of anthracite as a fuel. As early as 1839 the Cooper brothers had employed anthracite in the puddling process, and in 1840 Thomas Cooper took out a patent on a puddling furnace. Two years later Peter patented a mechanical puddling furnace consisting of a movable cylinder that was automatically fed by a turning screw.

While Cooper's iron establishment prospered during the early 1840s, it soon became clear to him that any further expansion of operations in New York City was impractical. In 1845 Cooper made a decision to move the ironworks and rolling mill from Thirty-third Street to Trenton, New Jersey, citing the complaints of neighbors who found the smoke from the burning of coal "objectionable." But there were other reasons behind the decision, including the fact that New York City was relatively far away from the anthracite coalfields and the rich iron ore sources of northern New Jersey. Trenton, however, was a near perfect site to fulfill Cooper's requirements. Not only was the New Jersey city located on the navigable Delaware River, but it had a canal that provided an excellent source of water-

power. Further, the Morris Canal linked the Delaware and Lehigh rivers, thus enabling anthracite to be easily transported to Cooper's new ironworks. Finally, the Camden & Amboy Railroad linked Trenton with the markets and supplies of New York and Philadelphia.

Cooper, who wanted to include family members into his expanding industrial empire, proposed that the newly formed Trenton Iron Works be placed in the charge of his son, Edward. Toward the end of 1844 Edward Cooper had returned from his European trip with his good friend Abram Hewitt, and he was ready to enter the family business. Edward had inherited much from his father in terms of inventiveness and mechanical skills; however, he was not forceful and decisive. On the other hand, companion Abram Hewitt had the drive, ambition, and business sense so desperately needed by the Trenton Works to succeed.

In an arrangement designed by Peter Cooper so as to avoid the pitfalls of partnership that he had experienced earlier in Baltimore, Edward and Abram Hewitt formed a company called Cooper & Hewitt in 1845. Cooper & Hewitt had the responsibility of managing the ironworks and held $149,000 of Trenton stock, while Peter Cooper controlled $151,000 of the ironworks capital. Thus, Cooper had the final say in any major decision. The Trenton Iron Works would soon emerge as the premier iron factory and rolling mill in antebellum America, and its shops were responsible for several innovations, including the rolling of iron beams for building construction.

As a result of the perseverance and hard work of Edward Cooper, Abram Hewitt, Charles Hewitt, and associate James Hall, the newly formed company began operations in spring 1846. The fledgling firm was quickly put on a secure basis with the successful negotiation of a contract to produce rails for the Camden & Amboy Railroad, and between 1845 and 1849 the company primarily produced 4½-inch rolled iron rails. Indeed, the close business relationship between the ironworks and the rail company was a primary reason for the former's financial profits during the 1840s. Because of the overproduction of rails by English mills in 1849 and the subsequent flooding of the American markets, the Trenton operation switched to the manufacture of wire for fencing, suspension bridges, and telegraph lines, as well as rivets and merchant bars made to order. Thus, flexibility and economic viability were demonstrated at a time of great crisis for the American iron industry.

By the early 1850s general prosperity had returned to the iron trade, and the annual report of the Trenton Iron Company stated that it was the largest rolling mill in the United States. Capable of making 400 tons of bar iron per week, including products like railroad iron, axles, chairs, forging bars, faggoting iron, wire rods, and brasier's rods, Cooper's Trenton Works were part of an integrated business that included three blast furnaces near Easton, Pennsylvania, and iron mines located at Andover, New Jersey. A railroad was constructed through eight miles of rough country to deliver iron ore at the rate of 40,000 tons per year, and the organization of this enterprise anticipated the integration which occurred after the Civil War.

Perhaps the greatest achievement of the Trenton Iron Works during the 1850s was the rolling of iron beams that were used in the construction of so-called fireproof buildings. This development was the product of years of empirical trials and machine design. As early as 1847 Cooper had directed shop foreman William Barrow to put together a rolling mill for the manufacture of 7-inch wrought iron beams. Initially unsuccessful, the attempt cost the company $30,000, and what was learned was that an ordinary rolling mill could not make such a product.

There were several reasons why a 7-inch, rather than a 4½-inch beam was desired. First, a larger beam was inherently stronger. Also, it was the smallest depth that could be structurally useful in 16- or 18-foot lengths. But the major stumbling block remained the design of efficient production equipment. A second attempt to make such a beam took place in 1852 as Peter Cooper was formulating his plans for the construction of his Cooper Union building in New York City. Once more William Barrow began work on the project, this time designing a new rolling mill. The cost of constructing the necessary apparatus—$150,000—was far more than originally anticipated. Before the work was completed, an exhausted Barrow died. Now in the hands of Abram Hewitt's brother Charles, the mill was completed and brought to perfection. The rolling equipment basically consisted of three vertical rolls rather than the typical horizontal configuration, and additional horizontal rolls were placed in a position where the vertical rolls came together at the top. Afterwards known as the "Universal Mill,"

the device made 7-inch I-beams rather than the rails consisting of a bulb at the top of the beam and a flange at the bottom.

The first beams from the mill were used in the Harper Brothers building in New York City rather than Cooper's own institute building, which had broken ground in 1853. Yet, the manufacture of beams for construction opened a new market for the iron trade, resulted in continued expansion for the Trenton Iron Works, and provided Cooper with additional wealth which by the late 1850s formed the basis of his philanthropic activities.

Without a doubt, Cooper's most enduring legacy was the consequence of his interest in technical education rather than the result of his success as a manufacturer. His idea of creating a unique educational institution for the working classes, and thereby acting as a patron of higher education at a time when it was uncommon for manufacturers to do so, was the result of his own personal experiences as a young man. As early as 1830, nearly thirty years before the opening of the Cooper Union, Peter had been struck by the comments of fellow New York common council member Dr. David Rogers, who had described to him the students and curriculum of a polytechnical school located in Paris. Cooper later remarked on Rogers's description:

> What made the deepest impression on my mind, was . . . that he [Rogers] found hundreds of young men from all parts of France living on bare crust of bread in order to get the benefit of those lectures. I then thought how glad I should have been to [have] found such an institution in the city of New York when I myself was an apprentice. . . . I determined to do what I could to secure to the youth of my native city and country the benefits of such an institution . . . and throw open its doors at night so that the boys and girls of this city, who had no better opportunity than I had to enjoy such means of information, would be enabled to improve and better their condition, fitting them for all various and useful purposes of life.

The fulfillment of Cooper's vision began in earnest in 1852, when he assembled land at the junction of Third and Fourth avenues between Seventh and Eighth streets; the next year a cornerstone was laid. By 1858 construction was completed, and a deed was executed that transferred the building to six trustees who were to use all income for "the instruction and improvement of the inhabitants of the United States in practical science and art." Cooper Union was a bold departure in the realm of education, progressive in spirit and response to the needs of a rapidly industrializing pre-Civil War economy. Its intended curriculum was clearly laid out and included

> [i]nstruction in branches of knowledge by which men and women earn their daily bread; in laws of health and improvement of sanitary conditions of families as well as individuals; in social and political science, whereby communities and nations advance in virtue, wealth and power; and finally in matters which affect the eye, the ear and the imagination, and furnish a basis for recreation to the working classes.

From the beginning Cooper Union's library, reading room, and classes were very popular. Admission was open to all with ability, and the school charged no tuition. Its first students ranged in age from sixteen to fifty-nine, and among its nineteenth-century graduates would be the sculptor Augustus Saint-Gaudins and physicist and inventor Michael Pupin.

But Cooper Union would prove to be more than an educational institution–it was a focal point for the dissemination of new ideas and a center for thought in New York City. In February 1860 presidential candidate Abraham Lincoln delivered a speech at Cooper Union that later was considered crucial to his winning the presidency. Others who spoke at Cooper Union included Susan B. Anthony, Horace Greeley, Mark Twain, Ulysses S. Grant, and Englishmen Thomas Huxley and John Tyndall.

As an old man in the 1870s, Cooper could look back with pride to his many business and philanthropic accomplishments. He was a most unconventional man, an individual whose qualities of generosity and concern for fellow men stood out in an age dominated by businessmen preoccupied with the amassing of large fortunes. It was in this context that civic-minded Peter Cooper, always interested and involved in local New York politics, would take a stand in 1876 for currency reform and support the working classes from which he himself had arisen.

Cooper had long been interested in "good" government and finance, and he had published several pamphlets on these subjects during the 1860s and 1870s. Beginning with the Civil War the federal government had issued greenbacks, or paper currency that was not redeemable with gold, and this cheap paper currency was favored by the working and agrarian classes as tender to pay off their debts. The greenback issue was a controversial one at the time of the election of 1876, and both major presidential candidates, Rutherford B. Hayes and Samuel Tilden, were unwilling to challenge the major financial and business interests and take up the cause of the common man. Thus in 1876 Peter Cooper, eighty-five years old, became the presidential candidate of the National Independent party, a party of protest that aimed at winning the support of farmers and laborers. Cooper, running on a platform that called for the regulation of railroads, supervision of land transfers in the West, and the control of federal currency, received only 100,000 votes or 1 percent of the total. But despite his poor showing, Cooper felt that he had achieved a moral victory, and that his ideas on paper money were voiced. He had become the champion of the masses, a role that Cooper had always coveted.

Until his death on April 4, 1883, the old man could be seen almost daily in the reading room of the institution that he so dearly loved. He was mourned by a city that had provided him with economic opportunity and that he had so ably served. Cooper was a mechanic and inventor, an entrepreneur and manufacturer, and above all a businessman who used his profits to fulfill what he considered to be the purpose of life, which was "to do good."

Selected Publications:

Deed of Trust (New York: Wm. C. Bryant, 1859);

Mr. Peter Cooper's Address (New York: Cooper's Union, 1860);

The Death of Slavery (New York: Wm. C. Bryant, 1863);

Letter of Peter Cooper on Slave Emancipation (New York: Wm. C. Bryant, 1863);

Address of the Graduates to Peter Cooper, Esq., and His Reply (New York: Whitehorne, 1871);

Protection to American Industry (New York: Baker & Godwin, 1871);

A Communication to Show the Dangers of a War of Commerce on All the Great Interests of Our Country (New York: Baker & Godwin, 1872);

Letter From Peter Cooper, to the Delegates of the Evangelical Alliance (New York: Baker & Godwin, 1873);

Letter to the Episcopal Church Congress (New York: Evening Post, 1874);

Thoughts Presented to the Pupils of the Cooper Union (New York: Baker & Godwin, 1874);

Currency (New York: Evening Post, 1875);

The Unmeasured Importance of an Unfluctuating Currency Over which the Government Has Entire Control (New York: Mercantile Journal, 1875);

A Sketch of the Early Days and Business Life of Mr. Peter Cooper (New York: Murray Hill, 1876);

The Dangers of a War of Commerce and the Necessity of a Tariff and of an Unfluctuating Currency to National Prosperity (New York: Trow's, 1877);

A Letter on the Currency (New York: Trow's, 1877);

An Open Letter to the Citizens and Voters of Maine (New York: Trow's, 1877);

An Open Letter to President R. B. Hayes (New York: Trow's, 1877);

The Nomination to the Presidency of Peter Cooper and His Address to the Indianapolis Convention of the National Independent Party (New York: Trow's, 1877);

An Open Letter to the Hon. John Sherman, Secretary of the Treasury of the United States (New York: Trow's, 1878);

The Appeal of Peter Cooper, Now in the 88th Year of His Age (New York: Advocate, 1878);

Good Government (New York: J. J. Little, 1880);

Address by Peter Cooper (Philadelphia: American Iron and Steel Association, 1882);

Ideas For a Science of Good Government (New York: Trow's, 1883).

References:

Gano Dunn, *Peter Cooper (1791-1883): A Mechanic of New York* (New York: Newcomen Society, 1949);

Robert A. Jewett, "Solving the Puzzle of the First American Structural Rail-Beam," *Technology and Culture*, 10 (1969): 371-391;

C. Edwards Lester, *Life and Character of Peter Cooper* (New York: John B. Alden, 1883);

Peter Lyon, "The Honest Man," *American Heritage*, 10 (1959): 4-11, 104-107;

Edward C. Mack, *Peter Cooper: Citizen of New York* (New York: Duell, Sloan & Pearce, 1949);

Rossiter W. Raymond, *Peter Cooper* (1901; reprinted, Freeport, N.Y.: Books for Libraries Press, 1972);

Esmond Shaw, *Peter Cooper & the Wrought Iron Beam* (New York: Cooper Institute, 1960);

C. Sumner Spaulding, *Peter Cooper: A Critical Bibliography of His Life and Works* (New York: New York Public Library, 1941);

J. C. Zachos, *A Sketch of the Life and Opinions of Mr. Peter Cooper* (New York: Murray Hill, 1876).

Archives:

Materials concerning Peter Cooper are located at Cooper Union, New York, New York; the New-York Historical Society, New York, New York; and the Library of Congress, Washington, D.C.

Thomas Dickson

(March 26, 1824-July 31, 1884)

by John A. Heitmann

University of Dayton

CAREER: Machine shop and foundry owner (1856-1862); president, Dickson Manufacturing Company (1862-1867); coal superintendent (1859-1867); vice-president (1867-1869); president, Delaware & Hudson Canal Company (1869-1884).

Thomas Dickson, a manufacturer and civic leader, was born in Leeds, England, March 26, 1824, and died in Morristown, New Jersey, July 31, 1884. Despite his lack of formal education and rather humble origins, Dickson rose to a preeminent social and economic position in the Scranton, Pennsylvania, area between 1850 and 1880 because of his persistence, business acumen, and ability to draw on his family for crucial support.

One of the six children of James and Elizabeth Linen Dickson, Thomas immigrated along with his uncle John Linen and family to Canada in 1832. In search of opportunity, the Dicksons subsequently settled in Dundaff, Pennsylvania, and in 1836 relocated to Carbondale, Pennsylvania, where Thomas's father, a millwright, gained employment with the Delaware & Hudson Canal Company. During the early 1830s Pennsylvania experienced a frenzy of canal building activity in the wake of the success of the Erie Canal, and great hopes were placed on the Delaware & Hudson Canal to tap the rich mineral resources of the Northeast Pennsylvania hinterland. The Delaware & Hudson would prove to be one of the few economically viable Pennsylvania Canals, and James Dickson quickly rose within the transportation firm to the position of master mechanic, a title he held until his death in 1880.

As a child Thomas Dickson briefly enrolled in one of the local primary schools in Carbondale, but his time as a student ended abruptly after he be-

Thomas Dickson

came involved in an argument with a teacher. Rather than apologize, Thomas Dickson chose to work as a mule driver for the Delaware & Hudson Canal Company, hauling coal out of one of the company's anthracite mines. He soon attracted the attention of Carbondale merchant Charles T. Pierson, however, and was offered a position as a clerk. Like many other businessmen of his day, the skills and personal contacts gained in clerking served as the starting point in Dickson's future career path.

After working with Pierson for a short time Dickson accepted an offer to work with prominent Carbondale businessman Joseph Benjamin. Ultimately this decision was pivotal in Dickson's life, since Benjamin was not only involved in the mercantile trade but also in a fledgling foundry business. Eventually Dickson assumed responsibility for the foundry, a thriving activity in the Scranton-Carbondale area that was experiencing rapid industrial and economic growth. Beginning in the late 1840s the region emerged as a leader in the manufacture of iron products, especially railroad rails, and was rapidly being linked to northern and eastern markets by the expanding railroad network. This linkage expanded the markets for the foundry's

machinery and fabricated iron equipment; Dickson was in an excellent position to meet that demand.

In 1856 Dickson dissolved his partnership with Benjamin, gained family support for his own foundry venture, and established a machine and foundry shop in Scranton near the Lackawanna Iron & Coal Company. Initially Dickson's firm supplied the area's two largest companies—the Lackawanna Iron & Coal Company and the Delaware & Hudson Company—with steam engines and boilers. But once established and with the coming of the Civil War, his foundry company expanded both its product line and market. In 1862 the operation was organized as the Dickson Manufacturing Company, and the new firm's family control was reflected in its first officers: Thomas Dickson was elected president, George L. Dickson was appointed secretary and treasurer, and John L. Dickson was named master mechanic.

During the 1860s the Dickson Manufacturing Company grew steadily, and by the end of the decade it had more than 150 employees. As a result of its purchase of a local locomotive manufacturer, the Cliff Works, the company began making steam locomotives, and with the acquisition of a neighboring planing mill the manufacture of railway boxcars was initiated. By the 1880s Dickson Manufacturing Company equipment, including a diverse number of industrial products, was distributed within an expanding international market. The firm produced a large variety of locomotives for mines, rolling mills, sugar plantations, and logging operations, as well as vertically and horizontally configured steam engines ranging from 4 to 75 horsepower. Perhaps because the foundry was located in the anthracite coal region and had its beginnings in supplying the local mines with equipment, a broad spectrum of mining machinery was also advertised in the company's 1885 catalog, including coal screens, ventilating fans, coal elevators, air compressors, and ore-roasting ovens.

Concurrent with Dickson's success in the foundry trade came business diversification. As early as 1859 Dickson had held the important position of coal superintendent within the ailing Delaware & Hudson Canal Company—a responsibility he no doubt viewed as being necessary to the maintenance of his position within the foundry business, since he could ill afford to lose one of his biggest customers. Subsequently Dickson would be named the transportation organization's president. Dickson also emerged at this time as one of a handful of civic leaders who would organize the First National Bank of Scranton in 1863. By the mid to late 1860s these diverse business activities intensified, as Dickson was not only a leading figure in organizing the Moosic Powder Company in 1865 but also involved in activities related to the Crown Point Iron Company and the Mutual Life Insurance Company of New York. During the 1870s Dickson, now one of the wealthiest men in Scranton, increasingly began to turn over his responsibilities to his son, James Pringle Dickson.

James, born in 1852, was educated at Lafayette College and gained valuable experience by working with an engineering company engaged in railroad construction and as a commission merchant. In late 1872 James was hired as a clerk at the Dickson Manufacturing Company, and three years later he became the firm's representative in Wilkes-Barre, Pennsylvania. In 1882 he was elected

vice-president, and in 1886 he became president, two years after the death of his father, thus maintaining family control of a business during an age when it was increasingly difficult to do so.

References:

Gerald M. Best, *Locomotives of the Dickson Manufacturing Company* (San Marino, Cal.: Golden West Books, 1966);

Burton W. Folsom, Jr., *Urban Capitalists: Entrepreneurs and City Growth in Pennsylvania's Lackawanna and Lehigh Regions, 1800-1920* (Baltimore: Johns Hopkins University Press, 1981);

Horace E. Hayden, *Genealogical and Family History of Wyoming and Lackawanna Valleys* (New York: Lewis Publishing, 1906);

Samuel C. Logan, *The Life of Thomas Dickson: A Memorial* (Scranton, Pa.: Devinne Press, 1888);

Portrait and Biographical Record of Lackawanna County, Pa. (Chicago: Chapman Publishing, 1897);

Benjamin H. Troop, *A Half Century in Scranton* (Scranton, Pa.: The Scranton Republican, 1895).

Joseph Dixon

(January 18, 1799-June 15, 1869)

by Bruce E. Seely

Michigan Technological University

CAREER: Chemist, Hall's Dye House, Lynn, Massachusetts, (1822-?); founder and owner, factory producing graphite stove polish and pencils (1827-1867); works manager, Adirondack Iron & Steel Company (1848-1850); organizer and owner, Joseph Dixon Crucible Company (1867-1869).

Joseph Dixon was one of the talented mechanics and inventors who played an important role in the development of industry in this country during the first half of the nineteenth century. Dixon's interests were never confined to a single field, but ranged widely in seemingly unrelated endeavors. Although we know that the successful initiation of manufacturing and industry in this country was not a simple matter of "Yankee ingenuity," the career of Dixon suggests why Americans widely believe that they possess some special ability to harness technology for the nation's needs. In addition to playing a central role in the creation of a domestic cast crucible steel industry, Dixon held equally significant posi-

tions in the establishment of the photographic, lithographic, and graphite industries. Unafraid to tackle any project that caught his attention, the word *inventor* truly describes Joseph Dixon.

Like so many of the pioneering entrepreneurs and engineers whose efforts placed this country on the road to industrialization, Joseph Dixon had only a rudimentary formal education. He was born in Marblehead, Massachusetts, on January 18, 1799; his father was a ship captain. His mechanical and inventive aptitude became apparent at an early age, for by the time he was twenty-one years old, he had invented a file-cutting machine. At this time, almost all files, a vital tool in the metalworking industry, were imported from England, either in completed form or as blanks for finishing. It is unlikely that Dixon profited from this device, or from his next invention—a superheated steam engine he built in 1821 in Lynn, Massachusetts, to power a boat. But these diverse projects were an indicator of the young man's breadth of interests.

The Joseph Dixon Crucible Company works in Jersey City, New Jersey, circa 1870

Dixon's early career remains hazy, and we know only that he continued to invent. But inventing was hardly a way to support oneself. Dixon made ends meet as a chemist for Hall's Dye House in Lynn, starting the job in 1822. Dixon's biographers claim that this chemical work stemmed from his interest in medicine, although it is not clear how that interest emerged or was pursued. The length of time he remained in Lynn is as unclear as the origin or extent of his qualifications for being a chemist.

It is clear, however, that Dixon's pattern of holding multiple interests continued through the 1820s. He moved in several directions, first following a youthful interest in printing. Apparently, he lacked the funds to purchase metal type, so he carved wooden letters. From this beginning Dixon went on to develop several ideas and innovations, including a new matrix for casting type. He also seems to have grown interested in lithography—a process for printing illustrations using a carefully prepared stone slab as the plate—during the late 1820s. Dixon initially adopted this technique for printing banknotes, and by 1832 he had developed and pat-

ented a process that used colored inks to prevent the counterfeiting of banknotes.

Dixon's interest in printing had, however, an even more significant side effect–he began searching for containers to hold molten metal at high temperatures. Pots made from various kinds of clays had been used as crucibles for melting metal, including type, but Dixon set out to find a durable material that withstood higher than normal temperatures. The quest was of some significance, for English success in fashioning crucibles for the metal trades constituted one of their most important advantages over American manufacturers during this period. By 1826 Dixon was testing plumbago–now called graphite–from a New Hampshire farm, and finding that this material was wonderfully resistant to high temperatures. He also recognized that markets for the crucibles were very limited at this time. But he had found that graphite possessed several characteristics that suggested other products. In 1827 Dixon built a factory in Salem, Massachusetts, for producing pencils and stove polish–the two uses of graphite most immediately useful in America at the time.

111

Dixon himself, not surprisingly, invented but did not patent the production machinery. He initially imported the raw material from Ceylon, although at some point he discovered a source of graphite near Fort Ticonderoga, New York. A further indication of the nature of American society in about 1830 was Dixon's marketing mechanism, for he sold his products to "Yankee peddlers" who traveled the country.

Dixon's graphite venture was an immediate success because his graphite and clay mixture was much more satisfactory as a pencil "lead" than was real lead. More important, from Dixon's viewpoint, the works were not difficult to oversee, and the profits supported his inventing. One of his enthusiasms in the 1830s was his production of the first locomotive in the country with a double crank. The machine ran on wooden wheels and only worked on level ground, suggesting the design had its flaws.

Dixon had much better luck in another field that appeared in the late 1830s—photography. Frenchman Louis Daguerre began to develop in 1826 a technique that was well suited only for still lifes and portraits because of the lengthy exposure required by the chemical emulsions used. These daguerreotypes quickly won fame in the United States and in Europe. Dixon has been identified as perhaps the first American to take a portrait using Daguerre's process. By 1839 Dixon had invented a process for making collodion, the chemical emulsion used to coat the plates exposed to light in a camera. In addition, Dixon introduced a reflector to prevent the picture from being reversed. Samuel F. B. Morse was so taken by this idea that he attempted to patent it for Dixon in Europe. Dixon also developed a method of grinding true lenses for camera tubes. Finally, in the early 1840s Dixon combined his interests in printing and photography, introducing photolithography, a process he used originally for printing banknotes. Eventually, this technique permitted rapid production of large numbers of art prints and illustrations, especially as used later in the Vandyke process.

These successes in the realm of photography did not exhaust Dixon's inventive talents. In 1845 he jointly patented with I. S. Hill an antifriction metal for lining bearings, but the Babbitt metal introduced a short time later seems to have been superior. Biographers suggest there may have been a connection between Babbitt and Dixon, but the evidence on this point is not clear. More successful, it seems, was Dixon's invention of metal cylinder rollers for calico transfer printing.

These developments indicate that Dixon was moving by the 1840s toward an interest in metals. Other evidence of his work in this direction came in 1839, when he received the Massachusetts Charitable Mechanic Association Medal for making a large steel mirror. Even low-quality blister steel remained a very expensive commodity in America at this time, and the British completely monopolized production of the top grade tool steels. Dixon's effort with the mirror, although probably just a demonstration, was only one of many attempts to end British domination. His interest in steel must have continued through the 1840s, and a good reason existed for it—Dixon's graphite crucibles solved one of the most pressing shortcomings of American steel manufacture. The cast crucible process—then the only way to produce uniformly high-quality steel—required converting good wrought iron bars into blister steel, melting blister steel in crucibles, and pouring the molten steel into ingots. American efforts to make high-quality steel had always failed because Sheffield manufacturers sold a superior product, in part because they alone made crucibles from Stourbridge clay, which was perfectly suited for steel melting. Dixon's crucibles eliminated this basic advantage of British steel makers.

In 1847 Dixon moved his graphite factory to Jersey City, New Jersey, deciding that the time had come to manufacture the crucible he had developed 20 years earlier. Apparently he wanted to be nearer to an expanding market. This move may also, however, have been due to his steel experiments, for Dixon immediately established a connection with a group of Jersey City businessmen who hired him to build a cast crucible steelworks. The company was the Adirondack Iron & Steel Company, which had grown out of an attempt in the 1830s to exploit a rich iron deposit in New York's Adirondack region near Lake Placid. The ironworks had a troubled technical history, but the owners were convinced that their iron was suited for steel production. After contacts with a steel maker in Sheffield, England, fell apart, the owners turned to Dixon. Bars of steel he had made were tested by a local blacksmith in January 1848, and as a result of a positive evaluation, the Adirondack Iron & Steel Company hired Dixon to build a cast crucible steelworks and teach the process to the company.

Dixon erected the steelworks in Jersey City at the outlet of the Morris Canal, proceeding in the traditional fashion of Benjamin Huntsman. The initial facilities were rather modest—a cementing furnace for making blister steel, 16 melting holes fired by anthracite coal, three or four tilt hammers, and a steam engine. From the start, however, the works encountered problems. First, construction proceeded slowly, with the hammers especially proving difficult to acquire. By early July the first blister steel was made, but the hammers were still not in place in November. So little progress had been made by March 1849 that a report by Dixon to Archibald McIntyre, the chief investor, was decidedly apologetic in tone. A further complication may have been the process that Dixon developed for making steel, for the inventor in him prevented simple adoption of Huntsman's process. Instead, he introduced a process that made blister steel from pig iron rather than wrought iron. The advantage in eliminating this early step in the process of steel making was a savings in fuel, and many others pursued this strategy. Dixon may have experienced some success because of the special quality of the company's iron, for he received a patent for his idea in April 1850. But the process may also have accounted for the inconsistent quality of the steel.

Exacerbating these technical difficulties was a feud between McIntyre and Dixon over the latter's contract. The company had initially agreed to pay Dixon a flat fee for building the works and teaching the process, but then decided to hire him as superintendent for $1,200 a year plus a share of any profits. Dixon had been paid only $500 of the $1,000 promised in the initial agreement when the new contract took effect, and he believed he was owed the difference. McIntyre, however, assumed the new agreement superseded the first. Each man accused the other of cheating, with Dixon's threat to quit and work for another steel company stymied only by his contract. A $500 bonus paid Dixon for erecting a new furnace ended the dispute, but McIntyre never trusted him again. By 1850 Dixon had left the company, perhaps at the termination of his contract. McIntyre did not grieve.

In spite of these technical and personal problems, Dixon succeeded in making the best crucible steel yet produced in this country. Testimonials from several users in 1849 and 1850 suggested that samples of Adirondack steel matched the quality of English products, while the Franklin Institute

awarded Dixon its Elliot Legacy Premium for the encouragement of American manufacturers. Unfortunately, consistently high standards of quality proved hard to maintain, and Dixon's departure did little to solve this difficulty, for his successor seems to have paid more attention to output than standards. In the end Dixon left behind him a viable steelworks, extensive know-how, and his crucibles. Yet, in 1853 the partners abandoned their efforts, although subsequent attempts to make steel in this plant succeeded both technically and commercially. No small amount of the credit for the successful conclusion of the struggle to make cast crucible steel in the United States belonged to Dixon.

Even after leaving the Adirondack Iron & Steel Company, Dixon retained an interest in steel making. In 1858, for example, he received another patent for improvements in his steel-making process. But he also showed his usual interest in other directions as well. Even while directing the steelworks, he had taken out a patent on the use of crucibles in pottery making—an idea that may have stemmed from contacts with the Adirondack company. The heirs of David Henderson, leading investor in the steel company and also owner of a Jersey City pottery, also worked at and invested in the steelworks.

Most of his time, however, Dixon devoted to his graphite manufacturing operation, which remained in Jersey City. His crucibles were recognized as the best and were widely used in the cast crucible steel industry. Although competing producers eventually appeared, Dixon remained the largest producer for a long period. Moreover, the United States government used Dixon crucibles exclusively for melting gold and silver. He had solved the crucial problem in fabricating these containers—the proper mix of the proper type of clay with the graphite—for graphite by itself had no adherence properties. The formula long remained a closely guarded secret. Dixon also expanded his product line after 1860, adding graphite greases and paints to the firm's production. But in terms of numbers, pencils always dominated Dixon's operation. He continued to work on manufacturing improvements as well, in 1866 inventing a special wood planing machine for shaping wood for pencils. By the time he incorporated as the Joseph Dixon Crucible Company in 1867, in order to protect the company after his death, he had founded the largest graphite firm in the country, and it continued to flourish after he died in 1869.

Joseph Dixon was a stereotypical American inventor, stricken with the obsession that infected many Americans in the nineteenth century—the need to find new ways to perform old tasks. Always inventing, in 1866 he took out a patent on a galvanic battery, a wholly new line of interest. Yet, beyond his inventing, we know very little about Joseph Dixon. He was married to Hannah Martin of Marblehead on July 28, 1822, and had one daughter, whose husband later played a small part in the business. We know nothing of his personality, and his civic and social interests remain a blank. We are left with the impression that Dixon was consumed by his passion for inventions. It seems safe to say that he found working for others a trying experience; he seems to have preferred being independently employed and free to invent. Altogether, Dixon resembled many of the businessmen and engineers who developed American technology during the early 1800s. Surely, there was more to this talented mechanic than we know, but in the absence of more information, we are left with Elbert

Hubbard's hyperbolic analysis. Hubbard, who wrote many inspirational biographies, labeled Dixon one of the 20 most influential men in history, ranking him alongside Gutenberg, Watt, Arkwright, Edison, Westinghouse, Darwin, Jefferson, Lincoln, and Pericles, among others. This is high praise, even if exaggeration is admitted, for a self-educated man who single-handedly created a new American industry—graphite products—and made possible the development of another—cast crucible steel.

Unpublished Document:

Bruce Seely, "Adirondack Iron and Steel Company: New Furnace, 1849-1854." Report NY-123 U.S. Department of Interior, Heritage Conservation and Recreation Service, Historic American Engineering Record, 1978.

References:

Elbert Hubbard, *Joseph Dixon* (N.p., 1912);

Mechanical Engineers in America Born Prior to 1861: A Biographical Dictionary (New York: American Society of Mechanical Engineers History and Heritage Committee, 1980);

Hamilton S. Wicks, "American Industries—No. 2: The Utilization of Graphite," *Scientific American*, 40 (January 18, 1879): 31, 34.

William Franklin Durfee

(November 15, 1833-November 14, 1899)

by Ernest B. Fricke

Indiana University of Pennsylvania

CAREER: Engineer, architect, and city surveyor, New Bedford, Massachusetts (1853-1862); member, Massachusetts General Assembly (1861); builder, engineer, Kelly Pneumatic Process Company (1862-1865); consulting engineer (1865-1899).

William Franklin Durfee was an engineer who helped introduce the age of steel in the United States by building the country's first successful Bessemer steel mill between 1862 and 1865. He was also one of the first to conceive of smelting as a chemical process; and in so doing he laid the foundation for analytical laboratories to be an integral part of iron and steel operations.

Durfee was born on November 15, 1833, in

New Bedford, Massachusetts, to William and Alice Sherman Talbot Durfee. New Bedford was at that time in the midst of the period of economic growth which would turn it from a whaling town into a textile center. Durfee's father was a carpenter with a good reputation as a builder of large structures. Durfee received from his father a practical training which was crowned by the completion in 1853 of a course of study in the Lawrence Scientific School at Harvard. In 1853, at the age of twenty, he began a career in New Bedford as an engineer and architect. Shortly afterward he also accepted the position of city surveyor. His primary work in this latter position was to survey and plot the community's undeveloped tracts of land. He also had the opportunity to augment his income by acting as the selling agent for the land he was surveying.

In those days prior to civil service examinations, the office of surveyor was a political appointment. The Whig party was dominant in New Bedford, and Durfee, like most Northern Whigs, became a Republican in the national crisis over the extension and abolition of slavery. In 1861 he was a member of the Massachusetts House of Representatives. As secretary of its military committee, in part charged with equipping troops, Durfee led the fight for a resolution pleading that Congress repeal "all laws which deprive any class of loyal subjects of the Government from bearing arms for the common defense." The resolution is said to be the first formal proposal for the arming of black troops. Early in 1862, as a private citizen, he further demonstrated his commitment to the success of the war by designing a new naval gun. Although Durfee was praised for his effort, the gun was never manufactured for use.

After the failure of his weapon design Durfee, through the help of his cousin Zoheth Durfee, moved to Michigan in mid 1862 and turned his attention to the manufacture of iron and steel. Zoheth and two others–Daniel Morrell of the Cambria Iron Company of Johnstown, Pennsylvania, and Eber Ward, who had iron interests in Wyandotte, Michigan, and Chicago–were interested in perfecting William Kelly's pneumatic process of making steel, a technique nearly identical to that developed by England's Henry Bessemer. In the process, air was forced through molten iron, thereby greatly reducing the use of fuel and the costs for converting pig iron into wrought iron and steel. In 1861 the group had decided to secure the Kelly patent and had sent Zoheth Durfee to Britain to acquire the American rights to Bessemer's patent on machinery. Their goal was to erect a large commercial-size converter and manufacture steel. This last project was to be William Durfee's job.

Durfee was not the only one who was attempting to perfect the pneumatic process for making steel, as others in America, in addition to Bessemer in Europe, were working toward this goal. But Durfee, unlike his competitors, had no background in iron and steel. His drive to be successful was buoyed by the fact that he was still a young man in his late twenties, who had already experienced success at home, in politics, and in naval design, and who now was becoming a scholar in the sciences.

William Franklin Durfee (from Cassier's Magazine, *May 1895)*

Without having seen either Kelly's or Bessemer's equipment, but after being briefed on Kelly's work, Durfee studied the patents of both and scrutinized William Fairbairn's *History of the Manufacture of Iron.* Thus "his own sense of the internal fitness of things" led Durfee to decide that Kelly's stationary converter was not as good for commercial production as "the highly original and ingenious apparatus invented by Bessemer especially the tilting converter. . . ." By the time Zoheth returned from England–without the Bessemer patent–Durfee's version of a mobile 1 1/2-ton converter awaited him.

Durfee was unable to follow through with other rapid strides toward a commercial works, largely because Ward diverted his efforts from the converter in Michigan to problems he was having in his rerolling rail mill in Chicago. However, Durfee did design and build other pieces of apparatus, such as a casting ladle and the cranes necessary to move the solid and molten iron. Some of the equipment, machinery, and implements were complex and huge, requiring long periods for their completion. For instance, the engine to provide pressure for the air blasts was not completed until spring 1864, and the new furnace to provide pig iron of a consistent quality to thrust into the converter was not finished until that summer.

In addition as work progressed, Durfee's mandate was expanded beyond what was originally conceived. Zoheth returned from England with the additional information that exactly the right amount of carbon had to be added to the blasted molten iron in order to create a good quality steel. This required that Englishman Robert Mushet's spiegeleisen compound be incorporated into the process. In fall 1863 Zoheth returned to England to make another effort at obtaining the rights to the Bessemer patent. As it became obvious that Zoheth would fail again, Durfee prepared to abandon his Bessemer-type converter and to build a large Kelly converter.

Everything that had occurred so far had frustrated Durfee's efforts. In spring 1863 a German chemist was hired to analyze the American iron ores which were to be placed into the converter. He left in December before completing this project, and Durfee was left to finish it. Again, when Zoheth returned from his last trip in spring 1864, Durfee had to incorporate his technical advice into the task. More beneficial to Durfee was the simultaneous arrival of Llewellyn Hart as an aide; Hart had experience working at the Bessemer works in France.

The two men worked well together as the new stationary converter was in the process of being built. But often the fruit of experimentation is one failure after another. During the testing they were plagued by the problem of proper pace, either working so quickly that the molten metal was blown out of the converter or so slowly that the metal chilled and solidified in the vessel. In any case Durfee decided by fall 1864 that he was ready for a public demonstration. The "public" who were present on September 6 were the owners of the works, and despite some apprehension, things went smoothly. Molten iron was poured into the converter, a successful blast was made, the spiegeleisen was added, and the molten result was poured into the molds. In a half hour the first steel made in the United States by the pneumatic or Bessemer process was finished.

There was jubilation among the workers, many of whom had spent nearly three years working with Durfee toward this goal. Some of the steel was sent to Bridgewater, Massachusetts, to be tested, some was formed into tacks, and more was welded into pipe. Durfee had two jackknives and a razor made from the steel. The razor was given to his father, who used it proudly for 15 years.

Durfee's personal celebration was brief as his task was still unfinished. The steel next had to be rolled into rails for the railroads, which were viewed as the best potential customers for commercially produced steel. In addition he was faced with a working environment which was not completely harmonious. He sensed a hostile attitude among the young men assigned to him as assistants. Although he was a capable engineer and builder, and he held a position of great responsibility, Durfee was not a skilled manager of men. It seemed to Durfee as though his assistants were more interested in pursuing their careers through Ward than in doing what Durfee asked of them. Another problem was an older man who headed the furnace which provided him with pig iron. He would modify the process only with the greatest reluctance. William described him in German as "Herr Unkunde Unheilschwanger" or as "Mr. Ignorance Pregnant with Disaster." Some of the workers were difficult as well; most were poorly educated, and the best of the men left for the war. Durfee once discovered that some pipe had been disconnected, and he could not help but wonder whether "mischief or murder" was planned. When one of the assistants, despite Durfee's orders, agreed to demonstrate the converter at Ward's request when Durfee was away, an ignored safety device caused an explosion which almost killed two United States senators and blew Ward out of the building.

The final outrage Durfee suffered at Wyandotte was the raid on the "apothecary shop." Wanting to see that "sound principles are established in place of old empiricisms" in the manufacture of iron and steel, Durfee had recommended in 1862 that an analytical laboratory should be built at Wyandotte. It was to be one of the first in what by the mid 1870s would become general policy in the industry. The laboratory was called by the workers the apothecary shop. When the German chemist left, supervision of the shop became another of Durfee's duties. In January 1865, on his return from a trip, Durfee found that the shop had been stripped of all equipment, chemicals, and notes. He wrote years later, "I manifested no surprise, of it I made no complaint; but then and there I mentally resolved that as soon as the first rail was rolled from steel made at Wyandotte, I would leave. . . ."

The steel was rolled into rails in May 1865. Ward refused to carry out the original plan to build a commercial plant using the perfected process, and Durfee left. Though he was one of the men most responsible for the birth of the steel age in America he was glad to say farewell to Wyandotte and what he labeled the "syndicate of sin."

Durfee never returned to Bessemer steel operations. His reputation as a designer of machinery and as a builder established him as a consulting engineer, and work was plentiful. Upon leaving Wyandotte he built the Bayview Merchant Mill at Milwaukee, Wisconsin. In 1869, at the rolling mill of the American Silver Steel Company of Bridgeport, Connecticut, he applied the puddling of iron to the new Siemens regenerative gas, or open-hearth, furnace. He was the first to do this in the United States. While he was with the Wheeler & Wilson Company of Bridgeport in 1878, he constructed the first successful gas furnace for refining copper at Ansonia. In 1886 he joined the United States Mitis Company as general manager and developed a new process for making wrought iron and steel castings. He also worked on a new machine for the manufacture of horseshoe nails.

Outside of his prosperity in the workplace, his life was serene. He married Annie Swift of Boston in 1880, and they made their home in New Brighton on Staten Island, New York. The esteem in which he was held as a scholar was manifested in his being consulted as an expert in patent cases. In 1876 he served as a judge of machine tools at the Centennial Exposition in Philadelphia. He wrote many articles of a technical or historical nature for the journals of the American Iron and Steel Institute, the American Society of Mechanical Engineers, the American Institute of Mining Engineers, and for a larger audience, the *Popular Science Monthly*.

Upon the outbreak of the Spanish-American War the United States planned to increase the size of the navy. Durfee was asked for his advice on how to increase the supply of steel for ships. It was while he was working on this project that he became ill. After a two-week stay at the State Hospital in Middletown, New York, he died, one day shy of his sixty-sixth birthday, on November 14, 1899.

Selected Publications:

Balanced Vertical Machines (Bridgeport, Conn., 1871);

An Account of a Chemical Laboratory Erected at Wyandotte, Michigan, 1863 (New York, 1884);

"In Memoriam: Alfred Charles Hobbs," *Transactions of the American Society of Mechanical Engineers*, 13 (1892): 263-274.

References:

Henry Howland Crapo, *The Story of Henry Howland Crapo, 1804-1869* (Boston: Privately printed, 1933);

Jeanne McHugh, *Alexander Holley and the Makers of Steel* (Baltimore: Johns Hopkins University Press, 1980);

Elting E. Morison, *Men, Machines, and Modern Times* (Cambridge, Mass.: M.I.T. Press, 1966);

James M. Swank, *History of the Manufacture of Iron in All Ages, and Particularly in the United States from Colonial Times to 1891*, second edition (Philadelphia: American Iron and Steel Association, 1892);

Peter Temin, *Iron and Steel in Nineteenth-Century America: An Economic Inquiry* (Cambridge, Mass.: M.I.T. Press, 1964);

Theodore A. Wertime, *The Coming of the Age of Steel* (Chicago: University of Chicago Press, 1962).

Zoheth Shearman Durfee

(April 22, 1831-June 8, 1880)

by Ernest B. Fricke

Indiana University of Pennsylvania

CAREER: Member (1856-1857), partner, J. & T. Durfee (1858-1859); proprietor, Zoheth S. Durfee (1860-1862); partner, Kelly Pneumatic Process Company (1863-1866); secretary, Pneumatic or Bessemer Process of Making Iron and Steel (1866); secretary-treasurer, Pneumatic Steel Association (1866-1877); secretary-treasurer, Bessemer Steel Association (1878-1880).

Zoheth Shearman Durfee was born in Fall River, Massachusetts, on April 22, 1831, the son of Thomas and Delight Shearman Durfee. Later the same year the family moved to New Bedford, Massachusetts, where Durfee was raised. The Durfees were members of Middle Street Christian Church, and Durfee was educated at the Friends' Academy. Rather than pursuing a college education he learned blacksmithing, the family trade practiced by his father, grandfather, and uncle James. In New Bedford their trade had a maritime bent, servicing the whaling industry. At Durfee's Wharf and other locations they were "ship smiths" who also made whale boats and acted as agents for items like anchors, chains, seashore fences, and shutters. By 1856 Durfee was a member of the firm of J. & T. Durfee, and in 1858 he became a partner. A year later the firm dissolved, due to his father's retirement and Durfee's involvement with business interests in Michigan. Despite the firm's demise Durfee remained on the town's books as a blacksmith until the end of 1862.

From the time of Durfee's birth the whaling industry had been in a decline. Rather than becoming sailors on long voyages into the Pacific, young men from the Massachusetts whaling coast moved further inland to farm or to follow trades, or they traveled further west into the Northwest Territory. New Bedford men who remained in the town and who had made money in whaling invested much of this wealth in the community, helping New Bedford to become a textile center. Those who sought higher returns on investments and were willing to take higher risks looked elsewhere for investment opportunities, often to the interior of the country. For many New Bedford residents Michigan was the focus of their investments.

In this age before investment banks, stock brokers, stock exchanges, or central banks, investment decisions were made according to the advice of someone knowledgeable in the industry, preferably someone who had visited the area and knew the "regional exchange" rates. Because the United States lacked a central bank there was not always a reliable currency. In Michigan merchants occasionally issued their own notes called "skin plaster." Wages were often paid in commodities: in Michigan a barrel of flour became the medium of exchange. The geographical area open for speculation was vast: the Northwest Territory, alone, was larger than any European state but Russia. Mining, railroads, timber, and iron making were all beckoning investment capital.

Sometimes the wealthy would give their money to a pioneer and trust him to find a good investment. It was in this way that the special tie was developed between New Bedford and Michigan. In 1835 "Michigan fever" hit the town when New Bedford residents acquired land warrants from veterans of the War of 1812. These were given to a young man destined for Michigan, who bought a good deal of land for the investors near Lansing. This link between New Bedford and Michigan would play a crucial role in Durfee's career.

Durfee was well suited to be an investment seeker, as he was young, skilled, and from an established family. Most New Bedford residents had no qualms in placing their confidence in him. Durfee was asked to travel to Jersey City, New Jersey, to investigate a new steel-making process being used by Joseph Dixon. Dixon, originally from Salem, Massachusetts, had made a number of inventions which had proved to be commercially successful. In 1849 he had improved crucible steel by lining the pots with graphite.

While in New Jersey Durfee broadened his research to include the Kelly-Bessemer controversy. Both William Kelly, in Kentucky, and Henry Bessemer, in England, claimed the invention of a process turning pig iron into wrought iron or steel. Bessemer received American patents in 1856 on his machines for working the process, and Kelly received the American rights to the process itself. Durfee became convinced that Kelly deserved the patent as the true inventor of what was known as the pneumatic process. This conviction led to Durfee's direct involvement in the iron and steel industry.

Eber B. Ward, a Detroit, Michigan, entrepreneur, was involved in many industries, including steamships, coal, copper, timber, newspapers, railroads, and iron ore. In 1855 he entered the iron-making business with Detroit backing, building a furnace and later a rolling mill ten miles south of Detroit at Wyandotte. In 1857, with funds from New England, he expanded his operation by erecting an iron rail rerolling mill in Chicago, where there were already six railroads in operation. Durfee, because of his research and experience, was hired as the technical adviser for the two iron-making operations.

Both works began to make money, especially the Chicago operation, and they continued to prosper once the Civil War began. Ward's attention had been drawn to investigate steel making using the pneumatic process by a good friend, Daniel Morrell, the general manager of the Cambria Iron Company in Johnstown, Pennsylvania. Morrell had hired Kelly to demonstrate his pneumatic process at Cambria from 1857 through 1859. In addition to Morrell, Ward consulted Durfee, who added his support not only for the Kelly process but also for the Bessemer machinery.

The three men—Ward, Durfee, and Morrell—agreed to a testing program to see whether the pneumatic process being used successfully in Europe would convert iron into steel when American ores were used, particularly the Lake Superior ores in which Ward had a financial interest. Durfee was the managing partner, or at least the mobile one, of the enterprise. Early in 1861 he made a gentlemen's agreement with Kelly for control of his patent. With Kelly's patent in hand Ward provided space at Wyandotte on which to build the converter and other apparatus necessary for the test. Durfee's cousin, William Durfee, was hired to conduct the test. Meanwhile, Morrell brought Kelly back to Cambria Iron to see if he could develop his process further. Kelly's last effort in 1862 included the trial use of a Bessemer converter sent from England by Durfee, who had made a trip there in fall 1861.

The purpose of Durfee's journey was to acquire the American rights to the Bessemer patents on machinery and to study the efforts being made in Europe to perfect the steel-making process. Bessemer refused to transfer the rights to his patents. The reason for his refusal is unknown but it could have been his uncertain situation. Bessemer was then in the midst of a number of legal cases; he was having trouble adjusting his process to making steel; and his character was being questioned because of his treatment of Robert Mushet, an English metallurgist who had received a patent on a compound known as spiegeleisen, which was added to the molten iron to produce steel. Perhaps the fact that Durfee had sided with Kelly in the matter of the American patent added to his reluctance; or possibly he refused because Durfee did not offer enough money. Whatever the reason, Durfee was unable to convince Bessemer to sell the rights to use the technology.

In his second task Durfee was more successful. He visited companies actually making steel in the Bessemer converter and others making finished products like rails and axles for the railroads. In addition to England, Durfee traveled to Sweden, where the Bessemer process had first been successful, and to France, near Bordeaux, where the process had been recently implemented. In his travels he learned of Mushet's patent on spiegeleisen. Caught up in the legal muddle with Bessemer, the patent had lapsed. Durfee believed, as did a number of continental steel makers, that it was the use of spiegeleisen that was the prime reason for the success of the Bessemer process in making steel. It was Bessemer's refusal to acknowledge this with words or money that was the cause of his public condemnation.

Durfee returned from Europe in 1862 with an estimate of the kinds of machines, such as cranes and blast engines, which would be needed to build a new Bessemer steelworks, and their cost. Money would be saved because of the smaller number of workers needed to run it. The ironworks in Michigan were prospering, largely due to the increasing or-

...ed by the Civil War. The three partners, following discussion of the value of both Durfee's ...opean venture and Kelly's Cambria efforts, de...ded to continue their plans for steel manufacturing at Wyandotte.

William Durfee requested, and his cousin approved, plans for an analytical laboratory to be added to the works, and a German scientist came to Wyandotte in spring 1863. In May the Kelly Pneumatic Process Company, as the Wyandotte firm was called, took shape. Formally, the partners or members were Ward and Zoheth Durfee, the Cambria Iron Company, and two new companies, Lyon, Shorb & Company from Detroit, and Park Brothers & Company from Pittsburgh. William Kelly was excluded from the partnership. Despite this, Kelly's patent was held in trust on behalf of his four daughters, and Kelly was guaranteed a portion of the profits.

Durfee went abroad again from fall 1863 to April 1864 in another attempt to obtain the patents for the Bessemer machines. The negotiations were long and tedious, and there was another bidder as well. Durfee, fearing that he would not reach an agreement, began to send suggestions to William Durfee on machine designs that would avoid patent infringement. His most drastic recommendation was to abandon the Bessemer mobile converter and replace it with a stationary one in the Kelly design, earlier used by the Swedes. It turned out that his fears over the negotiations were well founded, for he lost to his rival, Alexander Holley, who was working for a group from Troy, New York. Holley acquired the American rights to both the process and machines.

Durfee did much better in obtaining the American rights to the Mushet patent. These talks were conducted concurrently with the Bessemer meetings and were also long and tedious. Initially it was unclear as to who owned the patent rights. Having resolved that, it became a matter of their wanting cash advances which Durfee was not able to give. Finally a deal was struck, and on October 24, 1864, Durfee received the license for the use of Mushet's American patent. In exchange for the patent Mushet, Thomas Clare, and John Brown—the rival claimants to ownership—were admitted into the Kelly Pneumatic Process Company.

There was yet another mission to be accomplished during this last trip abroad. Durfee acted as the purchasing agent for the American government,

the Pennsylvania Railroad, and several midwestern railroads for replacement rails made of longer lasting and much cheaper steel. Since steel made with the new process was not yet available in the United States, Durfee arrived in Europe with over $300,000 in orders, $100,000 from the Pennsylvania Railroad alone. His purchases for the rail lines were more responsible for the rapid rise of British steel imports into the United States in 1863 than were the purchases of arms used to carry on the Civil War.

On his return to Wyandotte, Durfee brought along Llewellyn Hart, who had worked at a French works, to be an assistant to his cousin. In Michigan Durfee supported the construction of a new furnace to produce a better quality of molten iron, and he encouraged the company to continue to erect the unfinished stationary converter. On September 6 he and the other members of the Kelly Pneumatic Process Company witnessed the first successful making of Bessemer steel in the United States. The next spring, on May 24, 1865, he and George Fritz of the Cambria Iron Company watched at the Chicago works as Wyandotte steel was rolled into rails.

The Kelly Pneumatic Process Company grew with the success of their steel making. But the Civil War was just ending and the postwar economic adjustments meant the loss of government orders for iron. Ward was willing to expand the Wyandotte operations but was unwilling to follow the original program, which would have used William Durfee's improved process, to go into commercial production.

Another Bessemer steel plant opened in 1865, this one operated by Griswold & Winslow of Troy, New York, the firm with which Alexander Holley was associated. In its effort to hurry the production of steel, Griswold & Winslow had used Kelly's and Mushet's ideas before they were legally able, thereby infringing on patent rights. The two firms, to avoid legal problems, began to design replacement machinery. This was extremely difficult for the Wyandotte works because William Durfee had left the firm in mid 1865. Three successors were able to finish and effectively use the stationary converter, but it was not nearly as efficient as the Bessemer converter legally owned by Griswold & Winslow. Holley, working with a chromium derivative, also was unable to find a substitute for spiegeleisen, owned by the Kelly Company. The duel ended in 1866 when the two companies merged the patent rights into a new company, thereby avoiding a legal battle. A trust called the Pneumatic or Bessemer Process of Making Iron and Steel was established to avoid the delays of a court case.

The trustees of the merger were Winslow, Griswold, and Morrell, and Durfee was made secretary. The trustees then created a joint stock company in New York, known as the Pneumatic Steel Association, in which the patents were vested. Durfee became secretary-treasurer of the new company. His new position did not entirely remove Durfee from the workday world of iron and steel. As all licensees could, he went to Troy to study the steelworks and process. Throughout his career he used his ingenuity to make a number of improvements in steelmaking machinery and in steel products, obtaining 16 patents between 1862 and 1876. Of these, the most important came in 1866 for the use of a cupola on the iron furnace. The use of the cupola, rather than the reverberatory furnace as in England, became common in Bessemer plants in the United States.

Starting in 1867 it was to Durfee, acting as the general agent for the association, that entrepreneurs had to apply if they were interested in using the Bessemer process. He also collected the $5,000 license fee for plans, information, engineering help, and royalties. There were few takers, as the patents were to expire soon, and the salesmen for English steel had educated railroads on the advantages of buying their proven product now that the war was over. The new open-hearth furnace also provided strong competition to the Bessemer technique.

In 1870 William Kelly, with Durfee's help, applied for and received a seven-year extension on his patent. Durfee's honest accounting and good relations with Kelly led to the rights being assigned to the Pneumatic Steel Association, which now began to make small amounts of money. By 1880 there were still only 11 Bessemer steel plants in opera-

tion. This small number was a disappointment, but in retrospect the reason is clear. The vast quantities of pig iron that could be made quickly into steel using the Bessemer process led to the development of large plants, where furnaces and rolling mills were located side by side. This required more capital than had previously been needed to enter the iron and steel industry.

Profits were also less than had been anticipated because of the panic of 1873. The depression which followed the crisis slowed the growth of the railroads, the most avid customers for steel. By the time the Kelly renewal expired the licensees had begun to look upon the Pneumatic Steel Association as a means of limiting competition rather than of expanding it. Members began to use it as a pool to try to divide the market and set prices among themselves. As this situation developed Durfee's health deteriorated, and in 1876 he entered Butler Hospital in Providence, Rhode Island, for the treatment of a mental illness. During his absence the Pneumatic Steel Association was reorganized and renamed the Bessemer Steel Association in 1878. Durfee remained the official secretary-treasurer. On June 8, 1880, at the age of forty-nine, he died of paralysis.

References:

Henry Howland Crapo, *The Story of Henry Howland Crapo, 1804-1869* (Boston: Privately printed, 1933);

Jeanne McHugh, *Alexander Holley and the Makers of Steel* (Baltimore: Johns Hopkins University Press, 1980);

Elting E. Morison, *Men, Machines, and Modern Times* (Cambridge, Mass.: M.I.T. Press, 1966);

James M. Swank, *History of the Manufacture of Iron in All Ages, and Particularly in the United States from Colonial Times to 1891*, second edition (Philadelphia: American Iron and Steel Association, 1892);

Peter Temin, *Iron and Steel in Nineteenth-Century America: An Economic Inquiry* (Cambridge, Mass.: M.I.T. Press, 1964).

Durfee

.rges

by Paul F. Paskoff

Louisiana State University

The word "forge," as it was commonly used during much of the nineteenth century, connoted an installation at which pig iron was refined or worked up into bar or wrought iron, a metal with a lower carbon content and, therefore, greater strength than pig iron. Forges were sometimes called refinery forges to distinguish them from bloomery forges in which wrought iron was made directly from iron ore. Almost all forges during the nineteenth century were refinery forges.

Most forges consisted of two hearths, one called a finery or refinery and another called the chafery. The initial forging was executed at the finery where the pig iron was reheated to a plastic state and was then repeatedly pounded by a trip hammer. This sequence of heating and hammering was repeated until the iron had been worked up or wrought (hence the name) into thick bars shaped somewhat like attenuated dumbbells, called anconies. An ancony was usually subjected to further heating in the chafery and further pounding by the trip hammer to form it into the bar shapes demanded by wheelwrights, coopers, and blacksmiths.

As late as the mid 1800s, only half of the forges in eastern Pennsylvania, the part of the country with the largest concentration of forges, used steampower to generate the air blast for their hearths and the motive power for their hammers. Well over half of all forges were owned by individuals or 2-man partnerships, an arrangement permitted by the low capital requirements of this sort of iron-making enterprise. Forge size and, therefore, capitalization varied from one part of the country to another and reflected the local and regional character of demand for forge products. The average level of capital investment in forges in 1850 ranged from a low of about $2,300 in Alabama and New Hampshire to more than $17,000 in major iron-producing states such as Pennsylvania and New York.

Although few forges turned out specialized bar iron products for manufacturers, most forges produced bars and blooms. Unlike bars, which served as either finished goods or as intermediate goods for further processing at a slitting mill, where they were cut into narrower strips, or a wire mill, where they were drawn into rods, wire and nails, blooms were useful only as an input for rolling mills. The low heat of charcoal-fueled forge fires and the small size of their hearths and labor forces—typically 20 to 30 men and boys worked at a forge—limited their annual output to between 200 and 300 tons. This rate of production was, by 1860 and even as early as 1850, dwarfed by the rolling mills which were by then rapidly making forges obsolete survivors of a bygone era.

References:

J. P. Lesley, *The Iron Manufacturer's Guide to the Furnaces, Forges and Rolling Mills of the United States* (New York: John Wiley, 1859);

Paul F. Paskoff, *Industrial Evolution: Organization, Structure and Growth of the Pennsylvania Iron Industry, 1750-1860* (Baltimore: Johns Hopkins University Press, 1983);

Charles E. Smith, "The Manufacture of Iron in Pennsylvania," *Hunt's Merchants' Magazine*, 25 (November 1851): 574-581 and tables following 656;

United States Bureau of the Census, *Seventh United States Census, Compendium* (Washington, D.C.: GPO, 1854);

United States Bureau of the Census, *Eighth United States Census*, volume 3: *Manufactures of the United States in 1860* (Washington, D.C.: GPO, 1865);

United States Bureau of the Census, *Ninth United States Census*, volume 3: *Wealth and Industry* (Washington, D.C.: GPO, 1872).

Henry Clay Frick

(December 19, 1849-December 2, 1919)

by John N. Ingham

University of Toronto

CAREER: President and chairman, H. C. Frick Coke Company (1871-1900); chairman, Carnegie Brothers & Company (1889-1892); chairman, Carnegie Steel Company (1882-1894); honorary chairman, Carnegie Steel (1894-1900); director, United States Steel Corporation (1901-1919)

Known as the "Master Manager," Henry Clay Frick was one of the most dynamic and important industrialists of the "heroic age" of American business in the late nineteenth century. Frick consolidated the vitally important bituminous coal and coke fields of the Connellsville region of western Pennsylvania and managed the massive Carnegie steel complex during its years of great expansion and controversy in the 1890s. During the last 20 years of his life Frick established precedents for art collectors and benefactors by his actions.

Frick was born on his father's hardscrabble farm in West Overton, Pennsylvania, on December 19, 1849. Both his father's and mother's (Overholt) families were of Germanic origin. The Fricks came from Switzerland to America in 1732, and the Overholts came from the Palatinate on the Rhine in 1767. Henry Clay Frick's father, John W. Frick, was born in Adamsburg, Pennsylvania, in 1822, son of a farmer in the area. John Frick was a classic ne'er-do-well who fancied himself a painter but was forced to depend upon his meager farming skills in providing a marginal livelihood for himself and his family. Henry Frick had nothing but disdain for his unsuccessful father but greatly admired his maternal grandfather, Abraham Overholt. The Overholts were Mennonites who came to eastern Pennsylvania in the 1760s, where they flourished. In 1800 the entire extended family of some 33 members crossed the Alleghenies where they purchased several hundred acres in East Huntington Township and established the village of West Overton. Abraham Overholt, then seventeen years old, soon came into

Henry Clay Frick

possession of the large homestead farm, on which was a small log distillery, which Abraham used as the basis of the largest fortune in that section of the country. Increasing the capacity of the distillery from 3 bushels to 50 bushels of grain per day, by 1859 it was a massive operation for its time. Capable of distilling 200 bushels of grain daily, the "Overholt" brand of whiskey became famous in the region. But Abraham Overholt was not content with being a distiller; he also developed coal lands and was intimately involved in the social and political affairs of the region. He was, in a word, a local notable–the regional "squire" to whom all other families in the area gave their fealty and admiration. In 1847 his fifth child, Elizabeth, chose to marry John W. Frick, much to the consternation of her father and family. The couple moved into the small spring house on her father's massive farm, where two years later their son Henry Clay Frick was born.

Henry Frick was educated in the village schools but, except for mathematics, at which he excelled, was an indifferent student. At an early age he developed contempt for his struggling father, transferring his admiration to his grandfather, "Grandpap Overholt." His grandfather and other members of the Overholt family soon took young Henry Frick under their wing. In 1863 the fourteen-year-old Henry was given a position in his uncle Christian Overholt's store in West Overton, and two years later he took a job in his uncle Martin Overholt's store in Mt. Pleasant, Pennsylvania. All of this tutelage and sponsorship took place under the watchful eye of Abraham Overholt. After a falling-out with his uncle in Mt. Pleasant in 1868, Frick went to Pittsburgh, where the Overholt family connections secured for him a position as a clerk in the large Macrum & Carlisle store. Five months later the eighty-six-year-old Abraham Overholt brought his favorite grandson back to West Overton, where he rewarded Henry with the position of chief bookkeeper at $1,000 a year. Henry flourished at the Overholt distillery, and in January 1870, when Frick was just twenty, his grandfather passed away.

It was at this same time that Henry Frick began to move into the rapidly developing coal and coke industry in the region. His cousin, Abraham Tinstman, had been involved in the industry on a small scale for about 12 years. Tinstman had bought 600 acres of coal lands in 1859 with Joseph Rist, a longtime business associate. In 1868 they had joined their holdings with those of Col. A. S. M. Morgan for the purpose of manufacturing coke for fuel. This was one of the pioneer coke operations in the country, but there was not enough demand for the product to make their operations profitable. Morgan soon dropped out of the partnership, but Tinstman was so deeply in debt that he could not. Tinstman, who was a managing partner in the Overholt distillery, came to talk to Henry Frick about his unfortunate business investment.

Frick, who had no money but much vision, foresaw opportunity where Tinstman could only conceive of continuing disaster.

Frick, who had spent only a few months in Pittsburgh, was nonetheless aware that the city's iron and steel industry was on the brink of a revolutionary transformation. Pittsburgh's iron industry did not depend upon coke, but the emerging steel industry did. Although Bessemer steel was still in its infancy, and Andrew Carnegie had not yet converted to Bessemer production, Frick was astute enough to recognize that it was the wave of the future. He was certain that Bessemer production would dominate the iron and steel industry and was equally determined that he would control the supply of coke so essential to its operation.

Frick therefore advised his cousin not to liquidate his holdings but to expand them as rapidly as possible. Tinstman was astounded by Frick's boldness but felt he had little to lose at that point. Frick and Tinstman, together with Rist and another Overholt cousin, formed a new company and, borrowing every cent they could, invested $52,995 in the purchase of 123 acres of coal land. They incorporated the new company under the name of H. C. Frick Coke Company, since the older partners did not wish their names and reputations to be so closely connected with Frick's impetuous gamble. Once started, however, Frick could not be headed. Wishing to supply ever larger amounts of coke, he bought more and more coal land and set up increasing numbers of beehive ovens. To do this Frick had to borrow greater amounts of money from relatives, even persuading his father to mortgage his small farm to provide cash for expansion. When this proved insufficient, Frick turned to wealthy area farmers for capital and, finally in late 1871, approached the wealthy Pittsburgh banker, Judge Thomas Mellon, for financial backing. Mellon, usually quite conservative in his business dealings, rather surprisingly loaned Frick $10,000 to construct 50 additional coke ovens. Mellon was impressed with the talents and ambition of the young Frick but was also convinced that the young man's long-range vision for the Bessemer steel industry in the area was correct. In the process Frick gained a valuable ally and associate in the Mellon family.

By 1873 the H. C. Frick Coke Company was a great success. Frick was able to sell all the coke he could produce at a good price and had already paid back his most pressing debts. Prospects looked good for the ambitious twenty-four-year-old. He had promised himself he would become a millionaire before he was thirty and was well on the way to achieving his goal. Then, like a thunderbolt, came the panic of 1873 and the ensuing five-year depression. A prudent businessman would have followed the dictates of sensible business practice: cut back on expenditures, ride out the worst of the bad time, and then expand facilities when prices began to rise again. But Frick, like his future partner Andrew Carnegie, had different ideas. The business collapse severely depressed the price of coal lands, and Frick was determined to buy out as many of his competitors as he could.

Frick's partners had gone along with his apparent recklessness before, but they could not afford to risk their broader capital concerns in such perilous times. Taking advantage of a loan of $37,000 from the trustees of his mother's portion of his grandfather's estate, Frick bought out his partners and also began buying additional coal land and coke ovens at rock-bottom prices. Just as Andrew Carnegie pushed ahead with construction of his Edgar Thomson Steel Works during the depression, at a savings of about 25 percent on construction costs, Frick knew he could gain control of the Connellsville coke fields much more cheaply if he had the courage to expand at this time. It seems quite evident in hindsight that the policies followed by Carnegie and Frick were correct, but it was far less certain at the time. Frick himself commented toward the end of his life that the 1870s were "an awful time."

Frick's next move was even bolder. Just before the crash the local community had furnished the capital for a 10-mile railroad from Broadford to Mt. Pleasant, where it connected with the Baltimore & Ohio (B&O) Railroad. With the economic downturn, and the consequent reduction in coke shipments, the line was facing bankruptcy. Frick, however, had a plan. He obtained a list of the railroad's stockholders, who were widely scattered around the area, and set out one night to obtain options from them for sale of their shares. Since they feared they might get nothing for their investment, they gladly signed the options. Frick then took them to the B&O and persuaded the railroad that the small line, although not making money at present, would be a shrewd purchase at its bargain price. Part of the purchase price included a $50,000 commission for Frick. The railroad accepted, and

Frick was able to use that money to purchase still more coal land.

In 1874 Frick returned to Judge Mellon for additional financing. Despite the fact that Mellon had restricted lending during the hard times, Frick's quick repayment of his earlier loan and his cunning exhibited during the B&O deal convinced the banker that Frick was a good risk. Mellon lent Frick $15,000 to purchase freight cars to ship coke over the railroad he had just sold and also gave him a $25,000 line of credit. Two years later Frick, still lusting after more land and ovens, which by now were even cheaper, persuaded the Mellons to accept a new mortgage as security for a $76,000 loan and also convinced them to discount business paper not exceeding $24,000. It was an extraordinary transaction on both parts and says much about why both the Mellons and Frick were able to amass such sizable fortunes. Extremely conservative and careful with money, both also had an uncanny eye for the main chance and were willing to take great risks to achieve that. Others often saw them as reckless, but it was calculated daring, based on much study and a firm conviction that good times would return. When they did, both Frick and the Mellons were in positions to reap enormous benefits.

By 1878 Frick had vastly expanded his holdings in the coalfields but had gone about as far as borrowed capital could carry him. Yet, Frick knew that if he were to achieve his goal of total domination of the Connellsville coke fields, he must not slow his efforts. As a consequence Frick decided to take on a new partner at H. C. Frick Coke—E. M. Ferguson, a wealthy Pittsburgh capitalist from an old and prestigious family. With this capital backing Frick was able to buy out the larger, but financially troubled, Morgan coalfields. Almost as soon as he completed that transaction, the depression ended and good times returned. Suddenly the railroads, which had lain dormant for five years, were feverishly expanding into new territory. As a result, orders for steel rails poured into Pittsburgh iron and steel mills in unprecedented numbers. Suddenly the Pittsburgh mills could not get enough coke, and prices, which for several years had stood at less than $1 per ton, skyrocketed to $5 per ton. H. C. Frick & Company controlled the Connellsville coke fields, and Henry Frick was becoming a very rich young man. By the time of his thirtieth birthday in 1879, Frick had achieved an important milestone. He went to his small office on that day and calcu-

Frick's home in Pittsburgh, "Clayton" (courtesy of Palmer's Pittsburgh)

lated his net worth. With 1,000 coke ovens and 3,000 acres of coal lands, Frick was a millionaire. He had achieved his goal, yet his career was only beginning.

At about this time Henry Frick made his only truly close friend–Andrew W. Mellon, son of Judge Thomas Mellon. The elder Mellon retired from active management of the banking firm in 1879, and Andrew assumed control. Although Mellon was four years younger than Frick, the two remote, cold men hit it off immediately. Although neither would ever make other close friends, they remained intimate companions for the rest of their lives. In 1880 they celebrated their friendship and financial success by taking a trip to Europe together and repeated the European cruise together nearly every summer for the rest of their lives. Frick's friendship with the Mellons, his business association with the Fergusons, and his grandfather's prestige opened the doors of Pittsburgh society to him in the winter of 1880-1881. Frick soon was attending many balls and parties and at one of these met Adelaide Howard Childs, daughter of Asa P. Childs, one of Pittsburgh's most prestigious businessmen. It was love at first sight for Frick, and evidently Miss Childs was equally enamored, as they were married on December 15, 1881. Their subsequent honeymoon also brought Frick into another partnership,

one which had great historic significance for America's steel industry.

On their wedding trip the young couple traveled around the East Coast, finally arriving in New York City. But Frick, whose mind was never far from business, had taken along with him a note of congratulation and invitation for dinner in New York from his biggest coke customer, Andrew Carnegie. Despite their business relations, the two men had never met, and Frick was pleased at the chance to finally do so at a dinner with Carnegie and a few of his friends and business associates at the Windsor Hotel. Yet Frick was also wary of what the canny Scot had in store for him. Carnegie had a reputation as a conniving and manipulative man, and Frick knew that something more than simple good wishes was behind the invitation. He soon found out what. Carnegie raised his glass to toast the newly married couple and then proposed an additional toast to Henry Frick and to the success of a Carnegie-Frick partnership. Frick was surprised by the audacity of the proposal but quickly acceded to it, since it fulfilled important strategic goals and visions for both men.

For Carnegie the advantages were clear and immediate. During the early 1870s, when there were a large number of coke producers in the Connellsville region, Carnegie had little concern about either price or supply. With a multiplicity of producers, he, as the largest purchaser, was assured of the upper hand. By the end of the 1870s, however, he was much more vulnerable. Frick now controlled 80 percent of the coalfields and ovens and was in a position to put a stranglehold on the Carnegie interests. There is no evidence that Frick had ever given the idea any serious thought, but Carnegie was not the type to take chances of this sort. And, besides, Frick, like Carnegie, had proven himself an opportunist of the first order, so how long could he resist an opportunity to squeeze the nation's largest steel company and his biggest customer? Carnegie did not wait to find out. He began buying shares of H. C. Frick Coke on the open market but wanted more than that—he wanted to control his supply of coke, and he could do that only through a partnership with Frick.

The advantages for Frick were less direct but just as important. For one thing, Frick needed more capital. His arrangement with the Ferguson family had helped but was not nearly enough to finance his ambitions. Secondly, his coke company would garner much prestige by its connection to the massive Carnegie interests. Finally, and probably most important, Frick would be able to enter the steel industry. Ever since he had unfurled his vision of a growing steel industry in Pittsburgh for Judge Mellon, Frick had wanted to become a player on the main field. It was fine to be a supplier of a vitally needed resource to the industry, but in Pittsburgh steel was king; coke was a sideline. Frick wanted a chance to prove his mettle in the city's dominant industry, and he realized his association with Carnegie might give him that chance. In some respects Frick was the ideal manager. He was a brilliant tactician, with a powerful and determined intellect. As Carnegie partner John Walker commented, Frick was von Moltke to Carnegie's Napoleon. That is, Carnegie was the "commander and intuitive genius," while Frick was "long-headed, deliberate, a great tactician." But in one respect Frick was not the ideal manager, the perfect technician—he did not like to take orders. Cold and imperious, Frick could fly into blind rages when he felt his judgment or reputation were slighted. And Carnegie was infamous for his cavalier treatment of his partners. Most did not survive long, and nearly all, including Carnegie's own brother, were driven virtually to the point of distraction by Carnegie's demands. So, de-

spite the advantages accruing to both sides, this was not to be a marriage made in heaven.

In any event the partnership between the two industrial tycoons was rapidly consummated. The Henry C. Frick Coke Company was organized with capitalization of $2 million and 40,000 shares of stock. Frick, who had owned one-third of his original company, had his share reduced to 29.5 percent in the new firm. The Ferguson family, which had owned two-thirds of the old firm, had 59 percent of the revamped concern. The Carnegie interests had just over 11 percent. During the next 18 months, however, Carnegie pressured the Fergusons into selling him a large block of their stock, so that by the summer of 1883 he had become the largest single stockholder. A short time later, by assuming all outstanding debts in the concern, Carnegie obtained a 50 percent interest in the company, which by then was capitalized at $3 million. Frick remained in charge of all operations, and Carnegie regarded the "coke king's" managerial ability so highly that he asked no questions about his management. What had Frick gotten out of this? The deal had enabled Frick to pay off all of his debts for the first time in his business life. No longer just a millionaire on paper, he now had actual cash to purchase items which were part of the good life of Victorian America. He also had at least limited access to Carnegie's massive capital resources for future expansion. Further, although Frick and Andrew Carnegie had, at best, a tempestuous relationship, Frick found he worked well with Andrew's younger brother, Thomas Carnegie, who was manager of the Carnegie concerns of Pittsburgh. With Frick allowed to handle the coke concerns without interference, and with Tom Carnegie the contact man in Pittsburgh while Andrew globe-trotted around the world, the arrangement seemed to work well for all concerned. In 1886-1887, however, the situation changed dramatically.

Most significant was Tom Carnegie's sudden death in 1886. This set in motion a surprising chain of events. To replace him as manager of all the Carnegie concerns, Andrew chose his longtime partner, Henry Phipps, Jr. But everyone, including Phipps, knew this was a stopgap measure. Phipps was good at many things, but management was not one of them. The crown prince—the "master manager"—was waiting in the wings, but the road to the head of the Carnegie steel interest was not smooth and even. In fact it almost never happened at all.

Although in many respects Andrew Carnegie and Henry Frick were quite similar, in terms of temperament they could not have been more different. Carnegie was a gregarious "glad-hander," who craved public attention and wanted to be loved by all, including his workers. Frick, on the other hand, was a rather dour man who had few friends and was reserved and formal in all his personal and business relations. Of greatest significance, Frick had learned heavy-handed tactics for dealing with labor in the coal fields. Unions were barely tolerated, strikes were ruthlessly suppressed, and violence was often employed by both sides. Carnegie, on the other hand, wrote articles extolling the "dignity of labor," and adopted a conciliatory attitude (at least in theory) toward unions, strikes, and workers. He abhorred the idea of labor violence and always felt he could win labor confrontations by simply waiting out the strikers. This temperamental difference between the two men in regard to labor was to plague their relationship over the years. The first great flare-up came in 1887.

In January of that year Carnegie had taken the first step toward making Frick head of his steel enterprises by offering him 2 percent of the total capital of Carnegie Brothers. Frick was not required to put up any capital for this share but would be allowed to pay for it out of accumulated dividends. In return Frick had to sign the so-called "Iron Clad Agreement," by which all partners agreed to sell their shares back to the firm at book value (which was far below market value) if demanded by interests representing 75 percent of the stock. Frick complied, but soon afterward events occurred which severely strained his relationship with Carnegie.

In spring 1887, while Carnegie was on his honeymoon in Europe, a strike was called in the coalfields by the local lodges. Frick took the leadership among the coal operators in vowing not to give in to the men's demands and promising to import strikebreakers if necessary. But Henry Phipps, as head of the steel firm, was concerned about the continuing supply of coke for the mills, which were running full and had thousands of tons of back orders. Phipps implored Carnegie to intercede, which he did, instructing Frick to accede to the striker's demands at once. Frick had no choice but to comply, since Carnegie held a majority of stock in Frick Coke, but he was mortified and furious. Frick, who

had a reputation as one of the toughest labor managers in the Connellsville fields and who had promised his cohorts leadership in this strike, had given in to the workers' demands without a fight. As a result, the day the strike was settled, Frick sent Carnegie his resignation as president of the coke company, saying that he objected to "so manifest a prostitution of the Coke Company's interests in order to promote your steel interests."

At this point Carnegie must have had some misgivings about Frick's skills in labor relations, but as he later wrote in his autobiography, he believed Frick "a man with a positive genius" in management, so Carnegie wrote him a series of flattering letters, urging him to remain as head of Frick Coke. Then, too, Frick saw that the coke company continued to make enormous profits, even with the wage increase. His ovens were simply more efficient than those of his competitors, so that by agreeing to the wage demands, which became the standard for all operators in the region, he was actually doing much to drive the other companies out of business. Most important, though, was that Frick retained an almost paternal love for his coke company. He did not want to give it up but also realized that he could protect it only if he played a more significant role in the management of the parent company. In this ironic manner the interests of the two headstrong men moved ever closer together. Six months later Frick accepted reelection to the presidency of Frick Coke. Then in January 1889 Frick replaced Phipps as chairman of Carnegie Brothers & Company. The master manager had arrived at the helm of the world's largest steel company. Over the next decade much of the daring and innovation at the Carnegie works came more as a result of Frick's actions and ideas than Carnegie's. In return for his new responsibilities Frick's share of the giant steel concern was raised from 2 percent to 11 percent, again, without having to put up a penny in cash.

Within a year after assuming the chair of the company Frick proved his worth to the Carnegie firm by taking over the troubled Allegheny Bessemer Steel Company, which became the Duquesne Works of Carnegie Steel. The most technologically advanced steel firm in the industry, the plant had been plagued with labor and management problems from the day it opened. But it provided the first serious competition to Carnegie's steel rail monopoly in the Pittsburgh region, and since it used the "direct process" for making steel, which eliminated an expensive step in the operation, it could potentially produce rails at a lower price. But the firm's labor problems proved insurmountable, and William G. Park approached Frick about selling the works to Carnegie Steel. Frick paid $1 million for the company, and the plant paid for itself within a year. It was a masterful acquisition and is generally considered one of the great "steals" in American industry.

Once Duquesne had been absorbed, Frick set about streamlining the entire Carnegie empire. The Carnegie interests had traditionally been divided into two units–Carnegie, Phipps & Company and Carnegie Brothers & Company. Carnegie had sound financial reasons for keeping the firms separate, since each one could discount the notes of the other to provide working capital. But Frick was convinced that the decentralized nature of the Carnegie concerns was keeping costs higher by duplicating operations and facilities. He also argued strenuously for a much larger capitalization which would more realistically represent the true value of the vast holdings. Carnegie remained skeptical, but when he found that profits had increased over 75 percent to $3.5 million during the first year of Frick's management, he was impressed with Frick's acumen. He therefore gave him the go-ahead to proceed with the reorganization. On July 1, 1892, Carnegie Steel Company, Ltd. took over all the assets of the Carnegie interests. With capital of $25 million, Carnegie Steel was 2½ times larger than the two former firms combined. The new concern had three complete Bessemer steel mills, Keystone Bridge Company, Lucy Furnace, Union Iron, Frick Coke, and a number of other properties.

Frick's consolidation and reorganization of Carnegie Steel went beyond simple paper manipulation. To connect the widely scattered and diverse Carnegie enterprises, Frick constructed the Union Railway, a company-owned railroad that served all the company's various plants, steel mills, and furnaces in the Monongahela Valley and Pittsburgh areas and which connected with the Pennsylvania Railroad. The Union Railway freed the firm from dependence on the Pennsylvania and B&O railways for this service, and the savings to Carnegie in switching charges alone was enough to pay the interest on the cost of building the railway. With less than 100 miles of trackage, the railway in 1899 carried as much freight as many of the major railway systems in the country–16 million tons.

The future looked bright for Carnegie Steel and Henry Frick in 1892. By far the largest steel firm in the industry, Carnegie's serious competitors had been routed and the steel trade was booming. Profits averaged $4.5 million a year. But trouble loomed on the near horizon at the firm's Homestead plant, long a seedbed of union agitation and labor unrest. At Homestead the divergent philosophies and attitudes of Carnegie and Frick were put to the test, with disastrous results for all concerned.

Upon assumption of the chairmanship of Carnegie Steel, one of Frick's first accomplishments was to reassert his influence and hardfisted policies in labor negotiations. In February 1890 the unions in the coke fields called for a strike to force acceptance of a new general scale in the industry. Frick had calmly prepared himself for this siege, having filled the coke bins at Carnegie Steel and reduced orders to the plants. He refused to negotiate with the union and began bringing strikebreakers into the mines. The enraged strikers responded with violence and terrorism–but to no avail. Convincing county authorities to intervene on behalf of "law and order," the sheriff promised to shoot to kill and did in fact kill several strikers. As a result the strike was thoroughly crushed after three months, and the union was driven from the region. This time, Andrew Carnegie did not lift a finger to restrain Frick. Thus, when labor problems began brewing at the Homestead plant, Frick was certain he enjoyed the full confidence and support of the firm's majority stockholder.

The Homestead Works, like Duquesne, had fallen into Carnegie's hands as a result of troubled labor relations. When Carnegie took it over in 1883 he settled with the unions on generous terms, making his money in later years out of the plant's efficient operation and a booming market. In 1889, however, Carnegie management wanted to put the men on the same sliding scale as was already in place at the Duquesne and Braddock plants, a change that would result in a 25 percent reduction in wages. The union was furious and called the men out. Union leadership promised a battle to the death, and William Abbott, chairman of Carnegie, Phipps, intimidated by threats of violence and destruction of property, signed a 3-year agreement with the union which amounted to a nearly complete capitulation. Although the union accepted the sliding scale, it became the sole bargaining agent in the mill, and no men could be hired or fired with-

Frick with Adelaide Childs of Pittsburgh, whom he married on December 15, 1881 (courtesy of Miss Helen Frick)

out union approval. Frick and Carnegie were furious, feeling that Abbott had, in effect, turned management of the mill over to the union. As a result Abbott was forced out of the firm, and Frick inherited a nasty labor problem which festered until the union contract ran out in 1892.

Frick, for good reason, has long been regarded as the "villain" in the 1892 Homestead strike. His views on labor were certainly not liberal or progressive. James M. Guffey, a liberal New Deal senator whose father had been backed by and then dropped by Frick's allies the Mellons, was not an objective commentator. Nonetheless, his recollections of Frick are worth noting. He commented that Frick was "without question the most cold-blooded man I have ever known–absolutely without any trace of sympathy for the working man." One of Carnegie's workers later put the issue in clear terms when speaking to Andrew Carnegie: "It wasn't a question of dollars, the boys would have let you kick them, but they wouldn't let the other man stroke their hair."

But as recent scholars have emphasized, both Andrew Carnegie and the union must also share

the blame. Carnegie waxed eloquently about the rights of labor and the legitimacy of labor unions, but, truth be known, did not like them any more than did other steel manufacturers. He preferred to keep them out of his plants, but he was not an ideologue on the matter. Expediency was Carnegie's operating philosophy, and the key to his operation was to keep his mills running full at all times. So he often preferred to settle with unions if it meant he could keep the mills operating. But the whole matter was always counterbalanced and, indeed, dominated by the fact that Carnegie was a fanatic about costs. His favorite saying was "watch the costs and the profits will take care of themselves." As a result Carnegie was always badgering his managers continually to reduce costs. Ultimately the question of costs would run headlong into union demands, and the result would inevitably be conflict.

Then, too, the Amalgamated Association of Iron and Steel Workers had a virtual stranglehold on vital elements which affected the cost structure in the Homestead plant. Many older workers in the early twentieth century even admitted that prior to the Homestead strike, the union, in effect, ran the plant. Leon Wolff, in his exhaustive study of the strike, *Lockout*, reached a similar conclusion. What is obvious, then, is that some kind of confrontation between the Carnegie mills and the union was inevitable, whether Frick had been chairman or not. It is also clear from past history at Homestead that the union was quite prepared to use violence and terrorism if necessary to protect their position. But Frick brought to the confrontation an arrogance and intransigence that exacerbated the situation into one of the most tragic strikes in American history.

As the time approached for renewal of the Amalgamated rate at the Homestead plant in 1892, Carnegie and Frick had two demands. One was to lower the minimum on the sliding scale. The other, and more significant, demand was to eliminate the union as the exclusive bargaining agent in the mill—that is, to erase Abbott's costly blunder. Just before the contract expired, however, Carnegie sailed for Europe, leaving the taciturn Frick in complete command of the negotiations. Although Carnegie later claimed he was not aware of what Frick was planning to do, and that his aggressive, violent actions violated Carnegie's concept of peaceful, dignified labor negotiations, Carnegie in 1890 had been given a preview of Frick's labor policies in the coalfields, and he had not intervened in any way.

The Amalgamated, which had grown accustomed to a privileged position in the Homestead mill, had given clear indication in 1889 and after that they would not relinquish their power without a struggle. Further, if Frick tried to bring in strikebreakers, they were willing to respond with violence and terrorism if necessary. Thus Frick prepared the mill well for the impending confrontation. He erected a massive stockade around the works, bristling with watchtowers, gun slits, and barbed wire. So awesome and obvious were his preparations that workers and townspeople derisively dubbed the mill "Fort Frick." Next, Frick arranged with the Pinkerton agency to send 300 of their men into the plant, planning to have them take over the works and provide protection for the strikebreakers that Frick planned to use to reopen the works.

The strike began on July 1 and the Pinkertons arrived five days later. Coming down the river on barges during the night, lookouts spotted them and sent word ahead. When the barges arrived at Homestead, a pitched battle ensued between the Pinkertons and workers and townspeople which lasted all the next day. Finally the Pinkertons ran up the white flag and a truce was negotiated, after which the guards were forced to run the gauntlet through a howling mob. For several days the workers held the plant unchallenged, while Frick and the county sheriff appealed again and again for the governor, Robert E. Pattison, to send in troops. He finally relented and on July 12 sent hundreds of troops in to occupy the plant. The workers thought they could convince these good Pennsylvania workingmen to support their strike and met them at the train with a brass band and welcoming committee. But it soon became clear that the militia was there to protect property rights, not jobs, and the commander sided with the interests of the mill owners.

The strike dragged on, and workers at other Carnegie mills in the area joined the Homestead men in sympathy. In mid July Frick gave notice that workers had until July 21 to apply for rehiring, but not a single locked-out man applied. The tense struggle continued, and the Carnegie Company continued to bring strikebreakers into the mill but few highly skilled men could be found to do some of the most essential jobs in the plant. It was a classic standoff.

On July 23, however, an event occurred which did much to shift public opinion away from the strikers. On that afternoon, as Frick was sitting in his office, Alexander Berkman, a recent emigrant from Lithuania who was not a member of the Amalgamated, burst in and fired twice at Frick with a pistol at close range. Both bullets struck Frick in the neck. John G. A. Leishman, a vice-president of the Carnegie works, grabbed Berkman and wrestled him to the ground. Frick tried to assist and was stabbed three times by the assailant. At that point others entered the office, and a deputy sheriff raised his gun to shoot Berkman. Frick stopped him, saying, "Don't shoot. Leave him to the law but raise his head and let me see his face." They then discovered Berkman was chewing fulminate of mercury in an attempt to blow up the room. After Berkman was taken away, Frick remained in his office and, without accepting an anesthetic, had the surgeon extract the bullets and dress his wounds. Frick answered his mail and went about his duties as if nothing had happened.

The story of the attempted assassination of Frick made headlines around the world, and much public sympathy shifted to the side of the company, even though Berkman was in no way connected with the union. As an Amalgamated officer said,

Berkman's bullets "went straight through the heart of the Homestead strike." Frick remained in control of the strike and refused to budge an inch on his demands.

But the union was equally determined. As a result, the strike held solid for four months. Frick was amazed at the determination of the strikers, writing Carnegie that "the firmness with which these strikers hold on is surprising to everyone." But Frick was a tough, unyielding labor negotiator, and with the plant finally able to resume production in the fall and with winter approaching, the strikers' morale finally began to falter. On November 18 the unskilled workers asked to be released from their strike pledge, and two days later the Amalgamated called off the strike. The men were forced to return to work as individuals, and the union was broken in the Carnegie mills. Frick wired Carnegie: "Our victory is now complete and most gratifying. Do not think we will have any serious labor trouble again."

Carnegie was at first elated and sent Frick hearty congratulations. But Carnegie, ever sensitive to public opinion, soon found that the world press was highly critical of the way the workers had been treated in the strike. They particularly harped upon the discrepancy between Carnegie's professions of liberal labor policies and the severe tactics employed at the Homestead mill. As a result, although he continued to support Frick publicly, Carnegie privately began to have second thoughts. He told his friend, William E. Gladstone, the British prime minister, that it had been a mistake to try "to run the Homestead Works with new men . . . It was expecting too much of poor men to stand idly by and see their work taken by others." Carnegie never again trusted Frick and determined he would keep him under tight rein. He also decided he would ease him out of the chairmanship as soon as possible. His chance came two years later, when Frick, miffed over a relatively minor matter, sent in one of his frequent resignations as chairman. This time Carnegie accepted it, while allowing Frick to remain head of the coal concern and as a partner in the steel complex. He was also given the title of chairman of the board, a purely honorary position. Frick's 11 percent interest was reduced to 6 percent, but he continued to play a vital role in the development of the giant steel concern throughout the rest of the decade. He was able to continue his influence for two reasons. First of all, Carnegie under-stood that however much he disagreed with some of Frick's ideas, and however difficult it was for the two to get along, Frick was still an extraordinarily astute manager. Secondly, Frick, in effect, hand-picked his replacement, John G. A. Leishman, his trusted assistant. Leishman, who was not one of "Andy's young men," but had closer ties to the older Pittsburgh families, was a strong ally of Frick as long as he was president of Carnegie Steel.

Frick's most important contribution to the success of the Carnegie firm in the later 1890s involved his influence in the decision to buy large holdings in the Lake Superior ore regions. Carnegie, through his acquisition of Frick Coke, had gained control of sufficient supplies of coke at a reasonable price. He also purchased plentiful limestone deposits with his control of the Pittsburgh Limestone Company. The firm did not, however, control its own ore supply, and for some quixotic reason Andrew Carnegie was vehemently opposed to doing so. Like other steel firms Carnegie Steel relied on ore from the Marquette region of Michigan. Frick was convinced that the firm's lack of control over this vital raw material was a serious potential weakness. Just as he had garnered control of the Connellsville coalfields, so too could someone gain control of iron ore deposits.

A vast new deposit had recently been discovered in the Mesabi region of northern Minnesota. Just lying on the ground, the enormous fields of powdery ore were simple and inexpensive to mine, but many steel men felt it could not be used successfully in blast furnaces. By 1892, however, it had become clear at least to some that it could be employed successfully. At that point Henry W. Oliver, a rival Pittsburgh ironmaker and inveterate speculator, arrived upon the scene. He bought out the local owners with a worthless check and rushed back to Pittsburgh to persuade Frick of the importance of his holdings. Frick was greatly impressed and told Oliver he wanted Carnegie Steel to make a deal with him. Carnegie, however, was of a different mind. He thought owning ore fields was a waste of money, pointing out that John D. Rockefeller had controlled the oil industry in Pennsylvania not by ownership of raw materials but by his control of transportation and refining facilities. Also, Carnegie felt Oliver was a complete charlatan. They had worked as telegraph boys together in the early years, and he wanted no part of him. As Carnegie wrote to Frick, "Oliver's ore bargain is just like him—nothing in it." Frick would not back down, however, and ultimately persuaded Carnegie to purchase a 50 percent interest in Oliver's ore firm. No sooner had they made the deal than Frick's prediction came true. John D. Rockefeller moved into the Mesabi region, bought up all the rest of the holdings, and set up a line of ore boats on the lakes. But Carnegie, with his vast ore sources in the area, was now well protected and ultimately reached a deal with Rockefeller whereby he leased the latter's ore lands for 50 years and promised to ship most of his ore on Rockefeller's railroads and boat lines. *Iron Age* called this Carnegie's greatest achievement, but without Frick's vision and persistence it would not have happened, and the outcome of the confrontation between Rockefeller and Carnegie might well have been very different.

Despite these successes, the relations between Carnegie and Frick became more bitter as the years progressed. The situation became particularly difficult for Frick after 1897. In that year Leishman was fired by Carnegie for speculating. He was replaced by Charles Schwab, who was a firm Carnegie loyalist and less willing to listen to Frick's advice. As a result Frick wanted out of the Carnegie enterprises but was stymied by the Iron Clad Agreement he had signed years before. If he resigned he would only receive book value for his stock, at least $10 million less than it was worth on the open market. Further, since Carnegie paid notoriously low dividends, Frick did not have much money available for alternative investments. The situation, then, was vexing for Frick. Ignored at Carnegie Steel, restricted in his operation of the coke company, and unable to pursue outside business interests, he felt that he was being cut off in the prime of his life from pursuing his dreams.

Henry Phipps, Carnegie's longtime factotum and confidant, was similarly vexed. When Schwab announced in 1897 a vast expansion plan which would cut even more deeply into dividends, Phipps protested to Carnegie. Frick and Phipps thus became coconspirators in the late 1890s to find a buyer for Carnegie Steel who would enable them to emerge with their fortunes intact. They put out feelers on Wall Street in 1899 and found that William H. Moore and his syndicate were willing to offer $330 million to buy and reorganize the Carnegie firm. This was a staggering amount for the time and awesome for a company whose book value was only $50 million. Frick and Phipps stood to make

huge fortunes if they could pull if off. Further, Carnegie had been talking about retiring but had also made clear that he would never sell to a "speculator," and no one was more of a speculator than Moore. So, they knew they had to keep Moore's identity from Carnegie. Claiming they had been sworn to secrecy, Frick and Phipps would not tell Carnegie the name of the purchaser. The canny Scotsman was intrigued but suspicious, so he agreed to sell but demanded that $2 million in cash had to be put up for the option, which had to be picked up in 90 days. This meant $1,170,000 in cash had to be put up by Frick and Phipps (covering Carnegie's 58 percent of the company). The Moore syndicate was able to put up only $1 million, so Frick and Phipps had to contribute the other $170,000 out of their own pockets. They did so, confident that everything would be fine and also gloating over the fact that they were to receive a $5 million commission for the sale. Then disaster struck. First, Carnegie found out that Moore was the buyer. This infuriated him, but he had no way to cancel the proposed sale. Second, Moore found it impossible to raise the necessary cash for the deal on Wall Street in the required 90 days. Carnegie refused to grant any extensions, pocketed the $1,170,000, and proudly bragged to visitors at his Skibo Castle in Scotland years later that it was a "gift" from his "friend" Frick. But this was the last straw for Carnegie. He was now determined to force Frick from the firm.

Carnegie and Frick had verbally agreed that the steel company would pay Frick Coke $1.35 a ton for coke from 1899 to 1902. The price of coke began to escalate rapidly, and Frick soon tacked a "surcharge" on the coke, bringing the total price as high at $1.65. Schwab and Carnegie paid the Frick invoices only up to $1.35 a ton and marked the rest as "payments on advances only." Frick was furious, but his hands were tied. His revenge came over a minor matter. He learned that Carnegie Steel planned to buy a mill site at Peter's Creek and purchased the site himself. He then drove up the price to make a tidy personal profit. Carnegie was livid with rage and demanded Frick's resignation as chairman of the board, packed the board of Frick Coke with Carnegie loyalists, and forced the coke firm to agree to accept $1.35 per ton for coke. Then, in January 1900, Carnegie invoked the Iron Clad Agreement against Frick, demanding that he relinquish his shares. Frick was furious and flew into a tower-

Harper's Weekly *drawing by W. P. Snyder depicting the attempt on Frick's life by Lithuanian anarchist Alexander Berkman on July 23, 1892*

ing rage, screaming at Carnegie: "For years I have been convinced that there is not an honest bone in your body. Now I know you are a god-damned thief. We will have a judge and jury of Allegheny County decide what you are to pay me."

Frick then went to court and sued for a reevaluation of the steel firm's assets, and the whole matter exploded into a sensational court trial which fed details of the massive Carnegie operation to muckrakers for months. In March 1900 Carnegie decided he had had enough. He realized the harm to himself, to other industrialists, and to the Republican party as a result of the staggering revelations at the trial. As a consequence, he created a new holding company, called Carnegie Company, composed of Carnegie Steel and Frick Coke, capitalized at $320 million. With this new valuation it meant Frick's share would be over $30 million, considerably more than the $4.9 million he would have received under the Iron Clad Agreement. The new arrangement also automatically eliminated the old Iron Clad, so that these shares were far more liquid than was previously the case.

With this money Frick could now wreak further vengeance on Carnegie. Mellon had earlier pro-

vided backing to William H. Donner in building a small wire and rod mill which purchased steel billets on contract from Carnegie Steel. By 1900, however, the situation had become quite tense. Carnegie Steel, in response to the fact that J. P. Morgan's recently organized Federal Steel Company had stolen a number of finished steel customers, made plans to begin making wire and rods at a massive plant at Conneaut, Ohio. Further, now that Carnegie Steel and Donner Steel were competitors, would Carnegie continue to supply the small firm with billets? Donner and the Mellons decided they had to build their own blast furnace and open-hearth operations. But they needed more capital and they needed a manager—both of which were now available in person of Henry Clay Frick.

The confrontation with Carnegie Steel would be doubly sweet for Frick. Not only would he get retribution for past injury, but while he was at Carnegie Steel he had always counseled against moving into the finished product end of the industry and had frustrated any move in that direction. Now that Carnegie had the temerity to begin producing finished steel products, he would teach him an expensive lesson. Could any victory be more sweet? A new firm, called Union Steel, was organized, with Donner as president and Mellon as vice-president. Frick took no formal office in the company but held 25 percent of the stock and was clearly the major decision maker. The company bought two finishing plants in Johnstown, Pennsylvania, and Harrisburg, Pennsylvania, and Frick helped them acquire Republic Coke Company at bargain rates. Mellon then secured a favorable ten-year iron ore contract, and the company purchased 342 acres of land on the Monongahela River just south of Pittsburgh. In Donora, Pennsylvania, they would build a sparkling new steel plant.

This was a formidable challenge for Carnegie. To the west loomed the massive Federal Steel, controlled by the powerful J. P. Morgan, and now in Pittsburgh itself was a large new steel firm supported by the wealthy and powerful Mellon family and run by the "master manager," Henry Clay Frick. At this point Carnegie, who had wanted to retire anyway, let J. P. Morgan know that he would be willing to sell his huge firm. The price, however, had gone up since 1899. He now asked for a total of $480 million, and Morgan agreed. Morgan then began to put together the enormous U.S. Steel Corporation with Carnegie Steel as its centerpiece. But in order to do so, he needed the assistance of Frick.

To round out the creation of U.S. Steel, Morgan needed to get control of the vast Rockefeller holdings in Mesabi ore lands. But Morgan and Rockefeller did not like one another, and the negotiations were stalled. Morgan sent for Frick and asked him to act as an intermediary. Both parties put their complete trust in Frick, and he negotiated a satisfactory price, thus successfully culminating Morgan's plans. Frick received no direct compensation for his role in the negotiations, but when U.S. Steel was capitalized in 1901 as the first billion-dollar corporation, he reaped sufficient reward. His total holdings in stocks and bonds in the new corporation were over $60 million, and he was elected a member of the first board of directors.

But Frick was his own man. He was a capitalist of the first order, and the final two decades of his life demonstrated his uncanny ability to maximize his profits and provide astute advice to those firms with which he was connected. First of all, he very early became convinced that the future of U.S. Steel was less auspicious than it had seemed. Thus, in 1902 he slowly began unloading his shares on the market, disposing of his entire block of 218,324 shares of common stock and all but 10,000 of his 237,679 shares of preferred stock. He still retained some $15 million in bonds. A number of events occurred over the next few years, however, which induced Frick to become an active member of the steel firm's management.

The first event concerned Union Steel. While U.S. Steel was being created, Union continued building its mills and in 1901 had two huge 50-ton open-hearth furnaces producing their own billets. Frick and Mellon, watching U.S. Steel slide deeper into trouble, were determined to provide powerful competition and began buying other finishing mills. Many of U.S. Steel's mills were not technologically advanced, and they were having difficulty. Further, Andrew Carnegie's old billet contract with Donner Steel, which U.S. Steel inherited, obligated the corporation to provide Union Steel with all the billets it needed to compete with them. And these billets were sold at such a low price that Union Steel, because of its generally superior technical facilities, was able to easily undersell U.S. Steel on nearly every item and still make a profit.

Mellon and Frick had backed Morgan into a corner, just as they had done with Andrew Carnegie. Further, Morgan was at a loss as to what to do with U.S. Steel itself. It was teetering on the brink of collapse and was facing stiff competition from Union Steel. Morgan thus called on Frick for advice and assistance. Frick met Morgan on his yacht and freely gave him advice on reorganization of the giant steel firm. But when Morgan asked Frick to take a role in active management, Frick replied that he could not. He explained that since he owned 25 percent of a competing steel firm, it would not be ethical for him to engage in active management of U.S. Steel. But, of course, if U.S. Steel were to buy Union Steel, Frick would again have a financial interest in U.S. Steel, and there would be no conflict of interest. Morgan sent U.S. Steel's attorney, Judge James Reed, to negotiate with Mellon on the purchase of Union Steel. Since Reed was also Mellon's longtime intimate and erstwhile private attorney, it is not clear just how objective Reed was in this situation. In any event Mellon and Frick sold Union Steel to U.S. Steel for over $30 million in bonds, of which Frick received 25 percent.

Over the next few years Frick played an active and important advisory role in U.S. Steel, the most important coming in the purchase of Tennessee Coal & Iron Company. During the stock market panic of 1907 the Wall Street investment firm of Moore and Schley, which was in financial trouble, approached J. P. Morgan for help. As collateral they offered stock in the southern steel company. Morgan was eager to buy it, but Frick was opposed, feeling its production costs were too high. Morgan prevailed, however, and Moore and Schley were offered $12 million for their shares in the firm. Frick and Elbert Gary, head of U.S. Steel, were then dispatched to Washington to get approval from President Theodore Roosevelt for the purchase. They explained that unless the purchase was made an important Wall Street firm would go bankrupt and a serious depression might ensue. Roosevelt took their word as gentlemen for the situation and promised he would not prosecute U.S. Steel for the acquisition. In 1911, however, the Taft administration brought suit against the steel corporation for the purchase. Roosevelt exploded when he learned of the suit, viewing it as his final betrayal by the successor he had chosen. Nonetheless, Frick was charged by the Taft administration with misrepresenting many important matters to President Roosevelt in the suit. Nothing substantial, however, was done concerning Frick's involvement.

When Frick first pulled his money out of U.S. Steel, he invested much of it in the stocks of other corporations, particularly railroads. Frick once called railroads the "Rembrandts of investment." He invested about $6 million each in his seven favorite railroads–the Chicago & North Western, the Union Pacific, the Atchison, Topeka and Santa Fe, the Reading, the Pennsylvania, the Baltimore & Ohio, and the Norfolk & Western. He took his most active interest, however, in the Pennsylvania Railroad. He became involved in the road at a time when it was undergoing a major modernization and reorganization of its massive transportation system. Frick, as a director of the railroad, played a vital role in this reorganization. He also invested a great deal of money in Pittsburgh real estate, becoming the largest realty holder in the city. His money largely built the William Penn Hotel, and he also erected the Frick Building and the Union Arcade. He also remained intimately involved all of his life in Mellon National Bank and Union Trust Company (both Mellon institutions) in Pittsburgh.

In some areas, however, Frick was less successful in exerting his influence. In 1901 James Hazen Hyde had asked Frick to become a trustee of the Equitable Life Assurance Company, but he did not play an active role in the firm until 1905. In that year a great crisis engulfed Equitable and the insurance industry. Hyde had long indulged himself in a lavish lifestyle, and in the latter year it finally led a faction within the company, led by the president, James W. Alexander, to try to oust Hyde from control. This produced a terrible scandal of accusations and counter-accusations between the two groups which greatly blackened Equitable's standing and reputation. Further, the situation attracted so much notoriety that New York State established the Armstrong Commission to investigate the entire insurance industry and its practices. As a result of the crisis, Frick recommended in June 1905 that both Hyde and Alexander resign. But, contrary to Frick's experiences at U.S. Steel and other corporations, his suggestions were ignored, and Frick himself resigned from the board.

In the twentieth century Frick's attention turned increasingly to real Rembrandts; Frick became an art collector of the first order. Although Frick always retained his home in Pittsburgh's East End, in the early twentieth century he also bought a handsome home on Fifth Avenue in New York City and soon began filling it with the paintings of the old masters. He had begun the process in Pittsburgh, accumulating some 71 pictures there. But in New York Frick really came into his own as a collector. It was said that he spent between $30 and $40 million in building his collection, and the results were truly impressive. The first great work he purchased was Rembrandt's *Portrait of a Painter*, and this was followed by other valuable paintings and sculptures. Upon Frick's death the mansion, with its many treasures, along with an endowment of $15 million, was willed to the public as a museum. It continues today as one of the most important and impressive private collections of art in the world, a lasting tribute to Frick's wealth, his vision, and his taste in art.

Although Frick was not a philanthropist on the order of Andrew Carnegie or John D. Rockefeller, he did set important standards for benefaction in the American capitalist community. Besides his home and his art, he also gave liberal donations to Princeton University and over $300,000 for the improvement of Pittsburgh's elementary schools. Later he also left the public schools 10 percent of his estate. Frick also left many smaller bequests to educational and charitable institutions and donated 132 acres and an endowment of $2 million to Pittsburgh for the creation of a park. Called Frick Park, it is one of the largest and loveliest city parks in the nation.

When Frick died in 1919, he left behind his wife, a son, and a daughter. He also left a legacy as one of the most powerful and determined industrialists of his era. He was first of all the greatest of the partners Carnegie had in his company and was largely responsible for many of the revolutionary developments at the works in the late 1880s and 1890s. What is often obscured by his relationship with the Carnegie firm is that, as the "coke king" of America, Frick was an industrialist whose accomplishments were equal to that of Carnegie, Rockefeller, James J. Hill, or Henry Ford. Yet Frick is far less well known or well remembered today. A major reason for his relative obscurity probably lies in the fact that Frick accepted a partnership with Carnegie, which tended to dilute the significance of Frick's own accomplishments and forced him to attempt to march in lockstep with the flamboyant and quixotic Carnegie. In this framework the cold and imperious Frick was often overshadowed in public by Carnegie's showmanship and media manipulation. Like so many Pittsburgh steel men, Frick presented a dour, austere countenance to the public. He seemed interested in nothing but work and appeared to have no time for recreation, leisure, friends, or family. Yet this image was at least partially deceiving. Although Frick had few close friends, he was by all accounts a loving husband and devoted father, a trusted business associate, and respected public citizen. A ruthless competitor and a superior manager, Frick drove a hard bargain at all times and was particularly severe in his relations with his workers. Yet his associates, if not his workers, respected him for his honesty and rectitude, even if none appreciated him for his warmth. Whereas Carnegie and Rockefeller both tried to soften their images with the public, to appear as kind and benevolent public benefactors, Frick did not put on airs. He was tough, he could often be ruthless, but he was also honest and forthright. When he gave his monies away he usually did it quietly and with little fanfare, often embarrassed at the attention it might bring him.

Perhaps most significant in arriving at a historical judgment on Frick, he was the only man to stand up to both Andrew Carnegie and J. P. Morgan and win. Not only that, he retained at least the respect of both men. Morgan always accorded Frick great respect and deference as a manager and negotiator, and although Carnegie disliked Frick personally, he at least paid Frick his due as a businessman and manager many years later. Frick perhaps best typified a classic type of nineteenth-century businessman, one which has largely disappeared in the twentieth century. Disdaining public relations, Frick stood up to all challengers, whether they were labor unions, Carnegie, Morgan, or Rockefeller. If he did not win every encounter, neither did he compromise his principles. These are not principles that present-day historians and businessmen hold in high regard, but in many ways they represented the very essence of rectitude in late-nineteenth-century America. Frick was a tough, honest businessman, one who gave no quarter and asked none. He refused to mask his intentions behind flowery rhetoric and liberal pronouncements, and that is one of the reasons he is one of the least remembered of the great American businessman of the turn of the century. He deserves better. Henry Clay Frick was a giant of American business, consolidated the vastly important coke industry in America, and played a role second only to Andrew Carnegie himself in building and transforming the great American steel industry.

Unpublished Documents:

"Report of the Committee of the Judiciary on Employment of Pinkerton Detectives," House Report 2447, 52d Cong., 2d sess., 1892-1893;

U.S. House of Representatives, "Hearings," Committee on the Investigation of the United States Steel Corporation, 8 volumes, 62d Cong., 2d sess., 1911-1912.

References:

James Howard Bridge, *The Inside History of the Carnegie Steel Company* (New York: Aldine, 1903);

David Brody, *Steelworkers in America: The Nonunion Era* (New York: Harper & Row, 1969);

Herbert Casson, *The Romance of Steel* (New York 1907);

George Harvey, *Henry Clay Frick: The Man* (New York: Privately printed, 1928);

Burton J. Hendrick, *Life of Andrew Carnegie*, 2 volumes (Garden City, N.Y.: Country Life, 1932);

Burton Hersh, *The Mellon Family: A Fortune in History* (New York: William Morrow, 1978);

Gabriel Kolko, *The Triumph of Conservatism: A Reinterpretation of American History, 1900-1916* (New York: Free Press, 1963);

David E. Koskoff, *The Mellons: The Chronicle of America's Richest Family* (New York: Thomas Y. Crowell, 1978);

Joseph F. Wall, *Andrew Carnegie* (New York: Oxford University Press, 1970);

Leon Wolff, *Lockout, the Story of the Homestead Strike of 1892: A Study of Violence, Unionism, and the Carnegie Steel Empire* (New York: Harper & Row, 1965).

Archives:

Henry Clay Frick's papers are located at the Frick Art Museum, Pittsburgh, Pennsylvania. At this time, however, access to them is severely restricted.

John W. Fritz

(August 21, 1822-February 13, 1913)

by John W. Malsberger

Muhlenberg College

CAREER: Mill superintendent, Moore & Hooven Company (1844-1849); superintendent, Reeves, Abbott & Company (1849); superintendent, Kunzie Furnace (1852-1853); general superintendent, Cambria Iron Works (1854-1860); general superintendent and chief engineer, Bethlehem Iron Company (1860-1892).

A distinguished mechanical and metallurgical engineer, John W. Fritz contributed substantially to the growth and development of the nineteenth-century American iron industry. As an engineer who was largely self-taught, Fritz introduced many important innovations in the iron manufacturing process, most notably the three-high roll train, generally regarded as the last important improvement prior to the Bessemer process. Through Fritz's innovations, many of which became industry standards, the output and efficiency of American iron companies were improved dramatically. Under Fritz's management, in addition, the Cambria Iron Works and particularly the Bethlehem Iron Company, forerunner of Bethlehem Steel Corporation, were built up to become industry leaders in their day. Thus, as engineer and manager, John Fritz played a major role in the rapid evolution of America's iron and steel industry.

The first of the seven children born to George and Mary Meharg Fritz, John Fritz was born in Londonderry (Chester County), Pennsylvania, on August 21, 1822. George Fritz was born July 26, 1792, in Hesse Cassel, Germany, and as a boy moved to America with his parents in August 1802. Mary Fritz was a native-born daughter of Scotch-Irish immigrants, who had come to America in 1787. Like many youth of his era, Fritz received lit-

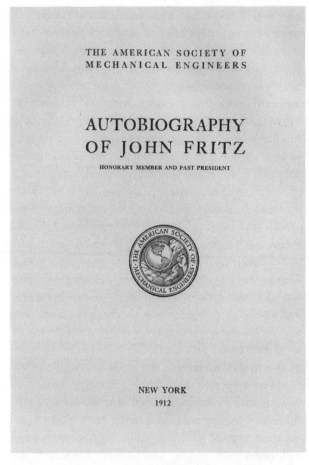

THE AMERICAN SOCIETY OF
MECHANICAL ENGINEERS

AUTOBIOGRAPHY
OF JOHN FRITZ

HONORARY MEMBER AND PAST PRESIDENT

NEW YORK
1912

Title page of Fritz's autobiography published in his ninetieth year

tle formal education, attending for several months each winter a variety of Quaker schools near his family's farm in rural Chester County. The bulk of his education prior to age sixteen came from performing the numerous tasks associated with farm life in early-nineteenth-century America and from his father, a self-taught millwright and me chanic.

In October 1838 Fritz was apprenticed to a Parkesburg, Pennsylvania, blacksmith named Thomas Hudders; the apprenticeship proved to be the turning point in Fritz's life. In Hudders's shop Fritz not only learned how to shoe horses but also how to manufacture and repair a wide variety of household and farm implements. Through this "country machine work," as it was then termed, Fritz became thoroughly familiar with farming implements, grist and saw mills, blast furnaces, and forges, as well as machinery used to manufacture cotton and woolen fabrics. As an apprentice, therefore, Fritz used much of the machinery that would form the basis of his later career in the iron industry. In Hudders's shop there were metal lathes, a primitive drill press, rollers for bending boiler plates, as well as shears, punches, and boilermaker's tools.

The experience that led Fritz most directly into a career in the developing iron industry resulted from a friendship he formed during his apprenticeship. Hudders's shop was adjacent to the Philadelphia & Columbia Railroad (later part of the Pennsylvania Railroad), and the railroad's repair shop was located nearby. Fritz spent many of his spare hours observing the techniques used in railroad maintenance and also developed a strong friendship with the superintendent of the repair shop, William Hardman. In 1839 when Hardman left Parkesburg to become superintendent of a railroad shop in Georgia, he was so impressed by Fritz's quick grasp of the technical aspects of repair work and by his industriousness that he offered the young apprentice the position of assistant superintendent. Fritz refused the offer because of his mother's objections to his moving so far away, but the introduction he had received to mechanics through these experiences whetted his appetite for more knowledge. As an alternative to working for Hardman, Fritz sought a job in the iron industry, thus fixing at an early age the path that he would follow in the future.

By the end of his apprenticeship in 1840 Fritz had already exhibited the two characteristics that were so essential to his great success. First, throughout his life Fritz repeatedly displayed an insatiable thirst for knowledge, a thirst which led him to change jobs frequently in order to further his education. The second distinguishing characteristic was his great industriousness, which made him willing to work as long as necessary to accomplish his goals. As he recalled late in life, "never shirking a responsibility and never missing an opportunity to acquire knowledge was at all times my guiding star."

Fritz's determination to find employment in the iron industry was delayed temporarily by the depressed economic conditions that followed the panic of 1837. At the end of his apprenticeship these conditions forced him to return to his family's farm where he worked until 1844 when, largely through his own efforts, he landed his first job in the industry on which he left such a strong mark. Each day during the fall of 1844, after his farm chores were completed, Fritz traveled to nearby Norristown, Pennsylvania, to observe the construction of a rolling mill for the Moore & Hooven Company. Striking up an acquaintance with the partners in this enterprise, Fritz so impressed them with his quick mind and determination that they hired the relatively inexperienced young man to assist the mechanics in erecting the new mill's machinery, boilers, and furnaces. After only several weeks as assistant Fritz was promoted to regular mechanic, and when the mill was put into operation, he was made superintendent of machinery at an annual salary of $1,000. The machinery Fritz oversaw consisted of three sets of rollers, through which he learned a great deal about rolling iron, knowledge that figured prominently in his later career.

The young superintendent of machinery advanced rapidly at Moore & Hooven largely because of his hard work and desire to learn. Fritz also worked hard to expand his knowledge of the other steps involved in the manufacture of iron, from puddling to heating to rolling, and he often stayed in the mill past 10:00 P.M. acquiring the skill of a common puddler. Such industriousness quickly caught the attention of the mill owners, and they promoted him first to night superintendent of the entire mill and later to day supervisor. In this new position Fritz displayed another hallmark of his career, the constant desire to innovate and improve. Both through his own and through other's inventions Fritz modernized the mill's machinery and improved its efficiency. One of his most important suggestions, for instance, was to raise the roof of the blast furnace some nine inches in order to improve the blast.

Fritz's entry into the iron industry in the 1840s came at a particularly fortunate time. Only four years before Fritz was hired by Moore & Hooven, the Welsh ironmaster David Thomas had

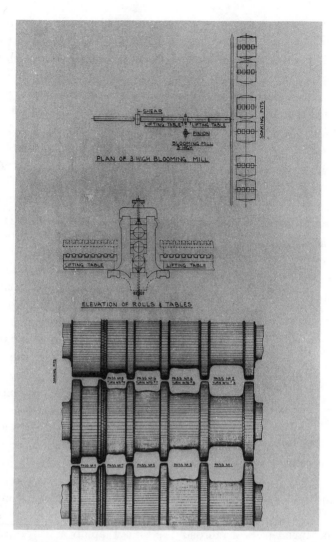

Diagram of the Fritz-designed three-high rolling mill (courtesy of Association of Iron and Steel Engineers)

demonstrated the commercial feasibility of manufacturing iron with anthracite coal at the Lehigh Crane Iron Company in Catasauqua, Pennsylvania. The advent of this new, more abundant fuel permitted the iron industry to expand tremendously over the next several decades, an expansion which afforded numerous opportunities for ambitious and talented young men like Fritz. In a sense Fritz and the American iron industry matured together in the mid nineteenth century.

After a little more than four years with Moore & Hooven, Fritz resigned his position in May 1849 to become the superintendent of Reeves, Abbott & Company, which then had a blast furnace and rail mill under construction at Safe Harbor, Pennsylvania, located about ten miles from Lancaster on the Susquehanna River. His new employers, because they manufactured iron rails, brought Fritz closer to his first love, the railroads. Be-

lieving that he had learned all that he could about rolling mills at Moore & Hooven, Fritz took the position at Safe Harbor, despite a pay cut to $650, in order to gain a greater understanding about blast furnaces and the manufacture of iron rails. After only several months at Reeves, Abbott, however, Fritz became ill with what was then diagnosed as fever and ague and was forced to resign his position and return to his family's farm to recuperate.

Fritz remained unemployed for the next two years, traveling part of the time with a friend to Michigan to inspect the iron ore beds around Lake Superior. Impressed by the abundance of the iron ore, Fritz tried unsuccessfully to persuade his former employers to import ore from Lake Superior, correctly perceiving that these beds would provide for the future expansion of the iron industry. Upon his return from the West in 1852, Fritz went back to work at Moore & Hooven for several months in his former capacity as superintendent, only to be lured away again by the owners of the Safe Harbor iron mill. Reeves, Abbott & Company had in early 1852 leased the old Kunzie Furnace on the Schuyl-kill River, twelve miles from Philadelphia, and hired Fritz to convert and rebuild the blast furnace to use anthracite coal to manufacture iron. When the reconstruction of the Kunzie Furnace was completed in 1853, Fritz and his brother George moved to Catasauqua to build a foundry and machine shop to service the rolling mills and blast furnaces, which were developing rapidly in that area. While at work in Catasauqua Fritz was again enticed to take a new position by Reeves, who had acquired part interest in the Cambria Iron Works of Johns-town, Pennsylvania. Fritz's appointment as general superintendent of Cambria represented an important turning point in his career by bringing him to the position that helped to establish his reputation as one of the foremost innovators in the American iron and steel industry of the mid nineteenth century.

When he arrived at Cambria in June 1854, the company was chiefly concerned with the manufacture of iron rails for the nation's burgeoning railroad industry. Bringing to his new employer all the knowledge he had earlier gained about the various aspects of iron manufacture, Fritz by 1860 had built the Cambria works into one of the most modern and efficient iron companies of its day. The changes that he proposed, however, often ran so contrary to the conventional practices in the iron industry that Fritz had to surmount bitter opposition from the company's directors and laborers before his ideas could be implemented.

The chief innovation introduced by Fritz at the Cambria works was the three-high rolling mill for manufacturing iron rails. The advantage of this technique over the rollers that were then used throughout the industry was that it allowed iron rails to be rolled in a continuous process, always with the same side up, thereby eliminating the time-consuming need of turning the rails. At the same time Fritz placed feeding rollers in the front of each machine to facilitate the passage of the bar between the rolls, cutting additional time from the manufacturing process. In constructing his three-high rolling mill Fritz also bucked conventional wisdom by eliminating most of the built-in safety devices that were widely regarded as necessary to prevent the breaking of the rollers. At critical points throughout the roll train, spindles and coupling boxes were deliberately made thin so that they would break first if trouble developed, thereby halting the rolling and preventing damage to the rollers. In Fritz's observation, however, these safety devices, because they were constantly breaking, were responsible for frequent and costly interruptions in production. Fritz decided to eliminate the safety features from his three-high rolling mill, instead risking major breakdowns periodically in return for steadier production.

Because Fritz's proposed innovations were so antithetical to the procedures then commonly used in the iron industry, most of the laborers at Cambria opposed the plans as dangerous and unworkable, and many refused to have anything to do with the three-high rolling mill. Many of the directors of the company also voiced strong opposition to the proposal, agreeing with the laborers that it was unlikely to succeed and arguing that it was far too costly. In the end Fritz's forceful arguments prevailed, and he was given grudging approval by the directors to proceed; most of the directors believed that the three-high mill would fail, thus ruining Fritz's career in the iron industry. Fritz proved his critics wrong. The three-high mill commenced successful operation on July 29, 1857, with only a skeleton crew of workers on hand. The innovation so increased the efficiency of rolling iron rails that the cost of manufacturing fell from an average of $160 per ton to $28. Fritz's method ultimately became the industry standard and is generally regarded as the first of the great advances that occurred in the iron and steel industry after 1857, advances which included the introduction of the Bessemer process and the Siemens-Martin open-hearth furnace.

Although the three-high mill proved successful, Fritz's difficulties at the Cambria Iron Works were far from over. The day after his invention was tested successfully, the Cambria mill burned to the ground but under Fritz's supervision was rebuilt and put back into operation by January 1858. In rebuilding the Cambria works, Fritz turned adversity into advantage by introducing additional innovations that further increased the efficiency of manufacturing iron rails. In traditional methods of rail manufacturing, rails came out of a roll train and were placed aside of it ("on the bank") to cool before being carted by hand to shears where they were cut to the desired length. In Fritz's mind this procedure was time-consuming, costly, and dangerous to the workers. He decided, therefore, to place a pair of shears in front of each set of rolls so that as a rail came out it was fed directly into the shears and cut immediately to the proper length. Once this innovation was implemented it meant that iron was not touched by human hand from the puddling furnace until it was completed, thereby greatly reducing the cost of manufacture and easing the burden on ironworkers.

As a result of all the Fritz-generated improvements, output at the Cambria Iron Works increased over 400 percent, the quality of the rails was enhanced, and the Johnstown company became one of the most efficient and prosperous iron companies of its day. Despite this success, however, contentious relations persisted between Cambria's directors and their general superintendent, finally leading Fritz in July 1860 to accept a new challenge as general superintendent and chief engineer of the newly organized Bethlehem Iron Company of Bethlehem, Pennsylvania. Hired to design and construct blast furnaces and rolling mills for this new firm, Fritz remained at Bethlehem Iron until his retirement in 1892, developing it into one of the foremost producers of iron and steel in America.

Under Fritz's supervision ground was broken on July 16, 1860, for Bethlehem Iron's first blast furnace, and early the next year construction began on the rolling mill. The unsettled economic conditions associated with the outbreak of the Civil War in April 1861 delayed completion of the facilities until 1863. The blast furnace commenced operation in January 1863, and the rolling mill turned out its first iron rails in September of the same year. Only several months after iron production began in full scale at Bethlehem, Fritz was summoned by the Union government to lend his mechanical skills to the war effort. In March 1864 President Abraham Lincoln, probably acting on the advice of Abram S. Hewitt of Cooper, Hewitt & Company, selected Fritz to design and construct a rolling mill in the South to allow the government to reconstruct southern railroads in order to facilitate the movement of Union troops. Given unlimited power to make whatever arrangements he deemed necessary, Fritz quickly brought to completion a rolling mill in Chattanooga, Tennessee.

Once the Chattanooga plant was operational, Fritz returned to Bethlehem Iron and directed his attention, as he had done throughout his career, to finding ways to improve the quality and efficiency of iron manufacturing. At the close of the Civil War the most talked-about innovation in the industry was the process of manufacturing steel developed separately though almost simultaneously in the 1850s by an Englishman, Henry Bessemer, and an American, William Kelly. Even before the Bessemer process was successfully introduced in the United States in 1865, Fritz had taken a strong interest in this new technology. Correctly foreseeing that the expansion of America's railroads would create a huge market for the more durable steel rails, Fritz was initially deterred from building a Bessemer furnace at Bethlehem because of Henry Bessemer's insistence that his process would succeed only if iron ore with a very low phosphorus content was used. Because most of the ore to which Bethlehem Iron then had ready access exceeded these phosphorus limits, Fritz believed that Bessemer steel was not commercially viable at his firm. A serendipitous accident, however, soon changed his mind.

In 1865 the Lehigh Valley Railroad, headquartered in Bethlehem, began, like many other railroads of its day, to import Bessemer steel rails from England. While these rails were being unloaded at Bethlehem, one of them broke and, upon analysis, was shown to have been manufactured from iron ore whose phosphorus content was higher than the limits established by Henry Bessemer. This revelation convinced Fritz that Bessemer steel could be manufactured in Bethlehem, and in the fall of 1868 construction began on Bethlehem Iron's first Bessemer furnace. Three years later Fritz installed another important technological advance, the Siemens-Martin open-hearth furnace, which greatly expanded his plant's steel-making capabilities. By the early 1870s Fritz's foresight and leadership had moved the Bethlehem Iron Company to the forefront of America's emerging steel industry.

Influenced by the many business failures that resulted from the depressed economy of the 1870s and devoted as always to innovation, Fritz began to argue in the late 1870s that product diversity was essential to the future success of Bethlehem Iron. Accordingly, he urged the company's directors to authorize construction of a new steel mill capable of manufacturing structural steel beams that could easily be riveted together. Despite his reasoning that the industrialization of America was creating a huge market for such structural steel, the more conservative directors vetoed the plan, preferring in the unsettled economy of that era to concentrate on the more reliable market for steel rails. For the remainder of his life Fritz continued to believe that Bethlehem Iron's refusal to build a structural steel plant was its most serious mistake.

In the early 1880s, several years after his plans for a structural steel plant was quashed, Fritz saw another opportunity for diversification when the U.S. government decided to modernize its navy. Arguing that most ships in the future would be built of iron or steel plate, Fritz, with the help of the company's largest stockholder, Joseph Wharton of Philadelphia, succeeded in persuading the directors to authorize construction of an armor plate mill. When the government asked for bids on armor plate in 1886, Bethlehem Iron was the only company able to bid on all the contracts; the company received $4.5 million worth of business. Fritz used $3 million of this money to modernize the plant, installing a series of open-hearth furnaces capable of producing steel in sufficient quantity to satisfy the government contracts. When these new furnaces were put into operation in August 1888, the foresight and determination of John Fritz had not only provided Bethlehem Iron with the means to grow and prosper during the difficult economic conditions of the 1890s, but also had built the Bethlehem firm into one of the most modern and diversified iron and steel manufacturers in America.

Fritz was married on September 11, 1851, to Ellen W. Maxwell of Whitemarsh, Pennsylvania. They had one child, a daughter Gertrude, who was born in 1853 and died at age seven.

Throughout his life John Fritz received many honors and awards in recognition of the great contributions he made directly to the iron and steel industry and indirectly to American life. At the 1876 Centennial Exposition in Philadelphia, for instance, Fritz was awarded a bronze medal for his innovations in the manufacture of iron and steel. In 1893 he was the recipient of the Iron and Steel Institute's Bessemer Gold Medal, and in 1902, in honor of his eightieth birthday, a fund was established through subscription to award an annual John Fritz Medal "for notable scientific or industrial achievement." Fritz himself was awarded the first John Fritz Medal. Later recipients included Lord Kelvin, Alexander Graham Bell, Thomas A. Edison, George Westinghouse, and Alfred Nobel. Finally, in 1910 he received the Elliot Cresson Gold Medal from the Franklin Institute "for distinguished leading and directive work in the advancement of the iron and steel industries."

Although he had little formal education, Fritz took a strong interest in it throughout his life. He served as one of the original trustees of Lehigh University, where he also endowed an engineering lab. Near the end of his life Fritz was also awarded many honorary degrees from some of the leading academic institutions in America. Columbia University acted first by granting him an honorary Master of Arts in 1895. In 1906 the University of Pennsylvania awarded Fritz the degree of Doctor of Science, and the following year he received a Doctor of Engineering degree from Stevens Institute of Technology. A second Doctor of Science degree came to the iron maker from Temple University in 1910.

Selected Publication:

The Autobiography of John Fritz (New York: J. Wiley & Sons, 1912).

References:

William J. Heller, *History of Northampton County, Pennsylvania*, 3 volumes (New York: The American Historical Society, 1920);

James Moore Swank, *History of the Manufacture of Iron in All Ages*, second edition (Philadelphia: American Iron and Steel Institute, 1892).

John Hill Garrard

(unknown)

by Bruce E. Seely

Michigan Technological University

CAREER: Partner, Cincinnati Steel Works (1831-1837); member, Ohio State Senate (1837-1838).

William Garrard

(October 21, 1803-unknown)

CAREER: Partner, Cincinnati Steel Works (1831-1844).

Until the development of the Kelly-Bessemer process, the production of steel was largely controlled by the English, primarily due to the invention of cast crucible steel in the 1740s by Benjamin Huntsman. By 1830 American manufacturers were beginning to produce blister steel, converting wrought iron bars by heating them in chests packed with charcoal for about ten days. Blister steel was suited for many uses, including springs, but tools and many other products required the more uniform quality obtained by melting blister steel to produce cast crucible steel. For 30 years after 1830 American steel makers struggled, largely without success, to copy the British method. The first serious attempt to do so was undertaken in 1831 by William and John Hill Garrard in Cincinnati, Ohio.

The Garrard brothers were born in Laxfield parish, Suffolk, England; William, the important figure in the firm, was born on October 21, 1803. In 1822 the boys' parents immigrated to the United States, settling near Pittsburgh. Apparently, William had been trained as a bricklayer; he also had acquired an interest in chemistry. At some point in the 1820s he and his father began experimenting with the manufacture of blister steel, experiments which convinced William that he could make steel equal in quality to that of the English.

In 1831 William and his brother launched their venture, calling it the Cincinnati Steel Works. William designed much of the machinery and the buildings and superintended construction. The completed facilities were not lavish, including only a converting furnace for producing blister steel from bar iron, a two-pot steel melting furnace fired by coke, and machinery for making saws and files. The works were modeled on English cast crucible steel plants, and in August 1832 the Garrards melted their first steel. The first saws and other tools were made in November. A key step was William's discovery of a high-quality clay near New Cumberland, West Virginia, for his crucibles; this material closely resembled the Stourbridge clay that was so important to Sheffield steel makers. For top quality steel Garrard relied on imported Swedish bar iron, while Missouri and Tennessee charcoal iron sufficed for lesser grades of steel. According to William's reminiscences, the material and the products were judged to be equivalent to English steel, and both found a market. Indeed, in the 1870s tests made on samples from the first batch of steel found it to be first quality.

John H. Garrard soon dropped out of the business, perhaps because he was elected state senator for Hamilton County in 1837 and 1838. In his place, W. T. Middleton joined the firm as a partner, creating the firm Middleton, Garrard & Company. Later, lawyer Charles Fox also became a partner in this firm. But the 1830s were a difficult time to begin a steelworks. First, Sheffield producers offered American importers much better credit

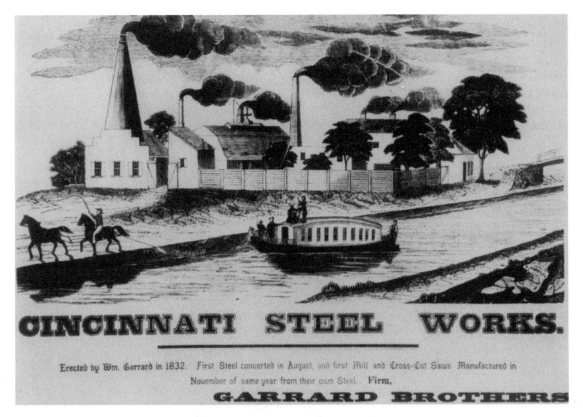

Advertisement for the Garrard Brothers' Cincinnati Steel Works (courtesy of the Historical and Philosophical Society of Ohio)

terms than Garrard, and American suspicions about domestic steel were another obstacle. Finally, the panic of 1837 caused the works to fail for lack of funds. After the panic Garrard managed to reopen the works, but he seems to have limited his efforts to the production of blister steel, a product that faced less competition from the British. One indicator of British dominance in the American market was the description of the works in the 1840 city directory, which identified William Garrard as English. Even so, Garrard's works succumbed in 1844.

The failure of this pioneering operation was repeated many times before commercial and technical successes in the production of cast crucible steel were combined by firms in Pittsburgh and Jersey City, New Jersey, during the Civil War. A combination of better processes, better materials, and the artificial market stimulation and protection provided by the war largely explained the timing of the eventual success. The Garrards' example, then, indicates that a pioneering steel manufacturing business needed to be more than technically correct to succeed in this country.

Unpublished Documents:
"Cincinnati City Directories," various years (1831-1846); located in Cincinnati Historical Society.

References:
State of Ohio, *Journal of the Senate . . . 36th General Assembly* (Columbus, 1837-1838);

James M. Swank, *History of the Manufacture of Iron in All Ages*, second edition (Philadelphia: American Iron and Steel Institute, 1892).

John Warne Gates

(May 8, 1855-August 9, 1911)

by Larry Schweikart

University of Dayton

CAREER: Partner, hardware store (1874-1878); salesman, Ellwood Wire (1878); partner, J. W. Gates & Company (1878-1880); partner, Southern Wire Company (1880-1882); partner, Braddock Wire Company (1882-1892); partner, Consolidated Steel & Wire Company (1892-1897); president, Illinois Steel Company (1894-1896); investor and stock speculator (1896-1911); partner, American Steel & Wire Company (1898-1902).

John Warne "Bet-a-Million" Gates, a promoter and capitalist, was born on May 8, 1855, in Turner Junction, Illinois (now a part of the greater Chicago metropolitan area), to Asel Avery and Mary Warne Gates. Gates was educated at the local public school and at North-Western College in Naperville. After receiving a degree of sorts in a six-month commercial course in 1873 he began working odd jobs and saving money, enough so that within a year he invested in a 50-percent share in a local hardware store. In 1876, at age twenty-one, he married Dellora Baker from St. Charles, Illinois.

While working in his hardware store Gates noticed that barbed wire for fencing had attained a sizable market. Assuming the vast tracts of western land could only cause the wire market to expand, Gates offered in 1878 to join wire manufacturer Isaac L. Ellwood in his wire business. Ellwood declined the partnership offer, but instead hired Gates at $25 a week as a traveling salesman. Gates immediately went to Texas, where he tried to impress upon local ranchers the usefulness of a wire that was unaffected by wind, fire, and, most important, wandering cattle. Although the ranchers resisted the wire at first, Gates won them over when he challenged them to test it with their best steers. He soon realized that the market was even larger than he had originally anticipated–large enough so that he abandoned his role of salesman and began to fill the orders himself. Lacking a manufacturing facil-

John Warne Gates

ity, Gates took a partner and built a wire plant in St. Louis. Ellwood was incensed and sought to file lawsuits to stop Gates from producing his own wire. But Gates moved his plant back and forth across the river and the Illinois-Missouri border, exhausting Ellwood's attempts to serve an injunction. Ellwood eventually settled with his former employee.

The ambitious ex-salesman concentrated on more than wire making. He attempted to consolidate the wire manufacturers beginning in 1880 and culminated his effort in 1898 with the formation of the American Steel & Wire Company of New Jersey (capitalized at $90 million in the state with the most relaxed incorporation statutes of the day). Gates detailed his maneuvers that led to this consolidation in his testimony at the *Parks v. Gates* case in 1902. With each consolidation the market value of the stock rose well above the par value, and Gates, in a classic case of "stock watering," issued new stock that exceeded in par value the assets that it represented. Nevertheless, the future of the wire industry was so bright that the market value of even the watered stock rose. In 1902 Gates sold American Steel & Wire Company to U.S. Steel.

Gates brimmed with energy and confidence and had a penchant for speculation, all backed with his smooth talk. He came to be known as "Bet-a-Million" or "Bet-You-a-Million" Gates because of his willingness to risk great amounts of capital with considerable confidence and conviction. In 1896 he was involved in a large Chicago gas investment, and he reportedly had cleared $12 million in wire profits alone by 1897. His speculations earned him a reputation as an early corporate "raider" on Wall Street, and it was rumored that his activities helped spark a panic in 1900. But he was not an ill-informed investor; he eagerly sought the advice of market technicians, such as William R. Walker, and also established a market network that involved private detectives.

During the decades following the Civil War the iron and steel industry had become an increasingly dominant sector in the American economy, and it was somewhat inevitable that Gates's wire concerns would bring him into business contact with the Carnegie Steel interests. By 1894 Gates had succeeded Jay C. Morse, at Morse's own request, as the president of the Illinois Steel Company, which permitted Gates to integrate backward his wire operations, then known as the Consolidated Steel & Wire Company. As usual Gates turned Illinois Steel into a high-yield operation which put Gates into more regular contact with the manufacturing aspects of iron and steel. By the late 1890s when Andrew Carnegie suggested he was ready to sell his steel interests, Gates saw an opportunity to further integrate steel production, iron manufacturing, and his own wire sales operations.

In 1899 Henry Clay Frick, the head of Carnegie Steel, had with Carnegie's approval attempted to arrange the purchase of Carnegie's interests by outside investors. Frick presented the package to a group of Chicago investors, including William and James Hobart Moore and Gates. These individuals toyed with the idea of forming a "supertrust" in steel, perhaps even creating a $1 billion corporation. Gates was active during the negotiations, but the deal collapsed. Undeterred, Gates helped to form the Republic Iron & Steel Company and in 1906 was a syndicate member of the group that took over the Tennessee Coal & Iron Company. Among the other companies he helped to organize were the American Tin Plate Company, the American Sheet Steel Company, the American Steel Hoop Company, and the National Steel Company.

Still, Gates suffered his share of setbacks, including a disastrous speculation in the Chicago Grain Exchange which wiped him out. He recovered completely. His experience with J. P. Morgan was not as pleasant, however. Gates had learned that Morgan needed the Louisville & Nashville Railroad for his rail consolidation plans, and Gates acquired it, reselling the line to Morgan in 1902 at a hefty profit. Morgan lured Gates into a huge speculation on credit extended by Morgan, then caused a plunge in the price of the securities that Gates had put up for collateral. As part of the settlement for Gates's debt, Morgan insisted that Gates retire from the New York Stock Exchange.

Gates was not finished, however. He organized the Texas Company, which had holdings in the Spindletop Oil Field and controlled real estate and industries in Port Arthur, Texas. He remained a financier to the end, if only on a smaller scale. John Warne Gates died on August 9, 1911, in Paris, France, where he maintained a castle. Gates predated the Wall Street "raiders" of the 1980s, but although he lacked social responsibility at times, he was hardly a ruthless "Robber Baron." Energetic and restless, Gates represented the archetypal "deal maker."

References:

William T. Hogan, S.J., *Economic History of the Iron and Steel Industry in the United States*, volume 1 (Lexington, Mass.: Lexington Books, 1971);

Joseph Wall, *Andrew Carnegie* (New York: Oxford University Press, 1970).

Gautier Steel Company

by Frank Whelan

Morning Call, *Allentown, Pennsylvania*

The Gautier Steel Company was one of many small-scale iron- and steel-making firms that were established in America during the post-Civil War era. Its origins were in Jersey City, New Jersey, where it began business under the name Gautier Mills as a division of Lafayette Steel Works of D. G. Gautier & Company. As was true of many manufacturing concerns of that era, it was a partnership. The firm was founded in the 1870s by Dr. Josiah Gautier, a Frenchman by birth, and his associates.

In late 1877 the owners of the Cambria Iron & Steel Works of Johnstown, Pennsylvania, invited Gautier into their partnership. Like many iron and steel firms of the day, the pull on Gautier to relocate was strong. The metals industries were still languishing in the depression that had followed the panic of 1873; and the collapse of the market for railroad rails left many smaller firms without a market. Moving out of Jersey City to Pennsylvania would put the firm closer to a primary source of raw materials. Coupled with the economy's depressed state, the potential savings on material made a move to the west seem irresistible.

Cambria was undoubtedly a good choice for Gautier. A long-established business–it was established in 1852–Cambria Iron & Steel was legendary in the industry. It was at Cambria that the steelmaster John Fritz had created his three-high rolling mill, a breakthrough which revolutionized the making of railroad rails around the world. Although John Fritz had long since departed for Bethlehem Iron Company, Cambria retained the image of his legacy.

The Gautier firm took up Cambria Iron & Steel's offer. On May 1, 1878, the new firm began its legal existence under the name of the Gautier Steel Company, Limited. Its capital was $300,000 and was increased in May 1879 to $500,000. The firm's members included Daniel J. Morrell, Cambria's general manager, George Webb, Cam-

bria's general agent, and Daniel Jones. The Gautiers were represented by founder Josiah Gautier and his family members, Thomas B. Gautier and Dudley G. Gautier.

For the next three years Gautier operated separately from the Cambria Iron Company. During this period the New Jersey Wire Mill, owned by Henry Roberts of Newark, New Jersey, was acquired by Cambria and combined with the new Gautier plant on the south bank of the Little Conemaugh River. Called the "Island," the plant was located at the junction of the western water route and the Allegheny Portage Railroad.

Steel and iron directories for 1879 list Gautier as a company that produced steel for plows. "Special attention given to the manufacture of all steel used in agricultural implements and tools," was one description of the firm. The same source describes the resources of the Gautier works as it was in 1880: "Rolling mill has 15 heating furnaces, 2 steam hammers, and 8 trains of rolls." The rolling mill had an annual output of 25,000 net tons. The wire mill made several grades of wire but specialized in galvanized wire, its output also reaching 25,000 tons. Other special products manufactured at Gautier were carriage springs and railroad coach springs. There was also a rake tooth shop.

In this period, Gautier retained offices at 93 John Street in New York and an office and warehouse on Commerce Street in Philadelphia. The products turned out by Gautier were known worldwide for their precision and high quality. The term "Gautier finish" quickly became "a synonym for the highest class of bar products with respect to accuracy, size, finish and quality."

For reasons that are unclear, by 1881 Cambria Iron had decided to buy out Gautier Steel Company, and in July of that year the partnership was dissolved. Cambria general manager Morrell, assistant general manager Powell Stackhouse, and W. S.

Robinson were the liquidating trustees. On December 12, 1881, Gautier Steel Company Limited became the Gautier Division of the Cambria Iron & Steel Company.

Under Cambria Iron the Gautier plant continued to manufacture the same products as before. According to an accounting done in 1888, the division had a brick building 200 feet by 500 feet, where wire was drawn and finished. Nearby was another building, the barbed wire mill, 50 feet by 256 feet, where the famous "Cambria Link" barbed wire was made. A merchant mill building, 725 feet by 250 feet, produced wire rods, shafting, springs, plowshares, rake and harrow teeth, and cold steel.

In 1889 a natural disaster altered this prosperous picture when torrential rains brought on the collapse of the dam at Johnstown, and a great wall of water swept down the valley. The Johnstown Flood wiped out almost all traces of the Gautier plant that had been built in 1878. When Cambria Iron started the task of rebuilding, its directors decided to discontinue the old wireworks. The rest of the facility was rebuilt. In 1923 Bethlehem Steel Corporation took control of Cambria Iron & Steel and, with it, Gautier.

Unpublished Documents:

"Report of Engineering Department of the Bethlehem Steel Corporation," Bethlehem, Pa., 1923;

"History of the Cambria Plant of the Bethlehem Steel Company," 1935. Manuscript at Canal Museum Archives of the Center for Canal History and Technology, Easton, Pa.

Miles Greenwood

(March 19, 1807-November 5, 1885)

by Marc Harris

Ohio State University

CAREER: Foundry worker (1827-1828); foundry manager, New Harmony, Indiana (1828-1829); foundry worker, T. & J. Bevan (1829-1832); partner and operator (1832-1836?), owner and operator, Eagle Iron Works (1836?-1885).

Miles Greenwood was a figure of great importance in southwestern Ohio in the mid nineteenth century. After an arduous youth and a brief involvement with utopian communalism, Greenwood developed an iron-manufacturing enterprise that was at one time the largest foundry in the West and an important Civil War supplier to the Union army. Of equal importance in his very active life was his vigorous civic leadership in Cincinnati, where he was largely responsible for creating a number of important civic institutions and helped ensure the city's loyalty to the Union cause during the Civil War.

Like many entrepreneurs of his era, Greenwood had roots in New England. His father was originally from Salem, Massachusetts, the descendant of an earlier Miles Greenwood who had fought with Cromwell in the English civil war. The elder Greenwood moved to Jersey City, New Jersey, early in the nineteenth century; Greenwood's mother was a Jersey City woman, and he himself was born there. It is not known what profession or trade his father followed. The following year the family moved to New York City where they lived for nine years before moving to Cincinnati in 1817. There Greenwood found himself having to work to support the family because of his father's invalidism. Over the next eight years he worked shining shoes, posting notices, cutting wood, and clerking in a small store. He had scant time, if any, left over for education and apparently felt the lack keenly.

In 1825, at eighteen, he went with his father to Robert Owen's colony at New Harmony, Indiana. The colony was one of many contemporary attempts to come to terms with the development and spread of newly productive manufacturing techniques and the restructuring of work and its rewards which those new processes entailed. Owen was already famous as the developer of a model factory village at New Lanark, Scotland; at New Harmony he attempted further reform by establishing communal ownership and a cooperative labor system. The colony introduced a number of other progressive reforms, including free public education and a free library. As a member Greenwood took advantage of the colony's educational opportunities and participated in the labor pool, and it may be that his involvement in New Harmony helped develop the very strong sense of communal responsibility he later evidenced.

In 1827, whether acting on his own or for the interests of New Harmony, Greenwood left for Pittsburgh. There he worked in an iron foundry for about a year, learning the trade of ironworking well enough to return to the colony and manage a foundry there. By this time New Harmony was falling victim to serious disagreements, and in 1828 the colonizers gave up the scheme of communal ownership and cooperative labor. Though many settlers remained, the colony as a common enterprise was dead. In 1829 Greenwood left for Cincinnati, where he was to live out the rest of his career.

For three years he worked for the foundry of John and Thomas Bevan, and in 1832 he opened the Eagle Iron Works in partnership with Joseph Webb. They began with borrowed capital and a complement of ten employees, and through good luck and astute positioning the company grew quickly. Early in its history Greenwood bought out Webb's interest and was responsible for the bulk of the company's growth. Its first product, an iron hinge ostensibly of Greenwood's own design, was a smashing success, and receipts from it allowed the works to expand its product line. The foundry eventually produced a wide range of foundry and forge products, including hydraulic presses, steam engines, stoves and other domestic ironware, tools, iron building fronts, and some fabricated hardwood pieces.

These goods served a variety of markets; Cincinnati had been for some time the main distribution center for the central Ohio Valley and capitalized on its position and trade connections to dominate a growing western trade in manufactured products. Throughout the pre–Civil War years steam navigation on the western rivers expanded greatly, along with the use of stationary steam power in the growing number of mills and factories in the Ohio Valley; when Louisiana and Texas sugar planters converted to steam-powered cane mills in the 1840s and 1850s many of their new engines were of Cincinnati manufacture. Additionally, the development of canals after 1827 and later of railroads opened up large areas of western Ohio, southern Indiana, and Kentucky to commercial farming, and increasingly dense population settlement provided a good market for domestic iron products

and tools as well as for architectural iron. By 1851 Greenwood's Eagle Iron Works employed 350 men and produced goods valued at $360,000. The elaborate growth of infrastructure in the West enabled Greenwood, carefully managing his foundry and shrewdly choosing his product line, to build a formidably prosperous business.

As his business grew, his drive to assume civic responsibility, whether inherited from New England traditions or impressed on him by his New Harmony experience, led him to deep involvement in public affairs. In the early 1840s he sat on the Cincinnati city council, where he lobbied for an expanded range of city services as well as for economy in government. He also took a special interest in education. In 1848 his contributions revived the Ohio Mechanics' Institute, which had grown out of the same intellectual ferment that encouraged Owen's experiments. Designed to offer a broad education to journeyman artisans who had no other educational opportunities, it had been in precarious shape for 20 years before Greenwood's contributions and solicitations enabled it to construct its own building. Ten years later he agreed to serve as a trustee of McMicken University, which later became the University of Cincinnati.

In the 1850s Greenwood became involved in the civic improvement for which he remains best known in Cincinnati history, the creation of a professional fire department. In the mid nineteenth century, Cincinnati, like other American cities, lacked virtually all the urban amenities now taken for granted. It relied on constables supplemented by posses for police protection and on volunteer fire companies to prevent the disaster of conflagration. Fire companies attracted ambitious, spirited, and pugnacious young men who used them as political and social instruments, and for many reasons the citizenry was deeply divided among supporters of the rival companies, which functioned as crosses between gangs and political clubs. They competed against each other for precedence at fires, which sometimes turned into brawls as companies fought each other while letting the fires burn. Greenwood participated eagerly in one of the volunteer companies from 1829 and, because of his strength and air of authority, was elected its president several times. He was an important man in the volunteer system.

But in 1853 some of the city's leaders and insurers, convinced of the overwhelming risk that fire presented to a crowded manufacturing city with a multitude of wooden buildings, decided to combine introduction of a powerful new steam-driven pumper with adoption of a full-time paid fire department to operate it. Greenwood was called on to support the new order against expected stubborn resistance, and he agreed to serve as captain of the city's first professional company. At the engine's first public appearance, a fire set by volunteers who intended to smash the machine, he led a hired contingent of rowdies and fought off the volunteer companies as the steam pumper demonstrated its superior performance. While resistance continued Greenwood put up $15,000 of his own money toward the new equipment and salaries, moved his family out of the city, and gave up running his business. He also donated his salary to the Mechanics' Institute and hired a foundry supervisor at his own expense. After 18 months of wrangling Cincinnati became one of the first cities in the country with a professional fire department.

Besides improving conditions in the city as a whole, the fire department reorganization ultimately benefited Greenwood's larger business interests, as did some of his other pursuits. In 1859 he was a charter director of a company formed to build a bridge over the Ohio River to Covington. After the war he acted in a similar capacity with other commercial leaders in promoting a railroad to Chattanooga. Both projects would be of strategic importance for the city's commerce; the bridge, however, was not built before the war, while the railroad was intended to reestablish a postwar commercial artery into the South.

Greenwood was equally active during the war years. The crisis of Abraham Lincoln's election in 1860, followed by the secession of seven states, had severe repercussions for the city, which largely depended on trade with the slave states for its economic well-being. Southeastern Ohio, Cincinnati's hinterland, was home to thousands of southern migrants sympathetic to the secessionists, and Kentucky, which could control the Ohio River, had yet to declare itself in early 1861. In such circumstances many were tempted to temporize or offer some support to the new Confederacy; Cincinnati's leaders were said to be divided. Greenwood, however, was a Union man and with others took drastic action to show his support. As grand marshal, he led a parade welcoming Lincoln's train into Cincinnati on its way to the inauguration in Washington.

When war actually began in April 1861 the potential danger for Cincinnati was enormous. Surrounded by Southern sympathizers, it was the nearest Union city in the West to any slave state and contained a temptingly large store of cash and supplies. In April several measures were taken to mobilize the area for the Union and ensure that the canals and railroads would remain loyal. Greenwood served on the committee of safety, which was charged with preventing supplies from heading downriver to the Confederacy. He again played an important role in the response to the first of two invasion threats in summer 1862, when the city was panicked by reports of John Morgan's raid into central Kentucky. Fearing that his cavalry would sack the ungarrisoned city, Cincinnatians elected Greenwood to a committee of vigilance to prepare defenses. The committee gained approval from Secretary of War Edwin Stanton to use cannons that the Eagle Iron Works had cast for the Union army.

The cannons were only a small part of Greenwood's production for Federal forces; Eagle Iron Works became virtually a Union arsenal. Employment nearly doubled to about 700 men, and Greenwood concentrated production on war matériel. The foundry produced 2,000 or more bronze cannons, along with gun carriages and caissons, entrenching tools, iron bedsteads for camps and hospitals, and some small ironclad naval vessels. Early in the war Eagle workmen converted some 40,000 old smoothbore Springfield muskets into percussion-lock rifles; "Greenwood rifles" were said to have an extraordinary recoil. And although good politics made good business, it also made enemies: Southern sympathizers set the foundry and works afire twice during the war years.

Greenwood continued his active life for another two decades. He was described as a huge and burly man, unafraid of a fight, and he certainly lived up to that reputation during the fire department reorganization when he was forty-seven. He was also said to be scrupulously just and honest even though, a rarity among the civic leaders and entrepreneurs of his time, he belonged to no church and professed no religion. Like most others of his position he assumed a place of public leadership, and some of his projects helped him in business as well as helping the city. But Greenwood, whose time in New Harmony marked him as one who had actively tried to create a self-consciously ordered community, was unusual in the degree to which he was willing to sacrifice money, time, effort, and business interests to improve his city and help his fellow citizens.

References:

J. Fletcher Brennan, ed., *Biographical Cyclopedia and Portrait Gallery of Distinguished Men of the State of Ohio* (Cincinnati, 1879);

Clara Longworth de Chambrun, *Cincinnati: Story of the Queen City* (New York: Scribners, 1939);

Gilbert F. Dodds, *Early Ironmasters of Ohio* (Columbus, Ohio, 1957);

Henry A. and Kate B. Ford, compilers, *History of Cincinnati, Ohio* (Cleveland: World, 1881);

Charles Frederic Goss, *Cincinnati: The Queen City 1788-1912* (Chicago: S. J. Clarke, 1912);

Charles Theodore Greve, *Centennial History of Cincinnati and Representative Citizens* (Chicago: Biographical Publishing, 1904);

Alvin F. Harlow, *The Serene Cincinnatians* (New York: Dutton, 1950);

Harry N. Scheiber, *Ohio Canal Era: A Case Study of Government and the Economy, 1820-1861* (Athens, Ohio: Ohio University Press, 1969);

Richard C. Wade, *The Urban Frontier: the Rise of Western Cities, 1790-1830* (Cambridge, Mass.: Harvard University Press, 1959).

James Harrison

(October 10, 1803-August 2 or 3, 1870)

by Terry S. Reynolds

Michigan Technological University

CAREER: Partner, Glasgow, Harrison & Company or Glasgow & Harrison (c. 1820-1840?); part owner and managing director, American Iron Mountain Company (1843-1870?); partner, Chouteau, Harrison & Vallé (1845-1870).

James Harrison, one of Missouri's pioneer iron manufacturers, was born in Bourbon County, Kentucky, on October 10, 1803, to John and Elizabeth McClanahan Harrison, both of northern Irish ancestry. He was their oldest child and was reared on the family farm.

The family moved to Fayette, Missouri, in 1819, and by 1822 Harrison was active in grain and livestock trade. In 1831 and 1832 he expanded this trade to Mexico, personally leading expeditions to Chihuahua to exchange produce for silver bullion. On one of these expeditions he was among the three survivors of a party of 13 attacked by Indians. In 1832 he married Maria Louisa Prewitt. That same year he began furnishing supplies to Indians under the terms of a contract with the United States government. Between 1833 and 1840 he successfully sold merchandise in several Arkansas towns while operating from a headquarters in Jonesboro, Arkansas. In 1840 Harrison moved to St. Louis.

In 1843 Harrison initiated the development of Missouri's iron ore resources. For some years

Missouri's iron deposits around Pilot Knob and Iron Mountain had been known, and in 1836 Van Doren, Pease & Company had attempted to develop them, but failed. Harrison was largely responsible for their successful development. In 1843 he purchased a one-third ownership in the Iron Mountain property, and in 1845 organized the American Iron Mountain Company in association with many businessmen, the most important being Pierre Chouteau, Jr., of St. Louis and Felix Vallé of Saint Genevieve, Missouri. That same year he became a partner in the firm of Chouteau, Harrison & Vallé. The American Iron Mountain Company established a blast furnace at Iron Mountain in 1846 and soon became one of the larger producers of iron in the country. By 1853 the company had two cold blast furnaces producing 4,300 tons of pig iron per year and one hot blast furnace under construction with a probable annual capacity of 2,500 tons. Harrison built both a church and schools for his employees at Iron Mountain, and the firm prospered through the 1850s despite a fire which destroyed its Iron Mountain furnaces and surrounding buildings in 1854.

In 1850 Chouteau, Harrison & Vallé purchased a large tract of land in the northern part of St. Louis and began erecting one of the largest rolling mills and nail factories (the Laclede Rolling Mill) in the United States to process the ores of

James Harrison (engraving by A. H. Ritchie, courtesy of Missouri Historical Society)

Iron Mountain. Harrison was, at the same time, instrumental in organizing the Iron Mountain Railroad Company and promoted the construction of the St. Louis & Iron Mountain Railroad to bring ore to the city. In the same decade he also negotiated a $7 million loan to purchase the property of the Pacific Railroad Company of Missouri from the state of Missouri. This company was later reincor-

porated as the Missouri Pacific Railroad Company which Harrison served as a director.

Harrison was politically conservative and, while opposing secession, was a Southern sympathizer during the Civil War. He reputedly had a horse saddled in his stable ready for any young man wishing to ride to Southern lines to join Confederate forces. Harrison's home in St. Louis was opposite a Federal prison during the war, and both Harrison and the ladies of his family ministered to Confederate prisoners there when permitted. Because Harrison had always been active in St. Louis charitable and civic causes and had a reputation for high integrity, President Lincoln granted him a pass in December 1864 to travel through Union lines in an attempt to negotiate a settlement to the war with Confederate officials in Texas and Louisiana.

In 1865 Harrison became a member of the Kelly Process Company, along with his partners Chateau and Vallé, with the apparent intention of initiating the production of Bessemer-Kelly process steel in the St. Louis area. Vallé succeeded Harrison as president of the American Iron Mountain Company and the firm of Chouteau, Harrison & Vallé after Harrison's sudden and unexpected death on August 2 or 3, 1870, in St. Louis. Although Harrison did not live to see the construction of a Bessemer-Kelly works, he was the prime mover in making Missouri, for a time, a major iron-producing state.

References:

Florence Harrison Bill, "Our Harrisons," *Bulletin of the Missouri Historical Society*, 6 (July 1950): 521-539;
Arthur B. Cozzens, "Chronology of Ironmaking (in Missouri)," *Missouri Historical Review*, 36 (1941-1942): 216-217;
J. Thomas Scharf, *History of Saint Louis City and County*, volume 2 (Philadelphia: Louis H. Everts, 1883): 1264-1265;
James M. Swank, *History of the Manufacture of Iron in All Ages* (Philadelphia: Privately published, 1884).

Henderson Steel & Manufacturing Company
by Brady Banta

Louisiana State University

The southern iron industry, having endured the economic slowdown of the mid 1870s, was prosperous and competitive until the late 1880s. By that time, however, more and more northern iron foundries were making the transition to steel production, a change that many southern companies were ill equipped, ill prepared, or ill disposed to make. This inability or unwillingness to stay abreast of rapidly changing technology cost southern plants dearly, as consumers increasingly expressed a preference for steel.

It should be acknowledged, however, that not all southern iron producers and capitalists were

oblivious to these developments. Indeed, some were receptive to change and attempted to produce steel from southern pig iron. A notable example of such an enterprise was the Henderson Steel & Manufacturing Company of Birmingham, Alabama. The central figure in this business was James Henderson, the holder of a patent, referred to by him as the "flame process," for making open-hearth steel. Having come to Birmingham to attend a conference of the International Association of Metallurgists and Mineralogists, Henderson touted his process and sold local entrepreneurs and capitalists on its commercial development. Their discussions led to the or-

The Henderson Steel & Manufacturing Company works in Birmingham, Alabama (from the Photographic Collections of the Birmingham Public Library, Birmingham, Alabama)

ganization in 1887 of the Henderson Steel & Manufacturing Company.

Having constructed a facility to test the "flame process," in early March 1888 the Henderson Steel & Manufacturing Company fired the furnace and in just under four hours produced its first ton of steel. Jubilant over their success and seeking capital for expansion, the owners reorganized the business as Henderson Steel Company. The expansion project moved rapidly toward completion, and before the year ended, a new and much larger facility was producing up to 16 tons of steel daily. Made from locally produced pig iron, often of a quality not desirable for other applications, the Henderson Company produced tool steel from which others fashioned razors and cutlery.

But Henderson Steel never overcame technical problems related to the "flame process" (i.e., the intense heat required rapidly consumed the lining of the blast furnace) and never negotiated dependable pig iron supply contracts. These difficulties, coupled with the greater efficiency of northern competitors, proved insurmountable. While the Henderson Steel Company could produce steel of acceptable quality, this technical achievement could not be translated into a commercial success.

Acknowledging their failure but not abandoning all hope of future success, in 1890 the company's directors gave the Henderson Steel facility to a committee from the Birmingham Chamber of Commerce. Despite the committee's promotional efforts, local investors could not be convinced to continue the steel-making project. The Tennessee Coal, Iron & Railroad Company eventually leased the furnaces for the purpose of conducting additional experiments involving the production of open-hearth steel. But the mechanical and technical problems inherent in applying Henderson's process to Birmingham pig iron could not be resolved. The project was abandoned, and in 1906 the Henderson Steel plant was dismantled and sold for scrap.

References:

Ethel Armes, *The Story of Coal and Iron in Alabama* (Birmingham, Ala.: Birmingham Chamber of Commerce, 1910; reprinted, New York: Arno Press, 1973);
Victor S. Clark, *History of Manufactures in the United States*, 3 volumes (New York: McGraw-Hill, 1929).

Abram Stevens Hewitt

(July 31, 1822-January 18, 1903)

by John W. Malsberger

Muhlenberg College

CAREER: Partner, Cooper, Hewitt & Company (1845-1903); president, New York & Greenwood Lake Railroad Company (c. 1870-1903); member, United States House of Representatives (1874-1886); chairman, Democratic National Committee (1876-1877); mayor, New York City (1886-1888).

As an iron and steel manufacturer, a politician, and a philanthropist, Abram Stevens Hewitt had a significant impact on American life in the nineteenth century. A partner in the noted firm of Cooper, Hewitt & Company, Hewitt was the first ironmaster in America to produce commercial quantities of steel through the introduction of the Siemens-Martin open-hearth furnace, one of the chief nineteenth-century advances in the manufacture of steel. Following his success in the business world Hewitt directed his energies to reform of New York City politics in the 1870s. His municipal efforts led to his election to the United States House of Representatives, where for five terms Hewitt was a tireless advocate of tariff reform and sound money. Hewitt also took an active part in promoting education, most notably as chairman of the board of trustees of Cooper Union, the school founded and funded by Peter Cooper. In all these capacities Hewitt left a major mark on the iron and steel industry and on the American nation.

The son of John and Ann Gurnee Hewitt, Abram Hewitt was born July 31, 1822, into modest circumstances in Haverstraw, New York, on the farm settled by his maternal ancestors, French Huguenots who had come to America prior to 1650. His father was born January 8, 1777, in England and was trained as a cabinetmaker and a machinist. As a machinist he worked at the noted firm of Boulton & Watt in Soho, near Birmingham. John Hewitt moved to America in 1796 and immediately found work in the Schuyler Foundry, where three

Abram Stevens Hewitt

years later he assisted in the construction of the first steam engine in the United States.

Abram Hewitt's youth was one of constant economic struggle, in part because there were seven in his family but also because his father repeatedly changed jobs. This experience apparently instilled in Hewitt both a high degree of industriousness and a strong determination to succeed. Early in his life Hewitt directed his energies to acquiring a good education; he attended public schools in nearby New York City and by the age of eleven had read all of Shakespeare's plays. When he was thirteen he enrolled in the Grammar School of Columbia College, then considered one of the finest schools of its type. After graduating from the Grammar School in 1838 Hewitt won a public school competitive scholarship enabling him to enroll at Columbia College. He excelled as a student at Columbia, winning the gold medal awarded for academic excellence four consecutive years. When he graduated with an A. B. degree in 1842, Hewitt began the study of law, supporting himself by serving as the acting head of the mathematics department of Columbia College's Grammar School and also by privately tutoring students. One of the students he tutored was his classmate at the college and future business partner, Edward Cooper, whose ill health had delayed his graduation.

The heavy burden of his law studies and his work caused Hewitt's eyesight to fail in fall 1843, and, when doctors ordered complete rest, he sailed with Edward Cooper to Europe in March 1844 for a ten-month excursion. It was this experience which directed Hewitt into the iron business. On the return voyage from Europe in December 1844 their ship, the four-masted *Alabamian*, already weakened by several storms, foundered 120 miles from New York City and was abandoned. Hewitt and Cooper were cast adrift in a lifeboat on the Atlantic Ocean and spent 12 hours on the water before being rescued. During this ordeal they pledged that if they survived they would go into business together. A business decision made in 1845 by Cooper's father, Peter Cooper, quickly enabled this pledge to be honored.

Early in 1845 Peter Cooper, a noted manufacturer and ironmaster, decided that if the anthracite iron mill he had established in New York City in the 1830s was to remain competitive, it had to be moved closer to the source of the raw materials. He proposed to establish an iron mill at Trenton, New Jersey, which was connected, by means of the Delaware River and its tributaries, with the anthracite coal mines of northeastern Pennsylvania. By 1845 there was also a water power company in operation at Trenton which could provide the power to run the mill, as well as railroads linking the town to Philadelphia and New York City. Peter Cooper offered to place Edward in charge of the Trenton mill, and at Edward's insistence, Abram Hewitt was included as a partner. As the Trenton Iron Company was initially established, Peter Cooper supplied the bulk of the start-up capital and Edward Cooper and Hewitt, through a separate partnership, assumed

the responsibility for the day-to-day management of the mill.

Initially, Cooper, Hewitt concentrated on the manufacture of wire, rods, and iron rails at the Trenton mill, purchasing their pig iron from other sources. By fall 1845 they had produced their first batch of wire and rods, and before the year had ended the first rails had been rolled. The success of Cooper, Hewitt's venture was assured when, shortly after its rolling mill had begun operation, the owners of the Camden & Amboy Railroad, one of whom was a close friend of Hewitt's father, offered the Trenton firm a contract for 2,000 tons of iron rails at $90 per ton. Largely as a result of this contract, the Trenton Iron Company grew until by the end of 1846 it commanded a work force of over 500. A year later it closed its books with a net profit of $67,500.

The rapid success of the Trenton mill led to several important internal changes in 1847. First, early in the year, the firm was formally chartered by the state of New Jersey and capitalized at $500,000. Chosen as officers of the new firm were Peter Cooper, president; James Hall, treasurer; and Abram Hewitt, secretary. Secondly, Edward Cooper and Hewitt formalized the partnership, Cooper, Hewitt & Company, to manage the daily operations of the firm. Although Peter Cooper was nominally the president of the firm, his son and Hewitt ran the business. Finally, in 1847 Trenton Iron also purchased land at Phillipsburg, New Jersey, located on the Delaware River north of Trenton, and began construction of two blast furnaces to enable the company to manufacture its own pig iron. Only two years after its establishment, Trenton Iron had embarked on a course of diversification that would bring it great success.

The establishment of the Trenton Iron Company in the mid 1840s came at a particularly propitious time in the development of the American iron and steel industry. Only six years earlier the Welsh ironmaster, David Thomas, had constructed in Catasauqua, Pennsylvania, one of the first blast furnaces in America capable of producing commercial quantities of iron using anthracite coal as fuel. In addition, by 1845 the American economy had begun to recover from the panic of 1837, encouraging railroad companies throughout the country to expand rapidly. The resulting railroad boom soon provided such manufacturers of iron rails as Cooper, Hewitt with more business than they could handle. The success of the Trenton Iron Company was rooted directly in technological innovations and economic developments that helped transform the American iron industry.

The technological and economic factors that boosted the fortunes of Cooper, Hewitt also benefited scores of other iron manufacturers throughout the nation. That the Trenton Iron Company was able to surpass most of its competitors and become one of the industry's leaders by the mid nineteenth century was largely the result of Abram Hewitt's efforts both to improve the quality and diversify the range of products manufactured by his firm. To improve the quality of their iron rails, which at this time tended to be rather brittle, Hewitt in 1848 hired an inspector whose sole duty was to follow carefully each step of production and report where any problems existed. Similarly, in an effort to manufacture more durable rails, Hewitt in the late 1840s scoured the New Jersey countryside in search of higher grade iron ore. He finally bought the Andover mine, one of the most famous iron mines in the colonial era, which by 1800 had been abandoned. The ore extracted from the Andover mine produced rails of such great durability that their fame spread rapidly throughout the United States, adding to the success of Trenton Iron. In 1852 Hewitt added a third blast furnace to the operations at Phillipsburg, giving the company an annual output of 25,000 tons of pig iron, and also greatly expanded the wire mill at Trenton, becoming for several years the sole supplier of wire to the notable bridge builder John Roebling.

Of all the steps taken by Hewitt to expand and diversify Trenton Iron, clearly the most important was to become the first American manufacturer of rolled wrought iron structural beams. Beams produced at Trenton in the 1850s were used in the construction of Cooper Union, the school established by Peter Cooper, and also in a new, multistory edifice for the publishing house of Harper & Brothers. The iron beams permitted the construction of large, sturdy, fire-proof buildings and thus contributed significantly to the urbanization of America. Trenton Iron's beams were soon in great demand throughout America, and in the 1850s the federal government alone let more than 100 contracts for iron beams to the Trenton firm. The great increase in earnings that these contracts generated allowed Hewitt, in turn, to expand the holdings of Trenton Iron still further. The most noted acquisi-

tion in the 1850s was of the famous Ringwood estate in New Jersey which, since the colonial era, had been the site of one of the largest iron ore deposits in the East. The future growth of Cooper, Hewitt's iron business rested heavily on Ringwood iron.

By the time the Civil War broke out in April 1861, Hewitt's reputation as a master iron maker was already so great that his services were twice relied upon by President Abraham Lincoln to aid the progress of the Union armies. In early 1862, as Gen. Ulysses S. Grant stood poised to attack Confederate forces at Fort Donelson, the president dispatched an urgent cable to Hewitt, whom he knew only by reputation, inquiring whether the ironmaster could manufacture 30 mortar beds and ship them to Grant at Cairo, Illinois, in less than 30 days. Without the mortar beds, Lincoln warned, the attack would fail. Because Hewitt had never seen a mortar bed, he immediately requested specific information from a Union officer at the Watertown, New York, arsenal who had recently completed one. Two days later when he received the mortar bed itself, shipped to New York on a boat of the Falls River Line, Hewitt disassembled it and subcontracted the manufacture of individual pieces to several different iron companies. In this manner Hewitt was able to complete his first mortar bed only 13 days after receiving Lincoln's wire and within 26 days had shipped all 30 beds to Grant's troops. Although Hewitt incurred a significant financial loss in this endeavor, his efforts helped make possible the success of Grant's capture of Fort Donelson, the general's first great victory in the Civil War.

Later in 1862 Lincoln again drafted Hewitt's services for the Union war effort. At the start of the Civil War Northern armories lacked sufficient weapons to supply the Union army because so many weapons had been sent to reinforce government troops in the South before secession. In an effort to relieve this shortage, Hewitt went to England in 1862 to purchase as many weapons as possible and to study the process by which gun barrel iron was manufactured. When he arrived in Birmingham, Hewitt found that the owners of the British mills guarded carefully their method for producing iron for gun barrels. The New Jersey ironmaster instead met frequently with the mill workers in local pubs at the end of their shift, extracting as much information from them as he could. After his return to America he relied both on this knowledge and on experimentation to perfect a process which yielded high-grade gun barrel iron from the ore available to the Trenton firm. For the last three years of the war Cooper, Hewitt supplied all of the gun barrel iron used by the Union forces. Again, Hewitt had made a substantial contribution to the war effort, but it was one which again resulted in heavy personal losses.

The end of the Civil War brought several years of unstable economic conditions that cut deeply into the profits of Cooper, Hewitt's Trenton mill. In an effort to reduce its operating expenses the company decided in late 1865 to sell the rolling mills at Trenton and concentrate exclusively on the production of pig iron. Because a buyer could not be found in the depressed economy, Hewitt, together with Peter and Edward Cooper, raised $500,000 to purchase the rolling mill from the stockholders. In April 1866 the rolling mill was reorganized as the New Jersey Steel & Iron Company. Thereafter the Trenton Iron Company continued to own the furnaces, mines, and wire mill which earlier had brought it great fame. The liquidation of the rolling mill, however, was not sufficient to calm the fears of Trenton Iron's stockholders, and throughout 1866 and 1867 the Hewitt and Cooper families steadily purchased the outstanding shares until, by the end of 1867, the two families had become sole owners of the firm.

The emphasis that Hewitt chose to give to steel in the name chosen for the reorganized rolling mill indicated that he understood the great significance that recent technological innovations in the manufacture of steel would have for the future of the industry. Initially, however, he eschewed the adoption of both the Bessemer process and the Siemens-Martin open-hearth furnace at Trenton because of the large sums that had been invested several years earlier to manufacture gun barrel iron. Hewitt was determined to offer as wide a variety of wrought iron products as possible in order to recoup his investment. His mind was soon changed by a European trip in 1867. Chosen as one of ten scientific commissioners representing the United States at the Paris Exposition of 1867, Hewitt observed a demonstration of the Siemens-Martin open-hearth furnace and was immediately impressed by the advantages this furnace offered over the Bessemer process. Not only was the open-hearth furnace cheaper to build than a Bessemer furnace but it also consistently produced higher quality steel and did so

using iron ore whose phosphorus content exceeded the tolerances established by Henry Bessemer for his method. Following this demonstration Hewitt sent a business associate, Frederick J. Slade, to Sireiul, France, to study the method more closely with its French inventor. After receiving a favorable report from Slade, Hewitt negotiated a contract with Pierre Martin whereby Cooper, Hewitt agreed to take out an American patent on the furnace at their expense and to pay Martin a royalty on each ton of steel produced by the open-hearth process. In return, Cooper, Hewitt was made the sole licensing agent for the furnace in the United States. Work on the Siemens-Martin furnace at Trenton began in spring 1868 and was completed in December of the following year. In early 1870 Cooper, Hewitt became the first American company to manufacture commercial quantities of steel. Hewitt's foresight was responsible for the introduction of one of the most important technological innovations in the iron and steel industry. The Siemens-Martin furnace was widely adopted by American steel manufacturers following its success at Trenton and, together with the Bessemer process, was largely responsible for revolutionizing the manufacture of steel, transforming a commodity which formerly had been produced in small, expensive batches, into one that was cheap and readily available.

By the early 1870s, through both innovation and integration, Cooper, Hewitt & Company had a work force of more than 3,000 men and had expanded to include the Trenton Iron Company; the New Jersey Steel and Iron Company of Trenton, New Jersey; the Durham Iron Works of Durham, Pennsylvania; the New Philadelphia Coal Mining Company of Pottsville, Pennsylvania; the Ringwood Iron Works of Ringwood, New Jersey; and the Trenton Water Power Company. These diverse holdings gave Cooper & Hewitt blast furnaces, rolling mills, steel mills, and machine shops, allowing them to manufacture a diversity of products including bar iron, bridge wire, telegraph wire, gun iron, and steel beams.

The depressed economic conditions of the 1870s dealt a severe financial blow to Cooper & Hewitt. Each year between 1873 and 1879 the company had annual operating losses of approximately $100,000 and was returned to profitability only through good fortune when it bought a large stock of iron just before prices soared in late 1879 and early 1880. One of the factors that certainly contributed to the hard times experienced by the firm was the labor policy implemented largely at Hewitt's insistence. Unlike many of his competitors in the iron and steel industry, Hewitt viewed his workers as more than merely one of the factors of production, believing instead that the company had a responsibility for its employees' welfare. Thus, when hard times hit in the 1870s, Hewitt kept as many of his workers employed as possible, often running the mills on reduced shifts and thereby contributing to the sizable losses the firm incurred. In 1887 Hewitt proposed a cooperative ownership plan to the employees at the Trenton Iron Company, which had lost $50,000 in the previous year. Hewitt offered the workers both the use of the Trenton mill rent-free and a loan of sufficient money to operate it, asking only for a return of 5 percent on his investment. This enlightened attitude toward the workingman was perhaps the chief reason why, under Hewitt's leadership, Cooper, Hewitt was renowned for its good relations with organized labor.

In addition to his prominent role in the iron and steel industry, Hewitt also influenced the business world in a number of other capacities. He served at various times throughout his career as the principal owner and president of the New York & Greenwood Lake Railroad (part of the Erie Railroad's suburban system) and as the director of many companies, including the Erie Railroad, Lehigh Coal & Navigation, United States Steel, New Jersey & New York Railroad, Susquehanna Railroad, and Shelby Iron Corporation.

Beginning in the early 1870s Hewitt also carved out for himself an important role in New York politics. A loyal Democrat in the Jeffersonian tradition, Hewitt was appalled by the corruption of New York City politics in the years after the Civil War. He joined with Samuel J. Tilden, Edward Cooper, and others to attack the administration of William Marcy "Boss" Tweed. When Tweed was finally ousted in 1874, Hewitt helped to reorganize Tammany Hall. As a result of his efforts, Hewitt earned the reputation of a reformer and forged the important political connections that led to his election in 1874 to the United States House of Representatives from New York's Tenth Congressional District, a position he held, except for the 1879-1880 term, until 1886. During his tenure in the House, Hewitt distinguished himself as an ardent and eloquent opponent of the various monetary inflation schemes then prevalent in American

politics, leading a group of 27 Congressmen in 1878 who voted against repeal of the Specie Resumption Act, and organizing the "gold bugs" who bolted from the Democratic party in 1896 and 1900 over the nomination of William Jennings Bryan as the free silver candidate. In Congress Hewitt was also an outspoken critic of the high protective tariffs upon which the country had relied throughout most of the nineteenth century. Contradicting the position supported by most members of the iron and steel industry, Hewitt argued that high tariffs hindered the nation's development by reducing the number of available jobs, increasing the price of most goods and services, and exacerbating the unequal distribution of wealth.

Hewitt was offered the Democratic nomination for governor of New York in 1876 but declined in order to avoid charges that he was a carpetbagger whose legal residence for the previous five years had been New Jersey. The same year, however, Hewitt agreed to serve as the chairman of the Democratic National Committee. In this capacity the New York ironmaster took charge of the presidential campaign of his friend Samuel J. Tilden in the infamous 1876 election. Convinced that if the Democrats presented their case fairly and fully the American public would not tolerate the selection in the House of the Republican candidate, Rutherford B. Hayes, Hewitt was initially determined to fight, but in the end he acceded to Tilden's wishes and accepted the GOP victory.

In 1886 Hewitt retired from Congress to run as the Democratic party's candidate for mayor of New York. In a three-way race that pitted him against the reformer, Henry George of the United Labor party, and Theodore Roosevelt, the Republican candidate, Hewitt was elected with the help of Tammany Hall. During his administration Hewitt devised a plan for New York's rapid transit system and reformed the city government by weeding out corrupt officials from all agencies, particularly the police department. His actions so alienated local Democratic party leaders that they threw their weight against his reelection bid, insuring his defeat in 1888. Following this loss Hewitt retired from politics and for the remainder of his life devoted his full energies to the numerous philanthropic interests he had long maintained.

Hewitt's philanthropy revolved mainly around educational institutions, chief among them Cooper Union, established by Peter Cooper between 1857 and 1859. Hewitt served as the chairman of the original board of trustees that drew up the school's charter and later became the secretary of the board, supervising for more than 40 years all of Cooper Union's financial and educational affairs. Throughout his life, moreover, Hewitt donated more than $1 million to the institution. In addition to his substantial role with Cooper Union, Hewitt served as a trustee of his alma mater, Columbia University, and also as chairman of Barnard College's board of trustees. Finally, Hewitt was also named to the original board of trustees of the Carnegie Institution and was chosen to be its first presiding officer.

Throughout his life Hewitt received numerous honors and awards testifying to the great success he had achieved in the business and political world. Columbia University awarded him an honorary LL.D. in 1857, and for his services promoting metallurgical science Hewitt attained the unique distinction of receiving a Bessemer gold medal awarded jointly by the British and American Iron and Steel Institutes.

He was elected president of the American Institute of Mining Engineers in 1876 and 1890, and one year before his death Hewitt was honored in 1902 with a gold medal from the New York City Chamber of Commerce for the great service he had rendered the city.

Hewitt was married in 1855 to Sarah Amelia Cooper, the only daughter of Peter Cooper, and together they had six children, three sons and three daughters. He died on January 18, 1903.

References:
"The Late Ex-Mayor Hewitt," *Nation*, 76 (January 22, 1903): 67;

"Man of the Nation," *Harper's Weekly*, 47 (February 7, 1903): 244-245;

Allan Nevins, *Abram S. Hewitt: With Some Account of Peter Cooper* (1935; reprinted, New York: Octagon Books, 1967), p. 1;

Edward M. Shepard, "Abram S. Hewitt, A Great Citizen," *American Monthly Review of Reviews* (February 1903): 164-167.

Alexander Lyman Holley

(July 20, 1832-January 29, 1882)

by Allida Black

George Washington University

CAREER: Mechanical draftsman, Corliss, Nightingale & Company (1853-1855); partner and editor, *Railroad Advocate* (1856); publisher, *Holley's Railroad Advocate* (1856-1857); coeditor, *American Engineer* (1857); chief engineer, Griswold & Winslow (1863-1866, 1868-1872); chief engineer, Pennsylvania Steel Company (1866-1868); editor, *Eclectic Engineering* (1869); engineer, Cambria Iron Works (1870-1872); chief engineer, Edgar Thomson Steel Works (1872-1875); inventor, Holley Vessel Bottom (1875); president, American Institute for Mining Engineers (1875); founder, American Society for Mechanical Engineers (1880); inventor, removable converter shell (1880); winner, Bessemer Award (1882).

Alexander Lyman Holley, mechanical draftsman, engineer, journalist, publisher, and inventor, was born July 20, 1832, in Lakeville, Connecticut.

His father, Alexander Hamilton Holley, received the finest secondary education available in his area but, just as his son would do some 20 years later, chose to enter the business world rather than attend Yale. Upon completing his schooling the senior Holley joined his father's business and traveled the Lakeville-Sheffield district, examining the family interests before taking a clerk's position in the family store. The Lakeville economy depended heavily on iron production and the manufacture and exporting of iron products for its growth. A factory-centered town attracted many types of workers with diverse backgrounds and skills. As he encountered different groups of workers and management as his clientele, Alexander H. Holley realized the shortcomings of his education, which he improved by rising before dawn to study. Thus he set a pattern of consistent study and diligent work habits for his son to follow.

Jane Lyman Holley gave birth to young Alexander the year following her marriage and died two months after giving birth. Three years later Alexander Hamilton Holley married Marcia Coffing, the daughter of one of his father's business associates and a friend from childhood. Marcia Coffing Holley was an extremely devout and pious person who disdained any sort of frivolity. She emphasized hard work, daily reflection, and piety. Consequently, Alexander Lyman Holley spent his childhood in a home governed by parental examples of strenuous self-discipline and self-improvement.

After attending public schools for three years, Holley enrolled in the Sheffield Academy. Upon completion of his course work at Sheffield, and as part of his preparation for Yale, he entered the academy at nearby Farmington. Common throughout these experiences was Holley's fascination with engineering. Before he entered his teens Holley had experimented with improving the design and construction of manufacturing equipment. First he constructed a replica of a rustic bridge to cross the stream near his house. A few years later, when he was eleven, he suggested improvements for his father's new cutlery machines and presented him with workable technical plans for the necessary changes. By the time he began classes at Farmington, his drafting skills were so developed that when his class visited a copper mine and observed a steam engine at work, Holley's teacher asked him to draw a plan of the engine from memory. Eager to do so, Holley produced a rendering of the engine which so impressed his teacher that the design was forwarded to the owner of the mine.

Farmington's classical curriculum could not compete with Holley's interest in railroad engines. The academy's restrictive policies limited the student's free time and thus prohibited the time Holley needed for his frequent trips to the machine shops and railroad yards in nearby towns. Holley's enthusiasm for the locomotive surpassed his interest in traditional subjects, and his academic record suffered. His father, realizing that his son would be better off at the technically oriented Brown University, abandoned his hopes of a classical Yale education for his son. Consequently, Holley enrolled at the more scientific Williams Academy where he was much more attentive to his class work. E. W. B. Canning, the head of the academy, encouraged his exceptional engineering and artistic talents:

Alexander Lyman Holley

Though excelling in all the branches of study required of him, his penchant for mechanics and invention developed itself markedly when he attacked Natural Philosophy and Physics. Dissatisfied with the meager description in textbooks of the steam engine, with which, he seemed to be better acquainted than the author of the treatise, he, at my request, made drawings in detail of a stationary engine and of a locomotive, with an accuracy and skill that would have done credit to a professional engineer or draftsman. These I used in demonstrating, in preference to the imperfect model among the school apparatus.

At Brown, Holley's academic performance suffered as his enthusiasm for engineering far surpassed his interest in the college curriculum. Nevertheless, after repeated disciplinary discussions with both his father and the university president, Holley graduated with a degree in philosophy in September 1853.

Holley remained in Providence and joined Corliss, Nightingale & Company as a mechanical draftsman and a repairer of drafting equipment. Soon after Holley joined the firm, George Corliss, impressed with Holley's enthusiasm for locomotives, asked him to sketch the Advance, the locomotive Corliss was constructing in his plant. Holley's detailed rendering of the locomotive and his passionate interest in its construction persuaded Corliss to turn over the failure-ridden project to Holley for supervision. When the Advance's design and its erratic performance ultimately convinced Corliss that the variable cut-off valve was too fragile for constant track use, Corliss scrapped work on the locomotive and tried to persuade Holley to remain with the firm to develop other nonlocomotive engines. But locomotives remained Holley's first priority, and on March 27, 1855, with an excellent recommendation from Corliss, Holley left Corliss, Nightingale & Company in search of a job working with railroad engines.

Railroads remained a fledgling industry in spring 1855. Canals and rivers carried much more traffic than the 20,000 miles of track then in operation. As Holley's experience with the Advance illustrated, the railroad was an industry troubled by problematic equipment and a volatile marketplace. Despite his enthusiasm and experience, Holley had great difficulty finding a job. Finally, 18 months into his search, Holley was hired by the New York Locomotive Works of Jersey City.

While continuing his interest in locomotives at the works, Holley perfected his drafting skills. His technique, a thick, rich application of watercolor to his initial ink rendering, attracted the attention of Zerah Colburn, editor of the *Railroad Advocate*. Colburn believed that by emphasizing the brilliance of watercolor, Holley would introduce a new technique of mechanical illustration. Yet Holley was interested in publishing more than just his mechanical sketches. Several articles on locomotive design appeared in trade journals, including the *Railroad Advocate*.

When Colburn and Holley finally met, Colburn had decided to leave New York and the *Advocate* to build a steam sawmill on recently purchased land in Iowa. Holley eagerly agreed to become Colburn's publishing partner, and on April 19, 1856, he left the locomotive works to assume his duties as editor and co-owner. Although Colburn was a regular contributor, Holley shouldered most of the writing and financial responsibility for the journal. In August 1856 the name of the journal was changed to *Holley's Railroad Advocate*. Soon after, a serious depression devastated the

economy, and Holley struggled to keep the journal solvent. Despite his effort to increase subscriptions and a willingness to accept other commissions to subsidize his publication, bankruptcy seemed inevitable by mid 1857.

Colburn returned to New York and purchased his stock back from Holley. The *Railroad Advocate*, however, ceased publication July 4, 1857. Convinced they could succeed in publishing if they could only promote their own work, Holley and Colburn began publishing the *American Engineer* a few days later. Even more short-lived than its predecessor, the *American Engineer* folded three months later. Holley appealed to his father for financial help, but the depression had also restricted the senior Holley's capital, even though he was the newly elected governor of Connecticut.

Colburn and Holley continued their partnership despite financial losses. Their publishing experience had brought them in contact with the leaders of the railroad industry, who were well aware of the European superiority in road construction and management. In fall 1857 seven American railroad presidents sent Colburn and Holley to Europe to observe firsthand European practices in order to suggest ways in which they could be adapted to improve American road performance. The two engineers spent three months abroad studying the British rail system. A year later Holley and Colburn published *The Permanent Way and Coal-burning Locomotives of European Railways; with a Comparison of the Working Economy of European and American Lines, and the Principles upon which Improvement must Proceed*.

The book clearly delineated the differences between American and European operational practices and in doing so highlighted the weaknesses of the American rail system. The press reaction to *The Permanent Way* was divided by the ocean. The British press praised Holley and Colburn, while American journals rationalized American procedures. Henry Varnum Poor, editor of the *American Railroad Journal,* interpreted the study as an insult to the American rail industry and assailed Holley and Colburn in the February 5, 1859, issue:

> The simple fact that Messrs. Holley and Colburn have done what they could to disparage our roads is the great reason why their report has been so warmly commended in England. They are held up as experienced and conscientious engineers; while, in fact, nei-

ther of them is, nor ever has been, a railroad engineer.

Enraged, Holley demanded a public apology from Poor and took legal action when Poor did not respond immediately. Holley assembled a 12-page booklet reviewing *The Permanent Way*; the pamphlet contained statements of support from American and British engineers and railroad executives. He then sent the endorsement to Poor with a letter saying he would initiate legal action for slander. Finally, Poor acceded to Holley's demands and printed a brief, and obscurely located, apology in his journal.

Despite its notoriety, *The Permanent Way* was not a financial success, and Holley again turned to his father for financial assistance. The elder Holley insisted that his son keep accurate financial records, a suggestion which Holley tried to follow. After a few months, however, his finances were again in disarray, and his relationship with his father worsened. Holley consistently tried to improve his income by taking whatever jobs were available to him. Consequently, Holley pursued Henry J. Raymond, editor of the *New York Times,* in search of commissions for editorials and science articles. Raymond, aware of Holley's expertise and appreciative of his enthusiasm, agreed to hire him. Holley, in turn, remained loyal to the *Times* and contributed to its engineering and science coverage until he died. Although most of Holley's pieces were printed without a byline, scholar Jeanne McHugh estimates that over 200 of Holley's pieces appeared in the *Times* for which he received a commission of $8 per column.

One of the major problems confronting the American rail industry was the short life span of the iron it used for its rails. Made predominantly from pig iron rich in carbon and phosphorus, the rails usually lasted less than two years. Very little was known in the 1850s about the science of metallurgy, and metallography was unknown. Consequently, men on both sides of the Atlantic struggled to perfect the production of high quality metal for use on the road. No one foresaw the durability of steel rails and focused instead on producing low-carbon, low-sulphur iron. It was in the making of durable cheap steel that Alexander Lyman Holley made his greatest contribution to the American railroad.

While Holley was struggling to complete his studies at Brown, Henry Bessemer was working to perfect his puddling process and furnace design in London. Familiar with the manufacturing process for both wrought iron and steel, Bessemer's goal was to create a more durable iron, a metal with all the durability of wrought iron or steel but which could be molded into shape when fluid rather than forged into shape after it hardened. Bessemer built and rebuilt furnaces in his struggle to create the perfect puddling process. From 1855 to 1857 the British patent office issued him 12 patents on this process. When finally convinced that his concept would work, Bessemer presented his findings to the August 1856 meeting of the British Association for the Advancement of Science. Entitled "The Manufacture of Malleable Iron and Steel without Fuel," reprinted in the *London Times,* the *Scientific American,* and reviewed in Holley's *Railroad Advocate,* Bessemer's paper became an immediate sensation on both sides of the Atlantic. Nevertheless, the initial effort to implement what would become known as the Bessemer process was marked by controversy, frustration, and disappointment.

Soon after presenting his findings, Bessemer sold the rights to his process to the world's largest iron manufacturer, the Dowlais Iron Works, and four other interested parties. In total Bessemer received £27,000 for his patents. None of the British purchasers, however, were able to re-create successfully Bessemer's initial findings. In the United States, where Cooper & Hewitt, the agents for Peter Cooper's Trenton Iron Company, had purchased rights to the Bessemer process, two small converters were built for two separate trial runs. Both tests failed. Cooper & Hewitt held onto the patents without any further attempts to implement the process.

Dejected, but confident in his initial findings, Bessemer isolated himself in an effort to discover his mistakes. He hired the renowned British chemist T. H. Henry to analyze the ingredients used in his tests. Henry determined that phosphorus was the agent which caused the iron to break apart. Bessemer's challenge was how to remove this chemical during the iron's malleable stage. After a year of contructing various converters, only to encounter more failures, Bessemer intensified his search for phosphorus-free pig iron. When he learned that Sweden was the source for the purest iron, he immediately ordered a sample to test. The results of the test were the same as his initial discovery, and Bessemer quickly set out to restore his reputation. Even after he secretly had tested the iron at the Galloway plant in Sheffield, no iron manufacturer expressed interest in his design.

Finally, Bessemer installed a converter at the Workington Iron Company in 1858. Having detected that a blast of air through the bottom of the converter prevented the molten metal from draining through the bottom of the vessel, Bessemer designed a movable converter which would receive the metal while lying on its side and then rotate to an upright position to receive the crucial blast of air. Although Workington successfully used this process for a year, Bessemer received no offers to purchase his patents. To prove the value of his discovery, Bessemer presented his work before the Institute of Civil Engineers. His paper, "Manufacture of Malleable Iron and Steel," was well received, and Bessemer's reputation as an engineer was restored. But Bessemer's professional worries were not yet over.

While Bessemer was struggling to implement his plans and his initial proposal was being discussed in American trade journals, William Kelly of Eddyville, Kentucky, wrote to the *Scientific American* that he, in fact, was the first person "to make powerful blasts of air do the work of the fire and the manipulation of the puddle's bar in the puddling process." Arguing that in November 1851 he perfected this process, Kelly cited the precedence of his work when he applied for his patent in 1857. The patent office had granted Bessemer a patent for his process in November of the preceding year. Kelly submitted several affidavits supporting his claim, and the commissioner of patents issued Kelly a patent for his process. Bessemer had the right to appeal this decision within 60 days but took no action against Kelly. Because Kelly used a stationary converter, the patent office ruled that Bessemer was entitled to the patent he received for the machinery he developed to tilt the converter.

The Bessemer-Kelly dispute raged for decades. Claims of espionage have been made by both sides, but whatever the intent or the actual conduct of the parties involved, the end results are clear. Kelly's claim is solidly argued and documented; his process produced high-quality iron and steel. The Kelly process, however, was not intended for mass production. The Bessemer process was, and his invention could rapidly produce large quantities of iron.

The contributions of these two men to the American metal industry cannot be overstated. But another man's contribution has been all but forgotten. Iron manufacturing was revolutionized by the contributions of Kelly and Bessemer. Their goal was to make a tougher, more durable iron which could replace the lower grade metal then used in construction. But Robert Mushet's contribution to the Bessemer process forced both men to consider steel, rather than iron, as their primary product. By further isolating the chemical breakdown of the puddling process, Mushet, a metallurgist, made it possible for steel to be produced by both the Kelly pneumatic process and the Bessemer process.

Fascinated by the poor qualities of "burnt iron," Mushet, after years of experimenting, discovered that when burnt iron was alloyed with high-grade, phosphorus-free iron, the iron was restored to its initial quality. The key to this conversion process was to release the oxygen locked within the burnt iron. Mushet achieved this by withdrawing manganese from the iron and creating manganese oxide. The carbon remained in the high-grade iron, and the iron was converted into steel. Mushet himself recognized the interdependence of his process with Bessemer's. As he wrote a friend, "Bessemer metal without Mushet = Iron; Bessemer metal with Mushet = Steel."

Alexander Holley did not immediately involve himself in the Bessemer-Kelly-Mushet dispute. Rather he spent the early part of 1862 in Europe studying armaments, especially ironclad ships, for Edwin Stevens. Stevens, an engineer, manufacturer, and railroad man, was well respected abroad, and many otherwise restricted doors were opened to Holley. With typical enthusiasm and thoroughness Holley gathered the most up-to-date information on armaments in existence, which he later published in 1865 as *A Treatise on Ordnance and Armor*. While visiting England, he, for an unknown reason, also went to Sheffield where he observed Bessemer's plant in operation. Immensely excited by the converters in action, Holley wrote in his report to Stevens:

> The wonderful success and spread of the Bessemer process in England ... within three or four years prove that great talent and capital are already concentrated on this subject and promise the most favorable results.

Arguing that steel was more advantageous than iron because "the number and quantity of its ingredients are better known at each stage of its refinement," Holley concluded that the new manufacturing process "based on chemical laws" was infinitely better than the old process of "tradition, trial, failure and guesswork."

While Holley was abroad for Stevens, Eber Ward and Zoheth Durfee planned to open a plant using the Bessemer pneumatic process at Wyandotte, Michigan. While Zoheth Durfee went to England to negotiate with Bessemer for the rights to his patents, Ward hired Durfee's cousin William Durfee to begin the construction of a converter based on the Bessemer and Kelly processes. Unknown to both Durfees, Holley was interested in negotiating with Bessemer for the American rights to his design. Bessemer refused Durfee's pleas and awarded the patent rights to Holley. Thus William Durfee's construction efforts at home infringed on Bessemer's patented rights. This does not imply, however, that William Durfee willfully violated Bessemer's patent. When asked about his use of Bessemer's designs in constructing his converter, Durfee openly acknowledged copying Bessemer's designs and justified this infringement by arguing that there had been no doubt in either his or his cousin's mind that Wyandotte would secure the American rights to the Bessemer process.

Holley, acting as an agent for Griswold & Winslow, a Troy, New York, firm, arrived with the authority to make all the financial arrangements necessary to complete the transaction with Bessemer. The arrangements were negotiated during fall and winter 1863. While completing the deal in Sheffield, Holley took advantage of the opportunity to study the process firsthand and to learn as much as he could about metallurgy. Holley turned to William Allen, Bessemer's brother-in-law, for instruction about pneumatics. Allen's agreement with Bessemer allowed him to use the Bessemer process free of any royalty as long as Allen shared his experience and information with those clients interested in purchasing rights to Bessemer's patents. After three months of study and arbitration, Holley returned to Troy, New York, with a license to Bessemer's steel-producing patents. Bessemer had agreed to give Griswold & Winslow a 3-year option on the American rights to his patent. During that time the partnership could manufacture as much steel with the Bessemer process as they opted to, as long as they paid Bessemer his proper royalty.

Just as Eber Ward had planned at Wyandotte, the Troy converter was to be an experiment for Griswold & Winslow. Instructing Holley to build a small converter which could be operated on a trial run, the partners adopted a wait-and-see attitude about the Bessemer process. Griswold & Winslow refused to purchase the patents until it was proved a solid investment. Holley's responsibility was to produce the evidence which they needed to make their decision.

Upon reaching Troy, his first challenge was to build the converter. Like William Durfee before him, Holley had numerous difficulties to overcome before steel production could begin. Unlike Durfee, however, Holley had no trouble constructing the converter using Bessemer's straightforward plans. The problems arose when the tuyeres and the stoppers, the fine points of the converter mechanism, were initially constructed. Holley, working alongside his crew, redesigned and built these intricate mechanisms himself. He also tackled the problem of lining the converter and its ladles with a material strong enough to resist the intense heat of the molten metal. Before processing any ore at the Troy converter, Holley had it sent to Sheffield for analysis. Thus, before the Troy plant was operational, Holley had served as an engineer, laborer, draftsman, and mechanic.

Despite numerous problems with boilers and poor-quality pig iron, the process was put into operation on February 16, 1865. Although Durfee had Wyandotte operational for five months, Holley's efforts to make the Troy converter successful ultimately improved the process for all American users of the patent. Holley continued to perfect the equipment throughout the summer, finally providing Griswold & Winslow with the evidence needed to purchase the Bessemer rights in August 1865. The following November Holley designed a steam engine to replace the waterwheel as the major force behind blast pressure, a change which made the converter more dependable and efficient.

Once the Troy plant was operational, the relations between the Griswold & Winslow operation and the Wyandotte plant became even more strained. In April the Troy plant circulated a flier stating that they were open for business and would accept orders for steel rails, axles, machine castings, pistons, boiler plates, and numerous other products made from steel. Promoting the advantages of Bessemer steel, the Troy circular also advertised that Griswold & Winslow would be happy to issue licenses to interested parties. Finally, its readers were warned against violating the Troy plant's patent rights. Clearly, this flier represented a challenge by Griswold & Winslow to Wyandotte to cease its infringement of the Bessemer license.

Yet Wyandotte was not the only company infringing on patents. Holley was also guilty of infringing on Kelly's and Mushet's patents, although he was desperately trying to develop techniques which would avoid them. All signs pointed to a major legal dispute. Yet suddenly, after both sides had developed their arguments, the case was settled out of court. Wyandotte agreed to merge with Troy and to give the Troy group 70 percent of the licensing fees. Although many legends have sprung up as to why this decision was reached, historians argue that Ward and his major backer, David Morrell, were apprehensive about a court battle, aware of the limited time remaining on the Kelly patent, and finally, appreciative of Holley's extensive engineering knowledge and expertise.

Winslow, Griswold & Morrell became the name of the new organization formed by the merger, and Zoheth Durfee was appointed general agent. All parties agreed that both the plant at Troy and the one at Wyandotte, as well as all future Bessemer works in the United States, would be subject to the combined Bessemer-Kelly-Mushet patents. They assessed royalties of $5 and $10 dollars per ton, the price depending upon the steel's final form.

Once the process was perfected, the company was free to capitalize on the railroads' desire for affordable, durable steel. Prejudice against steel was eroding as the railroad industry grew, and the costs of constantly replacing their iron tracks appreciated. At $120 in gold per ton, British steel was soon pricing itself out of the American market. Consequently, while Holley was putting the finishing touches on the Troy converter, Pennsylvania Railroad president Edgar Thomson and Samuel Felton, president of the Philadelphia, Wilmington & Baltimore Railroad, decided to finance the construction of another Bessemer plant near Harrisburg, Pennsylvania.

In January 1866 Winslow, Griswold & Morrell granted the Pennsylvania Steel Company a license to use both the Bessemer process and the improvements Holley had made on the converter. Recognizing that Holley's engineering expertise greatly facilitated the operation of the converter,

the new company also arranged to lease Holley's services from the Troy group so that he could design and construct the Harrisburg plant.

As chief engineer, Holley supervised the construction of the machine shop, the converter house, the water pumps, the inner-plant railroad operations, as well as all the other facilities needed to make the 100-acre plant run smoothly. Benefiting from his troubleshooting experience at Troy, Holley was able to open the Harrisburg plant more rapidly than he had been able to begin the Troy plant. Yet Holley was not alone responsible for the smooth operation of the new company.

Holley worked hand in hand with two other engineers, and together the three of them increased the Pennsylvania Steel Company's efficiency. George Fritz, the chief engineer at the Cambria Iron Works, and Robert Hunt, a chemist who had studied the Bessemer installation at Cambria for the Wyandotte group, soon joined Holley to complete the Harrisburg team. Since steel could not be produced in mills built to manufacture iron rails, new machinery was necessary, a requirement which initially increased manufacturing costs.

Neither Holley nor his associates yet envisioned producing cheaper steel; thus the problem confronting the three men was how to cut their overhead costs. From the outset, the relationship of the three men was a collaborative one. Holley contributed steel ingots made in the Harrisburg converters, and Fritz and Hunt turned his ingots into steel rails. Because none of the men thought that steel would completely replace iron, they envisioned a specific market for their product: heavily trafficked areas where iron rails were not durable enough to withstand constant use. Also, since they were not sure of the best way to manufacture superior steel, they pooled their ideas and carefully tested them. They discovered that by eliminating hammering of the steel blooms and raising the temperature at which the blooms were rolled, the quality of the steel was increased. Holley concluded that employing men without previous experience eliminated the difficulties associated with a skeptical and iron-oriented work force. To supervise this diverse and unskilled crew, Holley hired the legendary foreman, Capt. William "Bill" Jones. John Fritz, a skilled mechanic, completed the management team, and by mid 1868 the Pennsylvania Steel Company was operational.

In October a fire swept the plant at Troy and destroyed half the roof and the converter beneath it. Holley returned to manage the reconstruction of the plant and devoted the next two years to the project. While still in Troy, he consulted on the construction of the Cleveland Rolling Mill Company of Newburg, Ohio, and continued to keep a watchful eye on the Harrisburg plant. Yet, in keeping with his pattern of maximum use of time, in January 1869 Holley accepted an offer from the New York publishing house Van Nostrand to edit a new journal, *Eclectic Engineering*. After a year serving as both an editor and as the chief engineer at Troy, Holley recognized that he could not do justice to the two positions. In August 1870 Holley, with characteristic honesty, described his impatience to a friend:

> I have not got along far enough in life to look back on much work or much fruit from it; but I have lived long enough to conclude with certainty, that leisure is the hardest thing in life to get along with. I try to have as little of it as possible.

Observing Holley's success at installing Bessemer converters, David Morrell decided in late 1870 to install a converter at his Cambria Iron Works. Holley was asked to design the plant, and George Fritz was retained as chief engineer. With Holley's expertise in constructing the converter and Fritz's expertise in installing the blooming mill, the Cambria plant struck its first blow in record time. Fritz's changes in the blooming mill reduced the number of people required to operate the mill from eight to four. This improvement, coupled with Holley's changes in the converter itself, made Cambria the model iron and steel mill. For over a generation most of the major leaders in the iron and steel industry would be trained at Cambria.

Impatient with his past successes, Holley visited other plants using the Bessemer process, suggesting ways for improving efficiency. Consequently, each of Holley's designs reaped the benefits of his prior constructions. If there was one constant in Holley's theories, it was his belief that overhead costs must be reduced through increased efficiency and not by constructing less expensive machines. "One lost the most money," he once remarked, "from fixing a bad engine and made the most from melting that engine down."

By late 1870 there were ten Bessemer plants in operation in the United States, and Holley had been affiliated with all but one. He was clearly emerging as the leading expert in Bessemer installation and engineering. Therefore, when Andrew Carnegie, William Coleman, and Tom Carnegie decided to build a Bessemer plant on the 107 acres of Braddock's Field, outside of Pittsburgh, Holley was the natural choice to design the plant and to supervise its construction.

In September 1872 Holley began drafting plans for what would become the Edgar Thomson Steel Works. Yet this plant, unlike the vast majority of Holley's other designs, was not an iron-making concern. Like only two of his creations, these works were to manufacture steel exclusively. The plant's ideal location facilitated this process. The land was well served with transportation facilities. Bordered on the north by the Pennsylvania Railroad and on the south by the Monongahela River, the property also had the Baltimore & Ohio Railroad cutting through its center. Thus, moving equipment within the plant itself and exporting the plant's steel were not major problems. The plant could be constructed around existing transportation facilities, and Holley could concentrate on building the finest steel plant he could design.

Despite the ideal location for the new plant, there were problems. Holley and Carnegie were not free from personal and financial troubles. Late in the year Holley's three-year-old daughter, Alice, died. By mid 1873 the iron and steel industry suffered from the general economic depression. Only one-half of the mills which had produced iron and steel the year before were still in operation. The depression even affected Carnegie's cash flow, and construction on the new plant was temporarily halted. Carnegie decided to sell his shares in the Pullman Company to raise the necessary capital to complete the steelworks. After a few months delay construction resumed in Pittsburgh.

Designing the Edgar Thomson Works was Holley's greatest joy as an engineer; he was able to draw a design at the beginning rather than adapting an existing facility to his new technology. The railroads crossing the property increased the flexibility of his design, and the vastness of the acreage allowed the buildings to be placed near available transportation. He likened this design to a body shaped by its muscles and bones rather than a package into which the bones and muscles must be crammed. The separate buildings which housed the cupola, the blowing engine, the boiler, the converter, and the rail mill all had brick walls and iron roofs. These buildings, the muscles and bones of the plant, were linked to the body of the plant by connecting railroad tracks.

Over the years Holley learned that plant design, good management, and reliable machinery were equally important in increasing a plant's output. In an article in the *Metallurgical Review* published two years after the Thomson Works began production, Holley again compared British and American production techniques. Yet, unlike *The Permanent Way*, this time Holley praised the advancements of American engineers and managers:

> In the United States, while the excellent features of Bessemer's and Longsburn's plant have been retained, the very few first works, and in a better manner each succeeding works, have embodied radical improvements in arrangement and in detail of plant, the object being to increase the output of a unit of capital and of a unit of working expense. . . .
>
> The fact, however, must not be lost sight of that the adaption of plant, which has thus been analyzed, is not the only important condition of large and cheap production; the technical management of American works has become equally improved. Better organization and more readiness, vigilance and technical knowledge on the part of the management has been required to run works up to their capacity, as their capacity has become increased by better arrangement and appliances.

Three years after completing his design, Holley reviewed his goals for the Edgar Thomson Works. As continuous a flow as possible through the entire production process was central to the plant's design. From the initial shipment of raw material from the supplier to the production process itself and finally to the shipment of the finished product to market, fluidity was the key.

Holley's technological innovations on the Bessemer process and his focus on improved plant design were mixed blessings for plant owners. On the positive side, they did increase throughput—or production time—but they also created a more capital-intensive and energy-consuming process. In addition, managers found themselves suddenly having difficulty in coordinating all parts of production once the pace of the necessary components of manufacturing increased. Nevertheless, Alexander Lyman

Holley's contribution to the steel manufacturing process cannot be underestimated. Eleven iron and steel manufacturers built Bessemer converters between 1865 and 1876. Whether these converters replaced puddling or rolling mills or provided additional support to the mills, all shared one thing: Holley's engineering foresight. Less responsible for the technological innovations of the Bessemer process than other engineers, he did, however, excel in the design, construction, and placement of facilities within an existing plant to achieve his goal of assuring "a very large and regular output."

Holley became something of a proselytizer as he worked on the Thomson works. He had always shared his knowledge among his clients and published his findings in various trade journals. With characteristic vigor, he turned his attention to the training and education of future engineers. Between the mid 1870s and his death in 1882, Holley appeared before numerous American and British professional associations, where his speeches were meant to instruct, rather than impress, his audience. A gifted speaker, Holley was able to describe clearly and simply the complex improvements made to the Bessemer process. And typically, Holley carefully shared credit where credit was due rather than merely praising his own contributions.

Holley directed much of his praise toward the plant workers who operated the machinery, crediting increased expertise and knowledge as the prime reason for the Bessemer process's refinement. Holley argued that these steel men now had a more thorough understanding of their work, and they were able to recognize trouble spots before, rather than after, they occurred. For example, although Holley's most important contribution to the Bessemer process was the development of the Holley Vessel Bottom (a replacement lining which could be installed without interrupting the machine's operation), Holley postulated that the workers could detect when a converter lining was about to give way and thus decrease the converter's down time.

Yet Holley's lectures were not merely to educate his audience on the production techniques of steel and iron. He also strove earnestly to improve the quality of technical education engineers received. Here he drew on his observations from frequent travels in Europe. Firmly believing that technical instruction in the classroom alone was inadequate preparation for an engineer, he instructed students to spend time in the machine shop. Holley argued that a student was only qualified to discuss the new open-hearth process after he had spent a week or two observing the process firsthand.

To Holley professors presented secondhand information to their students. Consequently, students were deprived of the insights and personal contributions those familiar with the machinery could offer them. Recent graduates would therefore be unable to compete professionally with the engineers who had come up through the ranks:

> When one can feel the completion of a Bessemer "blow" without looking at the flame, or number the remaining minutes of a Martin steel charge from the bubbling of the bath, or foretell the changes in the working of a blast-furnace by watching the colors and structure of the slag, or note the carburization of steel by examining its fracture, or say what ore will yield from its appearance by weight in the hand, or predict the lifetime of a machine by feeling its pulse; when one in any art can make a diagnosis by looking at the patient in the face rather than by reading about similar improvements as theory may suggest, or to lead in those original investigations upon which successful theories shall be founded.

What better way to achieve this ideal, Holley argued, than to establish engineering schools within the plants themselves.

Having argued to change training of engineers, Holley then attempted to better the way improvements and technological innovations were communicated throughout the engineering profession. Again turning to his experience in Britain, Holley encouraged the efforts of fellow engineers who formed the American Institute for Mining Engineers in 1871. Holley was one of the first members of the institute and threw his unquestioning support and enthusiasm behind the new association.

In 1875, when Holley was elected president of the institute, his biographer Jeanne McHugh states that "he had become both the leader and the conscience of his profession." Nowhere was his dedication to the overall improvement of manufacturing and his earnest commitment to the improvement of engineering more apparent than in his presidential address to the members of the institute. Entitled "Some Pressing Needs of Our Iron and Steel Manufacturers," the speech detailed the problems confronting the iron and steel industry. After discussing each problem and the cost it incurred, Holley proposed a startling solution. Because these problems would not vanish overnight and required long-time study and scientific analysis, individual companies should not alone assume the financial and technological burdens of solving what were industry-wide problems. Rather, chemical studies and experiments in design should be a collaborative effort, jointly undertaken for the improvement of the industry itself.

Although a committee was established to investigate the feasibility of the suggestion, no collaboration occurred. Instead, the manufacturers tried to cope with the falling prices brought on by the depression of the 1870s. The major producers joined together in pools to divide up orders among themselves according to their size. Although Carnegie disliked pooling, the relationship worked for a while. The Pneumatic Steel Association still controlled the licensing of the Bessemer process, and Holley still remained its chief consultant. Even though the members of the association lowered the licensing fee, there was no demand for the process. Prevailing sentiment held that there were enough Bessemer plants in operation. Consequently, Holley's income once again plummeted, and he covered the costs his fees would not by consistently drawing on his savings account.

This does not mean, however, that the American steel industry was at a standstill. With 11 Bessemer plants in operation by 1876, and the improvements which Holley and others had initiated, the quality of American steel improved while the price drastically declined. Holley now urged all his corporate clients to buy only American steel. In the eight years since the Bessemer process had been introduced in the United States, annual production had increased from 8,000 to 470,000 long tons of steel. Plants were producing at approximately two-thirds of their maximum capacity. As a result, the price per long ton decreased from $158.50 in 1868 to $68.75 in 1876.

Despite the increased availability and decreased cost of the Pneumatic Steel Association's product, Holley became increasingly worried about his status within the organization. When the association decided not to build any more Bessemer plants, and when Holley had difficulty paying his staff, he decided to travel again to Europe to study the new open-hearth process he had discussed with William Siemens in Philadelphia the year before. Characteristically, he returned from Europe brimming with information and immediately composed detailed reports on the new developments for all his industrial clients. But April 1877 had seen a marked change in steel markets. While the demand for iron and steel had risen that year, the price had decreased so much that manufacturers made little, if any, profit. By the end of the year only one-half of the plants were in operation, and the price per ton had fallen to a record $40.40.

Holley realized that the only way he could stay afloat financially was to convince some of his clients to share his enthusiasm for the new open-hearth process. But the high royalties due to Siemens, the process patent owner, dissuaded many corporations from taking Holley's advice. Finally, in 1878 Holley convinced Cooper & Hewitt to enter into an agreement with the Cambria plant to construct two small open-hearth furnaces. In July, accompanied by his family, Holley again left for Europe to study the process and to determine what improvements had been made since his last visit. During his stay Holley attended meetings of the Iron and Steel Institute as part of his search for new information. There he met Sidney Gilchrist Thomas, who, although not a member of the institute, had developed a way to remove excess phosphorus from pig iron. This was exactly the information Holley needed, and he excitedly conveyed this information to his clients. A lukewarm response greeted Holley's efforts, and once more he found himself overwhelmed with financial worries.

American steel manufacturers increasingly shared Holley's economic problems. By the end of 1878 European rail prices had decreased substantially, and these low prices prompted sharp competition in the American market. The Pneumatic Steel Association responded to this challenge by asking Holley to return to the continent and determine the reason British steel had so declined in price. Although reluctant to meet the fee Holley demanded, association president Samuel Felton finally agreed that the need to have the information was more important than the cost of Holley's services. Holley agreed to leave in May 1879.

Keeping his usual frenetic pace, Holley delivered a series of lectures to the School of Mines of Columbia College while preparing for his trip abroad. In the week preceding his scheduled departure, Holley was also to attend the meeting of the American Institute of Mining Engineers in Pittsburgh. Exhausted, Holley almost decided to miss the meeting, until his friends encouraged him to go. Secretly

Holley's friends had prepared an evening tribute and reception in Holley's honor at a country home outside the city. The night of the surprise, however, Holley's exhaustion overcame him, and he decided not to make a journey in inclement weather. Distressed that their guest of honor would miss his celebration, his friends enlisted Mary Holley's help in persuading her husband to attend. When they finally arrived at the estate, cheers greeted their carriage. Holley then was presented with an elaborate silver pitcher and tray as a remembrance for his contribution to the profession.

The next week the Holleys sailed for Europe. Holley's excursion was short and marked by ill health. The high point of the trip was the time Holley spent with Sidney Gilchrist Thomas discussing his phosphorus-removing process for pig iron. Holley returned to the United States firmly believing that this new process would revolutionize the steel industry. When the steel industry began to recover from the depression in 1879, Holley, despite a personality conflict with the irascible Andrew Carnegie, was able to convince the magnate that Thomas's open-hearth process was superior.

In April 1880 Holley again sailed to Europe to negotiate the Thomas patents for the Bessemer Steel Company. Suffering from persistent chills and fever, Holley nevertheless kept up a pace that a healthy man would have found challenging. In his eagerness to study the new process, Holley journeyed from London to Dortmund, Essen, Rhurot, Liège, Paris, and Creusot. From France he traveled through Switzerland, Italy, Austria, and back to Germany. He was planning to visit Russia when he collapsed in Cologne. Hoping that a few days rest would be all he needed, Holley managed to return to his hotel in London. As word spread of his illness, his friends offered various kinds of aid. Holley, not wanting to be a burden to anyone, refused all offers. Ultimately, his friends won out. *Engineering* editor James Dredge suggested that a drive in the country on a pleasant day would help both his spirits and his health, and Holley agreed. Once out of London, Dredge then took his friend to his home in Clapham Common. Physicians diagnosed Holley's illness as a severe liver disorder, and Holley spent two months recuperating at his friend's estate.

When Holley returned to the United States, he immediately presented the results of his European study. This would be his last such effort. At a November 1880 meeting of the American Society of Mechanical Engineers, an association Holley had helped found, he delivered a paper describing how a converter's shell could be removed without interrupting the conversion process. With typical ingenuity, Holley had found the strengths of Thomas's process and improved its efficiency. Once the audience realized that major European plants had already successfully implemented his concept, they eagerly listened to his presentation.

Although still very weak from his illness, Holley continued to speak out on engineering issues. He considered, for example, the production of different rail sizes foolish and did not hesitate to speak out at the American Institute of Mining Engineers' meeting the following February. Although no one heeded Holley's advice immediately, in 1898 a committee finally met in Chicago to discuss the benefits of engineering cooperation among the railroads. This led to the formation of the American Railway Engineering Association, which would adopt a standardized rail size as its first project.

Holley was able to convince the Bessemer Steel Company to purchase the Thomas patent, but they refused to consider his own invention. This was a blow to Holley's finances, and he spent 1881 trying to recoup. At the end of the year, with his relationship with his Bessemer colleagues deteriorating, Holley returned to Europe to study new developments. By the time he reached Belgium, he was again racked with chills and fever. Nevertheless, Holley continued his fast pace until his daughter Gertrude contracted typhoid fever. Although he slowed down to be with his daughter, Holley's health also continued to decline. Concerned about the effects of a long winter, London physicians advised Holley to return home. After arrangements were made for his family in London, Holley sailed back to New York.

The New York climate did not help his condition. Friends notified his family that he was not improving and that they should return. Unfortunately, the ship carrying Mary and Gertrude Holley was delayed at the pier, and they reached the house 20 minutes after Holley died.

Alexander Holley was forty-nine years old at his death. His business account contained less than $1,000. Despite the investments he had made over the years, his estate was small. Finally, the Bessemer Steel Company agreed to appropriate $50,000

for Holley's shell patent. In April the most prestigious engineering award, the Bessemer Medal, was awarded to Holley posthumously. By the end of the year all the professional societies to which Holley belonged had held meetings praising his contributions. Two years later the associations memorialized their famous member by collecting $10,000 for a bronze memorial. The bust now stands in New York's Washington Square, a tribute to Holley's incomparable energy, professional devotion, and engineering creativity.

Selected Publications:

The Permanent Way and Coal-burning Locomotives of European Railways; with a Comparison of the Working Economy of European and American Lines, and the Principles upon which Improvement must Proceed: with Zerah Colburn (New York: Holley & Colburn, 1858);

American and European Railway Practices in the Economical Generation of Steam, including the Materials and Construction of Coal-burning Boilers, Combustion, the Variable Blast, Vaporization, Circulation, Superheating and Heating Feed-water, etc., and the Adaptation of Wood and Coke-burning Engines to Coal-burning; and in Permanent Way, including Roadbed, Sleepers, Rails, Joint-Fastenings, Street Railways, etc. (New York: D. Van Norstrand, 1860);

A Treatise on Ordnance and Armor: Embracing Descriptions, Discussions and Professional Opinions concerning the Material, Fabrication, Requirements, Capabilities and Endurance of European and American Guns for Naval, Sea-coast and Ironclad Warfare, and their Rifling, Projectiles and Breech-loading. Also, Results of Experiments against Armor, from Official Records with an Appendix Referring to Gun-Cotton, Hooped Guns, etc. (New York: D. Van Norstrand, 1865).

References:

Alfred Chandler, *The Visible Hand: The Managerial Revolution in American Business* (Cambridge, Mass.: Harvard University Press, 1977);

William T. Hogan, *Economic History of the Iron and Steel Industry in the United States*, volume 1 (Lexington, Mass.: Lexington Books, 1971);

Jeanne McHugh, *Alexander Holley and the Makers of Steel* (Baltimore: Johns Hopkins University Press, 1980).

Archives:

Materials concerning Alexander Lyman Holley are located at the Connecticut Historical Society, the Engineering Societies Library of New York City, the New York Public Library, the New-York Historical Society, and the American Iron and Steel Institute.

Hopewell Furnace

by Paul F. Paskoff

Louisiana State University

Hopewell Furnace was a charcoal-fueled, water-powered iron blast furnace in Union Township of Berks County, Pennsylvania. In almost every respect it was typical of the furnaces which, during the last half of the eighteenth and first half of the nineteenth centuries, turned out most of Pennsylvania's pig iron and, therefore, a large part of the pig iron product of the United States. More than simply a structure Hopewell was the center of a community or "iron plantation" which provided much of the required raw materials, housed most of the labor force, and supported a store from which the workers and their families were supplied with clothing, utensils, and sundries. Because Hopewell was representative, and also because its voluminous records survived its demise as an active iron-making establishment in 1883, the federal government purchased the facility and its surrounding lands in 1938 for a National Park Service Historical Site known as Hopewell Village.

Mark Bird, an ironmaster and the son of an ironmaster, built the furnace in 1771 on a site along French Creek, having first acquired the necessary parcels of land to provide ore and fuel. Bird chose Hopewell's location with care to insure the convenient availability of the raw materials required to make pig iron and access to the Philadelphia market. The land had more than an adequate supply of water which, when harnessed, provided the force necessary to turn the furnace's wheel which, in turn, powered the bellows for generating the blast of cold air. The several thousand acres that comprised Hopewell's domain provided ample supplies of wood for making charcoal, the fuel used in smelting the iron ore into pig iron. This acreage also contained substantial iron ore deposits which were located close to the surface, thereby obviating the need for extensive (and expensive) shaft mining, and considerable amounts of easily quarried limestone to be used as a flux to draw off impurities during smelting. A conveniently located hill against which the furnace was built readily permitted the construction of the necessary "charging bridge," the platform which led to the mouth of the furnace stack and from which the furnace was loaded with alternating layers of fuel, ore, and flux. The land also supplied another factor of production, labor, which was drawn from the population of the farms surrounding Hopewell.

Having selected a location that was rich in the necessary raw materials and that had a good water supply, Bird had only to assemble the capital required to build the furnace and its ancillary structures, get production under way, and find buyers for his iron. Fortunately for him his father, William Bird, who had died in 1761, had been a forge owner (the forge was also called Hopewell) and had left him not only a stock of practical experience but also an inheritance which provided the initial capital for Hopewell Furnace. Additional capital came from Mark Bird's sister and two brothers to whom he gave shares in the enterprise in exchange for their investments.

Bird was ambitious and also owned Gibralter Forge and the Birdsboro Slitting Mill & Steel Furnace in Berks County. When Hopewell Furnace began to run pig iron in the form of castings and to make stoves in late 1771, the outlook for the iron industry was clouded by a recent commercial depression and the deteriorating relations between Great Britain and its North American colonies, including Pennsylvania. Despite this uncertainty the local market for Hopewell's products, especially stoves, was strong, and the furnace did fairly well during its first few years. The outbreak of fighting between the colonies and Britain in the spring of 1775 dramatically changed the general business climate and Hopewell's prospects. Bird became active in the Revolution as a militia leader and also furnished the Continental Congress with considerable quantities of

The ruins of Hopewell Furnace before its restoration by the federal government

food and other supplies. The record of Hopewell's activities during this period is sketchier, and, although the furnace probably cast iron and made shells on contract for the army, the extent of its activities and their effect on Hopewell's ledger are unknown. And, although the furnace did stay active throughout the war despite the chaotic conditions which affected the supply of labor and currency dealings, the question of whether it made a profit in those years is more problematic.

However Hopewell and its owner fared during the Revolution, their fortunes declined with the war's end. The economic confusion that had accompanied the struggle outlasted the fighting and even deepened during the peace which followed. As prices collapsed and debts due the furnace—including amounts owed by the United States government—were not honored, Mark Bird found himself unable to repay his own substantial debts. Even nature seemed to conspire against him as floods disrupted production. By April 1786 he had put Hopewell up for sale. During the course of the next 30 years, he moved to North Carolina, probably to escape his creditors, returned briefly to Pennsylvania in 1796, and died in 1816 in North Carolina.

Hopewell Furnace was finally sold in 1788 to a partnership between James Old and Cadwallader Morris. Old, who owned two-thirds of the property, was an experienced ironmaster, having built Speedwell Forge in 1760 in what is today Lebanon County and Poole Forge in 1779 nearby in Lancaster County. The economic recovery that followed the ratification of the Constitution significantly helped the iron industry in Pennsylvania, and firms which enjoyed able management and sufficient capital resources thrived. Unfortunately, this was not Hopewell's situation. Cadwallader Morris sold his one-third share to his brother Benjamin, who, the following year, bought out James Old's two-thirds interest. Whatever Benjamin Morris's intentions, his involvement with Hopewell lasted scarcely two years, and in 1792 James Old bought the furnace back from him.

The return of Old to Hopewell did not, however, signal the return of continuity and stability to its affairs. Within only a few months of having purchased Hopewell, Old sold it again, this time to James Wilson, who had earlier purchased Birdsboro Forge and wanted the furnace's pig iron to feed the forge fires. Wilson's luck with the property was no better than that endured by his predecessors.

Floods in 1795 all but destroyed Birdsboro Forge, forcing him the following year to sell it. The forge gone, the rationale for Wilson having acquired the furnace went with it. Within a few months of Birdsboro's sale, he sold Hopewell. The buyer—for the third time—was James Old, who could neither stay away from Hopewell nor make a go of it. Hopewell went up for sheriff's sale January 1800 and was bought, again, by Benjamin Morris, who, only a few months later, sold it to a three-man partnership consisting of Daniel Buckley and his brothers-in-law, Thomas Brooke and Mathew Brooke, Jr. Buckley and the Brookes brought experience (they had built a forge in Lancaster County in 1785) and determination to Hopewell Furnace. Their descendants shared their resolve and retained title to Hopewell until the federal government purchased the property in 1935.

The years of chaotic management and revolving-door changes in ownership ended with Hopewell's acquisition by the Buckley-Brookes partnership at the turn of the century. The partners disdained absentee ownership and put in its place the practice of on-site owner-management. Hopewell's new owners immediately began to make significant improvements to the property, an undertaking that was all to the good but for the fact that it required the expenditure of the partners' fast-dwindling cash reserves. By 1807 a combination of misfortunes—some endemic to iron making, such as drunkenness among the workers, and some peculiar to Hopewell—had put the enterprise in jeopardy. As the partners' debt mounted they vainly hoped for relief through a contract with the U.S. War Department to cast cannonballs. This prospect had been the firm's last resort, and from late 1808 until well into 1816 Hopewell Furnace stopped running pig iron and making castings. During this hiatus the partners were able to retain title to Hopewell by settling its debts with money drawn from their other investments, notably nearby Hampton Forge. They also continued to operate Hopewell's stamping mill, the facility at which the slag from previous furnace blasts was pulverized to release trapped bits of iron which could then be worked up.

For Hopewell's owners the overriding problems were not the condition of the general economy and the fluctuations of the iron markets. The outbreak of war between the United States and Great Britain drove up prices of all commodities and sharply increased demand for iron products, espe-cially from the War and Navy departments. Even mediocre businessmen could profit handsomely in such circumstances. Yet Hopewell Furnace remained out of blast throughout the war. The reason was almost certainly the tangle of litigation over the ownership of much of the furnace lands which prevented the partners from cutting wood, making charcoal from the wood, and, therefore, making iron. Most of the suits were settled in 1816, although some worked their way through the Berks County Court during the next two years and were finally settled in the Pennsylvania Supreme Court in 1824. The partners did not have to wait until then, however, to begin to run pig iron again. By mid 1816 Hopewell Furnace was back in blast and casting stoves, in time to benefit from the prevailing high postwar prices and steady demand.

One of the partners' most astute decisions at this time was to hire Clement Brooke as Hopewell's resident manager, a position he was to hold until 1848 when, having become a partner 21 years earlier, he retired. Although new to the position Clement Brooke was hardly new to Hopewell Furnace. The son of Thomas Brooke, one of the partners, he had lived at Hopewell Furnace and worked there in a variety of capacities of increasing responsibility since 1800 and had become the furnace clerk in early 1804. He had also run the furnace's stamping mill during Hopewell's eight-year suspension of iron making. Thus, when the partners made him manager in 1816, their choice reflected not nepotism but their confidence in Clement Brooke's proven ability to do the job.

Brooke's talents as a manager were put to a brutal test within only a few years of his appointment when the panic of 1819 shattered both the general economy and the iron markets. The price of pig iron, which had already collapsed from its inflated wartime level of $55 per gross ton (2,240 pounds) to about $38.75, began to drop sharply again in the early spring of 1819 and did not bottom out at $35 until summer 1822. Business distress among Pennsylvania's iron producers was widespread, and the three considerations which probably counted most in determining a firm's survival were the availability of sufficient capital reserves to enable it to retain skilled workers and service its debt, a solid customer base, and, of course, competent management. Hopewell Furnace, thanks in great part to Clement Brooke, satisfied all three criteria for success, and the return of good busi-

Hopewell Furnace, rebuilt and restored by the National Park Service (courtesy of National Park Service)

ness conditions in the latter half of 1822 and early 1823 found Hopewell in fine shape to take advantage of the economic recovery.

Clement Brooke brought more than experience and sound operational judgment to the management of Hopewell Furnace. He also had an entrepreneurial flair which encouraged him to take judicious risks in pursuit of potentially profitable new lines of business and methods of production. At the same time, however, he made certain that the scale of the firm's capital and physical plant kept pace with the market's growth. Under his direction Hopewell began to emphasize stove production over pig iron and castings, a decision that was particularly well timed. Demand for stoves had increased substantially after the War of 1812 as the populations of the cities of the northeastern and the Great Lakes states grew. For example, New York City's population (including Brooklyn) of a little more than 100,000 in 1810 more than doubled by 1830 and had more than tripled by 1840. Similarly, Cincinnati's population increased about tenfold from 2,540 in 1810 to 24,831 in 1830 and then almost doubled during the next ten years. By the time Brooke became a partner in 1827, the firm (which, since its reorganization in 1816, had been called Daniel Buckley & Company) was doing a solid and ex-

panding business. That year the firm changed its name again to Buckley & Brooke to reflect Clement Brooke's new status. Four years later he and his brother, Charles, each acquired a one-third share of the firm and changed its name to Clement Brooke & Company.

The 15-year period that began in 1822 with the recovery from the panic of 1819 were the firm's best years. Buoyed by the general economy's prosperity, Hopewell and other furnaces and forges enjoyed rapidly expanding demand for their products and good profits. These halcyon days came to an abrupt end with the panic of 1837 and the ensuing depression which lasted from late 1839 to early 1843. At about the same time the anthracite furnace was introduced into eastern Pennsylvania's iron industry and, within only a few years, presented serious competition to the charcoal sector of the industry. Moreover, the rapid growth of railroad construction began to alter the character and extent of demand for pig iron. All of these influences brought unusual pressures to bear on Hopewell and played a role in the decision of its owners in 1844 to discontinue the making of stoves and concentrate instead on running pig iron for the region's forges and rolling mills.

Hopewell Furnace continued to make iron after its owners abandoned stove production, but the firm's best and, from the perspective of business history, most distinctive days were behind it. Although it also continued to make a profit for its owners well into the 1880s, it no longer did so as an innovative enterprise but, rather, as a stubborn survivor of a once-dominant and now increasingly outmoded charcoal iron industry. Clement Brooke's decision to retire as manager in 1848 further weakened the firm at a time when it was already being sapped by competition from domestic anthracite blast furnaces and British producers. In the decade that followed, Hopewell's owners, including Clement Brooke, attempted to adapt their firm's operations to new conditions. In 1853 they made a foray into pig iron production using anthracite coal but gave it up in 1857 when they encountered problems involving the quality of the ore and the expense of transportation to and from the Schuylkill Canal, the major coal shipment route. The lesson learned by the firm seems to have been that Hopewell Furnace was better situated and designed for making charcoal iron, an activity to which it returned in 1857, the year that a severe commercial and financial crisis began.

The depressed economic conditions associated with the panic of 1857 lasted well into 1861 when the stimulating effects of the Civil War increased demand and prices. Inflation of iron prices substantially eased the pressure on Hopewell and other charcoal iron furnaces from the more efficient anthracite producers by enabling the charcoal operations to cover their higher costs with higher revenues and, therefore, higher profits. The war also heightened demand for some of the products, notably iron for railroad wheels and other types of railroad iron, in which the charcoal producers specialized. Fortunately for Hopewell, demand from locomotive builders and railroads for its iron outlasted the war and helped to keep the firm afloat during the next ten years. Postwar economic conditions, however, were highly volatile, particularly within the railroad industry which, by 1873, was clearly overbuilt. The chaos within the industry following the panic of 1873 reverberated throughout most other industries, including the iron industry, and augered the demise of Hopewell Furnace in June 1883 when the furnace went out of blast for the last time.

Hopewell Furnace continued to function as an enterprise and community, though much diminished in population and level of activity, for about five years after it had ceased to run pig iron. By the mid 1890s, another period of severe general economic stress, Hopewell's domain supported only some charcoal making, stone quarrying, and ore mining. Meanwhile the furnace, its machinery, and associated structures had fallen into decrepitude. Hopewell's restoration began in 1935 when the National Park Service acquired the property to create a park for public recreation. Three years later the Department of the Interior, having declared the site and its buildings and machinery to be of significant value to an understanding of the nation's colonial history, established the Hopewell Village National Historic Site.

References:

Arthur C. Bining, *Pennsylvania Iron Manufacture in the Eighteenth Century* (1938; reprinted, Harrisburg: Pennsylvania Historical and Museum Commission, 1973);

Louis McLane, *Documents Relative to the Manufactures of the United States*, 4 volumes (1833; reprinted, 3 volumes, New York: Burt Franklin, 1969);

Allan R. Pred, *Urban Growth and the Circulation of Information: The United States System of Cities, 1790-1840* (Cambridge, Mass.: Harvard University Press, 1973);

James M. Swank, *History of the Manufacture of Iron in All Ages*, second edition (Philadelphia: American Iron and Steel Association, 1892);

Joseph E. Walker, *Hopewell Village: A Social and Economic History of an Iron-Making Community* (Philadelphia: University of Pennsylvania Press, 1966).

Robert Woolston Hunt

(December 9, 1838-July 11, 1923)

by Stephen H. Cutcliffe

Lehigh University

CAREER: Druggist, Covington, Kentucky (1855-1857); puddler and roller, John Burnish & Company (1857-1859); chemical metallurgist, Cambria Iron Company (1860-1861); military service (1861-1865); steelmaker, Cambria Iron Company at Wyandotte, Michigan (1865-1866); steelmaker, Cambria Iron Company (1866-1873); superintendent, John A. Griswold & Company (1873-1875); general superintendent, Albany & Rensselaer Iron & Steel Company (1875-1888); founder and consulting engineer, Robert W. Hunt & Company (1888-?).

Robert Woolston Hunt was born on December 9, 1838, in Fallsington, Pennsylvania, to Robert A. and Martha Lancaster (Woolston) Hunt. The family traced its ancestry back to John Hunt, a Devonshire immigrant, who had settled in 1712 in Hopedale, New Jersey. Robert Hunt's father was a medical doctor with a practice in Trenton, New Jersey, who then moved to Covington, Kentucky, where he ran a drugstore. He died in 1855 when Hunt was seventeen, thereby preventing any further formal education. Hunt took over his father's drugstore but in 1857 moved to Pottsville, Pennsylvania, with his mother, where he found employment at the iron-rolling mill of John Burnish & Company, of which his cousin was a senior partner. Here Hunt learned the practical side of the iron business as a puddler and roller. He subsequently took a course in analytical chemistry in the laboratory of Booth, Garret & Reese of Philadelphia, after completion of which he was hired by the Cambria Iron Company in August 1860 at a salary of $20 per month to establish an analytical chemical laboratory, the first formed and maintained as an integral part of an iron works in America.

The following spring Hunt moved to Elmira, New York, to become the night foreman of the Elmira Rolling Mill, a firm organized by his cousin.

Robert Woolston Hunt

However, Hunt's iron career was interrupted shortly thereafter by the Civil War; Hunt quickly volunteered for the Union army, eventually being mustered out with the rank of captain.

In 1865 Hunt returned to Cambria as chemist, whereupon the company sent him to the experimental Bessemer works in Wyandotte, Michigan, in which Daniel Morrell, Cambria's general manager, had a major interest. Morrell intended Hunt to observe the Bessemer process in the expectation that the Cambria management would eventually embark on steel production. However, within a month Hunt found himself in charge of steel making at Wyandotte due to the resignation and departure of several key individuals.

By his own account Hunt was reasonably successful, if somewhat lucky, in taking on this responsibility:

> As Mr. Hahn's retirement left the company without any practical steelmaker, and the works had thus far been conducted on an experimental basis, the proprietors determined on making the most hazardous experiment of all and put them in charge of the writer, who had gone there a few weeks before, in the interest of the Cambria Iron Company. In accordance with this arrangement the writer made his first "blow" and by some strange fatality happened to "turn down" at just the right time.

Hunt remained at Wyandotte for approximately a year, during which period he traveled to Troy, New York, in order to observe steel-making operations there and to familiarize himself further with the process. Although welcomed by Alexander Holley, he was politely but firmly refused entry to the plant itself and so left having failed to accomplish his immediate mission.

Following his return to Cambria in May 1866 Hunt was placed in charge of rolling the first batch of steel rails to be produced in America on a commercial basis. Cambria had not yet built its own Bessemer furnace when the Pennsylvania Railroad approached them to fill an order for rails with Bessemer steel produced by the Pennsylvania Steel Company. Because the Pennsylvania Steel Company, under the direction of chief engineer Holley, had not yet constructed its rail mill, the steel ingots were shipped to Cambria for final rolling. The occasion brought Holley and Hunt together, and despite the incident in Troy, the two men became fast friends and worked closely together for the remainder of their careers.

Following the successful rolling of the steel rails and the subsequent building of Bessemer plants by several other firms, Morrell convinced the Cambria management in 1870 to erect their own Bessemer converter, thereby becoming the sixth company to do so. Hunt worked closely with George Fritz, the chief engineer, and Holley, who had been called in to design and build the plant, and upon its completion a year later was placed in charge. He stayed in this position for two years until September 1873 when, following a labor dispute at Cambria, he moved to Troy as superintendent of the Bessemer works of John A. Griswold & Company. In 1875 the firm merged with Erastus Corning & Company to form the Albany & Rensselaer Iron & Steel Company (later the Troy Steel & Iron Company), and Hunt served as general superintendent.

Hunt remained at Troy for the next 14 years, during which period he conducted an extensive rebuilding of the works, including constructing a new large-scale blast furnace, and increased both the output and variety of finished products. He developed grades of Bessemer steel not previously produced in America, including soft steel suitable for drop forging as well as specialty steels for gun barrels, carriage axles, drills, and springs. In conjunction with August Wendell and Max M. Suppes he developed and patented a rail mill feed table subsequently adopted in the majority of American automated rail mills. While at Troy Hunt also improved and patented a process for handling and rolling wire rod blooms.

Hunt resigned from Troy in 1888, moved to Chicago, and established an engineering consulting firm, Robert W. Hunt & Company, which attained world recognition with offices and laboratories in London, Mexico City, and Canada, as well as other U.S. cities. Hunt was particularly interested in the establishment of standards and in materials testing, and he was associated with a number of organizations focusing on this last area, including the American Society for Testing Materials, of which he was president in 1912, and the International Association for Testing Materials. He proposed what later became known as the "Special Inspection," which involved close supervision of both steel manufacture and the subsequent rolling of rails, and the "nick-and-break test" for soundness of ingots. He also proposed and helped to design several new types of rail sections.

Hunt contributed frequently to the technical journals of the period. "History of the Bessemer Manufacture in America" in *Transactions of the*

American Institute of Mining Engineers (1877) and "Evolution of the American Rolling Mill" in *Transactions of the American Society of Mechanical Engineers* (1892), the latter Hunt's ASME presidential address, are probably his most well-known publications.

The career of Robert Hunt was long, varied, and notable. The importance of his role in the iron and steel industry is reflected not only by the award of the John Fritz Medal, which he won in 1912 for his early contributions to the development of the Bessemer process, but also by virtue of the fact that he was elected as an honorary member of almost every major American engineering society. This distinction included serving as president of the American Institute of Mining Engineers (1883, 1906), the American Society of Mechanical Engineers (1891), and the Western Society of Engineers (1893). He was also a member of the American Society of Civil Engineers, the United States Iron and Steel Institute, the Canadian Society of Civil Engineers, the Institution of Civil Engineers, the Institution of Mechanical Engineers, and the Iron and Steel Institute of England. He also served as a trustee of Rensselaer Polytechnic Institute. Final testimony to Hunt's contribution was made by the American Institute of Mining and Metallurgical Engineers, which established in his memory the Robert W. Hunt Award. The award, which consists of a certificate and silver medal, is currently awarded annu-

ally by the Iron and Steel Society to the author(s) of the best original paper(s) on iron or steel.

Selected Publications:

"History of the Bessemer Manufacture in America," *Transactions of the American Institute of Mining Engineers*, 5 (1877): 201-215;

"Evolution of the American Rolling Mill," *Transactions of the American Society of Mechanical Engineers*, 13 (1892): 45-69.

References:

John Fritz, *The Autobiography of John Fritz* (New York: John Wiley & Sons, 1912);

William Hogan, *Economic History of the Iron and Steel Industry in the United States* (Lexington, Mass.: Lexington Books, 1971);

Jeanne McHugh, *Alexander Holley and the Makers of Steel* (Baltimore: Johns Hopkins University Press, 1980);

Elting E. Morison, *From Know How to Nowhere* (New York: Basic Books, 1974);

Morison, *Men, Machines, and Modern Times* (Cambridge, Mass.: MIT Press, 1966);

Fritz Redlich, *History of American Business Leaders: A Series of Studies*, 2 volumes (Ann Arbor, Mich.: Edwards Brothers, 1940);

James M. Swank, *History of the Manufacture of Iron in All Ages* (Philadelphia: American Iron and Steel Association, 1892);

Swank, *Introduction to a History of Ironmaking and Coal Mining in Pennsylvania* (Philadelphia: J. M. Swank, 1878);

Peter Temin, *Iron and Steel in Nineteenth-century America: An Economic Inquiry* (Cambridge, Mass.: MIT Press, 1964).

Dr. Curtis Grubb Hussey

(August 11, 1802-April 25, 1893)

by Bruce E. Seely

Michigan Technological University

CAREER: Physician (1825-c. 1837); general-store owner (c. 1827-1860); member, Indiana House of Representatives (1829-1831); organizer and director, Pittsburgh & Boston Mining Company (1844-1879); organizer and president, C. G. Hussey & Company [Pittsburgh Copper & Brass Rolling Works] (1848-1893); founder and owner, Hussey & Wells (1852-1859); president, Pittsburgh & Boston Mining Company (1858-1879); organizer and president, Hussey, Wells & Company (1859-1876); president, Hussey, Howe & Company (1876-1888); president, Hussey, Brown & Company (1888-1893).

Dr. Curtis Grubb Hussey was a multitalented individual who guided American efforts at industrialization during the first half of the nineteenth century. Full of optimism and confidence, men like Hussey embarked, often successfully, on numerous industrial ventures; their careers seemed to prove the Horatio Alger myth of the self-made man and the booster and go-getter mentality identified by historian Daniel Boorstin. One could hardly find a better exemplar of these ideals than Curtis Hussey. Beginning as a doctor, he soon started a general retail business that stressed pork provisioning. From these origins he grew into a dominant figure in the industrial development of Pittsburgh after 1840. He pioneered and for a time controlled much of the copper business in the United States and later launched the first successful cast crucible steel firm. Befitting his business successes, Hussey was a leader in civic affairs in Pittsburgh throughout the second half of the nineteenth century.

Hussey was born in rural York County, Pennsylvania, but by 1813 he had moved with his parents to Mount Pleasant, Ohio. He began studying medicine with a local physician when he was eighteen and in 1825 opened his own practice in Morgan County, Indiana. His practice was immediately successful, and Hussey continued in it for over ten years. An indication of his stature in the community was his 1829 election to the Indiana House of Representatives; he declined to serve a second term.

Hussey's rejection of elective office was accompanied by his entry into the mercantile business, which he deemed more important than service in the legislature. With accumulated savings of perhaps $4,000, he opened a general store in Mooresville, Indiana, but soon moved it to Gosport on the White River. Working with a series of partners who operated the stores, Hussey opened branches in Monrovia, Columbus, Millvale, and Far West, Indiana. The doctor discovered, however, that many of his transactions involved barter payments, especially in pork, so Hussey invested in a packing house in Gosport. Moreover, he established connections in New Orleans to market the pork. This was a tricky business, but according to Erasmus Wilson's nineteenth-century account, because Hussey "Possess[ed] business ability of a high order, his enterprises prospered in a remarkable manner. . . ." Indeed, after 1835 he devoted himself fulltime to the provisioning trade with New Orleans. His skill as a businessman is clearly evidenced by his survival of the panic of 1837. After marrying Rebecca Updegraff of Mount Pleasant, Ohio, in 1839, he moved to Pittsburgh, Pennsylvania. Hussey then shifted his mercantile attention from New Orleans to Pittsburgh and cities on the Atlantic coast. He formed a business to market pork from the Gosport packing house.

After 1845, however, the pork business was a sidelight to several of Hussey's industrial ventures. The first of these involved copper; indeed, the interest in copper was the connection that brought Hussey and Calvin Wells together. Hussey was fascinated by rumors of huge copper deposits in northern Michigan, and in 1843 he dispatched a proxy to explore the Lake Superior region. His reporter

Dr. Curtis Grubb Hussey

bought Hussey an interest in the first three permits issued for mining copper on the Keweenaw Peninsula of Michigan. From this start emerged the Pittsburgh & Boston Mining Company, formed in 1843 but not chartered in Michigan until 1848, with Hussey and three other Pittsburghers holding a two-thirds interest in the company. In 1844 the company sent eight miners, furnace equipment, and a geologist to explore and test the claim near Copper Harbor. The following year they began work at the second parcel near Eagle River and found a significant copper deposit. This became the Cliff Mine, the first copper mine of the Lake Superior region and one of the most profitable. Almost from the start the Cliff produced metallic copper in masses weighing 1,000 pounds or more.

Unfortunately, the very richness of this discovery posed a problem. Removing the copper from the ground by breaking the masses into small pieces with hammers and drills was a trying and expensive experience; yet furnaces designed for large masses wasted too much copper. After experiments at Revere's copperworks in Boston and the Fort Pitt Iron Foundry in Pittsburgh, Hussey developed a re-

verberatory furnace with a removable top, into which cranes could lift the masses of copper. To exploit this design, Hussey and a partner founded C. G. Hussey & Company, built the Pittsburgh Copper & Brass Rolling Works in Pittsburgh in 1849 and 1850, and began smelting copper from the Cliff mine.

The smelter, however, was only part of Hussey's copper operations in Pittsburgh. The Pittsburgh Copper & Brass Rolling Works also included a rolling mill, the first such facility located away from the East Coast. The rolling mill enabled the company to market ingots, sheets, and bars, thereby competing with established eastern interests. Moreover, Hussey remained a leading figure in Michigan's Upper Peninsula. After 1858 and until the Cliff Mine was sold and all its affairs wrapped up in 1879, Hussey was president of the Pittsburgh & Boston Mining Company, which owned the Cliff Mine. During its years of operation the Cliff Mine produced more than $8 million in copper and returned $2,327,000 on an investment of $110,000. In addition, Hussey opened other Lake Superior copper mines, including the National Mine, and was president of the Adventure Mining Company, the Central Mining Company, and the Aztec Mining Company during the late 1850s.

Hussey was, in short, a leading figure in the copper industry throughout this period, the only individual to engage in all aspects of the copper trade. Even in the 1880s Hussey's was one of only three large-scale copper-refining works in the country, producing 300 to 400 tons of copper ingots annually. After 1870, however, the Pittsburgh Copper & Brass Works was replaced as the leader in the region by the Detroit & Lake Superior Copper Company, an operation controlled by the large mines on Michigan's Keweenaw Peninsula. After the Cliff Mine closed, Hussey had no access to copper ore from the Keweenaw, so his firm smelted mass copper from mines in Ontanogan, Michigan. Ironically, the Detroit & Lake Superior Copper Company owed some of its success to copying Hussey's furnace design. In spite of these problems, Hussey's rolling mill and brass foundry continued in operation long after its founder's death. The Copper Range Company acquired it during the Great Depression and as of 1988 continues to operate the Hussey Metals Division.

By the end of the 1850s Hussey's copper empire was firmly established, and his attention turned to another metallurgical pursuit–the production of steel. In the pre-Bessemer era that meant mastering the cast crucible process, monopolized by the British. American efforts to learn this process had begun in 1830 but had failed commercially, if not technically, because of the superior quality of crucible steel from Sheffield, England. Although his primary partner, banker Thomas Howe, initially opposed Hussey's initiative–numerous other Pittsburgh iron firms had already succumbed to British competition–Hussey eventually won his partner's support. An 1884 account claimed Hussey was motivated by a desire to prove his doubters wrong: "with a firm faith in himself, and a supreme conviction of right, that was in itself a sure prophecy of success, he persevered without a halt." It helped, however, that as a prominent member of the fledgling Republican party Hussey expected a protective tariff on steel to be enacted. Moreover, there were points of common technology between steel and copper production.

A central figure in the actual operation of Hussey's steel company was Calvin Wells, the young man who ran Hussey's pork marketing venture after 1852. Hussey first employed Wells as a clerk and bookkeeper in the copper rolling works in 1850, taking him into the pork company, Hussey & Wells, in 1852. In 1858 Hussey sent Wells to Jersey City to learn about steel making at the Adirondack Iron & Steel Company–then the most technically successful operation in the country. In 1859 Hussey abandoned pork marketing and created Hussey, Wells & Company, which bought the idle steelworks built by McKelvy & Blair in 1850. His primary concern was finding a way to produce cast crucible steel by melting wrought iron and charcoal in a crucible rather than using the traditional two-step Huntsman process of making and then melting blister steel. Other steel makers, including Joseph Dixon, introduced processes that eliminated a step in the process, and Wells may have brought the idea back to Hussey from his visit to Jersey City. Eventually Hussey and Wells mastered a procedure, but progress was slow. They produced only 10 tons during the first three months and 280 tons in the next year. Moreover, the firm had to fend off competitors' charges that their method produced inferior-quality steel. They survived all this, in part because they used charcoal iron from the Lake Champlain region, a step that other Pittsburgh firms soon followed. This high-quality iron was as

important to American producers as Swedish iron was to Sheffield manufacturers. As important in an indirect way, however, were the profits of Hussey's copper ventures, for they enabled Hussey to experiment by supporting the steelworks' losses for several years. Various accounts note that the partners spent between $300,000 and $400,000 in developing their production process.

By the mid 1860s, however, Hussey, Wells & Company was recognized as the first commercially successful American cast crucible steel maker. Production grew from 1,900 tons in 1863 to 6,167 tons in 1873. The crucial market Hussey, Wells & Company successfully targeted was the top-grade steel used for edge tools such as chisels; the firm produced about 1 ton per day during its first 15 years of existence. Hussey, Wells & Company also became specialists in plate manufacture. The first attempt to roll plates failed in 1863 due to poor equipment, but better rolls brought success in 1865, enabling the company to market plates for saws, locomotive fire boxes, boilers, and boiler flues. Plate manufacture also opened the market for agricultural implements, and during 40 years of existence the bulk of the firm's steel went to this use. Indeed, the company developed a rake tooth factory with special bending machinery. Finally, Hussey, Wells & Company was known for steel used by the railroads for locomotive slide bars, crank pins, axles, piston rods, and springs.

Hussey, Wells & Company built its success slowly with diligent effort. Historian Geoffrey Tweedale quotes a letter from Hussey to the Baldwin Locomotive Works that is typical of the way Hussey literally pleaded for orders. Eventually, however, the quality of the steel won a market. By the early 1870s the Chicopee Arms Manufacturing Company in Massachusetts, which used 200 tons of Hussey's steel for sabers, swords, and other uses, pronounced Hussey's product fully equal to British steel. And in the 1880s Hussey's firm provided the steel for what may have been the first all-steel bridge in the country, a railroad span across the Missouri River at Glasgow, Missouri.

With such success the works of Hussey, Wells & Company steadily increased in size. By 1865 the firm covered 3 acres, used 8 steam engines, had furnaces with more than 50 melting holes, and owned extensive rolling equipment. Daily capacity was 20 tons of cast steel, making this the largest and most recognized steelworks in the country. By 1876 annual capacity was 13,000 tons, although annual production during the previous five years had averaged only 5,000 tons. The facilities had been increased by the addition of 16 puddling furnaces, 6 modern 24-pot Siemens gas-fired furnaces, 100 traditional coke-fired melting holes, and 10 steam hammers. The 8 trains of rolls ranged in size from 9 inches to 28 inches. In the late 1880s the firm was no longer the largest in the city, although Hussey's firm remained the undisputed leader in rolling operations. New equipment added during these years included 2 newer 30-pot Siemens furnaces, 3 more roll trains, and 2 additional steam hammers. Nonetheless, the industry itself was changing. The greatest indicator was the construction of an 8-ton open-hearth furnace by Hussey in 1886; three other leading cast crucible steel manufacturers in Pittsburgh had acquired this new technology after 1873–Anderson & Woods; Park, Brother & Company; and Singer, Nimick & Company.

Changes were also affecting Hussey, Wells & Company during these years of growth; the firm went through a number of name changes as the various partners withdrew or retired. The constant, however, was Curtis Hussey, who remained in active touch with the steelworks until his death, making small improvements in the direct conversion process he had championed. The first change in the firm was occasioned by Calvin Wells's departure from the firm in 1876. Since about 1870 Wells had pursued other enterprises, becoming half-owner in the Pittsburgh Cast Steel Spring Works. Like Hussey, Wells juggled several enterprises, continuing as the Hussey, Wells & Company business manager until 1876. In that year the works were renamed Hussey, Howe & Company, reflecting the participation of Hussey's oldest associate, Pittsburgh banker Thomas Howe. In 1880 the firm incorporated, but Howe's death led to the introduction of a new partner and another designation in 1888–Hussey, Brown & Company, Limited. This name remained with the firm until Hussey's death in 1893. The works, which had an annual capacity of 20,000 tons at the turn of the century, became part of the Crucible Steel Company of America in 1900.

Physician, merchant, pork provisioner, copper magnate, and steel producer–this list still does not exhaust the business endeavors of Curtis Hussey. He apparently formed Hussey & Company in 1862 to manufacture crucibles, and he or his son established Hussey, Binns & Company in 1875 to manu-

facture crucible steel and shovels. Hussey was also involved in several Pittsburgh banks, especially in several joint ventures with Thomas Howe. He also had an interest in gold, silver, and copper prospecting in Canada, Mexico, and the western states, as well as in Michigan's iron ranges. According to his biographers, perhaps Hussey's only serious business misjudgment came in 1863, when he turned down the offer of the British steel firm Peabody & Company to develop the Bessemer patent in this country. But despite this miscalculation, Hussey was able to accumulate a fortune of between $10 million and $20 million.

As was typical of men of his position during this time, Hussey was active in Pittsburgh society outside of business. He helped found and was the first president of the Allegheny Observatory and purchased much of the facility's equipment. When the University of Pittsburgh later acquired the observatory, Hussey was named a trustee of the university, serving from 1864 until his death. His educational interests were not limited to this; he founded the Pittsburgh School of Design for Women in 1865 and supported schools in the Indian Territory, Wesleyan College for Women in Cincinnati, Earlham College in Indiana, and the Hussey School for Girls in Matamoras, Mexico.

Hussey was a member of the Society of Friends, an affiliation that accounted for his strong opposition to slavery and drink. His Quaker background may also be seen in other aspects of his life. In spite of his business and civic leadership, he did not pursue an ostentatious social life, preferring to spend time with his family of five children. The *National Cyclopedia of Biography* captured a sense of his character when it explained, "Personally he was retiring, dignified and affable, a man of sound business judgment and tact, clear brain and retentive memory and a lover of good literature."

Curtis Hussey's career offers an especially fine example of the spirit that propelled American industrialization forward at such a rapid pace during the nineteenth century. Even without formal education, Hussey mastered several difficult arts and trades and guided his businesses through a tumultuous environment in which failure was common. Yet he was also a man of high ethical standards and a civic and educational leader. He seems to have been genuinely liked by his partners and employees alike. More specifically, Curtis Hussey established two important industries in Pittsburgh and the Lake Superior region of Michigan—copper and cast steel—and founded or supported several important institutions of the city in which he lived. Historian of technology Eugene S. Ferguson has estimated that during the first half of the nineteenth century, as few as 5,000 people made the decisions and launched the efforts that made industrialization happen in this country. They tended to be confident, even brash, believers in themselves and their prospects, gaining support from a community of like-minded individuals who were convinced of the importance of industry and technology. They saw problems as challenges to be overcome, and this enthusiasm was the foundation on which America industrialized. Curtis Hussey was one of the most successful of these men.

References:

J. Leander Bishop, *A History of American Manufacture from 1608 to 1860* (Philadelphia: Galaxy, 1868);

Donald Chaput, *The Cliff: America's First Great Copper Mine* (Kalamazoo, Mich.: Sequoia Press, 1971);

Thomas Egleston, "Copper Refining in the United States," *Transactions of the American Institute of Mining Engineers*, 9 (1880-1881): 678-730;

Harrison Gilmer, "Birth of the American Crucible Steel Industry," *Western Pennsylvania Historical Magazine*, 36 (March 1953): 17-36;

History of Allegheny County, Pennsylvania (Philadelphia: L. H. Everts, 1876);

"Curtis G. Hussey," *Magazine of Western History*, 3 (February 1886): 329-348;

The Manufactories and Manufacturers of Pennsylvania of the Nineteenth Century (Philadelphia: Galaxy, 1875);

Pennsylvania Historical Review: Cities of Pittsburgh and Allegheny—Leading Merchants and Manufacturers (New York: Historical Publishing, 1886);

William P. Shinn, "Pittsburgh and Vicinity—A Brief Record of Seven Years Progress," *Transactions of the American Institute of Mining Engineers*, 14 (1885-1886): 665;

George H. Thurston, *Pittsburgh As It Is* (Pittsburgh, 1857);

Thurston, *Pittsburgh's Progress, Industries, and Resources* (Pittsburgh, 1886);

Geoffrey Tweedale, *Sheffield Steel and America: A Century of Commercial and Technological Interdependence, 1830-1930* (New York: Cambridge University Press, 1987);

Erasmus Wilson, ed., *Standard History of Pittsburgh* (Chicago: H. R. Cornell, 1898).

Charles Huston

(July 23, 1822-January 1897)

by Julian Skaggs

Widener University

CAREER: Physician (1844-1848); ironmaster, Brandywine Iron Works (1848-1855); manager, Lukens Rolling Mill (1855-1881).

Charles Huston was born July 23, 1822, in Philadelphia, Pennsylvania, the son of Dr. Robert M. and Hannah West Huston. His father was a member of the faculty of Jefferson Medical College and was active in community affairs. Most notably, he helped organize the Philadelphia Gas Works and ultimately served as president of the board of trustees of that institution.

Charles Huston enrolled at the University of Pennsylvania in 1836 and graduated in 1840. He then attended Jefferson Medical College and received his degree in 1842. Following graduation he went to Europe for a year and a half where he continued his medical studies. Upon his return Huston established a practice in Philadelphia where he met Isabella Lukens, the daughter of Rebecca Lukens, owner of Brandywine Rolling Mill in Coatesville, Pennsylvania. The subsequent courtship and marriage of Huston and Isabella Lukens were bitterly resisted by the elder Dr. Huston who gave his assent to the union only at the last moment. Dr. Robert Huston was quickly disappointed again, when his son abandoned medicine and became an ironmaster. Leaving Philadelphia in 1848 Huston became a partner of Abraham Gibbons, his brother-in-law, in the management of the Brandywine Rolling Mill.

The young physician quickly proved himself to be an adept student of the iron business. His pres-

Charles Huston

ence appears in the outgoing correspondence of the firm during this period, which shows him addressing the manifold problems of the trade. Prices of fuel and wrought iron blooms received almost daily written attention as did prices and varieties of plate that were to be shipped to commission agents in cities as distant as New Orleans and as close as Philadelphia. The problem of quality control, both of incoming raw materials and outgoing plate, were subject to Charles Huston's pen too, and his letters on these matters were ultimately reassuring, hectoring, or damning as his interests required. The correspondence also shows his periodic concern with maintenance and repairs of the mill and his relationship with the work force in the mill. Charles Huston immersed himself in his new job and mastered every facet of it.

It was well that he did, for after an apprenticeship of seven years, Charles Huston became the sole manager of the enterprise in 1855. His previ-

ous partner, Abraham Gibbons, left the partnership to take up a new and successful career in banking. The newly reorganized enterprise was called the Lukens Rolling Mill, in honor of Rebecca Lukens who died in 1854. The mill had been renovated in 1854 and had rolls 60 inches wide, two heating furnaces, and an annual capacity of 1,000 tons of plate iron.

The updated mill got off to a poor start in 1856. A series of breaks in the machinery wreaked havoc on delivery schedules and drove the owner and his customers to distraction. However, 1857 got off to a good start, and the troubles of the preceding year seemed to be safely past. The upturn in business at the mill proved to be brief. The panic of 1857 chilled the markets of the nation in general, and was especially harsh on the iron industry. At Lukens, the panic challenged the very existence of the firm. Charles Huston wrote that it was to be a "long night of Egyptian darkness."

Huston's principal agency in New York suspended payments in October, and the Lancaster Locomotive Works went bankrupt at the same time. Between them they owed Lukens $30,000. By mid October the mill was shut down because of lack of orders and remained idle for a month. In December of that year a slow recovery from the crisis began. The New York agency resumed its payments and successful negotiations were begun which salvaged the bulk of the debt owed Lukens by the Lancaster Locomotive Works. But it had been a frightening autumn for Charles Huston. Curiously, Dun and Bradstreet's report on Huston during this time was quite optimistic—in March 1858 their agent wrote of Huston, "Can't bust him."

The panic of 1857 cast a long shadow. Business continued to be slow for the rest of the decade. For part of that time Huston ran the business at a loss simply to preserve a place in the market. It was a mean, trying time and of it Huston said "blessed is he who holds out to the end." Huston survived it, however, and in late 1859 he took a junior partner, his cousin Charles Penrose, into the business.

The new partnership faced a new problem shortly after its creation. The iron market, fueled by the Civil War, was to test the upper productive limits of the business. By the middle of the Civil War the mill was producing 1,700 tons of plate annually, twice the average annual rate of the 1850s and 70 percent above its rated capacity of 1,000 tons per year as listed in the 1860 Census of Manufacturers. This burst in production was achieved by adding men to the labor force and by driving the mill harder. No expansion of the physical plant was done or considered during this time.

The war also drove its managers to address a moral issue. Charles Huston was a Quaker, as was his wife and as had been her mother, Rebecca Lukens. Rebecca Lukens had, during her stewardship of the business, been scrupulous about not selling iron destined for any possible military use. The war changed this. Quaker pacifism collided with a hatred of slavery and strong Union sentiment. This mix of inclinations, combined with the fact that the junior partner, Charles Penrose, was not a Quaker, led the firm to produce iron for military purposes for the duration of that war. After the war military orders were shunned as before.

By the late 1860s the managers of the mill came to realize that though their business was still profitable, they had to abandon the old plant and build a new factory to address the demands of the market. By that time the Lukens mill was decidedly antique. It was the only water-powered mill in Coatesville, and Lukens still used only a single set of rolls. The two other mills in Coatesville, Pennock & Company and Steel & Worths, were much larger and much more up to date. These competing mills were steam powered, had multiple sets of rolls, and employed four to five times more men than did Lukens. The other local mills also produced three and a half times more iron than Lukens.

In this situation Huston felt driven to modernize his mill, and by autumn 1870 the new mill was up and running. It was steam powered and had rolls of chilled steel, 84 inches wide. He was proud of it but as he said, it "was very like the old," but with "plenty of room for everything." In fact, by industry standards, the new plant was a nice, but rather modest affair. Both of the other Coatesville mills had greater annual production capacities.

This circumspect approach to the adoption of technologies and the expansion of production capacity was a hallmark of Lukens during most of the nineteenth century. Only once was Lukens a pioneer; that was in 1818 when the firm, under the hand of Dr. Charles Lukens, rolled the first boilerplate made in the United States. In the main Lukens survived by watching costs, identifying a niche in the market, and by watching over the quality of the

metal they rolled. Important too was the fact that the owners' fortune was never completely tied up in the mill. Thus, in bad times the family had financial reserves that enabled the business to meet its obligations until conditions improved

There was, however, one area in this business after 1870 where Lukens began to show superiority over its other competitors. This was in the field of quality control. Huston purchased tensile strength testing apparatus in early 1872 and began to systematically collect data on Lukens plate. Later, he evaluated the plate of other manufacturers. Moreover, he made his findings and testing procedures available to others. The editor of the *Nautical Gazette* was apprised of Lukens's testing procedures in 1874, and he was invited to come and observe the tests being made. The next year Huston notified the famous engineer Robert Thurston of the Stevens Institute of his research and offered to keep him informed of the results of ongoing testing. Professor Thurston was also invited to witness the testing process of Lukens. These investigations led to a series of essays that were published in the *Journal of the Franklin Institute* in 1878 and 1879. The work addressed the reaction of iron and steel to varying degrees of heat and stress, and Dr. Huston had the pleasure of seeing this work translated into French at a later time.

This work—careful, systematic, and tedious as it was—led the process of the manufacture of plate from an inexact art toward an exact science. It also led Huston toward national prominence. In 1877 the manufacturers of boilerplate were requested by the federal government to send a committee to Washington to assist in the definition of standards for boilerplate. Dr. Huston was a member of that committee, and his recommendations were generally adopted. Huston advised the government that testing modes had to be standardized and placed in the hands of a central authority in Washington. He pointed out that local federal inspection could be compromised by bias and by imperfect testing. Huston was adamant that manufacturers should not be allowed to test and approve their own plate. Here again, the series of tests made at Lukens was submitted to the government. The data was convincing and his advice accepted. Later, during the 1880s Huston advice on standards of boilerplate was sought out by insurance companies.

Huston's sons Abram and Charles continued their father's inquiries into the scientific aspects of

Workers at Huston's Brandywine Iron Works, circa 1880

the manufacture of iron and steel. Abram was sent to Europe to visit various mills there, and Charles was trained as an engineer at Haverford College. Both young men were introduced to the management of the firm during the late 1870s. Charles Penrose died in 1881, and that same year Dr. Huston's health became less than robust, forcing him to leave the daily management of the business more and more in the hands of his sons. Following the traditions of their father and their remarkable grandmother, Rebecca Lukens, the new managers pursued a policy of cautious expansion, cultivation of known markets, and scrupulous care for the firm's reputation. As Charles Huston gradually retired from active management of the firm, he increasingly indulged in an old enthusiasm for travel. He traveled widely in the United States, sometimes meeting with his sales representatives and sometimes traveling exclusively for pleasure. In 1887 he took his family, except for Abram and Charles, to Europe for several months. His private letters to his sons during the 1880s were encouraging and reassuring. He gave advice when it was requested but did not presume to meddle in the decisions of his successors.

Charles Huston was the last of a series of remarkable amateurs to assume the direction of affairs at Lukens. Dr. Charles Lukens, Rebecca Lukens, and their sons-in-law, Abraham Gibbons and Charles Huston, were not trained for the iron rolling business. They learned their craft from their foremen and by hard work. Charles Huston had surviving sons, and he raised them for the family's business. The heirs were obedient to his wishes, and they were talented and successful.

Dr. Huston was accomplished in fields other than medicine and the iron and steel business. He loved music, especially opera, and he played the flute and violin. He thought dancing a pleasant form of exercise and a harmless social grace but forbore its practice in his home because of his wife's strict Quaker objections. He liked bird hunting and handsome carriage horses. He tried his hand at architecture and designed his dwelling in Coatesville. He was a dedicated amateur botanist and gave his garden his own personal attention. Like his father before him, he helped direct his community's gas company. He died in January 1897.

References:

Stewart Huston, "The Iron Industry of Chester County," in *Southeastern Pennsylvania.*, edited by V. Bennet Nolan (Philadelphia: Lewis, 1943);

Clara Huston Miller, *Reminiscences* (N.p., 1930).

Archives:

The Lukens papers are located in the Hagley Museum and Library, Wilmington, Delaware.

Iron-Mining Machinery and Techniques

by Terry S. Reynolds

Michigan Technological University

There were three basic methods of iron mining practiced in the United States in the nineteenth century: (1) quarry mining [a primitive form of open-pit mining], (2) shaft mining, and (3) mechanized open-pit mining. Generally, quarry mining predominated until 1870, when it was replaced in importance by shaft mining. Shaft mining dominated the scene from the 1870s to the end of the nineteenth century but began to be rivaled in importance by mechanized open-pit mining around 1900.

At virtually all of the iron ore deposits exploited in the nineteenth century, including those in the southern and eastern United States, as well as in the Lake Superior iron ranges, quarry methods were the first used. Quarry mining required only the simplest machinery and techniques, was applicable where ore deposits were generally on the surface or very close to it, and could be used where there was no demand for production in vast quantities.

In quarry mining the process of ore extraction was uncomplicated, resembling rock excavation more than what we generally consider mining today. The overburden of trees, grass, and soil was stripped away, and the ore was removed from the exposed outcrop by breaking it off manually or by blasting with black powder. After the ore was reduced to manageable chunks with sledgehammers, additional blasting, or fire, it was then loaded by hand into wheelbarrows or carts.

At mines which primarily served local markets the ore was hauled, at first over flat paths, to an adjacent blast furnace for smelting. Rails were first used for moving ore at the Cornwall Mine in Pennsylvania in 1853 and at the Jackson Mine on Michigan's Marquette Range a few years later. For those mines where ores were exported rather than used locally, the ore was carried to an adjacent railroad siding.

The early Cleveland Mine on the Marquette Range provides a good example of quarry mining.

Opened around 1850, for almost two decades its ores were so close to the surface that little overburden stripping was required. The ore was simply quarried from the sides of the hills where the outcrops occurred. Mining was simply a process of blasting down segments of the ore ledge. After breaking the ore into small chunks, workmen loaded the chunks into one-horse two-wheel carts, which were pulled over inclined paths a few hundred feet to the Iron Mountain Railroad. There the ore was loaded into cars to be pulled by locomotives to Marquette.

The great advantage of quarry mining was the low capital and labor cost required. No heavy equipment was needed, only hand tools like picks, shovels, sledgehammers, and hand-held mine drills. Pumping and hoisting equipment, if needed at all, was required on only a small scale. Expensive labor was not required because skilled, labor-intensive processes like drilling and blasting were used at a minimum.

Before 1870 most iron mining, both in the Lake Superior district and in the rest of the United States, was largely carried on using quarry methods. By 1870, however, the days of quarry mining were numbered. As rich surface deposits were exhausted, it became necessary to follow iron veins increasing distances below the surface. As a result, the quarry mines began to require larger and more expensive pumping and hoisting plants, and the cost of stripping the overburden to allow miners to follow the ore deeper into the earth began to reach the point of diminishing return. Moreover, the increasing market for iron ore, due to the rapid expansion of the American rail system, among other things, quickly exhausted near-surface deposits and further accelerated the shift away from the hand-intensive methods which made quarry mining cheap. These circumstances forced increasing resort to the second basic type of mining operation–shaft mining.

Loading ore onto train cars at the Mahoning Mine in 1899

For relatively narrow ore beds, situated vertically, shaft mining became economically viable when depths reached beyond 40 feet. At that point, and in some cases even before then, it became cheaper to sink a shaft to the ore deposit than to attempt to strip off all of the earth covering it. Shaft mining, of course, had been widely used by miners for centuries and even in the United States had been utilized in certain locations. But after 1870 it became the dominant and unchallenged form of iron mining in all American mining districts until mechanized open-pit mining emerged on the Mesabi Range of Minnesota in the 1890s.

Shaft mining was generally more expensive than quarry mining because, as it was used mainly for deep iron deposits, it required more mechanized equipment as the shaft penetrated deeper into the earth. Shaft mining involved five basic steps: (1) exploration, (2) sinking a shaft, (3) mining proper, which involved drilling and blasting, (4) horizontal transportation, and (5) vertical transportation, or hoisting.

Because a shaft mine's ore deposits lay deep beneath the earth and required a rather extensive and expensive physical plant to reach, exploration was usually a prerequisite to the decision to construct a shaft mine. Sometimes test pits were sufficient to determine the presence of ores and to estimate their quality and quantity. But late in the nineteenth century one of the standard implements used for exploration was the diamond drill, which was steam powered and used a rotary action to produce a core for inspection and assay. It was capable of reaching depths as great as 1,000 feet to 2,000 feet. First introduced in 1869, it came into extensive use on the Michigan iron ranges in the late 1870s and 1880s. It was supplemented by the churn drill, introduced on the Mesabi Range in the 1890s. The churn drill, unlike the rotary diamond drill, relied on a chopping or pounding action to penetrate the earth and was more useful than the diamond drill in the soft iron ore regions.

Once ore bodies were discovered by test pits or diamond drilling and after analysis indicated commercial deposits, a shaft or shafts had to be sunk to reach the deposits. At first these shafts were often sunk at the angle that the ore deposits followed into the earth, but later mines tended to sink the shafts vertically and depend on horizontal tunnels to reach the ore. As iron deposits closer to the surface were exhausted, mines steadily penetrated deeper. By 1900 some mines on the Marquette and Menominee Ranges of Michigan were approaching 2,000 feet in depth.

Once the shafts had reached the depth of the ore deposit, mining proper began. Horizontal tunnels, called drifts, were dug from the shafts into the ore body, usually several at varying depths. After the ore body was reached, large rooms, called stopes, were mined out of the ore body. A number of different methods were used in shaft mining to break up the ore body. Open stoping, for example, required digging out the ore body and leaving large "rooms" with ore pillars to support the ceiling, a process useful for hard ores. For softer ores the room, or stope, required extensive timbering for support. Another method, block caving, involved undercutting large areas of ore and allowing it to break and cave in under its own weight.

Mining proper required two operations: drilling and blasting. To break iron ore from its outcrops and surrounding rock in the stope, or to drive through ordinary rock to an ore deposit, the rock first had to be drilled and then blasted. Both of these operations had been used in quarry mining but were much more extensively applied in shaft mining.

Drilling was one of the most labor intensive and skilled jobs in iron mining. Until 1880 drilling was almost universally done by hand, commonly requiring a three-man team. One man would hold a long iron or steel drill bit, commonly 4 to 5 feet long and 2 inches in diameter, with a chisel-shaped end. As he held the drill, two men with sledgehammers would alternately strike the drill, with the holder lifting and rotating it after each strike so that it would get a better bite. The depth a team could drill to per day depended upon the hardness of the rock, but 10 feet per shift would be typical for hard rock. With extensions, drill holes as deep as 20 to 30 feet were sometimes produced. Around 1880 the Rand Drill, a two-man machine operated by compressed air, was introduced into the Lake Superior mining district. Its drill bit was mounted on a piston, which was driven back and forth by compressed air. Capable of boring a hole at a much faster rate than the hand-drill team and requiring only two men for its operation, the Rand Drill became the standard drilling instrument in hard ore

mines for the rest of the century. Hand drilling, however, continued to be used extensively in mines where ores were soft.

After a number of holes had been drilled into the rock or ore at the face, or front, of a drift or a stope, the holes were cleaned and filled with an explosive. Before 1870 that explosive was invariably black powder. But black powder, while adequate for the dry conditions generally encountered in quarry mines, was not adequate for the wet conditions frequently found in shaft mines. Thus nitroglycerin was introduced on the Marquette Range as early as 1870. Because of the hazards of handling nitroglycerin, it was quickly replaced by dynamite when the latter became available after 1881. In the last two decades of the nineteenth century dynamite was generally the explosive of first choice in shaft mines, though black powder, because of its cheaper cost, continued in use where conditions permitted, either by itself or as a supplement to dynamite. The explosives were tamped into the drill holes, capped with clay, and linked to fuses of varying length. The differing fuse lengths staggered the explosion, making it more effective in breaking the body of the ore or rock loose from the face of the drift or stope.

Because the shaft which was used to hoist rock or ore to the surface was often several hundred feet from the stope or drift being worked, horizontal haulage was necessary in shaft mines. In most nineteenth-century iron mines the loose rock had to be hand loaded into tram cars mounted on rails which ran along the drifts to the shaft. However, most mines attempted to use gravity as much as possible to assist in loading. Hence, stopes were usually driven at an angle upward from the floor of a drift so that the ore would fall downward toward the tram cars which would be running along the floor of the drift. Just above the tram tracks barricades would collect and hold back the ore to permit easier loading of the tram cars from chutes. In other cases small shafts would be driven from a higher drift down to a lower drift to permit gravity loading of tram cars.

Once loaded, tram cars would either be pushed by hand or pulled by mules to the shaft. Mule haulage was common in the larger shaft mines, with dozens of mules stabled below ground and only taken up to daylight on Sundays. Only very late in the nineteenth century did a few mines begin experimentation with mechanical haulage.

A steam shovel with crew in the Biwibak Mine in 1895 (courtesy of the Nute Collection)

One of the earliest mines to apply electric haulage was the Lake Mine of Cleveland-Cliffs in 1892. Several mines on the Menominee Range used compressed-air locomotives, but these were the exceptions in the nineteenth century. Only after 1910 did electric haulage become common, and even then some hand tramming was necessary because exploratory and new drifts often did not have space for rails and electric lines.

Once the rock or ore was trammed to the shaft, it had to be hoisted to the surface. In the earliest shaft mines and in the quarry mines, hoisting, when it had to be done, was usually done with a swinging derrick or a whim, powered by man or horse. The first hoist at the Soudan Mine on the Vermilion Range, for example, was powered by a single horse. The whims lifted large cast-iron buckets containing a few hundred pounds of ore. By the 1870s, however, hoisting was generally being done in shaft mines with steam-powered winding engines and wire rope. The deeper the mines went, the larger the steam engines and associated hoisting machinery needed to be. By 1900 the special containers, called skips, which lifted ore to the surface in

the deeper mines, sometimes had a capacity of four to five tons.

Shaft mines had special problems not normally encountered in the early quarry mines, particularly pumping, air supply, and lighting. Since most shaft mines penetrated well below the local water table, large steam-powered pumping installations were needed to prevent flooding. By 1890 the Chapin Mine on the Menominee Range of Michigan was almost 1,500 feet deep and required a mine pump of 1,250 horsepower capable of pumping four million gallons of water a day to the surface to keep it dry. Smaller steam engines were required in some mines to provide fresh air and, by the end of the nineteenth century, in many more mines to provide compressed air to mine drills. For lighting, miners long relied on candles. When moving around, miners carried the candles on their hats. When working, miners placed the candles into special spiked holders which could be driven into mine timbers. Toward the end of the century some mines began to make use of "sunshine" lamps, which were filled with a solid fatty substance that supplied fuel to the flame as the wick became hot. Carbide lamps, battery operated lamps, and electric lighting systems began to come into use only around 1900 and did not become common until the early twentieth century.

While shaft mining was still the dominant iron-mining method in 1900, a more sophisticated and larger-scale form of quarry mining–mechanized open-pit mining–had become a major force in the 1890s. Modern open-pit mining began in 1892 and 1893 on the Mesabi Range of Minnesota. There were three reasons why large-scale open-pit mining initially emerged on the Mesabi. First, the Mesabi's relatively rich iron ore deposits were near the surface and lay flat or nearly horizontally, instead of vertically as on other ranges. Second, the Mesabi ore deposits were enormous. Third, Mesabi ores were relatively soft and so could be broken up and handled by mechanical equipment. By 1900 slightly over 50 percent of Mesabi Range ore was being taken out using open-pit methods.

Five basic steps were involved in open-pit mining: removing overburden, drilling, blasting, loading, and transporting. As several of these processes were identical to those used in early quarry mining and in shaft mining, they will be described more briefly. As noted above, quarry mining required removal of overburden, but its small scale required only pick and shovel work. The development of new technologies and the scale of the Mesabi operations, however, required the use of mechanized equipment. Among the new technologies available was the steam-powered shovel. By the 1880s American railroads were using large steam shovels mounted on rails for making railroad cuts. Steam shovels of this type had been used briefly in iron mining in 1884 at the Colby Mine on the Gogebic Range. But on the Gogebic Range iron ore deposits penetrated at a steep incline into the earth, and steam shoveling was not practical beyond removing the covering soil and the deposits very close to the surface. On the Mesabi, where ores were distributed horizontally and covered many acres, large-scale use of steam shovels for stripping overburden quickly became standard for open-pit mines. By 1900 the steam shovels being used for this purpose weighed from 40 to 60 tons, had a bucket capacity of 3 cubic yards, and could propel themselves on rails.

Once the overburden had been stripped away from the underlying iron deposits, the exposed iron ore had to be drilled and blasted to loosen it for loading. Drilling and blasting in open-pit mining in the

1890s differed little from the same processes in shaft mining, save that drier conditions and softer ores made hand drilling and black powder acceptable options even though compressed air drills and dynamite were widely available.

In the quarry mine the loose ore left in the pit after blasting was broken into manageable chunks, loaded by hand into either wheelbarrows or horse- or mule-drawn carts, hauled out of the mine, and loaded into cars on an adjacent railroad siding. By the 1860s small locomotives were used in a few mines to pull cars out of the workings. The open-pit mines of the Mesabi in the 1890s relied on much more mechanized equipment, such as steam shovels and standard gauge railroad cars, to load and transport ore from the mine. The steam shovels were similar to those used for removing overburden; they lifted the ore fragments and dumped them into standard ore-carrying railroad cars operating on temporary track leading down into the mine itself. Using conventional hand-loading and, in shaft mines, chute-loading methods, only a few tons of ore could normally be loaded in an hour. But in 1898 a 90-ton Vulcan shovel loaded a record 800 tons in an hour on an open-pit mine on the Mesabi Range. By 1900 a good open-pit shovel crew was normally capable of handling 3,000 tons a day. Once loaded, the train of ore cars would be pulled out of the pit by standard locomotives.

The open-pit mine, with its heavy use of mechanized, materials-handling equipment, could produce larger volumes of ore with a much smaller work force than the shaft mine. Studies in the early twentieth century indicated that the output per man-hour in an open-pit mine was roughly two to three times higher than in a shaft mine. Further, the larger scale of an open-pit operation brought other economies, such as reduced transportation expenses. These advantages were partially offset, however, by the high initial capital investment required for open-pit mining. This high capital cost was due to the expense of stripping the enormous amount of overburden from open-pit sites and the heavy cost of the capital equipment necessary to operate open-pit mines, particularly the steam shovels, railroad trackage, and locomotives. The high initial capital cost of open-pit mining permitted shaft mining to survive. But the high capital costs required for pit mining also gave an advantage to large corporations because of their ability to raise large amounts of capital and encouraged the emergence of the large consolidated companies which came to dominate iron mining throughout the United States in the 1890s.

References:

Harold Barger and Sam H. Schurr, *The Mining Industries, 1899-1939: A Study of Output, Employment and Productivity* (New York: Arno Press, 1972);

Nelson P. Hulst, "Methods of Mining Iron Ore in the Lake Superior Region," *Proceedings of the Engineering Society of Western Pennsylvania*, 15 (1899): 62-103;

Henry Raymond Mussey, *Combination in the Mining Industry: A Study of Concentration in Lake Superior*

Iron Ore Production, Studies in History, Economics and Public Law, Columbia University, 23, no. 3 (New York: Columbia University Press, 1905);

A. Tancig, "The Evolution of Equipment in Minnesota Iron Mining Industry, 1883-1953," *Skillings' Mining Review*, 42 (August 8, 1953): 1-2, 4, 6; 42 (August 15, 1953): 1-2, 4, 8;

Kirby Thomas, "Mining Methods in the Vermilion and Mesabi Districts," *Proceedings of the Lake Superior Mining Institute*, 10 (1904): 144-157;

N. Yaworski, et al., *Technology, Employment and Output Per Man In Iron Mining*, Works Progress Administration and Bureau of Mines Report No. E-13, (Philadelphia: Works Progress Administration and Bureau of Mines, 1940).

Iron Ore Ranges

by Terry S. Reynolds

Michigan Technological University

Iron ore deposits exist in practically every state in the United States but vary considerably in quantity and quality. The ores found and used in early New England, for instance, were collected from swamps and ponds and were known as either bog ore or pond ore. They had an iron content ranging from 20 percent to 30 percent and were generally found only in sufficient quantity to supply small-scale blast furnaces serving largely local needs. By the early nineteenth century larger and richer iron ore deposits had been discovered in New Jersey, upper New York, and eastern Pennsylvania. Pennsylvania ore contained as much as 50 percent to 70 percent iron. But these ores, because of lack of transportation, were used mostly to supply local or, at best, regional needs.

In the West several small mountains of iron ore–Iron Mountain and Pilot Knob–were discovered in Missouri, south of St. Louis, before 1800. First exploited in the mid 1840s, these iron ore deposits were considered, at the time, to be the largest in the world. Their high-grade ores were abundant enough to serve not only the needs of local ironworks, but to be exported in bulk to other regions, especially after railroads reached the area in the 1850s. By the 1860s a substantial amount of Missouri iron ore was being shipped to western Pennsylvania for smelting.

In the nineteenth century the regions with large ore deposits in states like Missouri, Pennsylvania, New Jersey, and New York were generally termed "iron districts." The term "iron range" was generally restricted to much larger ore districts in the Lake Superior region. These districts were called "ranges" because the iron ore deposits generally occurred in or along chains of hills, or ranges.

The ore production of Lake Superior iron ranges in the late nineteenth century quickly dwarfed that of any previous ore district and by 1870 enabled Michigan to surpass Missouri as the leading ore-exporting state in the United States. By 1900 almost 75 percent of American iron ore was coming from five of the six Lake Superior iron ranges: the Marquette, Menominee, Gogebic, Vermilion, and Mesabi. The sixth "range"–the Cuyuna–entered production only in the twentieth century.

The earliest of the Lake Superior iron ranges to be exploited was the Marquette Range, located in the north central portion of Michigan's Upper Peninsula, some 10 to 15 miles inland from Lake Superior. The Marquette Range extends for approximately 40 miles, from Marquette on the east to just south of L'Anse on the west. The most productive part of the range, however, ran for only about 25 miles from Ishpeming and Negaunee on the east to Lake Michigamme on the west. In width the

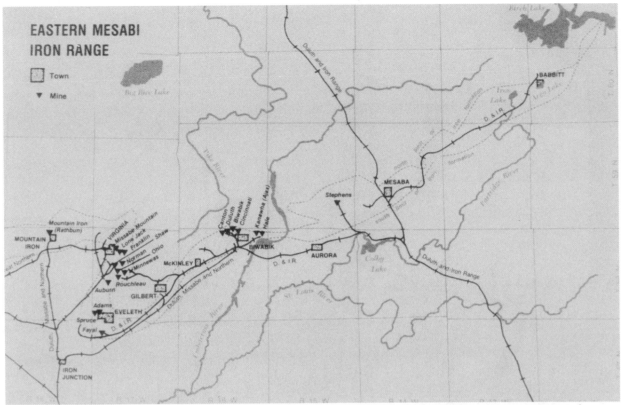

Courtesy of Alan Ominsky, Minnesota Historical Society

range varied from 3 to 10 miles. Marquette ores were of several varieties. Its hard specular ores were the first to be mined and were the standard for years, but its magnetic ores and red hematites had also become important before the end of the 1800s.

Iron ores were first discovered on the Marquette Range in 1844, when U.S. government surveyor William A. Burt noticed severe variations in his magnetic compass needle near what is now Negaunee, Michigan. Suspecting that a large iron ore deposit might be the source of the problem, he had his assistants search the area for ore specimens. Reacting to Burt's discovery, several companies had been organized to mine iron in the area by 1848. Initially, iron-mining companies on the Marquette Range hoped to smelt and forge bar iron locally and ship the finished product south. But these ventures failed. Instead, following the opening of the canal at Sault Ste. Marie in 1855, the region became a major shipper of iron ore. The Marquette Range remained the leading range in the Lake Superior district for nearly a half century. Until the 1870s all the mines in the district were on the Marquette Range, and not until 1892 did another range ship more ore.

The second of the Lake Superior iron ranges to be opened was the Menominee Range, located about 50 miles south of the Marquette Range along the Menominee River, which forms the boundary between Michigan's Upper Peninsula and Wisconsin. The Menominee Range covers the largest territory of any of the Lake Superior ranges save the Mesabi. Beginning at a point about 40 miles west of the Lake Michigan port of Escanaba, Michigan, the Menominee Range runs generally northwest for 50 miles. The bulk of the Menominee Range is in Michigan, but it overlaps slightly into Wisconsin. The Menominee ores were high in phosphorus and somewhat lower in iron content on the average than the ores of the Marquette Range, and some of its ores also had high silica content. These factors sometimes depressed sales of Menominee Range ores.

Iron ore was reported along the Menominee River several times in the late 1840s and at other points in the district as well. But the area was covered by dense forest and swamps and was too distant from any form of bulk transportation to attract serious exploration and investment until the growing demand for iron ore in the decade follow-

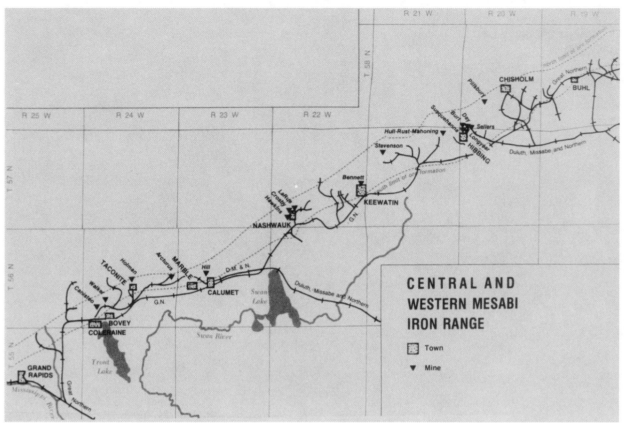

Courtesy of Alan Ominsky, Minnesota Historical Society

ing the Civil War. In 1866 two timber speculators, Thomas and Bartley Breen, discovered an outcropping of ore near Waucedah, Michigan, and four years later began serious exploration on the site. By the early 1870s the explorations by the Breens and others had indicated sufficient iron deposits for commercial mining. Commercial exploitation was delayed, however, by the panic of 1873 and by the lack of transportation facilities. In 1877, however, the Chicago & North Western Railroad extended a branch line from Escanaba to Quinnesec, and the first Menominee ores were shipped to market. By 1878 five mines were shipping from the Menominee Range. As the Chicago & North Western extended its lines westward, reaching Iron Mountain, Michigan, in 1880 and Crystal Falls and Iron River in 1882, more mines opened and began shipping. In 1882 the Menominee Range shipped more than 1 million tons of ore and had become a major supplier.

The third Michigan iron range to open was the Gogebic Range. Located about 100 miles west of the Marquette Range, it extends almost 80 miles from Lake Gogebic near the western end of Michigan's Upper Peninsula well into Wisconsin. Al-

though the Gogebic Range was 80 miles long, commercial mining occurred in a much more restricted area, extending about 25 miles westward from Lake Gogebic, reaching a few miles across the border into Wisconsin at Hurley. The band of ore on the Gogebic Range was exceedingly narrow, no more than a quarter of a mile wide. But it produced ore that was of high quality, generally a soft, red, somewhat hydrated hematite.

As with the Menominee Range, iron ores were discovered on the Gogebic Range long before they were exploited. On the Wisconsin side (where the range is called the Penokee), iron ores were noted as early as 1848 and again in 1858. But geographical isolation prevented development. A short-lived company attempted in 1873 to begin mining, but shipped no iron. Although first explored on the Wisconsin side, mining began on the Michigan side of the Gogebic Range following serious exploration in the 1870s and early 1880s. The first Gogebic Range ore was shipped by rail to Milwaukee in 1884 from the Colby Mine on the site of Bessemer, Michigan. The completion of the Milwaukee, Lake Shore & Western Railway line from the port of Ashland, Wisconsin, to the range in 1885 and the exten-

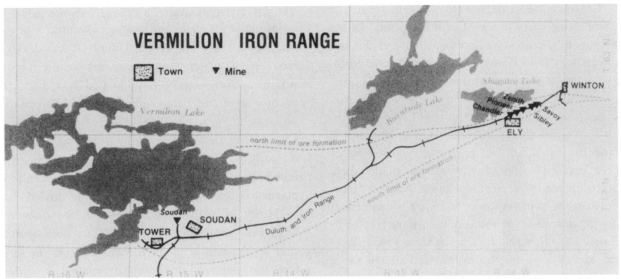

Courtesy of Alan Ominsky, Minnesota Historical Society

sion of the Wisconsin Central Railway from Penokee Gap to the range in 1887 promoted rapid development. In 1887 the Gogebic Range shipped over 1 million tons of iron ore, and in 1892 shipped more ore than the Marquette Range.

The first of Minnesota's iron ranges, the Vermilion, opened in 1884, the same year as the Gogebic. Located some 80 miles north of Duluth, Minnesota, the Vermilion Range extends for a distance of about 40 miles. Like the Gogebic, only about half the range was productive in the late nineteenth and early twentieth centuries. The ores of the Vermilion were of several types, but the one predominantly tapped in the nineteenth century was a very hard, granular, steel-blue hematite that was highly valued by furnace men.

Iron ore formations were reported in the Vermilion area as early as 1848. An abortive gold rush to Vermilion Lake in 1860 created new interest in the district and led to further reports of iron formations. But the area was isolated by dense forests and swamplands and was nearly 80 miles inland from Lake Superior, further inland than any of the previously opened iron ranges. This isolation meant that development would require railroad construction at considerable expense. As a result the early reports of iron ore on the Vermilion were not followed up. In the 1870s George Stone and George Stuntz persuaded Philadelphia entrepreneur Charlemagne Tower to investigate Minnesota iron prospects. Tower, in turn, engaged Albert Chester, who in 1876, working with Stone and Stuntz, uncovered rich ores at what is now Tower, Minnesota, on

the Vermilion Range. But the difficulties of mining the area initially intimidated Tower. But after having Chester conduct further explorations in 1879 and 1880, Tower plunged forward. In 1882 he organized the Duluth & Iron Range Railroad and financed construction of a rail line from what is now Two Harbors, Minnesota, to the Vermilion Range. The first Vermilion ore was shipped over the railroad in July 1884. Like the Gogebic, the Vermilion Range developed rapidly, shipping more than 1 million tons of iron ore in 1892.

The Mesabi Range (sometimes spelled Missabe or Mesaba) was the fifth Lake Superior range opened and the second of the Minnesota ranges. It is the largest of the iron ore ranges, both in terms of area and extent of ore deposits, extending from Grand Rapids on the Mississippi east-northeast about 100 miles to Birch Lake with a width varying from 2 to 10 miles. At its closest point it is about 50 miles northwest of Duluth. The ores of the Mesabi Range differed from those of the older Lake Superior ranges. First, they were laid down in nearly horizontal strata, rather than vertical or sharply inclined strata. This meant that they could be reached by open-pit mining methods. Second, while the ores first tapped were mostly very low phosphorous hematites of all varieties and textures, they were generally very soft and very fine, almost sandlike. This texture was to cause some problems in blast furnaces (cloggings and explosions), before methods were developed for handling fine ores.

Around 1850 there were scattered reports of iron-bearing rocks in the Mesabi area, but, like the Vermilion, the area was unpopulated and difficult to reach. In 1866 Henry H. Eames, first state geologist of Minnesota, reported seeing immense bodies of iron ore in his explorations of the western part of the Mesabi Range. The reports of Eames and others led several groups to begin claiming land on the range in the 1870s, but no commercial mining or even detailed exploration work was undertaken, probably due to the area's isolation. Another factor delaying the exploitation of the Mesabi was probably the distraction of the Vermilion Range. In 1876 Albert Chester, acting for Charlemagne Tower, studied Mesabi ores on the eastern end of the range and was disappointed with their low iron content. He thus reported negatively on the range, assuming that the lean ores of the east were typical of the entire range, and recommended that Tower develop instead the much richer ores he and his associates had found on the Vermilion Range.

The key figures in opening the Mesabi Range were the Merritt brothers of Duluth. After several years of preliminary exploration, in November 1890 they struck a rich deposit of iron ore just north of what became the Mountain Iron Mine on the western portion of the Mesabi Range. By 1892 they had found other rich deposits and had begun construction of a railroad to bring the ores to market. Their Duluth, Missabe & Northern Railroad was completed in October 1892, and the Mesabi shipped its first ores to market late that year. Because Mesabi ores could be mined using mechanized open-pit methods, the range's production rapidly passed that of all the other Lake Superior ranges. Two years after its opening, over 1 million tons were shipped from the Mesabi, and in 1895, three years after its opening, the Mesabi became the leading producer of iron ore in the United States. By 1902 the Mesabi was producing half of all the iron ore shipped out of the Lake Superior district.

The last of the great Lake Superior ore ranges, the Cuyuna, was not developed until the early twentieth century, even though it was discovered very late in the nineteenth. The Cuyuna Range was not, topographically, a "range." Its ore deposits lay under the plains and swamps of central Minnesota, some 100 miles west of Duluth. It was discovered later than the other ranges, despite the fact it was in a more accessible, populated area, because its ores were completely subterranean. There were no protruding rocks to reveal its presence. The Cuyuna was thus discovered by geophysical means. As early as 1869 surveyors working in the area noted abnormal magnetic variations. More detailed magnetic exploration of the region was carried out by Henry Pajiri in 1892 and by Cuyler Adams in 1895, both of whom concluded that possibly significant ore deposits lay some distance beneath the surface and mapped out the approximate boundaries of the deposits. In 1903 the Orelands Mining Company, using a diamond drill, confirmed the existence of iron ore underground near Deerwood, Minnesota. Adams, who directed the exploration work, named the new "range" the Cuyuna, combining the first three letters of his first name with the last three letters of his dog's name (Luna). In 1906 and 1907 several companies attempted to sink operating shafts, but it was not until 1911 that the first shipment of ore was made from the district.

References:

Donna Stiffler Bollinger, "The Iron Riches of Michigan's Upper Peninsula," *Michigan History*, 62 (November-December 1978): 9-13;

Burton H. Boyum, *The Saga of Iron Mining in Michigan's Upper Peninsula* (Marquette, Mich.: John M. Longyear Research Library, 1977);

Harlan Hatcher, *A Century of Iron and Men* (Indianapolis: Bobbs-Merrill, 1950);

William T. Hogan, *Economic History of the Iron and Steel Industry in the United States,* volume 1 (Lexington, Mass.: D. C. Heath, 1971);

"The Iron Ranges of Minnesota," *Proceedings of the Lake Superior Mining Institute*, 13 (1908): 13-19;

David A. Walker, *Iron Frontier: The Discovery and Early Development of Minnesota's Three Ranges* (St. Paul: Minnesota Historical Society Press, 1979);

Carl Zapffe, "Discovery and Early Development of the Iron Ranges," in *Lake Superior Iron Ores, 1938* (Cleveland: Lake Superior Iron Ore Association, 1938): 13-26, 308.

Iron Ore Transportation

by Terry S. Reynolds

Michigan Technological University

Transportation of iron ore was not a major problem in most of the early iron-producing districts of the United States. Ores were produced and smelted locally. When the Lake Superior ore districts opened, however, transportation quickly became a major concern. Because the districts were geographically isolated, without coal and without any major local markets, iron ore could not be smelted to pig or bar iron and sold at a profit. The only profitable way to exploit the ore deposits was to transport ore in bulk to distant blast furnaces on the lower Great Lakes.

The transportation system created to transport Lake Superior iron ores in the nineteenth century consisted of several distinct steps. First, ores had to be hauled from mine to port. Second, they had to be loaded onto vessels for shipping to ports like Cleveland, Chicago, and Ashtabula, Ohio, on the lower Great Lakes. Finally, the ores had to be unloaded for transportation to blast furnaces. Each step in this transportation system had to be balanced with regard to efficiency. Many of the changes which the system underwent in the nineteenth century came as a result of attempting to correct or eliminate bottlenecks which had arisen in the system.

The importance of an efficient and economical ore transportation system became clear when the Marquette Range, the first of the Lake Superior ore districts, began operations in the late 1840s. The first mines were located approximately 15 miles inland from Marquette Harbor, but the 15 miles were covered by thick forest. Ores could be moved from mine to port only in winter when heavy snows made it possible to use sleighs. Because this method was operable only part of the year and handled only small ore volumes, two of the leading companies on the range, the Cleveland Iron Mining Company and the Jackson Iron Company, jointly constructed a plank road, covered

with strap rails, which ran from the mines to the port. Completed in 1855, the road proved insufficient; the planks quickly began to rot, and the mule-drawn carts could make only one trip per day.

Economical transportation from mine to dock came only with the locomotive-powered railroad. In 1851 banker Herman Ely laid plans and began to raise funds to build a railroad from the mines to Marquette. His initial attempts failed for lack of capital, but by 1855-1856 work was renewed. Herman Ely died in 1856, but his two brothers, George and Samuel Ely, completed the Iron Mountain Railroad in 1857, operating it with a 25-ton steam locomotive.

The Ely brothers' railroad demonstrated the crucial importance of rail transportation between mine and dock. The same lesson was to be repeated on virtually every other range in the Lake Superior district. Commercially viable iron deposits on the Menominee Range were widely known as early as 1866, but active production began only in 1877–after the Chicago & North Western Railroad built a branch from the port of Escanaba to Quinnesec. The opening of rail connections from the Gogebic Range to Ashland, Wisconsin, in the mid 1880s was critical to the commercial success of that range. In Minnesota, Charlemagne Tower's Duluth & Iron Range Railroad was the key to successfully opening up the Vermilion Range, and the Merritts' Duluth, Missabe & Northern Railway was essential to opening up the Mesabi's ores.

The next potential bottleneck was at the lakeshore docks. The first ore-carrying vessels were loaded by ramp and wheelbarrow from simple piers. It took two dozen laborers three to six days to load 200 to 300 tons. The coming of the iron ore railroad made this a serious obstacle. In 1857, in anticipation of the completion of the Iron Mountain Railroad, Samuel P. Ely built a radically different type of dock for the Lake Superior Iron

Ore clippers operating on the Great Lakes, circa 1880

Company at Marquette. The new dock was built a full 25 feet above lake level and was equipped with pockets to store ore and with chutes to drop the ore by gravity directly into the holds of vessels moored beneath the dock. Ely's pocket dock was quickly imitated.

The most important geographical bottleneck in the water portion of the transportation system was the rapids at Sault Ste. Marie on the St. Mary's River, the link between Lake Superior and the lower Great Lakes. The first shipments of iron ore from the Marquette Range mines had to be carried by boat to the Sault, unloaded, portaged around the falls, and then loaded on a different vessel to be carried to ports on the lower Great Lakes. In 1853, however, the state of Michigan began construction of a canal around the rapids. Completed in June 1855, the Sault Canal made the export of Lake Superior ores economically feasible. Ore shipments from the Lake Superior region increased sharply. In 1856 less than 1 percent of the pig iron made in the United States was smelted from Lake Superior ores. Four years after the opening of the canal made uninterrupted water transportation possible the figure was 9 percent, and in 1869 it had risen to almost 24 percent.

Most of the early ships that carried iron ore were not specifically designed for that purpose and had very small capacities, often less than 300 tons. Many relied on sails for power and had to be towed by tugs through rivers. The *R. J. Hackett*, launched in 1869, was the first steamer designed exclusively for the ore trade. Its design became standard for lake freighters–pilothouse forward, engine aft, decks in between free of gear, hatches unobstructed and spaced for easy cargo handling. The *Hackett* could carry 1,200 tons and was 211 feet long. In 1882 the first iron-hulled freighter was launched at Cleveland. The 287-foot *Onoko* was built with a capacity of around 2,200 tons. The first steel freighter built for Great Lakes service was launched in 1886, the 310-foot *Spokane*. The steam-powered *Pontiac* and *Frontenac* of the Cleveland Iron Mining Company, both built in 1889, were more than 300 feet long and had a capacity of 3,500 tons. By 1900 the sailing vessel was virtually extinct, and steel ore freighters were capable of carrying 7,500 tons.

While the growing demand for iron ore and the productive capacity of the mines pushed the size of lake freighters upward, limits were placed on their size and carrying capacity by several other fac-

Workmen unloading wooden ore cars into dock pockets in the Minnesota ore ranges (courtesy of Library of Congress)

tors. One was the size of the locks at Sault Ste. Marie. The 1855 locks were only 250 feet long and 12 feet deep. As early as 1870 it was clear that the 1855 lock was inadequate, and a new lock, 515 feet long and 16 feet deep, was opened to traffic in 1884. It was sufficient to take the larger iron and steel vessels constructed in the 1880s, but continued growth in vessel size and lake traffic led to the opening of an 800-foot-long lock in 1896.

At the far end of the transportation chain was unloading. As traffic increased in the 1860s and 1870s and boats grew in size, this process became a major cause of delays. In the era of low ore shipments, unloading was done by hand with ramps and wheelbarrows and often took days. With growing volumes, the wheelbarrows and ramps were replaced with hoisting buckets operated by horsepower or, after 1867, steam power. But even these were inadequate as vessels grew in size. In 1882 Alexander E. Brown, a Cleveland-born engineer, developed a new form of unloader, a machine capable of removing cargoes of 5,000-7,000 tons in hours instead of days. Brown's unloader consisted of a movable truss bridge situated so that it extended both over a ship's hold and an onshore stockpile or railroad cars. A bucket traveling on the bridge enabled ore to be moved directly from a vessel's hold to the

stockpile or to cars. Brown later improved the system by using self-loading clam buckets. By 1893, 75 percent of the Great Lakes ore was being unloaded by Brown hoisting machinery.

In 1899 George Hulett, a chemist, made the process more efficient by designing an improved unloader. The Hulett device was steam powered with a clamshell bucket mounted on a movable arm. The bucket operator and the bucket controls were placed in a "cabin" on the arm just above the bucket. The operator could, therefore, move with the bucket down into the hold of the ship being unloaded, grab a shovel-load of ore, lift it up, carry it above a stockpile or waiting railroad car, and dump it.

The transportation system which had emerged by 1900 for transporting iron ores from the Lake Superior district thus consisted of steam-powered ore railroads to move ore from the mines to the docks on Lake Superior, large pocket-docks with chutes to gravity-load steam-powered steel vessels of increasingly larger capacity, an all-water route through the Sault Ste. Marie Canal to ports on the lower Great Lakes, and steam-powered Brown and Hulett unloaders at the docks there. By 1900 this system was annually shipping 20 million tons of ore, enabling the Lake Superior district to become the major ore supplier for the entire nation.

References:

H. R. Harris, "Transporting Ore from Mine to Docks," in *The Cleveland-Cliffs Iron Company and Its Extensive Operations in the Lake Superior District* (Cleveland: Cleveland-Cliffs Iron Company, 1929), pp. 39-43;

Harlan Hatcher, *A Century of Iron and Men* (Indianapolis: Bobbs-Merrill, 1950);

Walter Havighurst, *Vein of Iron: The Pickands Mather Story* (Cleveland & New York: World Publishing, 1958);

John Hearding, "Early Methods of Transporting Iron Ore in the Lake Superior Region," *Proceedings of the Lake Superior Mining Institute*, 29 (1936): 174-179;

William T. Hogan, *Economic History of the Iron and Steel Industry in the United States*, volume 1 (Lexington, Mass.: Lexington Books, 1971);

John Sampson, "Transportation of Lake Superior Iron Ores From Mines to Furnaces," *Proceedings of the Lake Superior Mining Institute*, 25 (1926): 160-163;

Kenneth Warren, *The American Steel Industry, 1850-1970: A Geographical Interpretation* (Oxford: Clarendon, 1973).

Joliet Steel Company

by Alec Kirby

George Washington University

The Joliet Steel Company was the ninth Bessemer steelworks in the United States at the time of its construction at Joliet, near Chicago. The firm became part of the initial merger that created the Illinois Steel Company in 1889, a steel conglomerate exceeded in capacity only by the group headed by Andrew Carnegie.

The Joliet Steel Company grew out of the Joliet Iron & Coal Company, which had experienced financial difficulty in the panic of 1873. In the late 1870s the receiver of the firm appointed Horace Strong Smith, the master mechanic of the Chicago & Alton Railroad, as manager. Under Smith's leadership the firm expanded and undertook several innovative projects, including installing automatic machines for rolling rails and billets. The firm was also the first to roll steel rails directly from the converter, without reheating from ingot. By 1887 the company was producing more steel rails than any other in the world. This was due in part to the firm's strategic location, Chicago, which during this period was the location of six major railroad companies. Through the mid 1860s these companies used iron rails because of the high cost of producing steel. Iron rails lasted, however, only an average of two years.

In the 1860s the Bessemer process for making steel was perfected, an inexpensive technique that removed carbon and silica from molten pig iron by blowing air through a converter. For the first time large amounts of durable yet malleable steel could be produced at a relatively low cost. By the 1880s steel was replacing iron as the standard alloy for railroad use, and the Joliet Company was prepared to meet the demand. The firm established a steelworks in 1870 and produced the company's first steel rail on March 15, 1873. The steelworks was a massive facility, with an annual capacity of 235,000 net tons of Bessemer steel ingots. The steelworks had six heating furnaces.

Under Smith's leadership the Joliet company expanded into other steel industry fields, and by the mid 1880s it was the largest single producer of wire rods in the United States. Much of the firm's expansion can be credited to Horace Smith, who was in his time a highly regarded innovator. In 1889 Orrin W. Potter of the North Chicago Rolling Mill Company and Jay C. Morse of the Union Iron & Steel Company initiated negotiations that led to a merger of their companies with the Joliet Company. The three companies together created the Illinois Steel Company. The Joliet works were further absorbed when the Illinois Steel Company was acquired by Federal Steel in 1898.

References:

American Iron and Steel Association, *Directory to the Iron and Steel Works of the United States* (Philadelphia: American Iron and Steel Association, 1880);

Herbert N. Casson, *The Romance of Steel* (New York: A. S. Barnes, 1907);

William T. Hogan, *Economic History of the Iron and Steel Industry in the United States*, volume 1 (Lexington, Mass.: Lexington Books, 1971).

Benjamin Franklin Jones

(August 8, 1824-May 19, 1903)

by John A. Heitmann

University of Dayton

CAREER: Clerk, Mechanics' Line (1842-1845); partner, Kier & Jones (1845-1853); partner, Jones, Kier & Lauth (1853-1856); partner, Jones & Lauth (1856-1861); partner, Jones & Laughlin, Ltd. (1861-1903).

Benjamin Franklin Jones, a businessman and iron manufacturer, was born on August 8, 1824, in Claysville, Washington County, Pennsylvania, and died in Allegheny, Pennsylvania, on May 19, 1903. During the span of his long career the American iron and steel industry was transformed in terms of technology, organization, locus of activity, as well as scale of production, and Jones played a significant role in creating and sustaining these changes.

Jones was the son of a farmer, merchant, and occasional surveyor, and at age fourteen he moved with his family to New Brighton, Pennsylvania, where he received a thorough and liberal education at the New Brighton Academy. Even as a young man, however, Jones preferred practical business enterprise to scholarship, and at age eighteen he left home, determined to make a fortune on his own. In 1842 he moved to Pittsburgh, which became the center of his business activities for the remainder of his life.

His first job was as a receiving clerk at the Mechanics' Line, a transportation company owned by Samuel M. Kier and doing business between Pittsburgh and Baltimore, Philadelphia, and New York. Jones's talent, organizational ability, and attention to detail soon won Kier's confidence, and at age twenty-one he was offered an equal partnership. The Mechanics' Line operated a transportation system consisting of canal boats and railroad facilities, yet its success was short-lived. The Pennsylvania Railroad was chartered in 1846 and by 1854 would control the routes once so profitable to Jones and Kier. The partners recognized this predicament and by the late 1840s began searching for new opportuni-

Benjamin Franklin Jones

ties. In 1847 the firm of Kier & Jones purchased an iron furnace and forge in the southeastern corner of Indiana County, Pennsylvania; much to Jones's credit, the mill was operated without loss for several years. Yet, Jones realized that the location of the works, two miles from the nearest canal, resulted in high transportation costs and that the low tariffs of the time depressed prices. The profit outlook for the venture was dismal. Despite the adversity this marginal business taught Jones much about the iron trade, and when he learned of Bernard Lauth's newly established American Iron Works on the Monongahela River in Pittsburgh, he immediately became interested.

In June 1853 Bernard Lauth and his brother John, both of whom had previous experience in the iron trade in Ohio, purchased land on the south bank of the Monongahela River in what was known as "Brownstown." Soon they began erecting an ironworks initially consisting of four puddling furnaces, two heating furnaces, a guide mill, muck rolls, and a crocodile squeezer. The Lauth brothers began operations in September 1853, and the output of the furnaces was about seven tons per day. This operation was reorganized in December 1853 with a capitalization of $20,000 and was held by a partnership of Jones, Kier, and the Lauth brothers. The agreement stipulated that the venture was to have a duration of five years and that if a partner wanted to sell his shares before the expiration of the contract, others in the group would have first option at purchase. Jones was placed in charge of the warehouse, accounts, and finances, the Lauth brothers supervised production, and Kier became an inactive partner, having no day-to-day responsibilities.

In part because of Jones's organizational talents the firm expanded steadily during the 1850s. On the eve of the Civil War the Monongahela (later known as South Side) works consisted of 5 steam-driven roller mills, 25 steam-powered nail machines, and 31 furnaces. In 1856 the firm was reorganized, as Kier and John Lauth removed themselves from the business and were replaced by financier and merchant James. In 1861 the firm of Jones & Lauth was dissolved, Bernard Lauth leaving the enterprise. The company was renamed Jones & Laughlin, Ltd. reflecting a new group of partners that included Jones and his two brothers along with Laughlin and his two sons.

Over the next 20 years Jones's business sense would play a major role in the growth of the iron and steel company that bore his name. Although not a man well versed in technical knowledge, Jones's strength lay in his organizational ability, particularly in his recognition of the advantages of vertical integration. Jones's firm was among the first, if not the actual pioneer, in buying coal lands and making coke in the Connellsville region. When coal was used as a fuel in the Pittsburgh mills, Jones & Laughlin had one of the most extensive mining operations in the local area. And when natural gas was used to heat the furnaces and fire the boilers of Jones's operation, the firm drilled their own wells

and laid their own pipelines. With the erection of two large-capacity blast furnaces in 1860, Jones spearheaded a move to purchase land in the Lake Superior ore region. The company would also own its own limestone quarries, in addition to a warehouse in Chicago. From the mines to the rolls to sales networks, Jones & Laughlin emerged by the 1860s as a vertically integrated business.

In 1860 and 1861 the Eliza furnaces were built by Jones & Laughlin on the north side of the Monongahela River, directly opposite the South Side plant. Boats were used to interchange materials until the construction of the Monongahela Connecting Railroad, which not only gave greater facilities to Jones & Laughlin, but also tied together every manufacturing plant north and south of the Monongahela River. One contemporary remarked that "there are no works in the country that run with greater regularity" than the Jones & Laughlin operation. The efficiency of the Jones & Laughlin operation was tied in part to its labor system. Jones advocated a system of paying wages that he insisted was based on the true value of labor, namely that the workers were to be paid out of product rather than capital. At a time when most manufacturers espoused the so-called wage-fund theory, in which wages were paid out of capital, Jones argued for a sliding scale in which labor rates fluctuated according to market price.

By 1876 the works on the South Side had grown to 75 puddling furnaces, 30 heating furnaces, 18 roller mills, and 73 nail machines. The plant's annual capacity had increased to 50,000 tons. Ten years later Jones & Laughlin would begin steel-making operations as a pair of 7-ton Bessemer converters were installed at the South Side works, with a third Bessemer unit made operational in 1890. And by 1896 the transition from iron to steel making would be complete; in that year 42 basic open-hearth furnaces were put into production. The puddling furnaces so prominent during B. F. Jones's early years in the business were abandoned. On the eve of World War I, the South Side works would produce Bessemer and open-hearth steel, steel bars, sheet piling, power-transmission machinery, cold rolled steel, shafting, rope drives, concrete reinforcement bars, steel-wire nails, barbed-wire nails, fence and special screw wire, template, railroad spikes, light rails and connections, steel-mine ties, structural steel, and plates.

Benjamin Franklin Jones had thus witnessed dramatic changes in the nature and scale of production during his lifetime. Perhaps because he was an organizational innovator, his company, although consolidated with other interests in 1900, remained an independent manufacturer. Thus Jones & Laughlin did not participate in the wave of mergers that climaxed with the sale of Carnegie Steel. Instead, Jones & Laughlin emerged as America's major independent steel producer in the decade before World War I and provided Jones with an enduring legacy.

After the Civil War Jones became increasingly removed from day-to-day plant operations. Always interested in the transportation business and obviously acting out of his own self-interest, Jones was a director of the Pittsburgh & Connellsville and Allegheny Valley railroads and was for some time president of the Pittsburgh, Virginia & Charleston railroad. Upon reaching middle age Jones took on the role of a civic-minded business leader, serving as a director of several leading banks, insurance companies, and hospitals; as an active member in clubs and engineering and scientific societies; and as a participant in the hub of social life in western Pennsylvania.

Jones also found more time to participate in politics, although in part he had a vested interest in campaigning and lobbying for a protective tariff. A Democrat before the war, Jones switched affiliation after the shelling of Fort Sumter and became an ardent unionist and Republican. He helped to organize troops and wrote letters to newspapers and congressmen, arguing for the issuance of greenbacks that would be convertible to bonds. This political aspect of Jones's career took on more significance during the 1880s when he became chairman of the Republican party's national committee. As a staunch supporter of James G. Blaine, Jones met disappointment in the close election in 1884. Later he attempted to heal political wounds among Republican party members, and he campaigned vigorously for the inclusion of a protective tariff stance in future party platforms. His views on the tariff were best expressed in an 1888 article published in the *North American Review.*

Jones's argument was conditioned by his own experiences in the iron and steel trade, particularly the hard times and depressed prices after the tariff of 1846. He asserted that the American people escaped the full economic consequences of that tariff

law because of the demand for surplus grain brought on by the Irish famine, the economic stimulus resulting from the Mexican War, and the discovery of gold in California. But the potential of disastrous consequences from such legislation still existed. The protective tariff, on the other hand, was a source of high wage rates. It is a policy "which inspires labor with hope, and crowns it with dignity; which gives safety to capital, and protects its increase; which secures political power to every citizen, comfort and culture to every home. . . ." Jones felt that President Grover Cleveland's linking the protective tariff with the rise of combinations was erroneous. While Jones asserted that the protective tariff was or should be a permanent fixture in American fiscal policy, he considered trusts as transitory and as a reflection of fierce competition brought on by the lack of adequate tariff rates.

Jones's death in 1903 coincided with the end of one phase in the history of the American iron and steel industry. With the formation of U.S. Steel in 1901 the business was no longer in the hands of a few prominent individuals but rather guided by an emerging corporate bureaucracy. In his obituary in *The Iron Age* it was stated that he was "the most highly respected man in the iron trade," respected by workingmen for his fair treatment and by competitors for his honesty. Jones's life had witnessed a fundamental transition in the American economy, and his business efforts played a major role in the remarkable rise of the nineteenth-century American iron and steel industry.

Selected Publication:

"Iron and Steel," *North American Review*, 146 (1888): 437-439.

References:

American Historical Society, *History of Pittsburgh and Environs*, volume 3 (New York: American Historical Society, 1922);

American Iron and Steel Association, *Directory to the Iron and Steel Works of the United States* (Philadelphia: American Iron and Steel Association, 1884);

Arthur G. Burgoyne, *All Sorts of Pittsburghers* (Pittsburgh: Leader All Sorts, 1892);

Thomas Cushing, ed., *A Genealogical and Biographical History of Allegheny County, Pennsylvania* (1899; reprinted, Baltimore, Md.: Genealogical Publishing, 1975);

Sarah H. Killskelly, *The History of Pittsburgh: Its Rise and Progress* (Pittsburgh: B. C. & Gordon Montgomery, 1906);

Ben Moreell, *"J&L": The Growth of An American Business (1853-1953)* (New York: Newcomen Society, 1953);

George Irving Reed, ed., *Century Cyclopedia of History and Biography: Pennsylvania*, volume 1 (Chicago: Century, 1910).

William R. Jones

(February 23, 1839-September 28, 1889)

by Larry Schweikart

University of Dayton

CAREER: Apprentice (1849-1853), machinist, Crane Iron Company (1853-1855); member, U.S. Army (1862-1865); ironworker, (1865-1872); assistant superintendent, Cambria Iron Company (1872-1873); chief assistant, Edgar Thomson Works (1873-1875); general superintendent, Edgar Thomson Works (1875-1889); consulting engineer, Carnegie, Phipps (1888-1889).

William R. "Bill" Jones, an engineer and supervisor at the Edgar Thomson Steel Works was born on February 23, 1839, and died on September 28, 1889. The son of a Welsh patternmaker and intellectual spokesman for the Welsh community in Pennsylvania, Rev. John G. and Magdalene Jones of Catasauqua, Pennsylvania, Jones was in poor health as a boy. His mother died when he was eight, and his father died six years later. Jones lacked an extensive formal education, although he did acquire a taste for literature.

Jones entered the work force at the age of ten as a molder's apprentice at the Crane Iron Company at Catasauqua, Pennsylvania, a firm managed by David Thomas, known as the "father of the American iron trade." He soon worked his way up to the machine shop and at the time of his father's death was a journeyman machinist. He then entered the employ of James Nelson at Janesville, but soon began to drift from job to job, working for a short period in the machine shop of J. P. Morris & Company in Philadelphia, and at various craftsman and farm positions.

During the panic of 1857 Jones worked as a lumberman in Tyrone, Pennsylvania. But this work did not suit him, preferring as he did the hot blast furnaces. As Jones would later tell visitors to the deafening, searing mill furnaces of the Carnegie works, "This is my home, a good preparation for the next world." He arrived in Johnstown, Pennsylvania, in

William R. Jones

1859, where he worked for the Cambria Iron Company and found his calling in iron and steel. He became a master mechanic and in 1860 went to Chattanooga, Tennessee, to erect a blast furnace. Before Jones could complete his assignment, however, the sectional tensions that would lead to the Civil War increased to such a degree that he returned to

Johnstown. On April 14, 1861, William Jones married Harriet Lloyd of Chattanooga, with whom he had four children, Ella, William M., Cora M., and Charles.

Fearless and tough, Jones enlisted as a private on July 31, 1862, in A Company, 133rd Pennsylvania Volunteers. He soon advanced to the rank of sergeant and fought at Fredericksburg and Chancellorsville, where he served with distinction and though wounded, insisted on staying with his unit. Although Jones's enlistment expired, he reenlisted during the Gettysburg campaign after a short trip back to Johnstown. Jones raised a company of the 194th Pennsylvania Regiment of Emergency Men and was promoted to captain on July 31, 1864, under the command of Gen. Lew Wallace. He left military service on June 17, 1865, and again returned to Johnstown.

After his return to Johnstown, Jones again worked for the Cambria Iron Works and in 1872 was appointed assistant to George Fritz, the general superintendent of the mill. While at the Cambria Company Jones met Alexander Holley, who at the time was the greatest authority on Bessemer steel in the United States. Holley had signed a contract with Henry Bessemer for the exclusive right to use the Bessemer process in the United States, and Jones watched Holley work at close range. Through this experience Jones decided that he wanted to lead the world in producing steel.

Although Jones expected to be named Fritz's successor when he died in 1873, he became the victim in a power struggle, largely over the wage structure at the firm, between Fritz and Daniel Morrell, the general manager. Fritz and Jones favored high wages as an incentive to increase production. When Fritz died, Morrell bypassed Jones and named another Jones–Daniel, a subordinate–as the new mill supervisor. Jones resigned from Cambria with dignity, then met with Holley, who was preparing plans for a new mill, the Edgar Thomson plant. Holley gladly took Jones with him to the new project as chief assistant. Holley expressed to Carnegie how valuable Jones had been, and Carnegie appointed Jones general superintendent of the Edgar Thomson Steel Works in 1875. Carnegie had obtained one of the most important employees in his empire.

Due to Morrell's policies of cutting wages at the Cambria works, Jones found his former employer a perfect recruiting ground. Jones knew all the foremen and workers and their families, and he visited the most able men, luring them to Edgar Thomson with little difficulty. In essence Jones took with him the command structure of Cambria—superintendents Thomas H. Lapsley, the head of the rail mill section; converting works chief John Rinard; machinery supervisor Tom James; top furnace builder Thomas Addenbrook; transportation director F. L. Bridges; and even the main clerk, C. C. Teeter. Jones's corporate raid thus stocked the Edgar Thomson Company with talented and qualified leaders at every level. One anecdote summarizes Jones's contribution to Carnegie's plants. Carnegie, several years after Morrell had shunned Jones, invited Morrell to inspect the Edgar Thomson mills. Morrell gazed upon the impressive output, all manned and supervised by his former employees. When Carnegie asked what Morrell thought, the simple answer paid tribute to Jones: "I can see that I promoted the wrong Jones."

With an army of his best men, Jones immediately set records for output at the Edgar Thomson plant. The first month's profit amounted to $11,000, and at the end of three months Jones had doubled the output of similar mills.

In 1888 Jones became a consulting engineer to Carnegie, Phipps & Company, and he was, like another Carnegie employee, Julian Kennedy, a mechanical genius. He patented a number of devices including new hose couplings, new designs for Bessemer converters, an apparatus for compressing ingots while casting ingot molds, a new method of operating ladles in the Bessemer process, new washers for ingot molds, hotbeds for bending rolls, a device for handling, setting, and removing rolls, a feeding apparatus for rolling mills, a method of making railroad bars, and the Jones mixer, a device capable of holding for the converter 250 tons of molten iron from many different blast furnaces. Jones's knowledge earned him $15,000 a year in royalties alone, and he received an invitation from the Krupp firm in Germany to visit its famous steelworks at Essen; he was one of the first Americans so honored.

The value of the Jones mixer alone would have been important to Carnegie. An 1895 letter from Carnegie to the president of his company, J. G. Leishman, criticized the attempt of the Illinois and Pennsylvania Steel Companies to use the mixer without paying patent royalties. Carnegie estimated its worth to the two companies at $350,000. Carnegie filed suit over this infringement, won a preliminary round, lost an appeal, then had the suit dissolved as his companies were sold to J. P. Morgan in 1901.

As critical as Jones's technological discoveries were, his daily observations and common sense, hands-on engineering also proved important to the company. The short Welshman would march about the plant, giving orders in his colorful accent. Making slight adjustments—supporting a tube here, changing an angle there—Jones instilled in his men the creative interest in their work that made producing record levels of steel a challenge they eagerly sought.

In addition to Jones's royalties his salary was a point of interest, because he was one of the few exceptions to Carnegie's policy of rewarding valuable managers with company shares. When W. P. Shinn threatened to resign as general manager of Edgar Thomson unless Carnegie promised to make him chairman (threatening also to take Jones with him to his new position at the Vulcan Iron Company), Carnegie moved quickly in order to keep Jones. Carnegie offered to make Jones a partner, but Jones did not want to be bothered with the business aspects of steel making. He did allow that he wanted Carnegie to give him "a hell of a big raise." Carnegie told the Welshman to name his price, but thought the figure Jones set—$15,000—was too low. Carnegie gave Jones the same salary as the president of the United States, $25,000. Jones was so excited that he bet Carnegie that his mills could outproduce Cambria; they did, making almost 8,000 more tons of rails than the Cambria mill in 1881.

Jones, along with Carnegie, instituted the "scrap heap" theory at Edgar Thomson, whereby he taught managers to discard equipment, no matter how new, if other processes or machinery rendered it obsolete. In this process Jones threw away millions of dollars worth of machinery a year and yet saved the company still more millions. This practice drew concern from the partners (other than Carnegie) who saw Jones throwing their dividend checks out with the discarded machinery. But Carnegie insisted on the policy, and Jones, who was not a partner, could only benefit from having the best and newest machinery. Carnegie was also convinced by Jones's arguments on behalf of the Thomson basic process, instituted at the Homestead works by Julian Kennedy. Carnegie ordered the presi-

dent of Carnegie, Phipps & Company, W. L. Abbott, to implement the Thomson process at four open-hearth furnaces.

As strongly as Jones and Carnegie agreed on installing new equipment, they disagreed over labor costs. Jones had always fought to keep his laborers' scale high, while Carnegie was the archetypal cost-cutter. Jones, who had demanded a high salary for himself, also persuaded Carnegie to keep wages high for his men. After Jones took over management of the Edgar Thomson plant, he ended, over Carnegie's protests, the two twelve-hour shifts in favor of three eight-hour shifts. Jones estimated that the savings from reduced absenteeism and fewer accidents would compensate for a one-third increase in the labor force. In 1881 Jones credited the increased production of the Edgar Thomson works to the new labor policy. In the end he convinced Carnegie that higher wages bought better workers, and Carnegie adopted Jones's plan in all his mills. Jones's idea was the norm for labor policy in the Carnegie plants until 1888.

On January 1, 1888, when no other steel manufacturer adopted the three shift policy, Carnegie ordered Jones to return to the two twelve-hour shifts. Carnegie appeared in person before Jones's men and persuaded them to adopt the old shift with a sliding scale over the three shift policy at lower wages. But the trouble had not ended. At the Homestead works in 1889, the Amalgamated Association of Iron and Steel Workers of North America went on strike over the sliding scale. Carnegie, in Britain for his annual vacation, left orders for Jones to control his men. Abbott, however, attempted to break the strike and failed; shortly thereafter he signed a compromise that recognized the union in return for the union's acceptance of the sliding scale.

Even passing acquaintances of Jones proved to be assets to Carnegie's company. At a grocery store in Braddock, Jones met a young clerk named Charles Schwab, who asked for a job at Edgar Thomson. Jones found a spot for Schwab, and within six months he was superintendent in charge of construction of the new blast furnaces. Carnegie liked Schwab so much that a report delivered by Schwab stood a better chance of being approved. It was a relationship that foretold the future of Edgar Thomson, as Carnegie soon promoted Schwab to superintendent of Homestead and his stock rose after the 1888 strike. Less than six months later, on September 26, 1889, Jones was fatally injured. He had been standing near one of the Edgar Thomson blast furnaces when it suddenly exploded, hurling him from the platform on which he was standing to the tracks below. He smashed his head on an ore car, never regained consciousness, and died two days later, on September 28, 1889.

The Carnegie companies could not measure Jones's contributions. In addition to his numerous patents, which "Dod" Lauder convinced Mrs. Jones to sell to Carnegie Brothers for $35,000, Jones had implemented his inventions at the plants. These meant untold millions ultimately in the production of steel, and the Carnegie companies got them for small yearly royalties. Perhaps the most significant of his contributions was the Jones mixer, which revolutionized steel making. More than his patents, Jones had given the Carnegie companies great energy, enthusiasm, loyalty to, and from, his men, and a grasp of iron and steel that few others shared. Carnegie's biographer, Joseph Wall, appropriately labeled Jones "the greatest steelmaker in America." On the day of his funeral, Braddock's stores, schools, and mill shut down while the city mourned. At Johnstown a memorial was held for him.

Jones had been a humanitarian, actively supervising the flood relief at Johnstown, where 3,000 people had perished. Immediately after the flood he delivered three boxcars full of provisions: then, as other workers arrived to aid the victims, he saw to their feeding and care at his own expense. He stayed in Johnstown for two weeks helping to repair the damage.

Carnegie honored Jones by calling him virtually irreplaceable. Indeed, except for Carnegie himself, no American made such an impact on the iron and steel industry. Jones worked hard, fought for the welfare of his men, applied his intellect and engineering skills to devices that would improve the quality and quantity of steel making, and donated his charitable energies to his community. William R. Jones was in all ways an American hero.

References:

Andrew Carnegie, *The Autobiography of Andrew Carnegie* (Boston: Houghton Mifflin, 1920);

William T. Hogan, *Economic History of the Iron and Steel Industry in the United States*, volumes 1 and 2 (Lexington, Mass.: Lexington Books, 1971);

Joseph Wall, *Andrew Carnegie* (New York: Oxford University Press, 1970).

Jones, Lauth & Company

by John A. Heitmann

University of Dayton

An organizational precursor to the Jones & Laughlin Steel Corporation, the history of Jones, Lauth & Company reflects the nature of business partnerships in the iron industry prior to the Civil War and demonstrates the significance of finance in fostering the growth of manufacturing enterprise.

The origins of Jones, Lauth & Company can be traced to June 1853, when Bernard and John Lauth purchased land on the south bank of the Monongahela River on what would become the site of the South Side Jones & Laughlin plant. The Lauth brothers had acquired iron-making experience in Ohio but for some reason had returned to their home of Pittsburgh in 1852. After securing the property they began erecting what was called the American Iron Works, initially consisting of four puddling furnaces, two heating furnaces, a guide mill, muck rolls, and a crocodile squeezer. Operations began in September 1853, and the output of the furnaces was approximately seven tons per day.

The Lauth brothers' firm soon attracted the attention of Benjamin Franklin Jones, a Pittsburgh entrepreneur and businessman who quickly saw the commercial and manufacturing potential of the newly organized American Iron Works. Jones was born in Claysville, Pennsylvania, in 1824 and at age eighteen was employed as a receiving clerk in the Pittsburgh office of the Mechanics' Line, whose principal owner was Samuel Kier. This venture operated a line of canal boats between Pittsburgh and Philadelphia and employed a mechanical lift to carry the barges over the Allegheny Mountains. Jones so impressed Kier with his abilities that in 1845 he was offered an equal partnership in the business. Jones and Kier soon became interested in iron making and in 1847 purchased an interest in an iron furnace and forge near Armagh, Pennsylvania. This operation was worked for several years with little financial success; profits lagged for several reasons, including its poor location, high transport

charges, lack of fuel, poor quality of its iron, and the general depressed state of the iron business during this time. By 1853 Jones was familiar with the intricacies of the iron trade and realized that the Lauth brothers' firm had many of the advantages that his first venture in manufacturing did not.

On December 3, 1853, a partnership was formed between Benjamin Franklin Jones, Samuel Kier, Bernard Lauth, and John Lauth to manufacture and roll iron. The agreement was to bind the parties together for five years, and it put Jones in charge of the warehouse, accounts, and finances. Bernard Lauth was to supervise the roller mills, and his brother John was given responsibility over the small mills and turning department. Kier had no day-to-day function in the operation of Jones, Lauth & Company, which was capitalized at $20,000. One stipulation of the partnership agreement was that all profits would be returned to the capital stock, enabling the company to expand steadily during the 1850s. By the end of the decade operations consisted of 5 trains of rolls, 25 steam-driven nail machines, and 31 furnaces.

In March 1855 Kier sold out his one-fourth interest to Jones, and a year and a half later James Laughlin entered the partnership. Laughlin, born in Ireland in 1807, had immigrated to the United States in 1828 and with his brother had formed a provision business in Pittsburgh and Evansville, Indiana. Laughlin had been instrumental in establishing the Fifth Ward Savings Bank in 1852, which later became the First National Bank of Pittsburgh and by 1856 had loaned more than $12,000 to Jones & Lauth Company. In August 1856 Laughlin became a partner in the firm, and John Lauth dropped out of the venture, selling his share to his brother and Jones. That same year the partnership was extended five years.

In 1861 the firm of Jones & Lauth dissolved as Bernard Lauth left the business. The partnership

The ironworks of Jones & Laughlin in 1880 (courtesy of Jones & Laughlin Steel Corporation)

was restructured to include James Laughlin and his two sons along with B. F. Jones and his two brothers. The firm was called Jones & Laughlin Limited and later became an integral component of one of the largest, independent steel producers in the United States.

References:

J. P. Lesley, *The Iron Manufacturers' Guide* (New York: John Wiley, 1859);

Ben Moreell, *"J & L": The Growth of an American Business (1853-1953)* (New York: The Newcomen Society, 1953);

George Irving Reed, *Century Cyclopedia of History and Biography: Pennsylvania*, volume 1 (Chicago: Century Publishing, 1910).

William Kelly

(August 21, 1811-February 11, 1888)

by Ernest B. Fricke

Indiana University of Pennsylvania

CAREER: Junior partner, McShane & Kelly (c. 1829-1846); senior partner, Kelly & Company (1847-1858); senior partner, W. C. Kelly (c. 1863-c. 1878).

William Kelly was a mid-nineteenth-century American inventor noted for the development of the pneumatic process of iron and steel making. This process was almost identical to, and is best known as, the Bessemer process. Kelly and Henry Bessemer each held patents, which were united into the Pneumatic Steel Association, a corporation which, along with its successors, controlled to a large degree the sales and prices of Bessemer steel in the late nineteenth century.

Kelly was born on August 21, 1811, in Pittsburgh, Pennsylvania, the son of Irish immigrants John and Elizabeth Kelly. His father had arrived in Pittsburgh as a nineteen-year-old in 1801. Pittsburgh was already at the center of a major migration route west of the Appalachians, sitting as it did at the wellspring of the Ohio River, which was the leading mode of transportation to the West. Pittsburgh was also a rising manufacturing center, providing westward-moving migrants with items difficult to haul over the mountains from the East. Some of these products included rafts, steamboats, wagons, wagon wheels, and glass; after 1805 iron foundries appeared, followed by the establishment of several nail factories. Even in this dynamic environment John Kelly possessed at least two advantages: he could speak English, and he had money. He quickly began to make money, most likely by real estate, and is credited with building the first brick house in Pittsburgh.

On March 25, 1805, John Kelly married Elizabeth Fitzsimons, another Irish native, and William was born six years later. Kelly received a typical Pittsburgh education in the common schools, but due to his family's financial standing he was

William Kelly (courtesy of American Iron and Steel Institute)

also able to attend the Western University of Pennsylvania (now the University of Pittsburgh). Family tradition maintains that he learned metallurgy at Western, but it is obvious that he knew very little about the subject when he first began experimenting many years later. He did, however, evidence a strong interest in chemistry. He also demonstrated mechanical talent, constructing a waterwheel and a rotary steam engine. Kelly continued his interest in science throughout his life and read widely in it, but he never joined any scientific society.

When he left the university Kelly moved into the world of business, becoming the junior partner with his brother-in-law in the dry goods and commission firm of McShane & Kelly. They were later joined by Kelly's younger brother, John F. Kelly. The company was headquartered in Philadelphia, but Kelly worked out of Pittsburgh, making annual visits to collect money from customers and, if possible, to secure more orders. In this era before the railroad, travel was mainly by horseback and stagecoach, and by steamboat and occasionally by canal. His trips took him into western Pennsylvania, Ohio, Indiana, Kentucky, and Tennessee. He survived the hazards of constant travel, remaining in good health, and was described as being openhearted, unselfish, optimistic, and self-confident. The fact that his firm survived and even prospered during the panic of 1837 and during the volatile economic swings brought on by the death of the Second Bank of the United States contributed to his confidence.

A series of occurrences in the mid 1840s led Kelly to change his vocation. On a trip to Nashville in summer 1846 Kelly met sixteen-year-old Mildred A. Gracey, whom he found quite attractive. Her father, James N. Gracey, was a tobacco merchant and steamboat owner, who lived near Eddyville, along the Cumberland River, in Kentucky. While visiting Mildred and her father in Kentucky, Kelly was exposed to the iron foundry business which had been in operation there since 1837. When a fire destroyed the McShane & Kelly warehouse, Kelly took his brother John to Kentucky to survey the area, no doubt comparing opportunities there to what was available in Pittsburgh. Liking what they saw, the two brothers entered the iron business in 1847 near Eddyville. Kelly also established a family in Eddyville, marrying Mildred Gracey.

Acting as Kelly & Company, the brothers purchased the old Cobb furnace and some 14,000 acres of land up the Cumberland River from Eddyville. The land contained both surface iron ore and timber for charcoal. That same year they built Union Forge so that they could manufacture finished products. When these initial efforts proved profitable, they constructed an additional works, the Suwanee Furnace, which was completed in 1851.

It was William Kelly's job to operate the furnaces while John Kelly handled the finances. Kelly followed the usual frontier method of iron making by using charcoal. In 1847 charcoal produced 75 percent of the nation's pig iron. Trees for the charcoal were readily available and cheap, and the smelting process, polished over the centuries, produced a consistently good quality of pig and wrought iron. The smelting process was relatively simple. About 1,500 pounds of iron ore were packaged between layers of charcoal. As the fuel burned, more charcoal was added until the ore was smelted. Good quality iron was increasingly in demand as the center of the American population moved west of the Appalachian Mountains. The newer smelting methods, which originated in Great Britain and used coal or coke as fuel with new furnace designs, were slowly being adopted in the eastern United States, primarily the eastern part of Pennsylvania. The rapidly changing industry was a challenge to the relatively inexperienced Kellys.

Out of the pig iron and blooms produced by the Suwanee Furnace, two widely marketable products were made. Large kettles of cast iron were sold in the Deep South and Cuba for use in sugar making. (James Gracey had a warehouse in New Orleans which might have been used for the Kellys' overseas trade.) Also from the forge came boilerplates to be used on Ohio River steamboats.

The firm was prospering, but not at the highest possible level. William was new both to the everyday life of the region and to the mysteries of the iron-making craft. The traditional charcoal method of iron making was labor intensive (one furnace requiring at least 40 workers), causing Kelly to rely on the still-thriving Kentucky slave economy. Kelly professed to feel guilt at using slave labor but argued that labor conditions (whites willing to work at an iron works were scarce) prevented any alternative. The rising national tension over slavery and the increased costs associated with the "peculiar institution" led Kelly to try a different solution to his labor problems.

Through a Philadelphia business acquaintance Kelly contracted to hire ten Chinese laborers and workmen on a trial basis. For a year they were treated on the books and at the workplace as blacks—they were paid $6 per month and provided with room and board. Kelly was satisfied with their work, and he arranged to replace the entire work force with the Chinese by hiring 50 more. This never happened due to circumstances of China's foreign policy at the time. Kelly's problems continued; his lack of experience in handling slaves caused anguish, the trial use of the Chinese lowered profits, and the purchase of slaves continued to eat into his investment capital.

In addition to labor problems Kelly's inexperience in the iron-working trade led to some poor business decisions and to mismanagement. Errors grew out of his lack of understanding of different types of iron ore and the amount of timber required to provide enough charcoal for his furnaces. The surface ore on the land the brothers had purchased was hematite, a good-quality ore. The business was prospering, which led them to build the new Suwanee Furnace, further increasing the company's output. Within a year they began to run out of conveniently located surface ore. Greater productivity also led to the faster than anticipated thinning of the timber near the furnace, raising the specter of increased transportation costs for charcoal. To avoid moving the furnace to another location the company decided to mine the nearby, but deeper, ore, which turned out to have more impurities than the surface ore. The impurities in the ore, little particles of flint, did not melt in the furnace. (These particles were called shadrach by the local iron men, for like Meshach and Abednego in the Old Testament, Shadrach did not burn when cast into the fire.) As the particles fell to the bottom of the furnace, they clogged it, making it difficult to pour the pig iron. Costs were rising, but only a poor-quality iron was being produced. The firm was also heavily burdened with debt from local banks and individuals. Faced with the possibility of losing the business, Kelly searched for new solutions and thus became the inventor of a new way of making iron and steel.

What Kelly brought to the search that other ironmasters of his day did not possess was an interest in science and some knowledge of chemistry. The problems that confounded Kelly were the burn-resistant particles of flint and the increasing scarcity of charcoal. At a time of crisis he was able to recall an idea he first had when working at the forge in 1847. At that time he talked over what he termed "air boiling" with a local furnace builder. The idea was, as Kelly later phrased it, "of converting fluid pig metal into malleable iron, with the aid of a strong blast of air.... My object was to drive off the carbon in the iron...." While his understanding was not scientifically precise, his thesis was correct. The oxygen in the air would combine with the carbon in the iron mixture and in so doing would create a great amount of heat. The heat would convert the pig iron into wrought iron as the air bubbled through. He had made a brief experiment in October 1847, using large posted drawings as an aid for the forge workers, who were convinced that the cold air blast would ruin the pig. The brief trial was more or less successful but was shunted as his attentions were focused on the primary goal of keeping Kelly & Company prosperous.

In November 1851 and continuing for a period of 18 months, Kelly began a series of experiments designed to use less charcoal but to provide enough heat to burn all impurities, including the flint. The results were erratic. Kelly thought the problem was to discover the right amount of air pressure and the time and speed at which the air should be pushed through the molten iron. He changed from cold to hot air and built new converters and machines to create air pressure. He also tried a variety of iron ores available in the vicinity. He was not, however, a metallurgist and did not understand why some ores available locally in small quantities were better suited to the process than others. Today we know that it was because they contained manganese. The other, more readily available, local ores had high sulfuric and phosphoric contents and did not produce good results.

The experiments were not conducted in secret. In fact Kelly gave public exhibitions of the process in the hope of whetting interest among local iron makers, apprentices, and others in the trade. He needed help from skilled craftsmen who would be able to design and build the equipment he thought was necessary to continue his experiments. He did not have the money to import such men and equipment from the eastern United States, and he thought local craftsmen might be willing to help him with their insights, experience, and suggestions if he could convince them that his idea was viable.

Because of his willingness to demonstrate it to all comers, the "air boiling" process became well known throughout the Cumberland District. But conducting the experiments cost money. A family crisis soon ensued, triggered by the firm's already insecure financial situation, and bankruptcy became a real possibility. James Gracey worried about his daughter's and his grandchildren's future, tied as they were to a man who had already made a number of poor business decisions and who went about conducting ludicrous experiments while ignoring the expertise of experienced ironmasters. The local doctor was called in to judge whether he was completely sane and rational.

When the doctor found him to be rational, Gracey agreed to loan Kelly money in return for his promise to concentrate on applying good business procedures in the firm of Kelly & Company. His experiments were to be placed on the back burner and literally moved to the back part of the estate, out of the public eye. His occasional successes were put into finished products, usually boilerplates. As word of Kelly's air boiling experiments spread, customer response encouraged more secrecy. The Cincinnati firm of Shreve, Steele & Company threatened never to order from Kelly again if they learned that the goods were made by the "newfangled" process.

At this point in his life Kelly was a moderately successful businessman, senior partner in an obscure iron company having some financial difficulties. He had family responsibilities to his wife and children. But he also had his avocation as an inventor: "I flattered myself I would soon make it the successful process I at first endeavored to achieve...." He renamed his invention the "pneumatic process" and continued, without success, to seek support for his experiments when business trips took him out of the area. In 1854 he visited James Ward of Niles, Ohio, who had built the first rolling mill west of Pittsburgh. As the door closed upon Kelly's departure, Ward noted to a friend that Kelly was "crazy." Kelly persisted, for he knew that if he could get it right, if he could make wrought iron cheaper than anyone else—all it cost was air, not fuel—then he would have provided for the financial security of his family.

In 1856 Kelly finally kindled the interest of iron maker Daniel Morrell. In 1855 Morrell had been made general superintendent of the Cambria Iron Company in Johnstown, Pennsylvania. This firm had recently suspended operations, and Morrell's job was to protect the interests of the major Philadelphia Quaker creditors. Morrell sought technological changes to improve Cambria's production and product; he became interested in the work of John Fritz and William Kelly, among others. Fritz, Cambria's previous general manager, redesigned the mill for rolling iron rails in 1857, an extremely profitable change as the speed of the operation was increased, the labor costs were reduced, and the quality of the final product was improved.

Hoping for a second success, Morrell invited Kelly to come to Johnstown to build a converter for demonstration purposes. At about the same time as he received the welcome news from Johnstown, Kelly also learned of developments abroad which were to cause him to delay his response to Morrell's invitation. In August 1856 Henry Bessemer read a paper to the British Association for the Advancement of Science entitled "The Manufacture of Malleable Iron and Steel without Fuel." In a letter of response to the report, Kelly described his experiments which he had begun in 1851. The letter continued:

> I was surprised to notice in *Scientific American* of the 13th of September an account of a similar process of converting pig iron into malleable iron, claimed as the discovery of Mr. Bessemer of London, and made within the past two years, the process not differing in the slightest from that I had in practical operation nearly five years since.

Bessemer sought patents in Britain, in Europe, and in the United States for the process and for a number of machines he developed to make it practicable. When Kelly learned of the application for the American patent, he submitted one of his own. Bessemer's patent was granted in November 1856; Kelly's application led to a reexamination of the situation. In April 1857 the United States Patent Office granted an "interference" patent to Kelly on the basis of priority. In effect, this meant that the patent for the process was taken from Bessemer and given to Kelly. Bessemer's patents for the machines were upheld.

Happy with the results and basking in his enhanced reputation, Kelly went to Johnstown. At the Cambria Iron Works he was given a corner of the yard and the help of a young man or two to build a converter. Over 30 years later one of the helpers, John Fry, recalled the experience.

> The apparatus was assembled from scrap heap material and was indescribably primitive. The entire operation consisted of blowing air into the molten iron through a blast pipe thrust from above into the liquid metal. The only agreeable result was to change this very poor quality of iron for most purposes into a practically worthless one for any.

The experimental converter used by Kelly to improve the steel-making process

The metal chilled. Later the metal was kept hot, "but the blast could not be made to penetrate the iron and the attempt was promptly abandoned."

Kelly continued his efforts into 1859. However, his intensity faded; his visits to Johnstown were infrequent, and the duration of the trials was short. The results improved. But when one of the trials started a fire which burned down the building surrounding his work area, the experiments ceased.

The panic of 1857 helped to switch Kelly's interests elsewhere. The inability of the banks to lend money early in the year deteriorated into a depression; the price of iron ingots fell, placing Kelly & Company in a very vulnerable position. He had borrowed money from banks as distant as Louisville, Cincinnati, and Philadelphia, all of which were demanding the interest to be paid. Kelly was unable to collect the money owed to him and was again forced to borrow from his father and father-in-law. In addition, he began transferring ownership of some slaves and the stock in his store to those who would buy his property. There was no one to buy the big items—the furnaces, forge, and land. Finally in 1858 he sold the patent to his father for $1,000.

Kelly's efforts were to no avail. On April 20, 1858, he declared bankruptcy and handed over his assets to a friend. Lawsuits by creditors carried on into the Civil War. It was not until the end of December 1859 that the original furnace, Union Forge, Suwanee Furnace, and the land were sold. Kelly valued the assets at a little over $100,000 but was able to liquidate them at the rate of 2 1/2¢ on the dollar.

In order to remove himself from the jurisdiction of the Kentucky courts and to avoid the worst effects of poverty, he moved his family to New Salisbury, Ohio, 60 miles west of Pittsburgh. During 1858 and 1859 he made periodic visits to Johnstown. In 1860 economic troubles again hit the nation. Kelly's financial plight worsened when he lost control of the patent upon his father's death, at which time it went into the estate of Kelly's sisters. As they had serious reservations about their brother's business acumen, the patent was eventually placed in trust for his children.

In 1861 there was renewed interest in the pneumatic process. By spring America was irrevocably involved in the Civil War, which quickly increased the demand for high-quality steel. Bessemer was using the process in England for making steel. Kelly had the American patent on the process, and there were questions raised in the columns of *Scientific American* about the availability of the process to Americans in the iron trades. Kelly replied, "I would say, that the New England States and New York would be sold at a fair rate." He had already sold patent rights to iron men in Pennsylvania and Michigan and the war ruled out selling them in the South.

One of the men to whom Kelly had given patent rights was Eber Ward, a man of many economic interests who had recently opened ironworks at Wyandotte near Detroit and in Chicago. Zoheth Durfee, Ward's technical expert on iron, and Daniel Morrell were the other partners. Money did not pass hands, but the terms of the agreement were that the three would explore the steel-making aspect of the process. If they were convinced that it was possible, they would build a plant and begin to manufacture steel. Kelly, of course, would share in the profits.

In fall 1861 Zoheth Durfee sailed to England to investigate Bessemer's use of the process and to evaluate the Bessemer patents on equipment. His cousin, William Durfee, was hired to begin construction of a converter in Wyandotte, Michigan. By spring 1862 a Bessemer converter had been sent from overseas to Johnstown, and Kelly returned to western Pennsylvania to reactivate his experiments using the new equipment. In his old age James Geer, one of Kelly's young helpers at the time, remembered that the other workers at the Cambria Iron plant did not rejoice upon the return of the "Irish Crank." Kelly was nervous on the day of his first experiment with the newly acquired converter. Two hundred workers gathered to watch. Remembering past failures, the "Irish Crank" told the engineer in charge of the air pressure to make the blast as strong as possible. He shut off the safety valve, and the resulting pressure blew the hot, fiery, golden contents into the sky. Kelly's "fireworks" remained a joke for many years after.

He quickly made preparations for another attempt. Again a large audience was on hand. The pressure was less this time, but Kelly still did not know how long to leave it on. As sparks of molten iron flew out of the mouth of the converter, he dashed from one hot spark to another for a half hour, banging them with a hammer, until he found one that would not crumble. When he found one that could be hammered flat the blast was shut off. Later, when the metal had cooled off, he hammered part of it into a plate. The story of Kelly's hopping about and hammering became a legend among local ironworkers.

This final visit to Johnstown, where experiments "met with the usual number of encouraging failures," ended Kelly's work with the pneumatic process, but not his connection with it. The group which owned the patent rights was ready to use the process to make steel. Zoheth Durfee had discovered that the degree of success in steel making in England was due to the invention of another Englishman named Robert Mushet. By adding manganese to the wrought iron of the pneumatic process, enough carbon was restored to create steel. To produce steel what was needed was neither the Kelly nor the Bessemer process but, as William Durfee was to call it, the Bessemer-Mushet-Kelly process.

After consideration the group obtained the United States rights to the Bessemer and Mushet patents. They would continue to use and improve the facilities already built at Wyandotte and there would lay the groundwork for large-scale commercial production of steel, using the pneumatic process. Kelly never gave advice or worked at Wyandotte. His inventive genius for ideas had run its course and, at any rate, did not seem to extend in the direction of machinery.

The last step for the owners of the patent rights was to incorporate as the Kelly Pneumatic Process Company in May 1863. Even though it carried his name, Kelly was not a member of the corporation nor did he exert any control. He was, however, guaranteed a certain share of the anticipated future profits. The economy in the North was booming in the third year of war, and Kelly's financial problems in the aftermath of the bankruptcy were diminishing. Kelly no longer had an occupational incentive to remain in Ohio, so he moved his family back to Kentucky. They did not return to Eddyville, but instead settled in Louisville.

During this period of relocation the personal financial value of the patent deteriorated. The Kelly Pneumatic Process Company acquired the American rights to Mushet's patent, but not Bessemer's, which went to a firm in Troy, New York. In 1864 and 1865 each company succeeded with its experiments and produced steel commercially. In the process each infringed on the patents of the other. The desire of each to manufacture steel, to avoid litigation, and to collect royalties from others for the use of the patents before they expired led to the merging of the patents into a single company. Under the new arrangements the Kelly group would receive 30 percent of the profits, thereby reducing Kelly's interest by 70 percent.

The new company was known formally as "The Trustees of the Pneumatic or Bessemer Process of Making Iron and Steel." It organized a corporation, the Pneumatic Steel Association, which decided that the finished product should be sold under the Bessemer name rather than Kelly's. The reasoning was that the product was being sold successfully in Europe under the Bessemer name, which meant that the company could sell in the United States at the same high price being asked for this imported steel.

The use of the Bessemer trademark was difficult for Kelly to swallow. He could not even balance the loss of fame with riches, as the new arrangements did not help his financial pinch. In 1870 he claimed that expenses exceeded income by $9,100. In his 1857 letter to the *Scientific American* he also stated:

> I have reason to believe my discovery was known in England three or four years ago, as a number of English puddlers visited this place to see my new process. Several of them have since returned to England and may have spoken of my invention there.

He certainly suffered a sense of injustice, of being the victim of what one iron maker termed "the crime of the Nineteenth Century."

There is a family story, revealed well after Kelly's death, which describes him studying a picture of Henry Bessemer brought back from England by Zoheth Durfee. In the anecdote Kelly exclaimed that in the picture Bessemer closely resembled an English worker Kelly had briefly hired at the Suwanee Works. He did not want to endure the possible humiliation of losing a court contest to make good his identification, so he did not pursue the matter and asked that it be kept a secret.

The question remains: how did both Bessemer and Kelly create the air process? Was it by deduction, following observations, as both claimed, or did they steal the idea from another? Bessemer almost certainly never came to the United States. He had serious problems with seasickness, even when crossing the English Channel. However, his brother-in-law did come to New Orleans in 1854 on Bessemer business. He could easily have heard of "air boiling" and could have visited Suwanee. Both Bessemer and Kelly could have found written accounts of earlier attempts by iron men in England and America who had also noticed the burning of iron. A more recent theory claims that Kelly's borrowing could have been from the Chinese. If his Chinese workers had had any skill in iron making, they would have known of a decarbonizing process using cold air which had been in use in China since the second century B.C. In southwestern Kentucky the tale went round that Kelly got his idea from an engineer constructing a rolling mill nearby. The engineer was Simeon Norvell Leonard, Sr., from England, a classmate of Henry Bessemer.

Contemporaries of the two men were more likely to believe in coincidence, historical determinism, or God's will than in any of the above attacks on their character. As William Durfee expressed it, the answer was simple:

> When in the fullness of time the world is prepared for a decisive advance in the sciences or arts, an overruling power indicates simultaneously to minds separated oftentimes by continents and oceans some way to satisfy the growing needs of the world, and all to whom such revelations are given, who contribute to their promulgation and success, are entitled to an honorable recognition and reward. . . .

Kelly was to make one last impact on the steel industry. In 1870 the Bessemer and Mushet patents were to expire, and Kelly's the following year. Kelly applied for an extension, chiefly on the grounds that the process was valuable, but that he had received little financial return. Over protests of iron manufacturers who were awaiting the end of the patent rights, and of the railroads who were major purchasers of the steel, the patent commissioner issued a 7-year extension.

Against the advice of family and friends Kelly kept to his agreement under the contract and turned the extension over to the Pneumatic Steel Association. Subsequently the company began to prosper. The country was in the last years of an economic boom that would end with the panic of 1873. As the nation came out of the resulting depression, the railroads were a growth industry building new track and replacing the less durable iron rails with rails made of steel. Kelly's share in the association's earnings until the patent's final expiration on June 23, 1878, is unclear. One estimate, given by someone whose figure for Kelly's royalties before the extension was the same as that submitted to the patent commissioner in 1870 by Kelly, was $25,000. Another source, speculating that his share was much higher, placed it at over $450,000.

After 1878 Kelly's efforts went into creating a company to manufacture axes in Louisville, a company which he hoped would provide a business enterprise for his sons. The axes bore the trademark "W. C. Kelly," the initials of his eldest son, William. Under his sons the company moved into West Virginia and became in the early twentieth century one of the larger edge tool firms in the country.

As Kelly grew older and royalties came in he spent more of his time and energy on banking and investments. The most lucrative type of investment was real estate. Again his business acumen came into question. He had acquired land and houses along the riverfront in Louisville at a low price. Although he purchased the real estate from his uncle, the low price was not a sacrifice to help out a relative but reflected the decline in river traffic. However, the Chesapeake & Ohio Railroad decided to build a line along the river, raising the issue of the price Kelly should get when he resold the properties. Friends advocated the highest possible price. Instead he sold at a profit but at a lower price than the Chesapeake & Ohio could have been forced to accept. Kelly's actions may have reflected his upbringing, the religious influence on his life of faith and good works. In the case of the river land he said, "If I thought I would live always and had no accounting to make in the end, I might ask a larger sum for them than they are worth, but I do not want to close my life in possession of any property or value that is not justly my own." The same reasoning was applied when the Pennsylvania Railroad, through the efforts of Daniel Morrell, offered Kelly and his wife free travel passes. He would not accept on the grounds that he could do nothing for the road in return.

At the end of his life both public and private recognition came to Kelly. Publicly it occurred in fall 1887 when he was guest of honor at the Masonic Temple Theater. The speaker was another William Kelley, a national politician from Pittsburgh who lobbied for a protective tariff on iron and steel. In his speech, "Pig Iron," Kelley described William Kelly as one "who by his genius, gave the world the greatest invention" of the age.

In 1888 Andrew Carnegie paid Kelly a private visit shortly before Kelly's death. The two were not acquainted, but by this time Carnegie was the best-known user of the Bessemer process. His visit was recognition that the process was at least partly Kelly's. A brief illness ended Kelly's life on February 11, 1888. He was seventy-six years old.

References:

John Newton Boucher, *William Kelly: A True Story of the So Called Bessemer Process* (Greensburg, Pa.: Privately printed, 1924);

Richard A. Burkert, "Iron and Steelmaking in the Conemaugh Valley," in *Johnstown: The Story of A Unique Valley*, edited by Karl Berger (Johnstown, Pa.: Johnstown Flood Museum, 1985);

Joseph Green Butler, *Fifty Years of Iron and Steel* (Cleveland: Penton Press, 1922);

Jeanne McHugh, *Alexander Holley and the Makers of Steel* (Baltimore, Md.: Johns Hopkins University Press, 1980);

Elting E. Morison, *Men, Machines, and Modern Times* (Cambridge, Mass.: M.I.T. Press, 1966);

James M. Swank, *History of the Manufacture of Iron in all Ages, and Particularly in the United States from Colonial Times to 1891,* second edition (Philadelphia: American Iron and Steel Association, 1892);

Peter Temin, *Iron and Steel in Nineteenth-Century America: An Economic Inquiry* (Cambridge, Mass.: M.I.T. Press, 1964);

Theodore A. Wertime, *The Coming of the Age of Steel* (Chicago: University of Chicago Press, 1962).

The Kelly Pneumatic Process Company and the Steel Patents Company

by Ernest B. Fricke

Indiana University of Pennsylvania

The dominant method of steel production from around 1870 until the beginning of the twentieth century was the Bessemer process. The Kelly Pneumatic Process Company, organized in 1863, was an association of steel producers who introduced the process and, consequently, the age of steel into the United States. Through a series of formal associations, culminating in the formation of the Steel Patents Company in 1890, the group grew in size and for nearly 30 years and in various ways controlled entry into the steel industry. One of the methods used to limit participation in the steel industry was through the control of patents on machinery and methods.

Commercial manufacture of steel in the late nineteenth century was beset with many problems. There were technical problems, as the sciences of chemistry and metallurgy were not fully developed disciplines, and experimentation functioned more on trial and error than on solid scientific methods. There were also legal problems over patents, since experimentation was taking place on both sides of the Atlantic. Henry Bessemer, working in England in the mid 1850s, discovered how to blast air through molten iron to remove the carbon and produce wrought iron. An iron maker in Kentucky, William Kelly, had discovered the same principle a few years earlier, but it was Bessemer who introduced the method to the world when describing it in a paper read at a scientific conference in 1856. The manufacturing world learned of it when Bessemer applied for patents for the process and the machines he designed to utilize it. Thus, the process was identified with Bessemer's name. He patented the process and machinery in Europe as well as in the United States. Kelly also filed for the American rights to the process, claiming he had priority rights, and in 1857 he was granted the United States patent. However, Bessemer retained rights for his machinery. To

further complicate matters, another Englishman, Robert Mushet, patented a process to produce steel by replacing the carbon in molten iron by using spiegeleisen, a compound of iron, manganese, and carbon.

Neither Kelly nor Bessemer had perfected what Kelly called the pneumatic process, and both had concentrated their experimental efforts on wrought iron rather than steel. A small minority in the iron trade was concerned with steel making; among them were three Americans who were to form the Kelly Pneumatic Process Company. Daniel Morrell, general manager of the Cambria Iron Company of Johnstown, Pennsylvania, had allowed Kelly to use his plant to present the pneumatic process and to continue experiments on it from 1857 through 1859. Eber B. Ward, a friend of Morrell's, entered iron making in 1855 by building a furnace in Wyandotte, Michigan, near Detroit, and expanded operations with a rerolling mill in Chicago in 1857. The third in the group, and Ward's advisor in these enterprises, was Zoheth Durfee, a blacksmith and occasional commission merchant from New Bedford, Massachusetts.

As the Civil War was beginning in the United States, these three men were intrigued by British advances in making steel using the Bessemer process. They were determined to use the process successfully in the United States using domestic ores, manufacturing steel for use by both the military and the railroads. With this goal in mind, they acquired the rights to the Kelly patent in 1861 and had him renew his experiments in Johnstown. In 1862 they hired an engineer and builder, William Durfee (Zoheth Durfee's cousin), to construct a converter at Wyandotte. Finally, they attempted but failed to acquire the rights to Bessemer's patented machinery. Despite this last failure, they were optimistic that their efforts would eventually succeed, and in

May 1863 they formed the Kelly Pneumatic Process Company.

In addition to Ward, Zoheth Durfee, and the Cambria Iron Company, the new company included two other firms: Lyon, Shorb & Company and Park Brothers & Company. William Kelly was not a member but was assured a portion of the profits. Once established, the company again failed in an attempt to acquire the rights to the Bessemer patents. However, it did succeed in acquiring an option in principle on the American rights to Robert Mushet's patent on the use of spiegeleisen. The use of spiegeleisen was regarded by many as the key to the successful use of the Bessemer process in making steel. Ward also allowed William Durfee to set up an analytical laboratory at Wyandotte for the testing of iron ores. In September 1864 Durfee produced the first Bessemer steel in the United States at Wyandotte.

In October 1864 the company expanded to include three new members, Mushet, Thomas Clare, and John Brown, the last two Englishmen with claims to the spiegeleisen patent's ownership. In September 1865, following the successful rolling of ingots into rails in May of that year, the company expanded for the last time. Three more possible users of the Bessemer-Kelly-Mushet technology were enticed to join: Charles Chouteau, James Harrison, and Felix Valle, all of St. Louis. As the list of members in the firm expanded, the value of the original partners' shares in the Kelly Pneumatic Process Company declined.

Another group of iron men, located in Troy, New York, had also built an experimental works and were successful in producing Bessemer steel in February 1865. John F. Winslow, John A. Griswold, and Alexander Holley had obtained a license for the use of Bessemer's process in December 1863. After their first success they purchased the American rights to the original patent and to five more recent Bessemer patents for machinery. Neither group was able to continue to produce Bessemer steel without infringing on the other's patents, as the Kelly Company had used the Bessemer converter, and Winslow, Griswold, and Holley had utilized the Kelly and Mushet inventions. When each attempted to find substitutes and failed, legal action seemed to be the only solution. As William Durfee later described the attitude of Morrell and Ward:

The various bugabooks [sic] and hopgoblins [sic] which their terrified imagination con-

jured up of the horrors of the life to come among courts, judges, lawyers, experts, witnesses and obstinate jurors in case they ventured to assert in a court their manifest right, at last drove them into making a proposition to Messrs. Winslow, Griswold and Holley. . . .

An agreement was reached early in 1866 to merge the rights to all the patents into a single company. The trustees of the Pneumatic or Bessemer Process of Making Iron and Steel were Winslow, Griswold, and Morrell. Zoheth Durfee became general agent. Profits were to be divided on a 70 to 30 ratio—70 percent for the Troy group and 30 percent for the Kelly Company. Soon after, the trustees organized a joint stock company in New York, naming it the Pneumatic Steel Association. Members were to be licensees who paid royalties on the basis of tonnage. Holley became a consultant to all the members of the association, and Zoheth Durfee was named secretary-treasurer.

The goal of the Pneumatic Steel Association was to encourage use of its patents by increasing the number of Bessemer steel producers. By the early 1870s most of the technical problems had been overcome, and eight new mills entered production with the expectation of high profits. By the middle of the decade, economic conditions—depression, declining profit and steel prices, and a growing alliance between the railroads and Bessemer steel interests—led to the 1877 organization of the Bessemer Steel Company (also known as the Bessemer Steel Association). It seemed necessary to restrict the use of patents in order to slow the growth of plants and to reduce competition. Since the primary use of Bessemer steel was for new rails for the railroads, the Bessemer Steel Association also acted as a pool to set up quotas and to share the market for rails. The same motives in the oil industry produced the Standard Oil Trust, organized in 1882 under John D. Rockefeller.

The last important patents in the development of the Bessemer process in Europe came with the invention in England of a process to remove phosphorus, which was present in large amounts in iron ores in England and Europe. Sidney G. Thomas and Percy Gilchrist received the English patent for this process in 1878. As in the earlier Kelly-Bessemer controversy, there was an American claimant to the United States patent rights; Jacob Reese of Pittsburgh had simultaneously developed the phosphorus removal process, and he assigned some of his patents to the Bessemer Steel Company in 1879. Andrew Carnegie, one of the new generation of entrepreneurs who were beginning to dominate the steel industry, negotiated for the Bessemer Steel Company and acquired the Thomas process in 1881. This led to another period of confusion and contention over patents between Reese and the Bessemer Association. In 1888 the Supreme Court of Pennsylvania ruled that Reese was required to give up certain of his patents to the Bessemer Company.

Until the court decision, there were few requests for licenses, as producers did not want to be sued for patent infringement. Following the court decision the Bessemer Association was accused of restricting licenses in order to keep out new competition. The association's stance was that producers did not need their licenses because most American iron ore was without the high phosphorus content of the European ores. In any case, by the mid 1880s the open-hearth process of producing steel was growing in popularity. The final English form of the Bessemer process, which included the Thomas process, was never used to any great extent in the United States.

In 1890 the Bessemer Steel Company was reorganized into the Steel Patents Company. By this time the original Bessemer process patents had expired, but the company had acquired a number of non-Bessemer patents which became increasingly more important as use of the open-hearth process surpassed use of the Bessemer process in the twentieth century.

References:

Jeanne McHugh, *Alexander Holley and the Makers of Steel* (Baltimore, Md.: Johns Hopkins University Press, 1980);

James M. Swank, *History of the Manufacture of Iron in All Ages, and Particularly in the United States from Colonial Times to 1891*, second edition (Philadelphia: American Iron and Steel Association, 1892);

Peter Temin, *Iron and Steel in Nineteenth-Century America: An Economic Inquiry* (Cambridge, Mass.: M.I.T. Press, 1964).

Julian Kennedy

(March 15, 1852-May 28, 1932)

by Larry Schweikart

University of Dayton

CAREER: Blower and furnace superintendent, Brier Hill Iron Company (1876); furnace superintendent, Struthers Iron Company (1876-1877); superintendent, Morse Bridge Works (1877-1879); superintendent, Edgar Thomson Works (1879-1883); superintendent, Lucy Furnaces (1883-1885); general superintendent, Carnegie, Phipps & Company (1885-1888); chief engineer, Latrobe Steel Works (1888-1890); consultant and contractor (1890-1927).

Julian Kennedy, a mechanical engineer, inventor, and chief furnace designer at the Edgar Thomson Steel Works, was born in Poland, Ohio, on March 15, 1852, and died on May 28, 1932, the son of Thomas Walker Kennedy and Margaret Truesdale Kennedy. He was educated at Poland Union Seminary and later served as draftsman under his father, an engineer and furnace builder who supervised the building of the Struthers Iron Company furnaces in Youngstown, Ohio. In 1871 he entered Yale University's Sheffield Scientific School; he received his Ph.B. in 1875, but continued to teach at Yale as an instructor of physics. Kennedy continued his education by taking a postgraduate course in the chemistry of iron and steel processes. Kennedy demonstrated prowess in rowing, and he participated as a member of the Yale University crew from 1873 to 1876, winning a considerable number of rows in different boat and scull categories, and setting a record for a two-mile course rowed by a pair (12 minutes, 20.25 seconds in 1876 at Greenwood Lake). He also wrote an essay, "The Mechanics of Rowing," as his valedictory thesis. Even after he accepted his first position at the Brier Hill Iron Company of Youngstown in 1876, he could be seen sculling on a stretch of slack water near the plant.

Kennedy began his professional career in steel as a blower at Brier Hill and was quickly promoted to furnace superintendent. For the next three years Kennedy worked in the Youngstown area, first at Struthers Iron Company as furnace superintendent for construction and operation, then at the Morse Bridge Works as a superintendent. In 1879 he entered a long association with the Carnegie companies. As superintendent of the Edgar Thomson Works from 1879 to 1883, Kennedy helped the mills attain record outputs of pig iron. Kennedy had already given indications of his talent while at Struthers, where the furnace became one of the first in Ohio to attain large outputs. This efficiency was due to Kennedy's practices of regulating the quantity of the blast by the revolutions of the engine instead of the traditional method of using a pressure gauge regulator. The Struthers furnace (55 feet high x 16 feet diameter) improved its output from 1,602 tons of iron in December 1871 to 2,032 tons by March 1876, achieving an increased output of 400 tons in the first month under Kennedy's supervision. Kennedy brought these techniques to the Edgar Thomson Works, where the Lucy furnaces increased output from under 215 tons a day (which is what the leading furnace, the Isabella could produce in 1881) to 300 tons a day as of 1883. After the Lucy Furnace #1, then Lucy #2, could no longer keep up with the growing demand of Edgar Thomson for more pig iron, Kennedy supervised construction of two new blast furnaces at Braddock, which, although smaller than the Lucy furnaces, produced up to 442 tons a week each. Andrew Carnegie carefully pitted the four furnaces against each other, awarding a steel boom trophy to the furnace with the highest weekly production.

Once Carnegie, who was traveling in Scotland, sent Kennedy a daily production report from one of his furnaces, furnace C, asking Kennedy why furnace C fell below average on that particular day. Kennedy shot back a reply: to have an average you had to occasionally fall below it, because an average measured highs and lows. Kennedy's furnaces turned out an astounding average for even Carnegie's high standards. In the late 1890s a Lucy-type furnace could turn out 100,000 tons a year.

The Edgar Thomson furnace A, originally a charcoal furnace at Escanaba, Michigan, was moved and erected on its new foundations in 1879. Its dimensions—65 feet with a 7-foot 1-inch hearth diameter—utilized Kennedy's "blowing in" process.

Julian Kennedy

Blowing 15,000 cubic feet per minute into the furnace, or twice as much used by similar-sized furnaces, the results of fuel consumption were much lower than any obtained from contemporary furnaces. In 1890 Kennedy, commenting on a paper in *Iron Age*, recalled that when he first began his driving, or blowing in, theory he had used a wide top "to our own detriment," and Kennedy corrected it. For example, Kennedy changed the plans to the B furnace, making the inner walls of the hearth straight with an 11-foot diameter. This differed from other furnaces using coke for fuel by its large hearth that provided more space for combustion. In the month following its opening in April 1880, 3,718 tons of iron were made.

As records fell, Kennedy rose. In 1885 he was named general superintendent of the Homestead, Pennsylvania, furnaces under Carnegie, Phipps &

Company, Ltd. Within a year he had implemented the "Thomas basic process" in some of the hearths. The Thomas process, developed by amateur chemist Sidney Gilchrist-Thomas, extracted phosphorus from iron, making it suitable for the Bessemer converter by changing the lining of the converter from an acid material to lime or magnesium. Kennedy experienced immediate success with the basic process and opened up extensive new iron resources for steel making. Once Kennedy had demonstrated the success of this technique, Carnegie urged W. L. Abbott, president of Carnegie, Phipps & Company, to implement the process in all the hearths. Kennedy was also credited by Carnegie with finding a way to make ferro-manganese in the Edgar Thomson furnaces, and as a result Carnegie's company was the only one in the United States to produce this valuable material.

At Homestead Kennedy also erected a 10-ton ladle crane and a 16-ton ingot crane in 1892. Kennedy not only received the contracts for constructing them but also for erecting a new converter. At this time Kennedy also introduced "the Kennedy Universal," a universal mill that improved construction by setting the guide bars so that they could be removed laterally, thus adding power and compactness to the mill.

Kennedy was reassigned in 1888 to the Latrobe Steel Works as chief engineer, where he incorporated many of his own designs as patents. Indeed, Kennedy found a showcase for his talent in these furnaces. He opened his own engineering consulting office in 1890 in Pittsburgh, eventually employing his sons, Joseph and Julian, Jr., and an English engineer, Frederick McClain. Kennedy's reputation ensured the success of the business; iron and steel companies from around the world sought his advice. Among his projects abroad were the Tate Iron & Steel Works in Sakchi, India, and the Nicopol-Maripol Mining & Metallurgical Company in Sartana, Russia. He also directed projects in England, Europe, and the Far East, especially China, but the majority of Kennedy's mill design and construction involved American plants. He and his brother Thomas W. Kennedy, Jr., teamed to build the Adrian furnace at Dubois, Pennsylvania, which used a top air feed, and designed a number of plants and furnaces in the Mahoning Valley near Youngstown.

Despite Kennedy's service to the Carnegie companies, he was never brought into the inner circles,

as were other engineers such as W. L. Abbott or George "Dod" Lauder. It is likely that Kennedy's consulting kept him occupied elsewhere. But Kennedy's position outside management also spared him many of the troubles that affected the other associates including Abbott, Lauder, and Abbott's successor, Henry Clay Frick.

A noted inventor, Kennedy obtained patents on 160 inventions, most of them related to mill and furnace operations, and of which half were in general use in his lifetime. His first patent, issued in 1881, was for a hot blast oven. He made a number of improvements on steel tires for railway cars, including a rolling machine for shaping the tires from ingots and put into effect at the Latrobe works; a method of forging annular blooms for car-wheel tires; and a device that used a moving guide to tilt work on a table without the awkward and dangerous use of tongs by hand labor. This last patent, issued in 1903 and known as the Kennedy-Wellman manipulator, permitted workers to move hot materials over feed rollers. He also developed the Kennedy "beam-end shear," which was a machine that cut the ends off I-beams. This shear could cut I-beams without changing dies. Other patents included a machine for cutting flanged beams; a hydraulic shearing apparatus; a blast furnace charging apparatus which prevented swinging of the charging bucket; and a novel blast furnace. With such expertise in engineering and patents, it is not surprising that J. P. Morgan accepted Kennedy's equipment appraisals when he consolidated the properties that formed National Tube Company, a step in the foundation of the United States Steel Corporation in 1901.

At every opportunity Kennedy sought to generate more power, reduce human labor, and save time. He affixed electric motors to moving tables, changed work routines so as to minimize effort, and reduced costs. Through these changes Kennedy revolutionized steel mill production, in essence, replacing manpower with machine power. He lowered costs at every plant that he supervised, and yet his retooled plants soon showed profits. In just over ten years after the Lucy and Isabella furnaces began, American furnaces produced in a day what had once taken a week, in no small part due to the genius of Julian Kennedy. When Kennedy retired from consulting in 1927, he was, as Sir Lowthian Bell of the British Iron and Steel Institute claimed, the "greatest engineer in the world."

Kennedy's interests extended into fuel and mineral aspects of the iron and steel business. He was president of the Orient Coke Company, the Ontario Gas Coal Company, the Lowber Gas Coal Company, and the Poland Coal Company. He also served as a director for the Emerald Coal & Coke Company. Kennedy was also a member of the Engineer's Society of Western Pennsylvania, the American Institute of Mining Engineers, the British Iron and Steel Institute, and several golf clubs. Kennedy actively participated in professional conferences and journals. In 1928 the American Society of Mechanical Engineers honored Kennedy by awarding him its Gold Medal in recognition of his services to the iron and steel industry. He also received a gold medal for blast furnace and rolling mill construction at the 1904 Louisiana Purchase exposition in St. Louis, and the Gary Medal of the American Iron and Steel Institute in 1932. Among his honorary degrees were an A.M. from Yale in 1900 and the D. Eng. from Stevens Institute of Technology in 1909. Kennedy was a Unitarian church member. Jul-

ian Kennedy married Jennie Eliza Breneman on November 14, 1878. They had six children: Lucy, Joseph, Julian, Jr., Hugh (who died in infancy), Eliza, and Thomas. Julian Kennedy died in Pittsburgh of a heart attack on May 28, 1932.

Julian Kennedy was a superior engineer and inventor in the most practical sense. His innovations were immediately applicable and, as a result of his service, the Edgar Thomson Works repeatedly set levels for output. Carnegie's steel empire would not have been as dynamic without Julian Kennedy.

References:

Andrew Carnegie, *The Autobiography of Andrew Carnegie* (Boston: Houghton Mifflin, 1920);

Burton Hendrick, *The Life of Andrew Carnegie*, 2 volumes (Garden City, N.Y.: Country Life Press, 1932);

William T. Hogan, *The Economic History of the Iron and Steel Industry in the United States* (Lexington, Mass.: Lexington Books, 1971);

Joseph Wall, *Andrew Carnegie* (New York: Oxford University Press, 1970).

Andrew Kloman

(unknown-c.1880)

by Larry Schweikart

University of Dayton

CAREER: Partner, Kloman Brothers (1858-1860); partner, Kloman & Company (1861-1863); partner, Kloman & Phipps (1863-1865); superintendent and partner, Union Iron Mills (1865-1867); partner, Carnegie, Kloman & Company (1867-1874); manager, Pittsburgh Bessemer Steel Company (1879-1880).

Andrew Kloman, a partner in Carnegie, Kloman & Co. and superintendent and partner of the Union Iron Mills, came to the United States from Treves, Prussia, in the 1850s with his brother Anthony. In 1858 the Kloman brothers opened a forge shop at Gerty's Run in Millvale across from Pittsburgh, Pennsylvania, where they produced railroad axles out of scrap iron, using a small steam-driven wooden trip-hammer.

In 1861 Andrew Carnegie met Kloman through Tom Miller, one of Carnegie's neighbors

and a frequent business partner, who had in turn met Kloman and his brother Anthony while Miller acted as purchasing agent for the Fort Wayne, Pittsburgh & Chicago Railroad. Miller, like other customers, had been impressed with the quality of the Kloman brothers' axles, then considered the finest available. The Klomans' technique featured a twisting process, employed during forging, that gave a special strength to the iron, and it typified Andrew Kloman's mechanical genius. Andrew Kloman had, by now, taken the reins of the business. He was by nature distrustful and thrifty, as he feared his language inadequacies would put him at a disadvantage in every transaction.

But Miller was able to put Kloman at ease, and by 1859 the two men had entered into an agreement by which Miller advanced Kloman the funds to purchase another trip-hammer in return for 30 percent of the profits. This agreement between

Kloman and Miller led to Andrew Carnegie's entry into the world of steel. Miller, afraid that the appearance of his name on a contract with Kloman might smack of a conflict of interest with Kloman's main customer, the Fort Wayne Railroad, instead suggested that a third party be brought in–Carnegie. But Carnegie was already a well-known business figure to Kloman, and the cautious German resisted. Miller instead recruited yet another acquaintance, Henry Phipps, Jr., who was also a good friend of Thomas Carnegie. Phipps had first-rate bookkeeping skills, something the new partnership needed. Phipps agreed to pay Miller out of his dividends for putting up his half of the $1,600 for the new trip-hammer. By 1861 Kloman & Company was in business as the Iron City Forge Company, taking in more business than it could handle with the extra trip-hammer.

Kloman sought to expand further, asking Miller to front the construction of a new mill in Pittsburgh and to finance the lease of the land. A new company emerged named Kloman & Company of Pittsburgh (capitalized at $80,000 in 1862), again consisting of the Klomans, Phipps, and Miller, whose interest was still held by Phipps. Andrew Kloman did not like sharing control of the company, however, and when Miller sold some of his shares to yet other outsiders, Kloman grew increasingly uncomfortable. Moreover, with orders piling up, he needed all the help he could get, especially from Anthony. His brother, however, had proven unreliable, and in 1862, at Andrew's request, Miller bought out Anthony Kloman for $20,000, and Miller became publicly identified as a partner.

Even though Andrew had arranged his brother's ouster, he realized belatedly that he had negotiated away a majority interest in the firm. He complained rather bitterly, so much so that Miller offered to sell Kloman enough shares to give him a 50 percent interest in the company. Kloman refused and began a struggle with Miller and Phipps. He separated Phipps from his friend, and Miller turned to Andrew Carnegie for help, asking him to arbitrate. Whether Carnegie saw a business opportunity or merely a chance to reconcile former partners is unknown, but he effected a settlement that greatly reduced Miller's share in the company and on September 1, 1863, created a new partnership called Kloman & Phipps. A special clause in the contract permitted the other partners to purchase Miller's shares for book value ($10,000) with 60 days

notice. Miller, of course, signed this agreement only reluctantly, requiring reassurances from Phipps that Phipps would continue to support him. Phipps, who lacked the necessary $20,000 for his share, asked Thomas Carnegie to put up half.

Soon thereafter, Kloman and Phipps notified Miller that they were exercising their option to buy him out. Phipps not only made an enemy of Miller but temporarily found himself opposite Carnegie, who helped Miller in October 1864 to build a rival company, the Cyclops Iron Company. Whereas Kloman's plant was geared for producing war goods, Cyclops was aimed at producing bridge structures, a much more profitable enterprise with the war winding down. Cyclops intimidated Kloman so much that the German offered to merge with the Carnegie-Miller company in 1865. The Carnegies, working each side of the merger, completed the deal. Kloman had simply been outmaneuvered, and Miller had to be placated with promises that he would receive the largest share of anyone (even a larger share than Carnegie) and would not have to interact personally with Phipps. This company, Carnegie's first actual iron-producing mill, born in 1865, commemorated the reunion of the nation with the name Union Iron Mills. Its main source of demand would be another Carnegie company, the Keystone Bridge Works (chartered in 1862).

Miller, despite Carnegie's pleading, refused to make peace with Phipps. When a vacancy on the board occurred in 1867, Carnegie tried to fill it with Phipps. Miller resigned and insisted that Carnegie buy him out; as a result Union Mills was renamed Carnegie, Kloman & Company. Kloman, through it all, awed Carnegie with his mechanical inventiveness and energy. Carnegie, in his autobiography, marveled at "How much this German created!" Kloman introduced dozens of time- and labor-saving devices to the mills, including a saw that could cut cold iron to specific lengths, the first universal mill in the United States, and upsetting machines to make bridge links. He also succeeded in making couplings for the arches on the St. Louis bridge, at the time the largest semicircles ever rolled. Kloman had good luck in developing the universal mill—a machine in which plates of various widths and rolled edges could be made—in that one of his employees, Johann Zimmer, had seen such an apparatus in Europe. Zimmer gave Kloman the details from memory, and Kloman designed the mill. Carnegie called Kloman "a genius" and dispatched him to observe new processes and machinery, Kloman at once determining their effectiveness. He, like Captain William "Bill" Jones, the supervisor of the Edgar Thomson Steel Works, and Julian Kennedy, the chief furnace designer at the Edgar Thomson Steel Works, as well as Carnegie's cousin George "Dod" Lauder, gave the Carnegie companies an advantage in sheer engineering talent over competitors.

In early 1870 the Union Mills associates laid plans for a new furnace to relieve them of their dependence on various local furnaces for pig iron. Construction began in 1871, and the furnace was blown in in 1872. Named the Lucy Furnace #1, after Thomas Carnegie's wife, the furnace and its design reflected Kloman's influence. Kloman had convinced the associates that a large furnace that was higher and had a wider bosh (the point of greatest diameter) would increase production and use less fuel. The Lucy captured this design, standing 75 feet high with a 20-foot diameter bosh. By 1873 the Lucy Furnace averaged 593 tons of steel per week and in 1874 produced 100 tons in a single day, eventually reaching the 300-ton-per-day mark.

Typical of the good luck that Kloman brought to the Carnegie interests was his introduction of a relative, William Borntraeger, to the Union Iron Mills. Borntraeger began his career as a shipping clerk, then worked at Union Shipping Mills when the Siemens Gas Furnace was introduced to the United States from Great Britain. These furnaces were expensive, but the process saved tremendous amounts of material previously wasted. On his own, Borntraeger made an accounting of the savings generated by the new furnace. The form he designed was unique, serving as the basis for new cost accounting. He invested in Carnegie, Kloman & Company in 1871 and held an interest worth $4,532.

Kloman and Carnegie disagreed about Carnegie's profit-retention policies, and he looked for ways to invest his money outside the Carnegie companies. Using his Union Iron Mills, Lucy Furnace, and Edgar Thomson Steel Works shares as collateral, Kloman invested in the Cascade Iron Company and the Escanaba Furnace Company, neither of which was incorporated. After the panic of 1873 these businesses failed, leaving Kloman exposed to considerable debt. The Carnegie associates quickly bought out Kloman, although he was re-

tained as an employee. After Kloman had settled his debts, Carnegie offered to let him return with a smaller investment but Kloman refused. Kloman's protégé, Borntraeger, took over as superintendent of the Union Mill and made it the most profitable part of the Carnegie interests.

Meanwhile, Kloman, seeking revenge against the Carnegie associates, allied himself with William Singer and a syndicate of five manufacturing companies known as the Pittsburgh Bessemer Steel Company Ltd. ($250,000 capital). This group purchased land and planned a rail mill at Homestead, Pennsylvania, in 1879. Kloman was named manager of the plant. The German had already bought land adjoining the syndicate's Homestead works and had started construction on a 684-foot-long building that would hold a Bessemer rail mill, two universals, and other equipment, all dependent on the Edgar Thomson plant for ingots. By combining with the Pittsburgh Bessemer Steel Company, Kloman thought it possible to produce 50,000 tons of steel rails and 30,000 tons of other steel materials a year. Although the plant was completed in record time, 15 months, Kloman died in 1880 before it rolled its first rail. He was survived by a son.

Kloman also indirectly helped to steer Andrew Carnegie toward backward integration, insisting that Carnegie and his partners develop the Kloman Mine, an iron ore mine in Michigan's Marquette Range, in 1872. Although Kloman thought the mine capable of producing 50,000 tons annually, from 1873 to 1875 it delivered just 36,000 tons at a high cost. It was closed in 1875 due to the expense of extracting the ore. Carnegie sold the mine in 1880 to P. B. Shumway.

Andrew Kloman was typical of the expert engineers and inventive geniuses with whom Andrew Carnegie surrounded himself. He helped bring Carnegie into the world of iron and steel, then helped make him the unsurpassed leader in rail production. Kloman never shared Carnegie's goals, though. He left Carnegie's empire a bitter and vengeful man.

References:

Andrew Carnegie, *The Autobiography of Andrew Carnegie* (Boston: Houghton Mifflin, 1920);

Joseph Wall, *Andrew Carnegie* (New York: Oxford University Press, 1970).

La Belle Iron Works

by John A. Heitmann

University of Dayton

The history of the La Belle Iron Works, one of the important nineteenth-century iron manufacturers located in the vicinity of Wheeling, West Virginia, illustrates the market challenges which confronted those involved in the nineteenth-century iron industry and the product strategies pursued in attempting to achieve profitability and stability. The origins of the La Belle Iron Works can be traced to 1851, when a group of seven nail makers, previously associated with the Wheeling, West Virginia, firm of Bailey, Norton & Company, subscribed a total of $40,000 and organized Bailey, Woodward & Company. The partners organized a new venture, called the La Belle Iron Works, and four acres were purchased for the erection of a nail works south of Wheeling's city limits. Initially the plant was equipped with 8 boiling or puddling furnaces, 10 heating furnaces, a muck mill, a nail plate or skelp mill, and 25 nail-making machines. With the operation's yearly capacity of 75,000 kegs of nails, the owners made more than 10 percent annually on their investment during the 1850s.

As Wheeling emerged as the center of the "western" iron trade during the 1850s, it drew entrepreneurs from a much broader class spectrum than other iron- and steel-making communities. As John N. Ingham has demonstrated, some 20 percent of the elite of the Wheeling iron and steel trade, including those involved in the La Belle venture, had working-class origins. For example, Henison H. Woodward, an original subscriber of the La Belle mill, was born in 1812 and by age eleven was an accomplished nail feeder in Massachusetts. After working for some time in the textile industry and then making a series of moves west, Woodward arrived in Wheeling in 1847. Woodward became a partner in the Belmont works in 1849 before participating in the organization of the La Belle firm two years later. One of Woodward's associates, John Wright, had similar origins. Born in Pittsburgh in 1824,

Wright was the son of an engineer employed by Peter Shoenberger. Following in his father's footsteps, Wright came to Wheeling in 1849, and after serving two years with Bailey, Norton & Company at the Belmont mill, he severed his connection and was one of the founders of the La Belle establishment. Wright would remain in charge of the mechanical department until his retirement in 1877.

During the 1860s, 1870s, and 1880s the La Belle company's prosperity was entirely tied to the production of cut nails. Indeed, the leadership became so entrenched in traditional manufacturing operations and in the manufacture of one product that they failed to recognize the threat of wire nails to their business. By the early 1890s the La Belle cut-nail business was eclipsed by the wire-nail trade, and the company became increasingly unprofitable. Aware that the cut-nail trade would no longer serve as the basis of a profitable business and encouraged by the McKinley protective tariff, La Belle's directors decided in 1894 to erect a tin-plate plant. Consisting of four hot mills, the tin-plate operation was almost immediately profitable, and in 1897 authorization was given for the enlargement of the tin-plate facilities. While La Belle still stuck stubbornly to the manufacture of cut-iron nails, its viability as a business was tied to the manufacture of another profitable commodity, this one insulated from international markets by a high tariff.

During the late 1890s the iron and steel industry experienced rapid consolidation, and La Belle did not escape this frenzy of mergers. In 1898 La Belle sold its tin-plate operations to the American Tin Plate Company. As a result of the sale the company now had considerable financial resources. But its directors were still confronted with the same problems that challenged them in 1894, namely the lack of a profitable commodity. A decision was made in 1899 to purchase property in Steubenville, Ohio, and to erect a continuous plate mill. Concerned

Workers at the La Belle Iron Works in Wheeling, West Virginia, circa 1880 (courtesy of the West Virginia Department of Culture & History)

about obtaining a sufficient supply of steel for this project, in 1901 La Belle officers started work on an open-hearth plant consisting of six furnaces, each with a 50-ton capacity. The firm also rebuilt its blast furnace in addition to constructing a sheared-plate mill and a tube mill.

La Belle's ambitious expansion program came perilously close to bankrupting the firm. Indeed, for a time its officers had to take stock as part of their salary, and it was not until 1905 that a dividend was declared. The crisis would lead to personnel changes at the top, and in 1905 La Belle's directors elected a new president, Issac M. Scott. Scott di-

rected La Belle's operations during World War I and oversaw the incorporation of La Belle into Wheeling Steel Corporation in 1920.

References:

John N. Ingham, *The Iron Barons: A Social Analysis of an American Urban Elite, 1874-1965* (Westport, Conn.: Greenwood Press, 1978);

Earl Chapin May, *Principio to Wheeling, 1715-1945* (New York: Harper, 1945);

Robert L. Plummer, *Sixty-Five Years of Iron and Steel in Wheeling* (Wheeling, W. Va., 1938);

Henry Dickerson Scott, *Iron & Steel in Wheeling* (Toledo: Caslon, 1929).

Lackawanna Iron & Coal Company

by John A. Heitmann

University of Dayton

The history of the Lackawanna Iron & Coal Company and its several organizational precursors, including Scrantons, Grant & Company and Scrantons & Grant, contains many lessons for businessmen confronting the complex challenges of the present. Although the Lackawanna Iron & Coal Company emerged as one of the largest steel manufacturers in nineteenth-century America, its success was hard fought and often in doubt. Beginning in the early 1840s with problems associated with a new technology—the use of anthracite coal in iron production—the firm's story illustrates how tenacity and determination coupled with creative flexibility enabled its directors and leaders to overcome obstacles that on several occasions threatened to bankrupt the fledgling organization. The story also illustrates that it was the cohesiveness of the family, in this case the Scrantons, that proved crucial during the troubled times of start-up, as family members invested substantial amounts of time and capital to insure the success of the venture.

The history of the Lackawanna Iron & Coal Company begins not with the efforts of George W. Scranton, however, but with the work of the visionary promoter, engineer, and geologist William Henry. During the late 1830s Henry had become interested in the economic potential of an area of land in northeast Pennsylvania known as Slocum's Hollow, the present-day site of the city of Scranton. Henry's surveys indicated that all of the materials necessary for the production of iron—coal, limestone, and iron ore—were in abundance in this region. Thus Henry confidently took an option to purchase this farming area, and in his efforts to raise the capital necessary for the project he engaged his son-in-law, Selden T. Scranton of Oxford, New Jersey, who had extensive experience in iron manufacturing, and other members of the Scranton family, including George Scranton.

While Henry had vision, he ultimately was unable to turn it into reality. Henry's geological surveys soon were discovered to be seriously flawed, and his initial attempt to erect a workable iron furnace met with disastrous failure. At first Henry was assisted by George Scranton, who eventually took the lead in dealing with the formidable challenges of iron production. During 1841 two experiments using anthracite and local supplies of iron ore and limestone resulted in unexpected failure, and in each case the furnace had to be cleared using hammers and chisels before being relined. As the reason given for this cementation of reactants was the poor quality of the local limestone, a new source in Columbia County, Pennsylvania, was subsequently purchased by the firm, and the material was shipped by boat and wagon to the site.

With the success of a trial in January 1842 using the new source of limestone and a mix of iron ores, the firm had overcome its primary technological obstacle. Yet the partners in the venture quickly discovered that while pig iron could be produced in relatively large quantities, the market for it was sluggish, and the transportation costs incurred in getting it to the East Coast eroded much of the anticipated profits. While the company store operated under partner Sanford Grant's management was thriving, little else appeared positive in late 1842 and 1843, and a growing number of creditors in Pennsylvania and New Jersey began clamoring for payments.

Amidst these pressures George Scranton, now firmly in charge of the operation, made a crucial decision; rather than to continue making pig iron with little chance for long-term viability, the company would henceforth concentrate on nail making, thus producing goods with a stronger market, a larger markup of price, and lower shipping costs. The bold move, however, required the infusion of a large amount of new capital into the proj-

The Lackawanna Iron & Coal Company works at Scranton, Pennsylvania, circa 1875

ect. In part George Scranton's decision was based on his faith that many prosperous family members in the mercantile business—cousins Joseph Scranton of Augusta, Georgia, and Erastus Scranton of New Haven, Connecticut—would recognize the opportunity and render assistance. But capital outside the family network was also necessary, and thus Scranton called on a mercantile firm located in New York City, Howland & Company, for $20,000 in financial backing.

By May 1843 Howland & Company and Scrantons, Grant & Company had agreed to terms and had become partners in a corporation reorganized as Scrantons & Grant. With a capitalization of $86,000, the firm designated George W. Scranton, Selden T. Scranton, and Sanford Grant as active partners. The Scranton family, despite the infusion of a large amount of outside capital, remained in control of the organization's stock, holding 51.6 percent of its shares. The reorganization ensured that the iron-making venture had the money needed to purchase the equipment necessary to shift production from pig iron to nails.

Early in May 1843 George Scranton began supervising the erection of two new installations on the company's property. A rolling mill, which measured 110 by 114 feet and was driven by a 90-horsepower waterwheel, was set up, and five puddling furnaces were readied. In an adjacent building a nail factory with a capacity of 100 kegs per day was equipped with 20 nail machines and one spike machine. By early 1844 optimism prevailed, but this mood was quickly squelched when it was discovered that the nails produced were brittle and often broke when hammered. Indeed, in the terminology of the trade, the nails made were "cold-short," possessing a coarse fiber that was the consequence of the iron ore employed in the process. One out of three nails broke upon impact, and thus the company, already in jeopardy, had invested much capital in an effort that was nothing short of disastrous.

Again, George Scranton's cool head prevailed. As he weighed his options, Scranton, in a creative yet flexible way, decided to shift the product line again, this time to the manufacture of T-rails necessary for the burgeoning railroad industry. And as it turned out, his "cold-short" iron, while a poor material for nails, would be the ideal material for rails.

During the mid to late 1840s the United States was just beginning to experience a boom in railroad building, a boom that was well under way in Great Britain. Precisely because of the frenetic construction activities abroad, Americans, long dependent upon British T-rail products, were confronted with serious shortages in the construction of their own lines. One pioneer American line was the Erie Railroad, an organization that had received a charter from the state of New York and a $3-million grant for the purchase of land and equipment provided that track would be laid to Binghamton, New York, by a certain date.

Confronted with a material shortage and a deadline, William E. Dodge, a major figure at the Erie Railroad and a large stockholder in Scrantons, Grant & Company, made arrangements with George Scranton to manufacture and deliver T-rail to the Erie Railroad. The railroad advanced Scranton $100,000 to enable the firm to convert its operations from rolling nail plate to rail, and Scranton promised to furnish the rail necessary for the road between Port Jervis and Binghamton. Scranton accomplished this formidable task within the next year and a half, using every mule and draft horse in the region to move the rail to six major construction points. And as a consequence of this long awaited success the firm finally found an important niche in the market and subsequently prospered.

Reorganized as Lackawanna Iron & Coal Company in 1853, the firm was headquartered on the site formerly known as Slocum's Hollow, by the 1850s renamed Scranton in honor of the family that had supported a venture that was now the economic lifeblood of the region. This firm would retain the Lackawanna name throughout the nineteenth century, expand into the Buffalo, New York, area before World War I, and later be incorporated into the organization now known as Bethlehem Steel.

References:

Robert J. Casey and W. A. S. Casey, *The Lackawanna Story* (New York: McGraw-Hill, 1951);

Burton W. Folsom, *Urban Capitalists: Entrepreneurs and City Growth in Pennsylvania's Lackawanna and Lehigh Regions, 1880-1920* (Baltimore: Johns Hopkins University Press, 1981);

W. David Lewis, "The Early History of the Lackawanna Iron and Coal Company: A Study in Technological Adaptation," *Pennsylvania Magazine of History and Biography*, 96 (1972): 424-468;

Nicholas E. Petula, *Scranton, Once Upon a Time* (Scranton, Pa., n.d.);

Benjamin H. Throop, *A Half-Century in Scranton* (Scranton, Pa.: The Scranton Republican, 1895).

Archives:

Material concerning the Lackawanna Iron & Steel Company is located at the Syracuse University Library, Syracuse, New York, and at the Hagley Museum and Library, Wilmington, Delaware.

George Lauder

(November 11, 1837-August 24, 1924)

by Larry Schweikart

University of Dayton

CAREER: Instructor, Queens College (1864-1868); manager, Unity and Larimer Coke Works (1870-1881); partner, Carnegie Brothers (1881-1886); partner, Carnegie Steel (1886-1899); partner, Carnegie Company (1899-1901).

George "Dod" Lauder, a partner in Carnegie Steel, engineer, and Andrew Carnegie's cousin, was born in Dunfermline, Scotland, on November 11, 1837. "Dod" or "Doddy," as he was named by his father, George (known in the Carnegie biographies as "Uncle George"), became Andrew's boyhood friend. Uncle George, a local grocer who had lost his wife Seaton, spent a great deal of time with his son and Andrew, whom he named "Naig." He tutored the two boys in Scottish history, politics, and literature, and they learned a deep sense of Scottish patriotism from him. His influence on the boys was so great that Carnegie, in his autobiography, claimed that Uncle George could make him and Lauder "weep, laugh, or close our little fists ready to fight."

After Andrew Carnegie sailed for the United States in 1848, the two young men continued to correspond, Uncle George encouraging their transatlantic conversations and suggesting topics for discussion. In one series, for example, the two boys debated by mail the strengths and weaknesses of their two countries' respective parliamentary and federal political systems. These exchanges continued after Carnegie settled in Pittsburgh and began work in the telegraph office. But when the two were reunited in 1862 on Carnegie's trip to Scotland, Carnegie was disappointed to learn that Lauder disliked the ambition that filled his now nearly American cousin. Carnegie tried at length but failed to impress upon Lauder the opportunities available in America.

Lauder instead developed considerable scientific talents studying at Glasgow University under

George Lauder

Lord Kelvin, and he graduated with a degree in mechanical engineering in 1864. He then taught mechanical engineering at Queens College in Liverpool from 1864 to 1868. Lauder immigrated to the United States in 1868 and entered the world of iron and steel, taking a position as manager of the Unity and Larimer Coke Works in 1870. As manager he oversaw the remodeling of both mills. In 1881 Lauder became a partner in Carnegie Brothers with a $30,000 interest. During this period he again grew extremely close to Carnegie, so much so that, according to Carnegie biographer Joseph Wall,

"Only cousin Dod Lauder dared reprimand [him], and he only by a parable." Lauder, for example, admonished Carnegie to keep quiet during the tumultuous years of labor struggle in the late 1880s.

But Lauder did not always agree with Carnegie. He found himself siding against his cousin in the matter of the distribution of profits and dividends. Carnegie repeatedly ploughed profits back into the company; many partners lobbied for higher dividends. In their quest they relied on Lauder as a source of inside information. Yet Lauder remained close to Carnegie and was a member of the Iron Clad Agreement by which Carnegie associates promised to transfer or sell their stock to the other associates if three-fourths of the interest (that is, the stocks issued) or if three-fourths of the remaining associates should request such a transfer. The agreement had originally been drawn up out of concerns raised by the death of Thomas Carnegie in 1886 and the subsequent disposition of his shares. Lauder survived in the company long enough to see several partners, including William L. Abbott, J. G. A. Leishman, and eventually, Henry Frick, removed from the firm through the agreement. His was always the final word on technical matters. With admiration, Carnegie wrote that Lauder "never failed in any mining or mechanical operation he undertook."

In addition to his general expertise Lauder performed numerous other valuable services for the Carnegie companies. When Captain "Bill" Jones died suddenly in a furnace accident in 1889, Lauder and Henry Phipps checked Jones's papers and realized the importance of having the company own Jones's patents. Lauder negotiated the sale of the patents from Jones's widow for $35,000, an extremely low price for patents of virtually incalculable value.

In the Homestead Strike of 1892 Lauder found himself between Carnegie and Frick. Frick had hired Pinkerton agents to break the strike, and a bloody battle ensued in which several of the detectives were killed and the union appeared triumphant. When Carnegie, who was in Scotland, heard of the incident he wired Frick that he would return immediately. Lauder and Phipps, who were in England, feared this would insult Frick and cause his resignation, which in turn would signal the union that it had won. Lauder contacted Carnegie and urged him to stay out of the episode, or risk losing Frick. Lauder's cousin reluctantly agreed, but Carnegie did not forget Frick's failure.

By 1892 Lauder held a $1-million investment in Carnegie Steel Company, Ltd., and was a member of the board of directors of the Carnegie-affiliated companies. He also represented the steel company in Henry Clay Frick's Frick Coke Company directorate. Lauder played an important role in reorganizing the company and suggesting that Carnegie combine Carnegie Steel and Frick Coke Company into the Carnegie Company, Ltd. But during the course of negotiations Lauder cautioned that Frick and his partners were profiting too much from the reorganization. Carnegie, based on Lauder's recommendation, delayed the reorganization. Eventually, a syndicate headed by John W. "Bet A Million" Gates and William and James H. Moore of Chicago submitted a plan for reorganization; but they failed to meet the deadline Carnegie had set for accumulating cash, and Carnegie looked for other ways to streamline operations.

Lauder, based on his long association with Carnegie, was trusted for advice on the most critical matters, and his intuition was usually good. In 1899 Frick met with Lauder, who represented the Carnegie Steel interests on the board of directors, to present an ultimatum revoking previous price agreements that Frick had verbally concluded with Carnegie; Lauder fought strongly for the validity of Carnegie's verbal agreement with Frick. Frick and Carnegie, he argued, owned majority interest in their two companies, and they should correct any problems themselves. Frick's board rejected Lauder's plea that the principals resolve their differences personally, to which Lauder replied that Frick's ultimatum seemed "like a declaration of war." The matter hardly endeared Lauder to Frick, but Lauder had already urged Carnegie to sever himself from what Lauder termed "a disturbing element," meaning Frick.

Indeed, Lauder frequently acted as Carnegie's alter ego, passing along policy changes within the company to individuals before formal announcement. For example, Carnegie instructed Lauder to tell Charles Schwab that he would see his power as president expanded once Frick had been removed. This inside track gave Lauder innumerable advantages. Even in the Frick-Carnegie struggle of 1899 and 1900, others, such as Schwab, operated under a sword of Damocles in that a commitment to one party or the other might lead to ultimate destruction if Carnegie and Frick suddenly reconciled. Lauder, of course, had no such concerns. But his relative invulnerability inside the company rested on far more than the fact that he was Andrew Carnegie's cousin; he was a senior partner, a shareholder, a director, the steel company's representative on the Frick Coke Company board, and still the chief technical adviser.

The strongest evidence that Lauder's loyalties ultimately were with his cousin could be seen in Carnegie's plans to create a new board to oust Frick, outlined to Lauder in memos (circa 1899, but not dated). Carnegie's plan was particularly appreciated by Lauder, who saw in the plan elements of the British reform movement history that had been drilled into their heads by Uncle George. Although Charles Schwab ultimately negotiated the new board composition with Frick, it reflected the essentials of the plan outlined to Lauder. Schwab, in fact, recommended Lauder be a member of the new board, but, according to minutes of the special board of directors meeting on January 2, 1900, the list of new directors did not include Lauder's name.

Always the voice of caution, Lauder repeatedly tempered Carnegie's plans to expand out of concern for stretching the company's capital. He frequently reflected the attitudes expressed by those who were participating across America in the phenomenon known as the "managerial revolution." He sought to eliminate risk and preferred the stability of pools. When Carnegie and Schwab planned to expand the company into manufacturing various finished products in July 1900, Lauder registered his expected opposition. He insisted that the board delay a decision until his cousin returned from abroad to evaluate the proposal in person. Carnegie, however, had already given his approval by cable and mail to a similar plan, and Lauder soon found himself caught in a new sandwich, this time between Carnegie and Schwab. Lauder believed he was speaking for Carnegie's interests, and in some ways he was. He spoke out of a concern to maintain Andrew Carnegie's interests in, and control over, the company. In meeting after meeting Lauder argued that the board should contact its associates abroad, or wait until other associates were present. In each case it was clear that Lauder represented his cousin.

Equally as important as Lauder's contribution to the administration of the steel company were his significant engineering feats. He developed and introduced to the United States coal-washing machinery that utilized waste from the mines and supervised in-stallation of this equipment in Pittsburgh for many years. Although his engineering talent was mostly employed in the steel industry, he pioneered other inventions, including constructing the first newspaper-folding machine in the United States in the 1870s.

Reserved, restrained, and reflective, Lauder proved to be the antithesis of his cousin. Cautious but possessing a sense of humor, Lauder achieved high status in local Pittsburgh society. Among his club and organizational memberships were the Duquesne Club, the Pittsburgh Golf Club, the Pittsburgh Club, and the New York Yacht Club. Lauder was a member of the Carnegie Veteran Association and, next to Carnegie, the member with the longest tenure in Carnegie company service. He hosted the 1917 annual association dinner and was the association's second president, succeeding Carnegie. He was, however, conspicuous for his absence from references to the association's early annual meetings. Lauder married Anna Maria Varick on May 2, 1877, and they had three children (George, Harriet, and Elizabeth). He remained Carnegie's friend to the end, but he and his cousin drew further apart with age, especially after Carnegie sold his company. Lauder visited Andrew, but biographers differ over whether the two had renewed their friendship or seen it wither. "Dod" Lauder died on August 24, 1924, outliving his cousin by five years.

George "Dod" Lauder was a valuable engineer and technical adviser in the Carnegie organization. He exerted a strong, conservative influence on the Carnegie associates, especially after the death of Thomas Carnegie in 1886.

References:

Andrew Carnegie, *The Autobiography of Andrew Carnegie* (Boston: Houghton Mifflin, 1920);

William Dickson, ed., *The History of the Carnegie Veteran Association* (Montclair, N.J.: Mountain Press, 1938);

Burton Hendrick, *The Life of Andrew Carnegie*, 2 volumes (Garden City, N.Y.: Country Life Press, 1932);

Hendrick and Daniel Henderson, *Louise Whitfield Carnegie* (New York: Hastings, 1950);

Robert Hessen, *Steel Titan: The Life of Charles M. Schwab* (New York: Oxford University Press, 1975);

William T. Hogan, *The Economic History of the Iron and Steel Industry in the United States* (Lexington, Mass.: Lexington Books, 1971);

Peter Temin, *Iron and Steel in Nineteenth Century America: An Economic Inquiry* (Cambridge, Mass.: M.I.T. Press, 1964);

Joseph Wall, *Andrew Carnegie* (New York: Oxford University Press, 1970).

James Laughlin

(March 1, 1807-December 18, 1882)

by John A. Heitmann

University of Dayton

CAREER: Partner, Alexander Laughlin & Company (1828-1835); president, James Laughlin & Company (1835-1855); president, Pittsburgh Trust Company (1852-1863); partner, Jones, Lauth & Company (1855-1860); partner, Jones & Laughlins Company (1860-1882); owner, Laughlin & Company (1860-1882); president, First National Bank of Pittsburgh (1863-1882).

James Laughlin, banker and ironmaster, was born near Portaferry, County Down, Ireland, on March 1, 1807, and died at his home in Pittsburgh, Pennsylvania, on December 18, 1882. Laughlin played a major role in the establishment and growth of Jones & Laughlin Company, one of the largest iron and steel manufacturers in late-nineteenth-century America. His career and the history of Jones & Laughlin illustrate the importance of financiers to the development of modern heavy industry.

Laughlin's father was a prosperous Irish estate owner, and Laughlin received his formal education in Belfast. Upon graduation he returned home to assist his father in the management of family business affairs. In 1828 his mother died, and he, his father, and two sisters sold the family property and immigrated to America with the purpose of joining his brother Alexander in Pittsburgh. After arriving in Baltimore and then traveling to Pittsburgh, Laughlin formed a partnership with Alexander in the provision business, the new firm known as Alexander Laughlin & Company. A branch office was established in Evansville, Indiana, and Laughlin was placed in charge of a pork-packing operation located there. The partnership continued until 1835, when Alexander Laughlin & Company was dissolved. However, Laughlin continued the provision business and entrusted fellow Irishman James Orr with the Evansville branch, which was subsequently expanded to include an iron business. This relation-

James Laughlin (courtesy of the Carnegie Library of Pittsburgh)

ship between Laughlin and Orr would continue for some 20 years.

By the early 1850s Laughlin was not content to remain in the general provision trade. In 1852 he was instrumental in the organization of the Fifth Ward Savings Bank in Pittsburgh and was elected its president. This institution was reorganized as a state bank later that year and renamed the Pittsburgh Trust Company. This bank was the precursor of the First National Bank of Pittsburgh, which was established in 1863 under the new National Currency Act with Laughlin as its president.

Laughlin's banking activities during the 1850s involved him with the local manufacturing community. He had been a strong financial supporter of the American Iron Works owned by Benjamin Franklin Jones and the Lauth brothers, and in 1855 he became a partner in Jones, Lauth & Company. Five years later Laughlin and his two sons, along with Jones and his two brothers, took control of what had been Jones, Lauth. At the same time Laughlin formed Laughlin & Company, which built two Eliza blast furnaces on the north side of the Monongahela River for the purpose of supplying the Jones & Laughlin South Side plant with smelted iron ore for further processing. Jones & Laughlin Company and Laughlin & Company were consolidated in April 1900 during the late-nineteenth-century wave of mergers, and the unified and integrated firm that bore Laughlin's name became, by the early twentieth century, the largest independent steel producer in America.

Laughlin, as did many other successful bankers and manufacturers of his day, turned to politics and philanthropy in semiretirement. He served a term as a member of the Select Council of Pittsburgh and was appointed president of the board of trustees of Western Theological Seminary, a position that he held until his death. He was also a founder of the Western Institution for the Deaf and Dumb, located in Pittsburgh, and established and was the first president of the Pennsylvania Female College. Laughlin died on December 18, 1882.

Laughlin's preeminent place in the business affairs of Jones & Laughlin Company was taken by his son, James, Jr., during the early 1870s. The junior Laughlin had been educated at Princeton University where he received an A.B. degree in 1868 and an A.M. in 1871. After his years at Princeton the younger Laughlin became secretary and treasurer of Laughlin & Company and was particularly active in the development of the company's iron properties in the Marquette range in Michigan. Thus the Laughlins, James and James, Jr., played an important part in the growth of a nineteenth-century business that was owned by only a few partners. As the iron and steel industry became larger and more complex in its scale of operations, men like the Laughlins surrendered power that previously had been confined within the family to an emerging corporate bureaucracy directed by the professional managers.

References:

Ben Moreell, *"J & L": The Growth of an American Business (1853-1953)* (New York: Newcomen Society, 1953);

George Irving Reed, ed., *Century Cyclopedia of History and Biography: Pennsylvania,* volume 1 (Chicago: Century Publishing, 1910), pp. 259-260.

Bernard Lauth

(c. 1823-?)

by Marc Harris

Ohio State University

CAREER: Operator, iron furnace (?-1852); partner, B. Lauth & Brother (1852-1853); partner, Jones, Lauth & Company (1853-1861); manager, American Iron Works (1861-1864); promoter (1864-?).

Bernard Lauth is known for two major contributions to the technology of rolling mills: the cold-rolling of bar stock and the three-high finishing mill with a small center roll. He was also B. F. Jones's original partner in what became the Jones & Laughlin Steel Company. For all his undoubted contributions to the industry, however, little information is available regarding Lauth's career, and much about it must be interpolated. Born the son of a Pittsburgh taverner, probably in the early 1820s, Lauth went with his brother John to the Ohio Valley in the 1840s. At that time the region of southern Ohio along the great bend of the Ohio River was a major producer of pig iron, and much of the production was controlled by Pittsburgh interests; the city was then the West's major iron-working center but produced very little iron itself. The Mexican War put a premium on western iron production and stimulated the Ohio Valley furnaces in particular. It is possible that this demand allowed the Lauths to make favorable arrangements in Pittsburgh for operation of a blast furnace in the region.

In 1852 the brothers returned to Pittsburgh, where they established a partnership, B. Lauth & Brother, with facilities on the south bank of the Monongahela River. This site, known as the American Iron Works or South Side, later became the nucleus of Jones & Laughlin's Pittsburgh plant. On it the Lauths constructed and put into operation late in 1853 a puddling and milling facility capable of turning out about seven tons of wrought-iron ingots per day.

Perhaps needing capital, the business absorbed two more full partners at the end of 1853, Benjamin F. Jones and Samuel Kier. Jones and Kier were merchants who dealt in iron and brick and also owned a line of canal boats that plied the Pennsylvania Main Line Canal route between tidewater and the Ohio River. Each of the four partners put up capital valued at $5,000, but only three participated actively in the business. Bernard Lauth was overall manager and head of the rolling-mill operations, John Lauth oversaw the small mills, and Jones tended the books and warehouse. Kier had no responsibilities. The three active partners were on salary, and any profits were reinvested in the business. In 1855 Kier and Jones arranged a stock swap which traded Kier's interest in Jones, Lauth & Company for Jones's interests in their other joint ventures. At the same time James Laughlin, the firm's banker, took an option to buy into the operation. The next year the firm was reorganized, with Laughlin becoming a limited partner and John Lauth selling out. Still operating under the rubric of Jones, Lauth & Company, the mill produced a variety of iron goods, including nails, spikes, and rod and bar products.

It was during this phase of the partnership that Lauth made the first of his major contributions to the American rolling-mill industry, a cold-rolling process for bar iron. This development has been described as Lauth's invention, but it is also known that cold-rolling had been used in England and Germany to give sheet and strip iron a polished finish. Lauth purchased American rights from the British patent holder and adapted the technique to bars and rods. The process produced a stronger end product with a finer surface finish and was applied to a wide range of bar products, principally the overhead power shafts widely used in steam-driven textile mills. The company announced this development in 1860, and in 1861 Lauth received a patent for his innovation. He apparently made some efforts to license the patent to other mills.

In 1861, when the original 5-year partnership agreement expired, Lauth decided to sell his interest in Jones, Lauth's facilities. Jones, in turn, was joined by two brothers and Laughlin by his two sons, and the name of the firm was changed to Jones & Laughlin. Lauth's career path after he left Jones, Lauth is unclear. Possibly he had some thoughts of promoting his patent, but whatever his reasons for selling, it seems likely that he remained at least for a time as manager of the rolling-mill operations at the American Iron Works; he has been described as holding that position in 1864 when he introduced his second and more important innovation, the three-high plate-finishing mill with a small center roll.

The three-high mill had its genesis in a basic problem connected with rolling iron products. A mill with two rolls could either turn in only one direction, normally using a flywheel to smooth power flow, or be reversible. In the first case power was used more efficiently, but the workpiece had to be returned to its starting place before each pass, losing heat and putting work crews to backbreaking effort passing the hot material over the top roller or carrying it around the mill. The second option, while solving this problem, entailed tremendous stresses on the machinery that only increased as larger and more complex products were needed. The addition of a third roll enabled the return pass to be productive and the material to be worked fast enough to retain heat, while easing the crew's job. Three-high mills had been adopted by John Fritz at the Cambria Iron Works in about 1857 to roll rails, though they were not widely used for the purpose before 1865. Some of the advantages of three-high design disappeared with the adoption of electrically driven reversing mills in the twentieth century.

The smaller center roll, however, was the innovation that assured Lauth's reputation. Large rolls have advantages in strength that became important as mills needed to roll larger and heavier products. Their greater mass, however, demanded greater power. At the same time, it was generally recognized that small rolls could reduce the material more with each pass even while drawing less power, and would therefore make a mill much more productive. Small rolls, however, could not be made strong enough to do the work required of them. Lauth's mill, the first of the "baby-roll" mills, strengthened the small roll by having it run against each of the large ones alternately; the original design, a jump-stand, featured a fixed bottom roll, with the other two free to move up until limited by adjustable anchor screws. The center roll was thus backed up by the others on each pass. This principle was quickly extended to other types of mills which used various arrangements of small working rolls and larger backing rolls.

Another feature that Lauth described as crucial was a constant flow of water over the rolls. He claimed that the water offered many advantages in a cold-rolling mill designed to finish plate stock to a smooth surface. It allowed the rolls to polish and true each other through a mild abrasive action, and it kept them cool enough to stabilize their dimensions and prevent both the buckling and the characteristically thickened center to which previous designs were prone. The "Lauth mill," combining three-high construction with the small center roll and a constant water spray, he maintained, could produce finished plates of a uniform section at greatly increased speed. The new design was quickly adopted by American milling companies and allowed them to roll plates of unprecedented size for boilers, fire boxes, and other applications.

The course of Lauth's career after this development is not clear. In the early 1870s he aggressively promoted the use of his "Lauth mill" in England and Europe, and in 1872 he told an English audience that he had been doing so for the better part of a decade. This would indicate that he had left the Jones, Lauth works, possibly sometime shortly after he developed his new mill, in order to devote his efforts to promoting the patent. It is known, however, that he continued to live in Pittsburgh until at least 1876, when he renewed his patent on the cold-rolling process.

References:

Frank C. Harper, *Pittsburgh of Today: Its Resources and People*, volume 2 (New York: New York Historical Publishing, 1931);

Alexander Holley, "Three-High Rolls," *Transactions of the American Iron and Steel Institute*, 1 (1872): 287-292;

F. H. Kindl, *The Rolling Mill Industry* (Cleveland: Penton, 1913);

Mackintosh-Hemphill Company, *Rolling Mills, Rolls, and Roll Making* (Pittsburgh, 1953);

Ben Moreell, *"J & L": The Growth of an American Business (1853-1953)* (New York: Newcomen Society, 1953);

Francis P. Weisenburger, *The Passing of the Frontier, 1825-1850* (Columbus: Ohio State Archaeological and Historical Society, 1941).

Lehigh Crane Iron Company

by John W. Malsberger

Muhlenberg College

Chartered in January 1839, the Lehigh Crane Iron Company of Catasauqua, Pennsylvania, was one of the first iron companies in America to succeed in producing commercial quantities of pig iron using anthracite coal. Throughout the nineteenth century the company expanded its capacity steadily, becoming one of the most important iron manufacturers in the Lehigh Valley. It was also one of the longest lived of the early anthracite iron companies, surviving into the early twentieth century as a division of larger corporations. Although its last blast furnace was not taken out of production until 1930, the significance of Lehigh Crane Iron Company clearly rests in the technological change it helped to advance. By developing and promoting the manufacture of iron with anthracite coal, companies such as Lehigh Crane Iron helped to transform the iron and steel industry from one that relied on the small, primitive charcoal forges of the colonial era, to one dominated by the open-hearth furnaces of the large, integrated steel corporations of the late nineteenth century.

The origins of Lehigh Crane Iron Company date to a business decision made in the late 1830s by a Pennsylvania coal and canal company. The Lehigh Coal & Navigation Company, which owned and operated the Lehigh Canal as well as extensive anthracite coalfields around Mauch Chunk, Pennsylvania, north of the Lehigh Valley, began to fear in the 1830s that inadequate transportation might limit the market for their coal to their immediate region. In an effort to diversify their company's products, the directors of Lehigh Coal & Navigation planned to smelt the abundant iron ore that lay around their canal, using anthracite coal in furnaces that had been patented in America by Dr. Frederick W. Geissenhainer in 1833 and in Britain by George Crane in 1837. In July 1838 Lehigh Coal & Navigation offered to provide all the waterpower not necessary for navigation, on any one of its dams, to any company or individual who would spend a minimum of $30,000 to build an ironworks capable of producing commercial quantities of pig iron for at least three months using anthracite coal. The offer was accepted less than a week later when the Lehigh Crane Iron Company was organized by two Philadelphians, Josiah White and Erskine Hazard, both of whom were also major stockholders in Lehigh Coal & Navigation.

Rather than experiment on their own to develop a furnace, Erskine Hazard sailed to Wales in fall 1838, hoping to entice George Crane to come to America to work for Lehigh Crane Iron. Crane, however, was unwilling to leave his native land, but suggested instead that Hazard approach David Thomas, Crane's chief ironmaster, with whom he had collaborated on the anthracite iron process. Hazard was able to overcome Thomas's initial reluctance to leave Wales only by making him a lucrative offer, and on December 31, 1838, a contract was signed. In this agreement Thomas promised to come to America, construct an anthracite iron furnace for Lehigh Crane Iron, and serve as furnace manager for at least five years. In return, Lehigh Crane Iron agreed to pay Thomas's moving expenses, to provide him with a house and coal to heat it, and to pay him an annual salary of £200 until the first blast furnace became operational. Thereafter, Thomas's salary was to increase £50 for every additional blast furnace he brought into operation.

Up to the time this agreement was signed, Lehigh Crane Iron had operated only on an informal basis. Once Thomas's services as ironmaster had been secured, the company was officially chartered on January 10, 1839, and capitalized at $100,000, divided into 2,000 shares of $50 each. Chosen as directors of the new company were Robert Earp (president and treasurer), Josiah White, Erskine Hazard, Thomas Earp, George Earp, John McAllister (secre-

The Lehigh Crane ironworks in Catasauqua, Pennsylvania, circa 1860

tary), and Nathan Trotter. One of the first decisions made by the directors was to select land on the Lehigh Canal near present-day Catasauqua, Pennsylvania, as the site for their proposed furnace. The site selection was of great importance in these early days of iron manufacture and was understandably dictated largely by the location of natural resources. Not only was the site chosen for Lehigh Crane Iron located near large iron ore and limestone beds but it was also bisected by the Lehigh Canal. Thus, the canal served as both a source of waterpower to run the engines of the blast furnace and as a vital means of transportation to bring in raw materials and to connect the company with the rapidly growing urban markets in the Lehigh Valley and Philadelphia regions.

Before it could begin construction of its blast furnace, however, Lehigh Crane Iron was faced with a substantial technical problem. Prior to setting sail for America in May 1839, David Thomas had built a blowing engine in England for the new blast furnace. Unfortunately, the two cylinders of the engine were too large to fit through the hatch of Thomas's ship, hence upon his arrival in America in July, a search had to be made for a foundry capable of casting cylinders with a bore of 5 feet. After great difficulty, Lehigh Crane Iron finally located a foundry in Philadelphia that was willing to retool its shop in order to produce the necessary cylinders. Ground for the first blast furnace at Catasauqua was finally broken in August 1839, and the furnace was blown in on July 4, 1840.

The first blast furnace of Lehigh Crane Iron was lined with firebrick imported from Wales and was operated by a breast wheel 12 feet in diameter and 24 feet long, geared by segments on its circumference to a spur-wheel which drove two blowing cylin-

ders, each 5 feet in diameter and with a 6-foot stroke. The blowing cylinders were powered by changes in the water level of one of the locks on the Lehigh Canal. In its first six months of operation, the blast furnace produced 1,088 tons of pig iron, with 52 tons the largest weekly output, thus demonstrating the commercial viability of anthracite iron manufacture.

Encouraged by its initial success, Lehigh Crane Iron built four additional blast furnaces in the 1840s under David Thomas's supervision. When the Lehigh Valley Railroad commenced operation in 1855, it provided the Catasauqua firm with a more extensive and more efficient transportation system, and this, in turn, provided Lehigh Crane Iron with the impetus for further expansion. In the decade of the 1860s it added a sixth blast furnace, purchased two locomotives to facilitate the movement of raw materials to the plant, and constructed a foundry and machine shop capable of doing most of the casting and repair work for the blast furnaces. By the end of the decade Lehigh Crane Iron had become the largest producer of pig iron in the Lehigh Valley.

The financial uncertainty of the 1870s brought a temporary halt to the expansion plans of the Catasauqua firm, and at least one of its furnaces was shut down. The company was also recapitalized in 1872 and renamed the Crane Iron Company. When more promising business conditions returned at the end of the decade, the firm embarked on a substantial program of modernization. In 1879, for instance, it was decided to dismantle the three oldest furnaces built by David Thomas and to replace them with two larger, more modern blast furnaces. When these two new furnaces were put into operation in 1881, each was capable of producing 102 tons of pig iron weekly, nearly twice the capacity of Thomas's original furnace. By the end of 1881 the completion of the new furnaces gave the Crane Iron Company five operating furnaces with a combined annual output of 100,000 tons of pig iron.

Although the Catasauqua firm had expanded tremendously in the 40 years after its first blast furnace was blown in, it is clear in retrospect that the 1880s marked the beginning of its decline. Rapidly changing technology and business practices within the iron and steel industry combined in the late nineteenth century to make anthracite iron manufacturers such as the Crane Iron Company steadily more obsolete. By the 1880s, for instance, the advent of the Bessemer and open-hearth furnaces had so reduced the price and expanded the output of steel that it was rapidly displacing cast iron and wrought iron as the preferred choice in the business and industrial world. Similarly, by this time the discovery of huge veins of bituminous coal in western Pennsylvania and elsewhere, coal which could be made into coke, a vastly more efficient fuel for blast furnaces, had also begun to erode the position of the anthracite iron industry. Finally, as the iron and steel industry came to be dominated in the late nineteenth century by large, integrated firms such as Andrew Carnegie's company, which enjoyed significant economies of scale, small regional producers found it increasingly more difficult to compete. Thus, the rapid transformation of the iron and steel industry in the late nineteenth century forced producers to adapt or perish. By continuing to rely on the basic process for manufacturing iron with which it began in the 1840s, the Crane Iron Company inevitably followed the latter path.

The demise of the Crane Iron Company, though inevitable, came slowly. In the depressed economy of the 1890s one of the firm's furnaces was shut down and dismantled. New life was injected temporarily into its operations in 1899 when the Crane Iron Company was acquired by the Empire Steel & Iron Company of New Jersey, a holding company formed to buy up small iron mills, railroads, and ore fields. The continuing pressure of technological advancement and monopolization within the iron and steel industry inexorably eroded the position of small producers, however, and by 1914 the Crane Iron Company had only two furnaces still in operation. The Catasauqua firm was acquired in 1922 by the Replogle Steel Company, and with the onset of the Great Depression, iron production was halted permanently at the Crane Works in 1930 and its last remaining blast furnace dismantled.

The commercial success achieved by the Lehigh Crane Iron Company was clearly an important factor in the dramatic transformation of the nineteenth-century American iron and steel industry. Prior to its demonstration of the feasibility of manufacturing iron with anthracite coal, the chief fuel used in producing iron had been charcoal. Consequently, the early iron forges of America had generally been located near forests, the source of the wood for charcoal, meaning that the iron producers were often far removed from the markets they served. The cost of transportation that was thus added to the price of iron helped make it an expensive and relatively scarce commodity in the charcoal iron age. The method pioneered by Lehigh Crane Iron and other companies allowed iron to be manufactured in larger quantities and closer to the burgeoning urban markets. As a result, the efforts of Lehigh Crane Iron provided an important step by which iron became a cheaper, more readily available commodity.

References:

Craig Bartholomew, "Anthracite Iron Making and Industrial Growth in the Lehigh Valley," *Proceedings*, Lehigh County Historical Society, 32 (1978): 129-183;

M. S. Henry, *History of the Lehigh Valley* (Easton, Pa.: Bixler & Corwin, 1860);

Manufacturing and Mercantile Resources of the Lehigh Valley (Philadelphia: Industrial Publishing, 1881);

Alfred Matthews and Austin N. Hungerford, *History of the Counties of Lehigh and Carbon* (Philadelphia: Everts & Richards, 1884);

Charles Rhoads Roberts, et al., *The History of Lehigh County, Pennsylvania*, 2 volumes (Allentown, Pa., 1914).

Rebecca Lukens

(January 6, 1794-December 10, 1854)

by Julian Skaggs

Widener University

CAREER: Ironmaster, Brandywine Iron Works (1825-1847).

Rebecca Lukens was born on January 6, 1794, in West Marlboro Township, Chester County, Pennsylvania. She was the eldest surviving child of the six born to Isaac and Martha Webb Pennock. The Pennock family settled in Chester County in 1710 and made their living by farming. Isaac Pennock, not captive to tradition, abandoned farming for the iron business in 1793, a change made in the face of strenuous objections from his father and other family members. The new enterprise was called the Federal Slitting Mill, and its product was chiefly strips and rods of wrought iron which were subsequently fashioned into spikes, barrel hoops, wheel rims, and blacksmith iron.

In 1810 Isaac Pennock, in partnership with Jesse Kersey, a Quaker minister, purchased a tract of property along the Brandywine River in what is now Coatesville. There was already a water-powered sawmill on the site, and the partners converted it to an iron rolling mill. It was this mill that became the foundation of the Lukens fortune; it was named the Brandywine Iron Works.

Rebecca Pennock's early life and education were pleasant and conventional. Her father, "affluent, in his circumstances," had enrolled her in a boarding school in West Chester, Pennsylvania, when she was twelve. A year later she entered a boarding school for young ladies in Wilmington, Delaware, where she flourished. Of her childhood she said, "Every innocent amusement was allowed me" and that she was as "wild and happy and joyous as youth could make me." At school she found herself "a favorite with my teachers and at the head of all my classes." She read everything that "fell in my way" from wild dramatic stories to Shakespeare. She also gave careful study to chemistry and French, liked to take walks, and was fond of riding

Rebecca Lukens

horseback. She thought hers a nearly perfect upbringing, writing that "I was young, ardent and happy."

Shortly after finishing school, Rebecca Pennock visited Philadelphia with her father, and while there she met a young Quaker physician named Charles Lukens. They were married in 1813 and had six children, of which only three daughters survived to maturity. They were Martha (Mrs. Abraham Gibbons, Jr.), Isabella (Mrs. Charles Huston), and Charlesanna (Mrs. William Tingley).

Charles Lukens abandoned medicine after his marriage and became an ironmaster, successfully managing the Federal Slitting Mill for Isaac Pennock. In 1815 Isaac Pennock bought out Jesse Kersey's share of the Brandywine property, and Charles Lukens became a partner of Pennock and ran the Brandywine Mill. Pennock was a silent partner in the venture, and Dr. Lukens saw to the daily administration of the business. He immediately ran into difficulties. The mill's foundations were discovered to be too light for the machinery installed in it, and as a consequence the mill was always drifting off center, constantly causing expensive machinery to break. Charles and Rebecca Lukens were persuaded to keep at the business, and to invest in plant improvements by Isaac Pennock's promise that the mill was to be willed to Rebecca at his death.

This reassurance kept Lukens at the mill, which was reinforced and repaired several times. A severe flood almost wrecked it in 1822, but the mill survived. In spite of these various difficulties, Charles Lukens looked for new markets and began to experiment with new products. Most notably, he began to experiment with the production of wrought-iron boilerplate. His pioneering work paid off as his mill became the first in America to roll that product on December 30, 1818. There was another successful pioneering effort that Charles Lukens would claim. In March 1825 he received an order for sheet iron that was destined for the fabrication of the hull of the *Codorus*, the first iron-hulled vessel built in the United States. These two successes anticipated what would prove to be the Lukens company's principal markets for the next 150 years.

There were inevitably unanticipated difficulties that were to face the business in its future; two harsh blows fell in 1824 and 1825. In 1824 Isaac Pennock died leaving a muddled will, and the next year Charles Lukens died, leaving his wife Rebecca pregnant, in debt, and bound by his deathbed wish that she assume the responsibility of managing the business. Rebecca Lukens was devastated. She bore her husband's posthumous sixth child, a daughter named Charlesanna, and began to take up a responsibility which she had not anticipated and for which she had no formal training. She was thirty-one years old.

The young widow immediately ran into other trials. She found that there was a legal cloud on the use of and possession of the mill which she thought was hers. She found that the real property of her father's estate could not be divided until Rebecca

Lukens's youngest sister attained her majority in 1827. Moreover, an appraisal of all the property would have to be made before it could be distributed to the heirs. The verbal understanding that Rebecca Lukens would come into clear possession of the mill was worthless, and remained unsettled for another 25 years.

The property so wished for by Rebecca Lukens was in poor repair, and when the books of the estate were examined an "alarming deficiency" was shown. Martha Pennock did not encourage her daughter to go into the iron business. Indeed Rebecca Lukens wrote that "Mother wanted me to leave Brandywine and said it would be folly for me to remain" and that she "thought as a female I was not fit to carry on such a concern." Nor did her mother offer any assistance to the widow if she left the Brandywine. Years later Rebecca Lukens observed that "necessity is a stern taskmistress and my every want gave me courage," and ". . . where else could I go and live." Thus driven and finding that "grief made me eloquent," Rebecca Lukens obtained grudging assent from her mother to carry on the business, and she set about her new career.

Though sorely pressed, Rebecca Lukens was in possession of several assets. Her husband's friends were also hers, and they gave her encouragement and support. Among these were Charles Brooke and James Sproul, both of whom supplied her with iron blooms on credit. Her small labor force was experienced and loyal, and her husband's brother, Solomon Lukens, came to Coatesville to be the foreman at the mill. These people and their various talents made an effective combination. The mill began to make money, and Rebecca Lukens started to pay off her late husband's debts and made other payments to the estate to equalize the shares of her sister's claim.

This done, Rebecca Lukens felt that her claim to the property was clear, and she set about improving it. By the early 1830s the much repaired old mill "could no longer go and was ready to fall on the heads of the workmen . . ." By 1834 she had a new mill in operation. She described the job thus:

> The mill has been entirely remodeled and rebuilt from the very foundation. Dam entirely newly built, wheels put in, castings, furnaces, millhead, mill house much larger, all were built anew; not a vestige of the old remained and not a dollar of all this or any part of it came from my mother or father's estate. . . . I had built a very superior mill, though a plain

one and our character for making boiler iron stood first in the market; hence we had as much business as we could do. Prices were then good, I had few competitors and the opening of the Pennsylvania Railroad gave our iron ready access to market.

It was a good mill; it carried the family's fortune with little change until 1870 when it was converted to a puddling mill. The structure remained in hard use until the late 1880s.

Rebecca Lukens made other improvements to her property. Her own home was repaired, she built "good and substantial" tenant houses for her workmen, and she fenced and limed the fields of her farm. No sooner were these things done, than new misfortune struck. There was a lawsuit that threatened the water supply to the new mill, the settlement of which cost Lukens $800 and a year's worry. On the heels of this, the panic of 1837 gave Rebecca Lukens, and everyone else, a good scare. In May 1837 she wrote to her sister:

> Mother is returning this afternoon and I could not let such an opportunity pass without addressing a few lines to thee. The difficulties of the times throw a gloom on everything. All is paralyzed—business at a stand. I have as yet lost nothing but am in constant fear, and have even forbidden my agents to sell, not knowing who would be safe to trust.
>
> I have stopped rolling for a few weeks, and set my men to repairing the race dam, and having a heavy stock manufactured already. I do not wish to increase it until the times are more settled; but shall take advantage of the first gleam of sunshine to resume.
>
> We do not know how to do without a circulating medium. Every one that has a dollar in silver hoards it up as if he never expects to see another, and our cautious people are yet afraid of your small notes.

Rebecca Lukens's mother died in 1844 and left the estate in yet another ambiguous muddle. Once again Rebecca Lukens had to defend her title to the mill, and she made heavy cash payments to secure her own possession of it. In the end, all these trials were overcome, and she survived them handsomely.

In fact, from the late 1830s onward Lukens advanced admirably in the business. In the early 1840s she added a store and freight warehouse to her primary business and began to make money. In 1844 she celebrated her success in her personal journal by listing some of her income and some of the debts owed her. These sums were not listed in conventional bookkeeping style; they were written out in full, thusly:

> My rents are as follows: Mill $120 per year. Warehouse $200. Storehouse $150. Saddler shop $18. Also average $15 a year for ten dwellings. Also Geo. Gleming has $5500 of my money. Alban Hook owes me $5000. Solomon Lukens owes me $6300. John Mitchell owes me $6000.

The management of the business underwent a series of changes in the 1840s. In 1840 Solomon Lukens retired from the firm and left to do missionary work among the surviving Indians of Pennsylvania. Rebecca Lukens took a new junior partner named Joseph Baily, who was, like Solomon Lukens, her brother-in-law. Baily stayed on until 1842, when he left to open his own ironworks near Pottstown, Pennsylvania. Baily was replaced by Abraham Gibbons, Jr., a son-in-law to Rebecca Lukens. At this point the business was called Lukens & Gibbons.

This arrangement remained in place until October 1847, when Rebecca Lukens retired from active management in the firm now known as A. Gibbons & Company and became a silent partner entitled to two-thirds of the profits. A year later A. Gibbons & Company took another partner, Dr. Charles Huston, yet another son-in-law of Rebecca Lukens. At that time Rebecca Lukens left the business altogether. The Gibbons and Huston partnership lasted until 1855, when Gibbons left the iron business to take up a career in banking. After that the firm was guided by the Hustons with remarkable success into the twentieth century.

In spite of the requirement of running a business, choosing partners, and raising a family, Rebecca Lukens found time for other interests and concerns. She was a Quaker, and her surviving writings show her to have taken her God's implacable judgments and her inescapable duties very seriously. Thus she held to her Quaker pacifism even though it cost her money. By way of example, an order for boilerplate destined for the Boston navy yard was refused.

She was also active in charitable works, financially helping Solomon Lukens after he left to work among the Indians. He left with her blessing so that he might be "useful to this wronged and persecuted race." She was also an active abolitionist; the family, according to Lukens's granddaughter, hid a runaway slave on his passage northward.

To the poor near her, she felt an obligation. She said that her heart felt for them, but typically she went on to confess that:

> I am far from satisfied with what I have been able to accomplish. I must try to do more—be more active, energetic in the cause of humanity. There are positive duties incumbent on all—I must seek them out and never weary in the work.

Rebecca Lukens took an active interest in politics too. She supported Henry Clay in the presidential campaign of 1844, presumably because of his stand for high tariffs. Though she could not vote, her support for Clay was not passive. She was moved to compose the following bit of campaign verse in praise of the man:

> Hail to the omen of a brighter day
> When Freedom's bugle sounds the name of
>   Clay
> Arouse then Patriots with a general voice
> Proudly proclaim a Nation's choice.

Though the election went against Clay, this period was probably the most satisfying of her life. The business was thriving, her personal financial situation guaranteed a security that had been absent only 20 years earlier, and she was seeing to the orderly succession of management within her firm. But tragedy and remorse were not far off. Shortly after her retirement in 1848, Rebecca Lukens's youngest surviving child, Charlesanna, died in childbirth. After her retirement and the death of her daughter, everything turned to ashes. On December 1, 1850, she wrote in a diary kept intermittently that:

> I look over my life with regret—so many opportunities of being useful and doing much good utterly neglected and now at this late approach of time it seems almost a hopeless task to redeem the errors of the past or strike out a new path for the future.

A later entry in that diary shows that while she thought her past was an error in some unspecified way, it still attracted her and she missed her old life badly.

> I believe that having passed so many years in constant excitement has had a very deleteri-

ous effect on me—It was a stimulant to my existence and now I feel the want of something to give an impetus—a spur to the routine of everyday life—and not finding this I become apathetic and indifferent to all around me. It is an incubus I greatly desire to shake off.

These two black entries reveal how she felt about herself and her life during her retirement. The diary shows only a brief series of lamentations and beseeching prayers. The last entry in this document is dated January 7, 1852, and it contains this: "I feel so weak—that without His aid I can do nothing. Lord will thou not help thy poor unworthy servant?" Rebecca Lukens died two years later on December 10, 1854.

Her assessment of her life is harsh, sad, and inappropriate. Her efforts as a businesswoman were brilliantly successful in the face of grave difficulties. She had finished the pioneering work of her husband by establishing a market for high-quality boilerplate and boat iron. The new plant she built in 1834 ran well, sustained the family through good times and bad, and remained essentially as she had designed it until 1870. Two years after she retired from active work, the partners she took, the market she cultivated, and the mill she made combined to give Lukens the best year ever with a production run of 944 tons of plate. This iron was sold in a national market that extended from Boston to New Orleans. She had left the business in the hands of chosen, capable successors. At her death Rebecca Lukens left a personal estate valued at $107,000, and her real estate was probably worth at least the same amount. Judged as an entrepreneur, her achievements were undeniably remarkable.

She did not judge herself, however, simply as an entrepreneur. She had never planned to go into business, but circumstance and obedience to a dying husband conspired to put her in an unusual position, one that she occupied with unusual success. She saw herself as an interim trustee dedicated to increasing the estate of her children. There is no evidence that she ever thought her surviving children, all daughters, should be trained to run the business; she had no intention of imposing her lot on them. She did, however, impose her activities on her daughters' husbands. Those men were tested as co-partners, and as soon as they were found competent, Rebecca Lukens retired from active management. But she missed it all very badly as her diary shows.

Rebecca Lukens's sense of loss was born out of her acceptance of the gender stereotypes of her age. She had been raised and trained to be an accomplished young woman, a good wife, and a wise mother. Widowhood and necessity drove her out of a conventional niche and into a situation where she knew the excitement that comes from the assumption of great responsibility, and the pleasure of exercising independent executive authority. Given this it is no wonder that she was deeply unhappy when

she returned to the constrained world of mid nineteenth-century women.

References:

Isabella Huston, *Autumn Leaves* (N.p., 1873);
Stewart Huston, "The Iron Industry of Chester County," *Southeastern Pa.*, edited by V. Bennett Nolan (Philadelphia, 1943);
Clara Huston Miller, *Deminiscences* (N.p., 1930).

Archives:

Materials concerning Rebecca Lukens are located at the West Chester Historical Society, West Chester, Pennsylvania, and at the Hagley Museum and Library, Wilmington, Delaware.

Marquette Iron Company

by Terry S. Reynolds

Michigan Technological University

The Marquette Iron Company was a typical early company of the kind created to exploit the iron ores discovered on the southern shore of Lake Superior in the mid 1840s. Short on financial resources, plagued by the high costs of transportation between Michigan's Upper Peninsula and the eastern markets, and attempting to produce finished iron at or near the mine site instead of simply shipping iron ores elsewhere, Marquette Iron operated for only three years. But in its brief history it made the first commercial shipment of iron ore (six barrels) from the Lake Superior ore district.

The roots of the Marquette Iron Company date back to 1844 when U.S. Surveyor William S. Burt noticed a substantial deviation of his magnetic compass near what is now Negaunee, Michigan. Suspecting that this indicated the presence of large deposits of iron ore, he and his assistants searched for and quickly found specimens. Reports of Burt's findings spread quickly, and in the winter of 1844 and 1845 a group of Jackson, Michigan, businessmen formed the Jackson Iron Company. The area's plentiful supply of hardwoods and the high cost of transporting the castings and machinery needed to operate a blast furnace to the area, led the Jackson Iron Company to attempt to produce bar or wrought iron directly using traditional forging tech-

niques. Their Carp River Forge made its first bloom on February 10, 1848.

Burt's discovery and the Jackson Iron Company's activities soon attracted others to the area. Among these newcomers was Robert J. Graveraet, whose actions led to the formation of the Marquette Iron Company. Graveraet had been living at the far eastern end of Michigan's Upper Peninsula at the fur trading post of Mackinac. He visited the area claimed by the Jackson Iron Company around 1846 and in 1848 persuaded two friends to assist him in laying claim to an ore outcrop near the Jackson Mine which, unfortunately, was also claimed by the Cleveland Iron Company. In 1848 Graveraet met Edward Clark, who had been sent to explore copper prospects on the Keweenaw Peninsula of Upper Michigan by Waterman A. Fisher, owner of a large Massachusetts textile mill. After visiting the Jackson Mine and the area claimed by Graveraet and his associates, Clark returned east with iron ore specimens to talk with his sponsor. Graveraet followed on his own initiative and was successful in persuading Fisher to form an iron company.

Waterman Fisher, together with Clark, Graveraet, and Amos R. Harlow, the owner of a small Massachusetts machine shop, formally created the Marquette Iron Company on March 5,

The Marquette Iron Company in 1852 (courtesy of Marquette County Historical Society)

1849. The proprietors of the new company followed the course already taken by the Jackson Iron Company, laying plans both to mine ore and process bar iron in the Upper Peninsula. Harlow was placed in charge of ordering the components for a forge and transporting them to the Upper Peninsula. Graveraet returned to the Upper Peninsula in May 1849 to begin clearing a site for the forge, building housing for workers, and erecting a dock; the site was initially named Worcester, in honor of Fisher's hometown, but was soon changed to Marquette.

Harlow reached the site in July 1849 and began erecting the second forge in the region. The new forge went into operation on July 6, 1850, supplied with ore purchased from the Jackson and Cleveland mines. By the fall of 1850 the Marquette Iron Company employed around 70 men.

The Marquette Iron Company soon encountered problems. Its land claims were overturned in November 1850, the rights going to the Cleveland Iron Company. But the company succeeded in leasing half interest for 99 years in an adjacent ore property, and its forge continued to operate steadily. In May 1852, for example, the company shipped 1,905 blooms of iron, weighing more than 120 tons, from Marquette. Moreover, in July 1852 the company loaded six barrels of iron ore on the steamer *Baltimore* on consignment to Detroit, the first shipment of iron ore from the Lake Superior district.

The high cost of transportation, however, made the company's operations unprofitable. The only available transportation routes to market involved shipping iron by boat to Sault Ste. Marie, at the northeastern tip of the Upper Peninsula. Rapids there did not permit Lake Superior shipping to proceed further. Goods had to be unloaded, portaged around the rapids, and then reloaded aboard different vessels for transportation to ports on the lower Great Lakes. The distance of Marquette from markets and the cumbersome nature of the transportation system made the price of Michigan iron, both in ore and bar form, uncompetitively high. In the early 1850s, for example, it cost around $200 to produce a bar of iron at the Marquette Iron Company forge and to ship it to Pittsburgh, where wrought iron was selling for $80.

Waterman Fisher, whose funds had largely sustained the early operations of the Marquette Iron

Company, was ready to sell his interests by late 1852. In April 1853 the Cleveland Iron Mining Company purchased the assets of the Marquette Iron Company, thus ending its corporate history.

Unpublished Document:

Burton Boyum, "Cliffs Illustrated History," unpublished manuscript, c. 1986, in Michigan Iron Industry Museum, Negaunee, Michigan.

References:

H. Stuart Harrison, "The Cleveland-Cliffs Iron Company," *Transactions of the Newcomen Society in North America*, 44, no. 6 (1974): 8-10;

Harlan Hatcher, *A Century of Iron and Men* (Indianapolis: Bobbs-Merrill, 1950);

Kenneth D. LaFayette, *Flaming Brands: Fifty Years of Iron Making in the Upper Peninsula of Michigan, 1848-1898* (Marquette: Northern Michigan University Press, 1977).

Samuel Livingston Mather

(July 1, 1817-October 8, 1890)

by Terry S. Reynolds

Michigan Technological University

CAREER: Secretary, Cleveland Iron Company (1847-1850); director (1850-1853), secretary-treasurer (1853-1869), president, Cleveland Iron Mining Company (1869-1890).

Samuel L. Mather, one of the figures primarily responsible for opening the rich iron ore resources of the Lake Superior region in Michigan, was born on July 1, 1817, in Middletown, Connecticut, the son of Samuel and Catherine Livingston Mather. Samuel L. Mather's father owned both extensive local properties and large numbers of shares in the Connecticut Land Company, which held tracts of unexplored land in the Western Reserve of Ohio. He was also a prosperous businessman.

Mather spent his childhood in Middletown, Connecticut, attended Wesleyan University in Middletown, and graduated with its first class in 1835. He then worked for his father in Middletown and New York City in the commission business. After acquiring experience under his father he set up a commission business of his own.

In 1843 Mather moved to Cleveland to sell his father's Western Reserve holdings and to serve as real estate agent for other Connecticut families with Western Reserve lands. Although he studied law and was admitted to the bar in the late 1840s, he was soon diverted from legal practice by reports of large metal deposits on the southern shores of Lake Superior. In the winter of 1845-1846 Mather and several other Clevelanders organized the Dead

River Silver & Copper Mining Company of Cleveland, and they sent J. Lang Cassels to explore possible copper and silver deposits. Cassels, however, was persuaded while en route to look at iron deposits on the southern shore of Lake Superior, near what is now Negaunee, Michigan.

On his return Cassels persuaded his associates, including Mather, to transfer their interests from copper and silver to iron. In November 1847 the group created the Cleveland Iron Company with Mather as secretary. In 1850 the company reorganized itself as the Cleveland Iron Mining Company of Michigan to meet the requirements of Michigan law. Mather was named to the reorganized company's board of directors. In 1853 the Cleveland Iron Mining Company purchased the Marquette Iron Company, which had struggled for some years to forge iron in Michigan's Upper Peninsula rather than simply ship ore. To complete this transaction the company was reorganized again, and capitalization was increased. Mather served as secretary-treasurer from 1853 and, though not initially the leading stockholder, quickly became the driving force behind the company.

By 1860 the Cleveland Iron Mining Company had become a successful iron ore producer on Michigan's newly opened Marquette Range. By 1868 its annual shipments had reached 100,000 tons, and it had become one of the nation's leading iron ore producers. In 1869 Mather was elected president of the Cleveland Iron Mining Company. He

Samuel Livingston Mather

held both posts until his death in 1890, guiding the company to a preeminent position on the Marquette Range and playing a major role in making Cleveland a major iron and steel center.

In his last years Mather further consolidated the position of the Cleveland Iron Mining Company. In 1889 he contracted for a fleet of steel ore carriers, the first owned by any mining company. In 1890 he purchased a controlling interest in the Iron Cliffs Company, a mining company with extensive holdings on the Marquette Range. The formal consolidation of Iron Cliffs with the Cleveland Iron Mining Company to form the Cleveland-Cliffs Iron Company, long the dominant iron-mining company on the Marquette Range and a major independent iron producer in the Lake Superior district, did not occur, however, until 1891, following Mather's death in Cleveland on October 8, 1890.

In addition to the Cleveland Iron Mining Company, Mather invested in and held offices in many other iron-related concerns headquartered in the Cleveland area, including the Cleveland Boiler Plate Company, the McComber Iron Company, the Bancroft Iron Company, and the American Iron Mining Company. Mather was widely known and respected in late-nineteenth-century iron-mining circles. His personal integrity and his conservative policies provided a tranquilizing effect on a normally volatile industry. Moreover, according to Mather's biographer in the *National Cyclopedia of American Biography* he "did more than any other one man to bring Cleveland and the Great Lakes region to their position of supremacy in the iron industry." Two of Mather's sons, Samuel and William G. Mather, continued their father's eminent association with iron mining and the iron trade in the twentieth century.

Unpublished Document:

Burton Boyum, "Cliffs Illustrate History," unpublished manuscript, circa 1986, in Michigan Iron Industry Museum, Negaunee, Mich.

References:

"A Bond of Interest," *Harlow's Wooden Man* (Journal of the Marquette County [Michigan] Historical Society), 13 (Fall 1978): 5-8;

The Cleveland-Cliffs Iron Company and its Extensive Operations in the Lake Superior District (Cleveland, Ohio: Cleveland-Cliffs Iron Company, 1919).

Archives:

Samuel L. Mather's papers are in the W. G. Mather Library, Cleveland, Ohio.

Joseph McClurg

(unknown- 1825)

by Paul F. Paskoff

Louisiana State University

CAREER: Merchant (1798-1803 or 1805); owner, McClurg & Company (1803 or 1805-1815).

An Irish immigrant to the United States in 1798, Joseph McClurg was a pioneer in the development of the iron industry in Pittsburgh, Pennsylvania, where he first established himself as a wholesale and retail merchant. Because Pittsburgh then had a population of less than 1,600 and, therefore, could hardly have supplied McClurg with the store goods wanted by his customers in and around the city, he dealt regularly with Philadelphia and Baltimore merchants. This mercantile experience unquestionably helped him when he shifted his interests to iron making.

In 1803 or 1805 (the sources disagree as to which date is correct) McClurg built the first foundry in Pittsburgh which, only a decade earlier, had scarcely supported any semblance of an iron industry at all. In fact, the city's first venture in iron making, George Anshutz's furnace of 1793, lasted only a year before it failed. Due in large measure to the enterprise and persistence of McClurg, Anthony Beelen (another foundry owner), and a few others, Pittsburgh and its surrounding area quickly began to develop as an iron production center. This growth was furthered in 1810 when McClurg constructed at least two furnaces in Westmoreland County, just south of the city. However, the development of the iron industry in Pittsburgh and its environs received its greatest stimulus not from the efforts of individual ironmasters, but from the outbreak of war with Great Britain in 1812.

Joseph McClurg (engraving by A. H. Ritchie, courtesy of the Carnegie Library of Pittsburgh)

The War of 1812, which enjoyed considerable support along the nation's western frontier, found McClurg in a good position to serve the war effort and profit from it. Aware of the demand by the army and navy for cannon and shot, McClurg converted his business into a cannon foundry and operated it throughout the war under a substantial contract with the federal government, supplying the Lake Erie fleet of Commo. Oliver H. Perry and the army of Gen. Andrew Jackson at New Orleans with artillery.

The war had clearly been good for Pittsburgh's economy, including the iron industry, and especially McClurg's firm. The war's end found the city unabashedly proclaiming itself "The Birmingham of America." Although this assertion was as inaccurate as it was enthusiastic, the city's inhabitants had reason to be impressed by its iron-making enterprises. By 1815, about the time that McClurg retired and left his share in the foundry to his son, Alexander, these included rolling mills and a steam engine works, as well as the McClurg and Beelen foundries. Nearby, in surrounding Allegheny County, a vigorous group of blast furnaces and nail mills was in operation.

McClurg's retirement in 1815 enabled him to enjoy the considerable fortune which he had amassed during the war as one of only sixteen "Gentlemen" listed in James M. Riddle's *The Pittsburgh Directory for 1815*. His firm, now directed by Alexander McClurg, continued to operate under the name of McClurg & Company. By 1825, when Joseph McClurg died, the firm's partners had expanded its physical plant and range of products by building furnaces to complement its foundry and had contracted with the federal government to supply the army with artillery. The firm was still in business under the name McClurg, Pratt & Wade in 1832 when U.S. Secretary of the Treasury Louis McLane's report on manufactures listed it as a steam engine manufactory. It was by then steam powered, employed some 80 workers, and claimed an annual output of $95,000 worth of iron and machinery, of which about $50,000 represented steam engines. The firm's prosperity during the 1820s and early 1830s apparently was not enough to enable it to survive the panic of 1837 and the ensuing depression of the early 1840s. By 1850, when Pennsylvania's ironmasters conducted a census of their industry, no firm bearing the name of McClurg was listed.

References:

J. Frederick Byers, "Pittsburgh and the Iron Industry," in *Pittsburgh and the Pittsburgh Spirit: Addresses at the Chamber of Commerce of Pittsburgh, 1927-1928* (N.p., n.d.);

Samuel Jones, *Pittsburgh In The Year 1826* (1826, reprinted, New York: Arno Press, 1970);

Stefan Lorant, *Pittsburgh: The Story of an American City* (Garden City, N.Y.: Doubleday, 1964);

Louis McLane, *Documents Relative to the Manufactures of the United States* (1833, reprinted, New York: Burt Franklin, 1969);

Paul F. Paskoff, *Industrial Evolution: Organization, Structure and Growth of the Pennsylvania Iron Industry, 1750-1860* (Baltimore: Johns Hopkins University Press, 1983);

Catherine E. Reiser, "Pittsburgh, The Hub of Western Commerce, 1800-1850," *Western Pennsylvania Historical Magazine*, 25 (March-June 1942): 121-134;

James M. Swank, *History of the Manufacture of Iron in All Ages* (Philadelphia: American Iron and Steel Association, 1892).

David Cummings McCormick

(August 22, 1832-March 12, 1910)

by John A. Heitmann

University of Dayton

CAREER: Iron maker, Shoenberger iron interests (1852-1885).

David Cummings McCormick, an iron manufacturer, was born near Savannah, Georgia, on August 22, 1832, and died in Pittsburgh, Pennsylvania, March 12, 1910. McCormick's mother was the daughter of Dr. Peter Shoenberger of Pennsylvania. Shoenberger, known by his contemporaries as the "iron king," erected a number of iron furnaces during the 1820s and 1830s named after his daughters–the Sarah Furnace, the Martha Furnace, the Maria Furnace, and the Rebecca Furnace. Shoenberger had also owned the Juniata Forge in Huntington County, Pennsylvania, and erected the Juniata Works, the first rolling mill in Pittsburgh. It was in this family iron business that Shoenberger's son John H., son-in-law Pollard McCormick, and grandson David C. McCormick made their fortunes.

David C. McCormick was educated at Carlisle College in Carlisle, Pennsylvania, and at Yale. After his formal schooling he entered the family business at the Sarah Furnace located at Holidaysburg, Pennsylvania. It was the first operation to supply pig iron to the Pittsburgh rolling mills. The Sarah Furnace's stack was a tall stone structure approximately 50 feet high with a 70-foot-square base tapering to about 30 or 40 feet at the top. It was located on the side of a hill, which facilitated the charging of the huge furnace with alternate layers of charcoal, iron ore, and limestone. Once charged, the charcoal was ignited and a blast of hot air was introduced into the furnace to promote combustion; the fusion of impurities with the limestone formed a dross, which was subsequently separated from the molten iron.

Outside the furnace a sand bed was prepared in the casting shed. With the tapping of the furnace, molten iron flowed into channels shaped in-

David Cummings McCormick

side the bed. As soon as possible before the metal cooled men in wooden-soled shoes wielded sledgehammers to break the metal into pigs of a convenient size which were then transferred to wagons. At the forge this material would be heated almost to the point of fusion and then shaped at an anvil by the continuous battering of a trip-hammer. The resulting forged block of iron was called a "bloom," which was transported by mules and horses to Pittsburgh for further processing. The supervision of this type of manufacturing was McCormick's responsibility, and his success and that of the Shoenberger family iron companies resulted in the amassing of a considerable fortune by the early 1860s.

As in the case of many nineteenth-century manufacturers, McCormick spent the last 25 years in
semiretirement, promoting community activities
and being active in various philanthropies.

Reference:
Calvin W. Hetrick, *The Iron King* (Martinsburg, Pa.:
Morrisons Cove Herald, 1961).

Philip Louis Moen

(November 13, 1824-April 23, 1890)

by Alec Kirby

George Washington University

CAREER: Partner, president, Quinsigamond Iron
& Wire Works (1846-1868); partner, I. Washburn
& Company (1850-1865); partner, I. Washburn &
Moen Wire Works (1865-1868); vice-president
(1868), president (1868-1890), treasurer, Washburn
& Moen Manufacturing Company (1875-1890).

Philip Louis Moen, a manufacturer, was born
in Wilna, New York, on November 13, 1824. During Moen's childhood his parents, Augustus and
Sophie Moen, repeatedly relocated before finally settling in Brooklyn, New York, where Augustus
worked as a representative for British hardware
firms. The family was prosperous enough to afford
a private tutor for Moen, and he was apparently
ready to begin studies at Columbia College when
eye problems forced him to discontinue his studies.
Although the exact nature of the problem is not
clear, it apparently did not cause him great difficulty in his career as a manufacturer. Having abandoned his education, Philip began to work for his
father, from whom he learned the fundamentals of
business. In 1846 he married the eldest daughter of
Ichabod Washburn, of Worcester, Massachusetts,
the founder of a wire production facility. The following year Moen and his wife moved to Worcester
where he became a partner in a firm established to
construct a rolling mill.

In 1850 Moen took a decisive step by joining
his father-in-law's business. He purchased a half interest in the firm, I. Washburn & Company, which
was already a leader in the steel wire industry, utilizing innovations devised by Washburn. After creating the 1850 partnership, Washburn appointed
Moen vice-president. A talented inventor, Washburn continued his experiments, concentrating on
the production of piano strings, and developed proce-

Philip Louis Moen

dures for greatly expanding output. This enabled
the firm to break into a market which for years
had been monopolized by the British. Meanwhile,
sewing machines were creating a demand for crinoline wire and steel wire for needles. To raise capital
to increase capacity the I. Washburn and Company
incorporated under the name I. Washburn & Moen
Wire Works, and the company began a rapid expansion. The corporation constructed new plants in

South Worcester and in Quinsigamond and in 1868 again reorganized under the name of Washburn & Moen Manufacturing Company.

Moen continued to serve as vice-president in the newly reorganized firm. On December 30, 1868, Washburn died, and Moen became president, and in 1875 its treasurer as well. Under his leadership the firm continued its rapid expansion, and in 1890 the company established a rolling mill and wire-producing plant in Waukegan, Illinois, the first Washburn & Moen facility outside New England. By 1895 the firm was producing 100,000 tons annually of various wires.

Moen served on the board of several prominent corporations, including the State Mutual Life Assurance Company, and was the first president of the board of trustees of the Memorial Hospital, founded by Ichabod Washburn in memory of his daughters. Moen died on April 23, 1890, survived by Maria S. Grant, his second wife, a son and two daughters.

Reference:
J. D. Van Slyck, *Representatives of New England* (Boston: Van Slyck and Company, 1879).

Daniel Johnson Morrell

(August 8, 1821-August 20, 1885)

by Stephen H. Cutcliffe

Lehigh University

CAREER: Clerk, Trotter, Morrell & Company (1837-1842); partner, wholesale dry goods business (1842-1845); clerk and partner, Martin, Morrell & Company (1845-1855); general manager, Cambria Iron Company (1855-1884); member, U.S. House of Representatives (1867-1871); president, American Iron and Steel Association (1879-1884).

Daniel Johnson Morrell, following an early career in the dry goods business, was a leading pioneer in the iron and steel industry at the Cambria Iron Works where he served as general manager for three decades. Morrell was born on August 8, 1821, the seventh child of Thaddeus and Susannah Ayres Morrell. His father was the descendant of a New England family that had emigrated from England in the seventeenth century. Morrell's early youth was spent on the family farm in Berwick, Maine, in a settlement of Quakers where he was taught the values of economy and self-reliance. His early formal education was limited to a few years of elementary school studies, although he apparently later took a course in commercial studies after entering the business world. He also attended lectures at the Franklin Institute in Philadelphia.

In 1837, at the age of sixteen, Morrell traveled to Philadelphia, where he joined his brother David, who was a partner in the dry goods firm of

Trotter, Morrell & Company. After Morrell served as a clerk for five years, Trotter, Morrell dissolved; Morrell then joined with his brother to form a new company in the same location, only to have this firm dissolve also. In 1845 he joined Oliver Martin, a dry goods dealer, as a clerk and then later became his partner in Martin, Morrell & Company. Following Martin's death, Morrell continued the business of the firm until 1855, when he ended his mercantile career to begin his association with the iron and steel industry.

Morrell's active entry into the iron business developed out of his role as an investor in the Cambria Iron Company, which had been established in 1853 but had foundered financially the following year. Despite an infusion of additional funds the company had still produced no salable product by 1855. In that year a second group of six Philadelphia businessmen, including Morrell, each invested $30,000 to operate Cambria as lessees under the name Wood, Morrell & Company. Morrell was sent to take over direct management of the firm.

When Morrell arrived in Johnstown, Pennsylvania, to take over management of the Cambria works, he knew little of the iron industry; John Fritz, the noted iron maker, was later to say of him: "He was a very clever gentleman, but knew nothing about the iron business. . . . " If Morrell

Daniel Johnson Morrell

knew little about the iron business, he was certainly a willing learner and also kept an open mind regarding the development of the industry. In the year following his arrival at Cambria, Morrell was approached by William Kelly, the Kentucky iron maker, who requested permission to demonstrate his patented pneumatic process for producing steel. Morrell was receptive and not only provided him space for his experiments but also an assistant, James H. Geer, to help build the converter. Although hardly successful in any real sense of the word, Kelly's experiments must have piqued Morrell's imagination; despite the difficulties of the Civil War, in 1862 Morrell joined forces with Eber Ward, a Midwest shipping tycoon and iron producer, and Zoheth Durfee, a skilled iron maker, in an attempt to buy both the Kelly and Bessemer patents to the steel-making process. Despite failing in their quest for the Bessemer rights the group was able to acquire the American rights to the Mushet

spiegeleisen or recarbonization process. On the strength of these two sets of patents the Kelly Pneumatic Process Company was formed in May 1863 with the iron makers William M. Lyon of Detroit and James Park, Jr., of Pittsburgh as additional partners. Subsequent additions enlarged the partnership further during the course of the next two years.

The company had selected Wyandotte, Michigan, the location of a blast furnace and rolling mill owned by Ward, as the site for their experimental plant. Anticipating their success in obtaining the requisite patents, the company had instructed William Durfee, Zoheth's cousin, to begin constructing a converter as early as 1861, even though officially in violation of the patent rights. Completed in 1864, the converter was first successfully blown in on September 6; on May 24-25, 1865, the first set of six experimental rails was rolled at the North Chicago Rolling Mill Company from steel ingots produced at Wyandotte.

Despite Morrell's active involvement in the Wyandotte experiment, it would be eight more years before he convinced Cambria's management of the wisdom of adapting the Bessemer process. In the meantime he continued to play a leading role in Bessemer developments through his promotional and organizational roles within the controlling companies.

In 1864 a group of Troy businessmen (John F. Winslow, John A. Griswold, and Alexander Holley) obtained the American Bessemer rights, and in February 1865 they were successful in making Bessemer steel at their experimental works in Troy. Since both groups—Wyandotte and Troy—were in violation of each other's patent rights, they pooled all the American patents in 1866 as the Pneumatic Steel Association, a New York state joint-stock company in which Winslow and Griswold owned seven-tenths and Morrell held three-tenths in trust for the Kelly Process Company. In 1877 the company was reorganized as the Bessemer Steel Company and again as the Steel Patents Company in 1890. Through these companies, in which Morrell played an ongoing role, licenses for the Bessemer process were sold for an initial fee of $5,000 plus a charge of approximately $5 per gross ton of iron used to make the steel. In exchange for the initial fee the licensee received plans for a plant and information on the process. In return the licensee had to agree to keep the firm open to Winslow, Griswold, and Morrell to provide a check on the royalty payments. Of the 11 plants subsequently built and in operation by 1880, most were designed by Holley, while the free flow of ideas and consultation among the leading engineers–Holley, John and George Fritz, Robert W. Hunt, and Capt. Bill Jones–contributed to the solving of common problems. It was under these conditions that Morrell eventually convinced Cambria to construct the sixth integrated Bessemer steelworks, which conducted its first blast on July 10, 1871. Because of Morrell's leadership, Cambria remained as one of the larger integrated steelworks during the 1880s and 1890s.

Morrell played an active role as a community leader and on the national level. He aided the Union cause during the Civil War by helping to recruit volunteers for the army; following the war in 1866, he was elected as a representative from his district to the Congress and subsequently reelected in 1868, only to be narrowly defeated for a third term in 1870. While a member of Congress Morrell was an active supporter of American industry and labor upholding the Republican protectionist position. During his first term he was made chairman of the standing committee on manufactures. In 1870 he proposed the bill to provide for what became through his persistent support the 1876 Philadelphia Centennial Exhibition. He subsequently served as chairman of the centennial executive committee. Following the success of the centennial celebration and in recognition of his services, Morrell was selected a commissioner to the 1878 Paris International Industrial Exposition. Upon his return he submitted an extensive report to the secretary of state on the iron and steel exhibits and on the state of the industry worldwide.

Morrell's important contributions to the iron and steel industry were also recognized by his election on March 6, 1879, as president of the American Iron and Steel Association, a position which he held until December 15, 1884. Late in 1884 he resigned his active management of the Cambria Iron Works. Morrell failed to live out the year, passing away on August 20, 1885. He was survived by his wife, Susan Stackhouse Morrell, whom he had married on February 11, 1845, and by their daughter,

who married Philip E. Chapin, Morrell's successor as general manager of the Cambria works.

Daniel Morrell's name will always and rightfully be closely associated with the Cambria Iron Works. His nearly 30 years of service helped to stabilize the company and to force its transition to the manufacture of steel. However, his activities outside the company itself should not be overlooked, especially his role in developing and promoting wide-scale adoption of the Bessemer process throughout the industry. In addition, his promotion of the industry as a whole, as reflected by his term as president of the American Iron and Steel Association, places him in the forefront of nineteenth-century iron and steel businessmen.

References:

John Fritz, *The Autobiography of John Fritz* (New York: John Wiley & Sons, 1912);

William Hogan, *Economic History of the Iron and Steel Industry in the United States* (Lexington, Mass.: Lexington Books, 1971);

Robert W. Hunt, "Evolution of the American Rolling Mill," *Transactions of the American Society of Mechanical Engineers*, 13 (1892): 45-69;

Hunt, "History of the Bessemer Manufacture in America," *Transactions of the American Institute of Mining Engineers*, 5 (1877): 201-215;

Jeanne McHugh, *Alexander Holley and the Makers of Steel* (Baltimore: Johns Hopkins University Press, 1980);

Elting E. Morison, *From Know How to Nowhere* (New York: Basic Books, 1974);

Morison, *Men, Machines, and Modern Times* (Cambridge, Mass.: M.I.T. Press, 1966);

Fritz Redlich, *History of American Business Leaders: A Series of Studies*, 2 volumes (Ann Arbor, Mich.: Edwards Brothers, 1940);

James M. Swank, *History of the Manufacture of Iron in All Ages* (Philadelphia: American Iron and Steel Association, 1892);

Swank, *Introduction to a History of Ironmaking and Coal Mining in Pennsylvania* (Philadelphia: J. M. Swank, 1878);

Peter Temin, *Iron and Steel in Nineteenth-Century America: An Economic Inquiry* (Cambridge, Mass.: M.I.T. Press, 1964).

North Chicago Rolling Mill Company

by Alec Kirby

George Washington University

The North Chicago Rolling Mill Company was established by Eber B. Ward in 1857 and rapidly became a force in the iron and Bessemer steel industry. In 1889, under the leadership of Orrin W. Potter, Ward's protégé, the firm became the basis for the consolidated Illinois Steel Company, which became the largest steel conglomerate next to the Pennsylvania group headed by Andrew Carnegie.

The founder of the North Chicago Rolling Mill Company, Eber B. Ward, was the first of the American "iron kings." Ward began his career as a cabin boy on one of his uncle's Lake Superior schooners, a ship he would later captain. By 1855 Ward had earned a reputation as the "steamship king" of the Great Lakes, for he owned a fleet of 30 ships. In that year Ward entered the iron industry, establishing the Eureka Iron Company at Wyandotte, Michigan, one of the earliest rail mills in that region. He also constructed a furnace to smelt Lake Superior ores.

The main market for Ward's rails was the railroad, which was forced to use iron rails despite the fact that they wore out in an average of two years. Ward realized the business opportunities that the continual need for rails created and in 1857 took advantage of the presence of six leading railroad companies in Chicago to found the North Chicago Rolling Mill Company in that city. In addition to meeting the demand for rails, the rolling mill also provided nails, plate iron for boilers, and the simple "merchant bars" used by toolmakers and blacksmiths.

The Chicago firm grew rapidly, particularly after 1864 when Ward appointed Orrin W. Potter to become secretary and general manager. This represented a major promotion for Potter and a risk for Ward, who had first hired Potter as a clerk at his Wyandotte plant when Potter was only twenty years old. Under Potter the company rapidly expanded its iron- and steelworks in Chicago. These facilities included blast furnaces, Bessemer and open-hearth plants, rail, beam, plate, and slab mills, and facilities for the production of spiegeleisen, which consisted of a triple compound of iron, manganese, and carbon.

In the late 1880s Potter, who had assumed the presidency of the North Chicago Rolling Mill Company in 1871, began negotiations with Jay C. Morse to create the Illinois Steel Company. Under this arrangement consummated on May 4, 1889, the North Chicago Rolling Mill Company became the basis of a merger with the two other big steel companies in the region, the Union Iron & Steel Company and the Joliet Steel Company. The Union Company at this time was under the direction of Morse and H. A. Gray and specialized in Bessemer steel. The Joliet Company, under the direction of Alexander J. Leith and Horace Strong Smith, had the highest output of steel of any firm in the world.

The merger of the three firms helped to secure the position of Illinois as a major steel-producing state. Orrin Potter assumed the presidency of the firm and held the post until 1896. As the nucleus of the Illinois Steel Company, the North Chicago Rolling Mill Company secured a significant place in the history of the steel industry.

References:

Herbert N. Casson, *The Romance of Steel* (New York: A. S. Barnes, 1907);

William T. Hogan, *Economic History of the Iron and Steel Industry in the United States*, volume 1 (Lexington, Mass.: Lexington Books, 1971);

Glenn Porter and Harold C. Livesay, *Merchants and Manufacturers* (Baltimore: Johns Hopkins University Press, 1971).

Henry William Oliver

(February 25, 1840-February 8, 1904)

by Terry S. Reynolds

Michigan Technological University

CAREER: Telegraph Messenger, Atlantic & Ohio Telegraph Company (1853-1856); clerk, Clarke & Thaw (1856-1859); shipping clerk, Graff, Bennett & Company (1859-1861); partner, Martin, Oliver & Bickle (1861-?); partner, Lewis, Oliver & Phillips (1863-1880); president, Common Councils of Pittsburgh (1871-1872); delegate, Republican National Conventions (1872, 1876, 1888, 1892); partner, Oliver Brothers & Phillips (1880-1888); chairman of the board, Oliver Iron & Steel Company (1888-1904); president, Pittsburgh & Western Railroad (1889-1894); president, Oliver Iron Mining Company (1892-1901); chairman of the board, Oliver & Snyder Steel Company (1897-1904).

Henry W. Oliver was the creator of one of the earliest vertically integrated iron and steel companies and of an iron mining empire that provided the ores used by Carnegie and, later, United States Steel. He was born in Dungannon, County Tyrone, Ireland, on February 25, 1840, the third of six children born to Henry W. Oliver and Margaret Brown Oliver. In 1842 the family emigrated from northern Ireland for both religious and political reasons and settled in Allegheny, Pennsylvania. Henry Oliver's father had been a manufacturing saddler in northern Ireland and continued that trade.

Oliver received five years of education at the First Ward Public School in Allegheny and two years at Newell's Academy in adjacent Pittsburgh. At thirteen he left school to provide a supplementary income for his family. His manual skills were considered poor, and instead of taking up a trade, he secured a job as a messenger for the Atlantic & Ohio Telegraph Company, where his former schoolmate Andrew Carnegie was, for a time, similarly employed. Three years later, in 1856, Oliver began to work for Clarke & Thaw, a freight forwarding agency.

Oliver's introduction to the iron and steel industry began in 1859 when he took a job as shipping clerk for Graff, Bennett & Company, iron manufacturers, at the site of their new Clinton blast furnace. As his family's position improved, Oliver was able to save his earnings, which in 1861 enabled him and several partners to form the small firm of Martin, Oliver & Bickle. The firm bought an idle puddling mill at Kittanning, Pennsylvania, and restored it to operating condition. In January 1863 Oliver became even more entrenched in the iron business, forming with William J. Lewis and John Phillips the firm of Lewis, Oliver & Phillips to manufacture nuts and bolts. Oliver, the youngest of the three, was responsible for managing sales and the office. He was a "superb salesman" according to his biographer Henry Evans, and the firm grew steadily; in 1866 his brothers David Brown Oliver and James Brown Oliver were taken into the partnership.

The outbreak of the Civil War only briefly interrupted Oliver's business career. He volunteered and served in the Union army from April 24 to August 5, 1861, before being mustered out. During Gen. Robert E. Lee's invasion of Pennsylvania in 1863 Oliver again volunteered, serving from mid June until early July, mainly erecting forts and embankments in the Pittsburgh area.

On November 13, 1862, Oliver married Edith Anne Cassidy of nearby Minersville. His wife was a homebody caring little for social climbing and made his home life a contented one. In 1865 she gave birth to their only child Edith Anne.

Oliver entered the Pittsburgh iron business during a period of transition. Before 1860 Pittsburgh had not been a major iron producer, there being only six blast furnaces in Allegheny County in that year. Instead, the city's foundries had fabricated iron produced elsewhere into finished products, selling these products to small consumers like black-

Henry William Oliver

smiths, farmers, and mill owners. Between 1860 and 1880 the picture changed. By 1880 Allegheny County had 15 large blast furnaces and a host of iron and steel fabricating plants serving mainly industrial customers like the railroads. Henry Oliver, acting in several arenas, was both a producer and a beneficiary of this transition. In the political arena, he was an effective lobbyist for protective tariffs for the iron and steel industry. At the same time he was an active promoter of the transportation facilities which made Pittsburgh a major center for iron and steel production. And he was, of course, an active manufacturer of iron and steel products himself.

Politically, Oliver was, like most business leaders of the day, staunchly Republican, attracted to the party both by its antislavery stance and by its pro-

tectionism. He was active in Republican politics at local, state, and national levels. At the local level he served several terms on the Pittsburgh common council and in 1871 and 1872 served as president. In 1881 Oliver was nominated by organization Republicans for U.S. Senator from Pennsylvania; feuding within the Republican party created a deadlock within the Republican caucus, leading to Oliver's withdrawal. At the national level Oliver was a major spokesman for the iron industry in the movement for tariff protection. He was a delegate to four national Republican Conventions—1872, 1876, 1888, and 1892. At each of those conventions he served on the platform committee and played a leading role in producing a series of Republican planks advocating protective tariffs. In 1882 President Chester A. Arthur appointed Oliver to a special tariff commission on which Oliver played a prominent role in influencing the 1883 metal tariff schedule which afforded increased protection for American iron and steel.

Oliver's political success was partially attributable to his personality. He had a dignified manner, but one tempered by an irrepressible friendliness and sense of humor. He was unfailingly courteous, even to those opposed to him and even under the most adverse circumstances. He was generous, but in an unostentatious manner. This combination of traits brought him the devotion of subordinates and the trust of peers, both in politics and in business.

Oliver's active role in advocating protectionism for the iron and steel industries was paralleled by simultaneous activities in promoting and supporting efforts to provide Pittsburgh with improved transportation facilities. In 1875, for example, Oliver became an active promoter and leading stockholder of the Pittsburgh & Lake Erie Railroad, an attempt to bring a third trunk line into Pittsburgh. In 1883, when it became clear that the Pittsburgh & Lake Erie would probably pass under control of one of the two existing trunk lines (the Pennsylvania and the Baltimore & Ohio railroads), he engineered the sale of the road to the Vanderbilts of the New York Central Railroad, thus successfully bringing another trunk line into Pittsburgh. In 1879 Oliver was successful in buying control of the Pittsburgh & Northern Railroad, and through it obtaining for Pittsburgh an additional route to the West under the name of the Pittsburgh & Western Railroad. He served as president of the Pittsburgh & Western Railroad from 1889 to 1894 and later as chairman of its board. He also actively promoted construction of the Akron & Chicago Junction Railroad to provide Pittsburgh with better connections to the West. In 1882 Oliver aided the Pittsburgh & Lake Erie in building the Pittsburgh, McKeesport & Youghiogheny Railroad. He was similarly active in promoting his city's water transportation facilities. Between 1882 and 1885, for example, he played an important role in securing government construction of the Davis Island lock and dam to create a stable water level for Pittsburgh harbor and improve navigation on the Ohio.

In the meantime Oliver's involvement in iron and steel manufacturing continued to grow. In August 1880 William J. Lewis retired from Lewis, Oliver & Phillips, and the remaining partners, with James Smith, formed a new partnership–Oliver Brothers & Phillips. The firm had been steadily expanding since its creation as Lewis, Oliver & Phillips in 1863, erecting new rolling mills and gradually diversifying its offerings. By the time of its transformation into Oliver Brothers & Phillips, the company produced more iron and steel products than any manufacturer in Pittsburgh and, with 3,000 employees, had become one of the largest manufacturers of bar iron and iron specialties in the United States. Its diverse array of products included chains, structural shapes, wagon parts, agricultural hardware of all kinds, ratchets, and the ironwork for telephone and telegraph poles. In 1888 the company was reincorporated as the Oliver Iron & Steel Company. Henry Oliver served as chairman of the board, a position he retained until his death.

In the 1880s, using Oliver Brothers & Phillips as a base, Oliver began to expand his holdings, in the process becoming one of the first iron and steel manufacturers to create a vertically integrated firm. His first step was to diversify production of iron and steel products. In 1881 Oliver, his two younger brothers, and his two brothers-in-law formed the Oliver Wire Company, Ltd., and built a wire drawing plant in Pittsburgh. They also purchased a barbed wire plant at Joliet, Illinois, and moved it to Pittsburgh. Soon after Oliver extended the scope of his wire interests even further by creating the Pittsburgh Wire Nail Company to manufacture wire nails.

To supply his wire and nail production facilities, Oliver next sought to control raw materials by securing a supply of wire rods, previously imported primarily from Europe. In 1887 Oliver built a wire rod rolling mill in Pittsburgh. Soon after, all of the Oliver wire manufacturing concerns and plants were combined under the corporate name of Oliver & Roberts Wire Company, which became in 1894, on the withdrawal of his partner Henry Roberts, the Oliver Wire Company.

At the same time Oliver began to seek a supply of raw iron and steel to supply both the hardware manufacturing business and the wire rod rolling mill. Prior to 1881 Oliver had largely used iron puddled and rolled in his own plant, but the coming of Bessemer steel forced him to build new facilities. In 1882 Oliver erected a plant adjacent to and linked with his iron and steel plant to produce a malleable Bessemer steel using the Clapp-Griffiths process. When the process failed, Oliver closed down and scrapped the Clapp-Griffiths plant, absorbing large losses. This and the minor recession of 1883 plunged him momentarily into serious financial difficulties, but he quickly recovered.

In 1886 Oliver continued to develop his access to raw iron and steel by leasing the Rosena Furnace at New Castle, Pennsylvania. In 1889 he purchased an interest in the Hainsworth Steel Company in Pittsburgh and undertook remodeling and enlarging of its plants. In 1891 he leased and later purchased the Edith Furnace, also in Pittsburgh.

In the late 1880s and early 1890s Oliver sought increased control over the fuel supplies for his iron and steel production facilities. In 1889 he joined with E. C. Converse in establishing the Monongahela Natural Gas Company for the supply of natural gas to his plants and those of the National Tube Company. In 1891 Oliver organized a separate company which leased coal lands near Uniontown, Pennsylvania, and built three coking plants. These plants made Oliver a major coke producer. Oliver forged additional links with coal mining as an outgrowth of his association with the building and operation of the Pittsburgh & Western Railroad and the Fairport Docks on Lake Erie. In these activities he formed a close association with Francis L. Robbins, a large Pittsburgh coal miner. Oliver and Robbins formed the Pittsburgh Coal Company in 1899, which shortly became the largest bituminous coal company in the world.

Some of Oliver's previously mentioned activities in railroad promotion in the Pittsburgh area were, of course, partially designed to link the elements of his growing empire. These activities were supplemented by the construction in 1890 of a railroad from Pittsburgh to Fairport, Ohio, to receive and transport Lake Superior iron ores. To supplement this railroad, Oliver built docks at Fairport, purchased two small lake steamers, and organized the American Transportation Company to ship coal to the northwest and import ore from the Lake Superior district.

As his holdings in coal lands, coke production, transportation facilities, and iron and steel manufacturing plants grew, Oliver sought additional markets for his finished steel. Toward 1890 he aligned himself with bridge building and structural steel work through the Schultz Bridge Company. In 1891 Oliver and several of his acquaintances formed the Monongahela Tin Plate Company.

Oliver's interest in expanding markets for his steel and his associated interest in railroads led him in the 1890s to the steel railroad freight car. In 1891 Oliver met Charles T. Schoen, an inventor who had been developing steel springs and parts for freight cars and had taken out patents on an all-steel railroad car. Oliver decided to back Schoen's work financially, and in February 1891 Oliver moved the Schoen Manufacturing Company to Pittsburgh. In 1893 Oliver ordered the first all-steel gondola freight car from the Schoen Manufacturing Company. In 1895, after Schoen had patented a pressed-steel truck frame for railroad cars, Oliver and Schoen formed Schoen Pressed Steel Car Company with Oliver as one of its largest stockholders. In January 1899 the Schoen Pressed Steel Car Company and a rival company (the Fox Pressed Steel Equipment Company) were consolidated to form the Pressed Steel Car Company, long one of the largest steel railroad car builders and a heavy user of Oliver steel. Oliver was also involved in other steel car ventures. In October 1899 Oliver formed the Steel Car Forge Company for the purpose of manufacturing the forged parts used in the steel car; he placed his son-in-law, Henry R. Rea, in charge of it. And in 1901 Oliver with the Mellons established the Standard Steel Car Company at Butler, Pennsylvania. Through his financial support of Schoen's work, his creation of companies to manufacture steel railroad cars and parts, and the car orders directed to these companies, Henry W. Oliver played a vital role in creating a branch of manufacturing which substantially increased both the safety and the freight hauling efficiency of the American railroad network.

By 1893 Oliver had built a vertically integrated iron and steel empire that included coal

lands, coking facilities, blast furnaces, rolling mills, and plants that manufactured iron and steel into consumer products and controlled elements of the transportation systems linking these facilities. Oliver had just begun to sense the importance of also securing a reliable supply of iron ore when the panic of 1893 hit his empire. Overexpansion and construction cost overruns, coupled with the depressed market for iron and steel, forced the heart of Oliver's empire, the Oliver Iron & Steel Company, into receivership between 1893 and 1895.

The impact of the 1893 depression was, however, probably most severe on the Oliver Iron Mining Company, a new venture which Oliver had undertaken in 1892 to secure control of a long-term supply of iron ore by leasing iron mines on the newly opened Mesabi Range in Minnesota. This venture was a risky one, partially due to the nature of the Mesabi's ores. The ores of the older iron ranges in Michigan and Minnesota, as well as most of the iron ores of the northeastern United States, were hard and rocklike. All existing blast furnaces had been designed for these ores. Mesabi iron ores, on the other hand, were soft and fine, almost sandlike, and tended to clog blast furnaces.

Oliver's involvement in Mesabi iron ore had begun in 1892 when he sent George E. Tener, manager of his American Transportation Company, to investigate the prospects of the newly opened Mesabi Range. A few months later, following the Republican National Convention in Minneapolis, he visited the Mesabi himself. On September 18, 1892, Oliver formed the Oliver Iron Mining Company to operate several mines on which he had obtained leases, including the Missabe Mountain Mine, which he had leased from Leonidas Merritt in August. In 1893 the Missabe Mountain Mine paid the first stock dividend ever paid by a Mesabi Range firm.

The panic of 1893 almost destroyed Oliver's good start. Handicapped by the receivership of the Oliver Iron & Steel Company, the depressed market for iron ore, and the almost universal prejudice of blast furnace operators against the use of soft Mesabi ore, he was forced to seek outside investors, something that he usually avoided, preferring instead to raise capital from family members or close partners. Those investors initially attracted to the venture were quickly discouraged and in 1894 pressured Oliver, who was convinced of the potential importance of Mesabi's vast ore deposits and had no wish to abandon the venture, into repurchasing their stock. To keep the Oliver Iron Mining Company afloat, Oliver sought relief from Andrew Carnegie. Carnegie was very reluctant to get involved in the risky field of iron mining. He felt, moreover, that Oliver was prone to take too many chances and was a poor manager. He thus recommended to his friends not to invest with Oliver and was not inclined to do so himself despite a generous offer from Oliver in 1892. But Henry C. Frick, then the active head of Carnegie's steel company, recognized the importance of securing a reliable ore supply, and John D. Rockefeller's growing presence on the Mesabi further encouraged Carnegie to retreat from his earlier position. In early 1894 Frick proposed that in return for a loan of $500,000 to the Oliver Mining Company, the Carnegie Steel Company, Ltd., should receive 50 percent of the stock in the Oliver Mining Company. Oliver, in desperate need of funds, agreed. On May 1, 1894, half of the shares of Oliver Mining Company were transferred to Carnegie Steel Company, Ltd.

Carnegie's backing eased Oliver's financial troubles and provided the Oliver Mining Company with a guaranteed outlet for its ores. It also provided Oliver with the funds to continue expanding his mining holdings. In August 1894 Oliver leased a major mine from Lake Superior Consolidated Iron Mines, the Rockefeller ore concern. Oliver pushed production at both of his mines, and in June 1895 leased yet another major property. By the end of 1895 Oliver was shipping half of all ore mined from Mesabi. That same year the Oliver Iron & Steel Company emerged successfully from receivership.

The second leading ore producer on the Mesabi Range in 1895 was Rockefeller's Lake Superior Consolidated Iron Mines. In 1896 Oliver initiated negotiations with Rockefeller and his associates which led to an alliance between Carnegie, Oliver, and Rockefeller. Under the terms of the December 1896 alliance, the Oliver Mining Company leased all of the iron ore properties on the Mesabi held by Rockefeller's Lake Superior Consolidated Mines at a low royalty. Oliver Mining agreed to ship a minimum of 600,000 tons annually from the Rockefeller leases and another 600,000 tons from the Oliver properties and not to expand Oliver Mining holdings further. All ore was to be shipped via Rockefeller's Duluth, Missabe & Northern Railroad at set freight rates, with Rockefeller's Bessemer Steamship Company to then carry the ore

to Lake Erie ports at a rate determined by general market conditions. In keeping with Oliver's preference for long-term arrangements, the terms of the agreement were for 50 years.

The alliance's competitive advantage in transportation demoralized the remaining independent producers on the Mesabi. Taking advantage of this, Oliver began to acquire mining leases on the older Lake Superior ranges despite Carnegie's continued reluctance to support Oliver's acquisition program. But Oliver's acquisitions in the late 1890s gave Oliver Mining a good balance between soft Mesabi ores and the harder old range ores and enabled Oliver to keep down costs of freight by playing one range against the other. Among Oliver's purchases were control of the Tilden and Norrie mines, two of the largest producers on the Gogebic Range, and the Zenith mine on the Vermilion. In 1899 Oliver Mining also purchased control of the Lake Superior Iron Company, a leading property on the Marquette Range, which brought with it ownership of a fleet of six ore vessels. By the late 1890s the Rockefeller-Oliver-Carnegie combination controlled through lease or ownership 34 working mines on the 5 then-opened Lake Superior iron ranges.

Carnegie was understandably reluctant to allow control of the ores on which his company was increasingly dependent to remain in the hands of a company not under his control. Although he had secured half of Oliver Mining stock in 1894, in 1897 he insisted on firmer control. Thus in 1897 Carnegie Steel bought additional shares of Oliver Mining Company stock from Oliver, bringing Carnegie's share of ownership to 83.3 percent. Oliver retained 16.7 percent of the shares and a provision that he should remain president of Oliver Mining Company as long as he desired.

On March 31, 1897, with William P. Snyder as his associate, Oliver created the Oliver & Snyder Steel Company, a merger of three large coke plants near Uniontown, the Hainsworth Steel plant in Pittsburgh, the Edith Furnace in lower Allegheny, the Rosena Furnace at New Castle, and Oliver's Unity coal lands in Westmoreland County, Pennsylvania. Oliver then sold the new company and his Oliver Mining stock, making Oliver and Snyder Steel a fully integrated enterprise, with Snyder as head and Oliver as controlling owner.

In 1900 the informal Rockefeller-Oliver-Carnegie alliance began to come apart. Rockefeller was interested in selling his ore properties and was disturbed by Oliver's expanding Mesabi ore leases. After Carnegie reportedly refused a Rockefeller offer to sell all his mines, railroads, and steamships, Rockefeller cornered the market on lake shipping through his Bessemer Steamship Company and through chartering, purchasing, or leasing most additional bottoms. This step substantially raised the cost of shipping on the Great Lakes and taught Carnegie and Oliver the importance of transportation. They countered by chartering a new railroad to the Mesabi and forming the Pittsburgh Steamship Company. For the latter they ordered five large ore boats. Supplied with steel by the Carnegie company, the boats were rushed to completion for the next shipping season, giving Carnegie and Oliver the fourth largest fleet on the Lakes. With these counters, the conflict was quickly settled through a negotiated peace.

In 1901 Carnegie sold his iron and steel properties to J. P. Morgan and the emerging United States Steel Company. Oliver followed suit, selling his shares in the Oliver Mining Company and Pittsburgh Steamship Company. Oliver had some years earlier begun to sell some of his other holdings, for example, disposing of the Monongahela Tin Plate plant, the Oliver Wire Company, the Rosena Furnace, and a track of coal land in 1899. With the sale of his interest in the Oliver Mining Company and the Pittsburgh Steamship Company, all Oliver retained was Oliver Iron & Steel and the Oliver Snyder Steel Company, the last being the third largest producer of coke in the United States.

In 1901 and 1902 Oliver again entered the mining arena, first purchasing a large block of shares in the important Calumet & Arizona copper mine. He then reentered iron mining as a partner in the Chemung Iron Company. Chemung leased or purchased several mines in the Lake Superior district, but in 1903 sold out to U.S. Steel.

In his last years Oliver was also involved in Pittsburgh real estate. He had become one of the largest, if not the largest, individual owners of downtown Pittsburgh property by the time of his death on February 9, 1904. After his death his estate built the Henry W. Oliver Building, long the tallest structure in Pittsburgh, as a memorial.

Henry Oliver's contributions to the iron and steel industry were numerous. He was a promoter of improved processes in iron and steel production, particularly the modification of blast furnaces to take the softer Mesabi ores. With Schoen he introduced the all-steel railroad car. He was one of the first in the iron and steel industry to seek integrated, vertical organization. And, more clearly and consistently than any other in the business, he saw the importance of owning the raw materials used in steel making.

References:

Henry Oliver Evans, *Iron Pioneer: Henry W. Oliver, 1840-1904* (New York: Dutton, 1942);

History of Pittsburgh and Environs, volume 4 (New York & Chicago: American Historical Society, 1922);

John W. Jordon, *A Century and a Half of Pittsburgh and Her People*, volume 3 (Lewis, 1908);

Henry Raymond Mussey, *Combination in the Mining Industry: A Study of Concentration in Lake Superior Iron Ore Production*, Studies in History, Economics and Public Law, Columbia University, volume 23, no. 3 (New York: Columbia University Press, 1905);

David A. Walker, *Iron Frontier: The Discovery and Early Development of Minnesota's Three Ranges* (Minnesota Historical Society Press, 1979).

Archives:

The Evans biography listed above cites "Oliver Papers" but does not indicate a repository. An inquiry directed to the Western Pennsylvania Historical Society, Pittsburgh, the most logical repository for such papers, yielded negative results.

Open-Hearth Process

by Paul F. Paskoff

Louisiana State University

The open-hearth steel process was an alternative, and ultimately a successful competitor, to the Bessemer process of steel making. The technology originated in experiments performed during the late 1850s and early 1860s in England by William Siemens and in France by brothers Emile and Pierre Martin. Although first applied as a commercial process in 1866 in Europe and in 1868 by Abram Hewitt and Peter Cooper in the United States, open-hearth steel making did not come into widespread use in this country until the mid 1880s.

In its essentials the open-hearth process was similar to the puddling process, but the open-hearth process operated at considerably higher temperatures. The puddling process applied heat to pig iron by allowing only the flames from the furnace's burning fuel to touch the iron. This arrangement avoided the undesirable contact between iron and fuel that caused contamination of the iron, while at the same time bringing the pig iron to a boiling molten state. The open-hearth process achieved even higher temperatures and greater efficiency through the use of Siemens's regenerative gas stove. In the stove the fuel, natural gas, was burned in a chamber that had already been heated by exhaust gases which had been passed through a matrix of refractory brick arranged in an open-weave checkerboard pattern and called "checkers." The checkers heated the air flow forced into the furnace, and because there were two sets of checkers, one on either side of the furnace, the heating of one set of checkers could proceed while the other set sent its superheated air into the furnace hearth to raise the temperature.

The open-hearth furnace's high temperatures represented a significant advance over the puddling furnace. Unlike the latter's need of a puddler to work the impurities out of the molten iron, the open-hearth process substituted capital, in the form of the high-temperature furnace, for the labor of the puddler to drive impurities from the metal. The resulting product was steel which, together with the steel poured from Bessemer converters, began to transform the American iron and steel industry in the mid 1880s. The lengthy interval of time between the first attempts in the mid and late 1860s to make steel with the open-hearth furnace and the widespread adoption of the process almost 20 years later had many causes.

Initially the high operating temperature of the open-hearth furnace posed a severe problem because the furnace linings available in the 1860s and 1870s disintegrated too rapidly. The solution to this problem, as had been true of the solution to the similar problem with the Bessemer converter, lay in devising an improved refractory lining that could withstand the heat. Even with the development of better linings, on the eve of World War I the use of the open-hearth process still required the periodic rebuilding or relining of the furnace. Comparatively high labor costs were also an early obstacle to the early adoption of the open-hearth furnace because the first models were charged with solid rather than molten materials, an arrangement which required hand-charging and lower temperatures during charging, as opposed to the Bessemer process's use of machine-charging. These considerations made open-hearth steel more expensive than Bessemer steel and, therefore, limited the extent to which the open-hearth process succeeded in the market prior to the mid 1880s. There were, however, some inherent advantages to the use of the open-hearth furnace which encouraged some steel makers to adopt it.

Because licenses to use the Bessemer process were tightly controlled by the Pneumatic Steel Association, prospective steel makers who were unwilling or unable to meet the demands of the association could resort to the open-hearth process and face far fewer institutional obstacles. Also, unlike the high

YEAR	PER CENT OF TOTAL STEEL PRODUCTION MADE BY		
	BESSEMER PROCESS	OPEN-HEARTH PROCESS	ALL OTHER PROCESSES[1]
1869	35	3	62
1879	89	5	6
1889	86	11	3
1899	71	28	1
1900	66	33	1
1901	65	35	*
1902	61	38	1
1903	56	40	4
1904	57	43	*
1905	55	45	*
1906	52	47	1
1907	50	49	1
1908	44	45	*
1909	39	60	1
1910	36	63	1
1911	34	66	*

[1]Includes crucible steel process and, in later years, a small amount of steel made with the electric-arc process.

* Indicates less than 1 percent of total.

Source: adapted from Peter Temin, Table C.5, "Proportions Steel Made by Different Processes," *Iron and Steel in Nineteenth-Century America: An Economic Inquiry* (Cambridge, Mass.: M.I.T. Press, 1964).

Steel production by process, 1869-1911

level of capitalization—as much as $250,000 to $300,000—necessary for a Bessemer steel plant of standard capacity, an open-hearth steel furnace could be built on a smaller scale and for a cost of from one-fourth to one-half that of a Bessemer plant. The lower cost of the open-hearth furnace made it an attractive alternative to entrepreneurs and firms with limited capital resources. Of course, champions of the Bessemer process pointed out that because steel could be made from iron in a converter in only 20 minutes, compared with the six hours required for the operation in an open-hearth furnace, one got what one paid for. Still, the necessary license and large capital investment needed for Bessemer technology worked against that argument.

Until the early 1880s the refractory lining of both Bessemer converters and open-hearth furnaces was acidic and the steel-making process was called the acid process. This arrangement underwent a profound change in 1877 when an Englishman, Sidney Gilchrist Thomas, received a patent for a process known as both the Thomas process and the basic Bessemer process. The process provided for the installation of a basic (as opposed to the usual acidic) lining in the Bessemer converter, thereby permitting the use of ore high in phosphorus which, before Thomas's invention, if not removed, made bad steel. The older acid process did not remove the phosphorus and was therefore limited to using low-phosphorus ores—ores with a phosphorus content of less than one-tenth of 1 percent—which meant that Bessemer steel plants could not avail themselves of the high-grade, high-phosphorus ore from the Great Lakes ranges. The same problem beset open-hearth

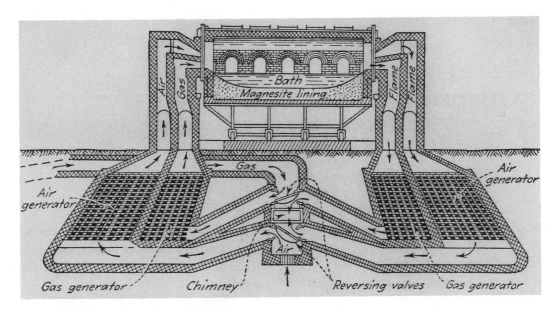

Diagram of an open-hearth furnace

steel producers until 1886 when Samuel T. Wellman, an American pioneer of the open-hearth process at the Otis Iron & Steel Company in Cleveland, Ohio, applied the open-hearth basic process that had been invented in the late 1870s by Percy Gilchrist, another Englishman and the cousin of Sidney Gilchrist Thomas.

With the development and widespread availability of the basic open-hearth process, the open-hearth furnace could make use of ores not suitable for Bessemer steel making. Even with the basic converter lining, the Bessemer process could not make use of most American ores, a fact which limited its adoption. The ability of the basic open-hearth furnace to use almost any phosphoric ore resulted in a significant cost advantage for the open-hearth process and was probably the single-most important reason for the gradual move by steel makers to it and away from the Bessemer process. Additional savings were realized in the 1880s when Wellman introduced mechanized charging of the open-hearth furnace, thereby greatly increasing operational efficiency and reducing labor costs. Another reason for the growing acceptance of the open-hearth process during the 1880s and 1890s was the fact that it could make use of scrap iron, permitting a reduction of the cost of input materials and, therefore, of

the final steel product. Bessemer converters, on the other hand, were as particular about the character of scrap that went into them as they were about pig iron. Still another inherent advantage of the open-hearth process after the mid 1880s was the perception by many steel makers and consumers that the open-hearth process permitted greater quality control. Accurate or not, market perceptions are not to be dismissed out of hand.

The growing acceptance of the open-hearth steel furnace can readily be seen by considering the quantitative record of steel production by the Bessemer and open-hearth furnaces from 1869 when the latter was first tried to just before World War I.

References:

William T. Hogan, S.J., *Economic History of the Iron and Steel Industry in the United States*, volume 1 (Lexington, Mass.: Lexington Books, 1971);

Jeanne McHugh, *Alexander Holley and the Makers of Steel* (Baltimore: Johns Hopkins University Press, 1980);

Fritz Redlich, *History of American Business Leaders: A Series of Studies*, 2 volumes (Ann Arbor, Mich.: Edwards Brothers, 1940);

James M. Swank, *History of the Manufacture of Iron in all Ages*, 2d ed. (Philadelphia: American Iron and Steel Association, 1892);

Peter Temin, *Iron and Steel in Nineteenth-Century America: An Economic Inquiry* (Cambridge, Mass.: M.I.T. Press, 1964).

Frederick Overman

(1803-January 7, 1852)

by John A. Heitmann

University of Dayton

CAREER: Mining engineer (1830s-1842); consultant (1842-1849); author (1849-1852).

Frederick Overman, a mining engineer and author, was born in Elberfeldt, Germany, in 1803 and died in Philadelphia, on January 7, 1852, as a result of a laboratory accident. While little is known of Overman's life, he was a prolific writer on subjects related to the iron industry. His significance in the development of a modern American iron and steel industry cannot be underestimated; Overman's works were published at a time when there was scant technological information available in print on mining and metallurgical operations. Among his treatises were the following: *The Manufacture of Iron* (1850); *The Manufacture of Steel* (1851); *Mechanics for The Millwright, Machinist, Engineer, Civil Engineer, Architect and Student* (1851); *The Moulder's and Founder's Pocket Guide* (1851); *Practical Mineralogy, Assaying and Mining* (1851); and *A Treatise on Metallurgy: Embracing the Elements of Mining Operations and Analyses of Ores* (1852). These works were republished on several occasions during the nineteenth century and became not only important reference works for those engaged in practical manufacturing operations but also served as textbooks in American colleges and universities.

Overman was born into a relatively poor family and received only a rudimentary education. Initially apprenticed in a mercantile business, he soon found the work distasteful and bound himself to a cabinetmaker. He subsequently became a journeyman craftsman and began traveling extensively, finally settling in Berlin. There he gained admission to the Royal Polytechnic Institute, where he soon won the attention of director Beuth, who introduced the young Overman to the leading artists and scientists of the day, including the sculptor Christian Daniel Rauch, the architect Karl Friedrich Schinkel, and the naturalist Alexander von Hum-

boldt. It was as a student in Berlin that Overman became imbued with the spirit of Romanticism, and he struggled to discover beauty, truth, and moral freedom. These ideals remained with him the rest of his life.

During the 1830s Overman applied his technical knowledge to the burgeoning European iron industry, and he introduced improved methods in the process of puddling. During this time the industrial revolution took a firm hold on the continent, in large part due to the efforts of German machinery manufacturer Fritz Harkort. Overman had the kind of expertise necessary to sustain the rapid industrialization that was spreading through much of Europe and that was being linked with a nascent nationalism by such writers as Friedrich List. For a time Overman supervised the erection of a large engineering establishment in Chemnitz, Saxony, and later he worked directly for Prince Klemens von Metternich on problems related to the Austrian trade and industry.

Yet, Overman's Romanticism and his dislike of autocracy led him to a fateful decision—he felt that he could no longer live in either Prussia or Austria. As a result he immigrated to the United States in 1842. According to bridge builder, close friend, and obituarist John A. Roebling, Overman attempted to introduce new ideas concerning manufacturing in his adopted country but met with widespread resistance because his talents and inventive genius were not wholly appreciated by mercantile interests. For the last three to four years before his untimely death in 1852 Overman occupied himself in his technological writings.

As mentioned previously, Overman's works were significant because they constituted an important source of scientific and technical knowledge at a time when little information existed in print in the United States. But perhaps just as important, these treaties represented a distillation of European ideas on the topics of mining and metallurgy. Thus Overman was an important link in the transfer of technology from Europe to America. His works are filled with illustrations of European machinery and process equipment, and the text borrows heavily from the studies of the German mineralogist K. J. B. Karsten, whose writings were unavailable in English translation. Secondly, in the mid nineteenth century, Overman was a solitary voice emphasizing quality rather than quantity in the American iron trade. As the century progressed and American iron and steel manufacturers came in direct competition with British and European producers, Overman's treatises contributed to the U.S. industry's competitive edge in an increasingly global international market.

Selected Publications:

The Manufacture of Iron, In All Its Various Branches (Philadelphia: H. C. Baird, 1850);

The Manufacture of Steel: Containing the Practice and Principles of Working and Making Steel (Philadelphia: A. Hart, 1851);

Mechanics For The Millwright, Machinist, Engineer, Civil Engineer, Architect and Student (Philadelphia: Lippincott & Grambo, 1851);

The Moulder's and Founder's Pocket Guide (Philadelphia: A. Hart, 1851);

Practical Mineralogy, Assaying and Mining (Philadelphia: Lindsay & Blackiston, 1851);

A Treatise on Metallurgy: Embracing the Elements of Mining Operations and Analyses of Ores (New York & London: D. Appleton, 1852).

References:

David McCullough, *The Great Bridge* (New York: Simon & Schuster, 1972);

John A. Roebling, "Frederick Overman," in *A Treatise on Metallurgy*, by Frederick Overman, sixth edition (New York: D. Appleton, 1868).

James Park, Jr.

(January 11, 1820-April 21, 1883)

by Bruce E. Seely

Michigan Technological University

CAREER: Partner, James Park & Sons (1840-1843); owner, James Park, Jr. & Company (1843-1862); president, Pittsburgh Common Council (1853); founder and president, Park, McCurdy & Company (1857-1883); president, Dime Savings Bank (1862-1865); founder and president, Park, Brother & Company (1862-1883); partner, Kelly Pneumatic Process Company (1863-1865); vice-president, American Iron and Steel Association (1873-1883); member, U.S. Tariff Commission (1882-1883).

As the center of the iron and steel industry in this country during the nineteenth century, Pittsburgh was home for many of the most important figures in the industry's development. One of those figures was James Park, Jr. Park spent almost his entire life working on various aspects of that industry, and the high points included opening one of the first successful cast crucible steelworks in the country, introducing a number of technical innovations to that branch of manufacturing, investing in the Kelly Pneumatic Process Company, and acting as champion of a protective tariff, one of the pressing issues of the day for steel makers. Successful in every endeavor, he was a leader in Pittsburgh's business and civic community.

James Park, Jr., was born on January 11, 1820, in Pittsburgh, one of seven children born to his immigrant parents. His Irish father had arrived in the city in about 1812, where he established a successful grocery and metal retailing business; in about 1835, he added queensware china to his interests. After a common school education, Park began working in his father's store in 1837; his initial responsibility was for the china business. He became a partner in the operation in 1840, and only three years later, James and a younger brother, David E., took control of the business after his father died, renaming the firm James Park, Jr. & Company. The two young men curtailed sales of china goods and focused on metal products, beginning Park's practical education in the iron trade.

Few details exist about Park's iron retail business during the 1840s and 1850s, but it must have flourished, for, from that base, Park expanded in sev-

James Park, Jr.

eral directions. First, he became part-owner of the Banner Cotton Mill, then acquired a similar stake in Smith, Park & Company's National Foundry. An indication of his increasing stature in Pittsburgh was Park's election to the Common Council during this period; he served for several years and was president in 1853. Finally, Park launched his own industrial project, choosing the copper industry.

After 1848 Pittsburgh was a center for copper smelting and marketing, thanks to Curtis Hussey's connection with the mines in the Lake Superior region. In 1857 Park followed Hussey's lead and erected the Lake Superior Copper Works. This facility had two 10-ton smelting furnaces, modeled on Hussey's design, two 3-ton scrap-melting furnaces, and a rolling mill to produce sheathing for ship hulls. With typical care Park had consulted with leading copper and brass firms in New England; he also hired Welshmen from Swansea, center of Britain's copper industry. John McCurdy—perhaps a

relative of Park's mother—joined his venture; Park, McCurdy & Company thrived because of the Civil War, in spite of being cut off from copper ore from Lake Superior by the establishment of the Detroit & Lake Superior Copper Company, a smelting and fabricating company controlled by the large copper-mining companies on Michigan's Keweenaw peninsula. Success came because Park, McCurdy was able to procure smelted copper elsewhere. A change in partners after the Civil War created Park, Scott & Company, but Park retained control of this firm throughout his life. As late as 1880 this operation was one of only three large-scale copper-working companies in the country.

In 1861 Park embarked on another, more risky operation—cast crucible steel manufacturing. Again, he followed the lead of fellow Pittsburgh native Curtis Hussey, who had started such an operation in 1859. In 1862, against the backdrop of the Civil War, Hussey's success in producing salable steel, and the protection of the Morrill Tariff of 1861, Park, Brother & Company began to build the Black Diamond Steel Works. This plant, located in the Lawrenceville section of Pittsburgh, was immediately adjacent to his copper rolling mill. The first ingots were poured on May 1, 1862. Unlike Hussey, who developed his own production method, Park relied on the traditional process developed by Huntsman—converting wrought iron into blister steel and melting the blister steel in crucibles. Like most other firms entering this trade, Park focused initial efforts on making top-quality tool steel. He benefited from the employment of several experienced steel hands from Sheffield, center of the British crucible steel industry. Perhaps as important, however, was the experience that his copperworks provided in metal working and furnace operation.

Within a short time the Black Diamond Works was producing five tons per day, although the firm lost money while struggling to master the complex cast crucible process. Part of the problem lay in convincing American consumers that Black Diamond brand steel could equal Sheffield steel in quality. Every earlier entrant in this industry, except Hussey, had failed to win American acceptance. Park adopted the strategy, at least by 1863, of supplying all grades of steel, surviving on sales of lower grade steels while patiently winning acceptance for its top-quality tool steels. In pursuit of this strategy of supplying all needs, Park patented in 1863 a process for making soft center cast steel for agricultural implements such as plows and for burglar-proof vaults and safes. By the time the firm reorganized in 1873, Park, Brother & Company offered all grades of cast steel: German Steel for agricultural implements, Black Diamond Homogeneous Cast-Steel for boiler and firebox plate, and various steels for railroad use.

The Black Diamond Works quickly won a reputation for technical innovation, largely due to Park's efforts. The initial facilities were traditional and included a cementing works to prepare blister steel, 72 coke-fired melting holes, 14 steam hammers, 4 helve hammers, 2 trains of rolls, shears, and 4 steam engines. Park, however, always sought out new methods and machinery. For example, he built the first Siemens gas-fired regenerative furnace for metal conversion in this country. In 1863 his works superintendent visited English steelworks, where he observed Siemens's newest furnaces and brought back drawings and a description. By August 1863 Park had constructed and successfully used such a furnace at the copperworks. Only problems in obtaining copper ore led Park to close this furnace. Attempts to adapt a Siemens furnace to the steelworks later that year, however, failed. Both were built only from printed drawings and constructed without a license. Not until 1867 would Park use the Siemens furnace for melting cast crucible steel, but then it spread rapidly through the American industry. By the late 1870s the Black Diamond Works had three 24-pot regenerative furnaces, fed by Siemens gas generators, in addition to the original coke melting holes. There were 6 puddling furnaces to provide wrought iron for 6 cementation furnaces, which had a capacity of 30 tons. The works also boasted 48 heating and annealing furnaces, and 6 trains of rolls ranging in size from 8 inches to 28 inches. A spokesman quoted in Thurston's *Pittsburgh's Progress, Industries, and Resources* (1886) commented, "We can cast an ingot weighing 30 tons, and can work such ingot under our hammer. Our mill has rolls 115 inches long, and we can finish plates, say 3/8 to 1/2 inch in thickness, 104 inches wide, and say 30 to 40 feet long." With a capacity of 12,000 tons a year, the Black Diamond Works was one of the largest in the country.

Park's interest in new technologies showed in other ways as well. In 1877 he cooperated with another large crucible steel maker in Pittsburgh, Miller, Metcalf & Parkin, in testing a process developed by Siemens in England for making wrought iron and steel directly from ore. Limited success halted the test in 1879, but in 1881 Park added open-hearth furnaces to his works. He employed Alfred E. Hunt, an M.I.T. graduate who had helped to erect the second open-hearth furnace in the United States in Boston, to superintend the new facility and the heavy forging department, and to act as metallurgical chemist. Hunt remained only two years before joining the Pittsburgh Testing Laboratory—an early industrial consulting firm. Even so, Hunt's tenure gave Park, Brother & Company a significant advantage in mastering the open-hearth technique. Park also was aware of the possibility of using Bessemer steel and in 1879 joined a combination of Pittsburgh crucible steel makers to form the Pittsburgh Bessemer Steel Company, which built the Homestead Works.

In short, Park was constantly following new technical possibilities in the steel industry. Not surprisingly, after his election to the group in 1872, Park was active in the American Institute of Mining Engineers. After 1880 he was a member, as well, of the Engineers' Society of Western Pennsylvania. Even in the 1880s Park spent more time roaming the works than in the office. It was typical, then, that the Black Diamond Works installed the largest steam hammer in the country in 1880, a 17-ton behemoth built by William B. Bement & Son of Philadelphia. The anvil alone weighed 160 tons, the heaviest iron casting in the country.

Perhaps the best marker of Park's enthusiasm for new technical processes was his involvement with the Kelly patent in the early 1860s. Along with E. B. Ward, Z. S. Durfee, Daniel J. Morrell, and William M. Lyon, Park became a partner in the incorporation of the Kelly Pneumatic Process Company. This chapter in the development of the Bessemer process for mass-producing steel culminated in construction of an experimental works at Wyandotte, Michigan. But the legal tangle caused by the Kelly Pneumatic Process Company's ownership of the Kelly and Mushet patents and another syndicate's license on the original Bessemer patents stymied development of the process. The Wyandotte group, moreover, was hampered by squabbling among the partners and the workers over such issues as Durfee's chemical laboratory. These problems apparently led Park to step aside from the partnership in 1865; Ward may have purchased Park's share. Even so, at least one source gives Park a role in the consolidation of the Kelly, Mushet, and Bessemer patents that occurred in an out-of-court settlement in 1866. Clearly, Park appreciated the possibilities of technological innovations in the iron and steel industry. As Harrison Gilmer explains, "James Park, Jr., was not an inventor, nor an inventive type of mind. But he had insight and could grasp new ideas instantly when they were presented to him. He constantly encouraged the introduction of new industrial process...." His sons William and David E. continued this legacy, making Park Brothers & Company, Ltd., the largest producer of special purpose open-hearth and crucible steel in the world. Moreover, they were the leading figures in the creation of the Crucible Steel Company of America in 1900. As Gilmer simply stated, "In many ways the Park Company dominated the special-purpose steel industry even more than the Carnegie interests dominated the production of common steel."

James Park's leadership extended well beyond the primacy of the Black Diamond Works in the cast crucible branch of the business. As a prominent member of the Republican party (again like Hussey), Park certainly encouraged the Morrill Tariff as protection for steel makers in 1861. This support for protective tariffs became a constant cause for Park. In 1873 he became a vice-president of the American Iron and Steel Association, in which capacity he steadfastly defended protectionism. In 1882 and 1883 he was a member of the United States Tariff Commission, which was created to reduce tariff rates. Park presided at an American Iron and Steel Association convention on the subject and presented the resulting information on the iron and steel industry's attitudes to the commission. More importantly, Park testified and lobbied effectively for his industry as a member of the commission in Washington. The results were a clever victory for iron and steel makers, who disguised existing high rates on steel imports in a new classification system.

Like many other businessmen of his day, Park was involved in a wide array of industrial ventures outside the steel industry. His obituary in the *Bulletin* of the American Iron and Steel Association noted that "He was, in a word, for at least the last thirty years of his life, one of the most enterprising, active, and influential business men in that remarkably busy community, which has Pittsburgh for its capital." He participated, for example, in the Suspen-

sion Bridge Company, MacIntosh, Hemphill & Company (one of the leading foundries and machine builders in the city), and the Penn Cotton Mill. He was the first president of the Dime Savings Bank, from 1862 through 1865, and an officer or director of several other banks. Earlier, he was a director of Allegheny Valley Railroad. Finally, Park helped organize utility companies for Pittsburgh, playing a leading role in forming both the Allegheny Gas Company and the city waterworks.

In another area—labor relations—Park may not have been typical of late-nineteenth-century businessmen. It is difficult to accept as accurate the comments of late-nineteenth-century observers in these matters; yet it seems that Park was not cut from the mold of Henry Frick. During the violent railroad strike of 1877, Park, according to *Appleton's Cyclopedia of American Biography*, "showed great courage . . . in facing the rioters during the labor troubles . . . and making an earnest appeal to them at the Union depot." Morever, he reportedly maintained good rapport with his workers, a matter of some importance in an industry that depended on their skill.

Park also discharged the civic responsibilities commensurate with his status in the Pittsburgh business community after the Civil War. In line with his antislavery position, Park served on the Subsistence Committee, the U.S. Sanitary Commission, and other movements to benefit soldiers during the conflict. He was involved in recruiting troops, and the Park Independent Battery, the Park Zouaves, and the Park Rifles reflected his efforts. In common with other leading industrial figures, he was a trustee of Western University of Pennsylvania (later the University of Pittsburgh), an organizing director of the Western Pennsylvania Hospital, a director of the Law and Order Society, and an Elder at the First Presbyterian Church. Park also was first president of the city park commission and took an active role in the creation of park facilities.

James Park, Jr., rose from humble beginnings to become one of the most influential men in Pittsburgh at a time when many individuals were staking their claim to the notice of history, especially in the iron and steel trade. The father of five children—three boys and two girls—he left an estate of between $2 million to $5 million. His success came less with developing original ideas than in copying closely the efforts of pioneers in the copper and steel trades such as Curtis Hussey or E. B. Ward.

Park had, nonetheless, a deep interest in new processes and technology. The makers of cast crucible steel had fewer chances to capture fame, for national attention went to Andrew Carnegie, Alexander Holley, and others developing huge steelworks after 1870. But Park was a dominant figure in an area of steel manufacturing that faced far greater problems of British competition than the Bessemer steel firms. From this fact came Park's connection to the giants of steel making, for he became their spokesman, through the American Iron and Steel Association, on protective tariffs. As *The Dictionary of American Biography* summarizes, "We wonder if the manufacturers of this country and its workingmen realize the sacrifices that a few willing and earnest men like James Park, Jr., have always made to secure to them the benefits and the blessings of a Protective tariff. . . ."

References:

"Death of James Park, Jr.," *Bulletin of the American Iron and Steel Association* (May 2, 1883): 116;

T. Egleston, "Copper Refining in the United States," *Transactions* of the American Institute of Mining Engineers, 9 (1880-1881): 678-730;

Harrison Gilmer, "Birth of the American Crucible Steel Industry," *Western Pennsylvania Historical Magazine*, 36 (March 1953): 17-36;

History of Allegheny County, Pennsylvania, volume 1 (Chicago: A. Warner, 1889), p. 668;

History of Pittsburgh and Environs (New York: The American Historical Society, 1922), pp. 69-70;

A. L. Holley and Lenox Smith, "American Iron and Steel Works: Park, Brother & Co.'s Works," *Engineering*, 23 (May 4, 1877): 337;

Henry K. James, "James Park, Jr.," *Magazine of Western History*, 4 (1886): 524-528;

The Manufactories and Manufacturers of Pennsylvania of the Nineteenth Century (Philadelphia: Galaxy Publishing Company, 1875);

"The Manufacture of Steel in Pittsburgh," *Scientific American*, 12 (April 1, 1865): 207-208;

Jeanne McHugh, *Alexander Holley and the Makers of Steel* (Baltimore: Johns Hopkins University Press, 1980);

James M. Swank, *History of the Manufacture of Iron in All Ages* (Philadelphia, 1884);

George H. Thurston, *Pittsburgh's Progress, Industries, and Resources* (Pittsburgh, 1886), p. 83.

Pennsylvania Iron Works

by Frank Whelan

Morning Call, *Allentown, Pennsylvania*

The Pennsylvania Iron Works was founded in Danville, Montour County, Pennsylvania, in 1844, but it was not until 1860 that it was known by that name. The business was founded as the Montour Iron Company, but it was not the first iron furnace to open in Danville. In 1829 the first iron manufactory was founded by John Theil, a maker of andirons, griddles, and plowshares. But it was not until the 1840s, when the use of anthracite coal as a fuel in blast furnaces was adopted in this country, that Montour County became a center of pig iron production.

At first, Montour Iron was just one blast furnace operation among many. But it was luckier than most. Even before 1839, when Welsh ironmaster David Thomas arrived in the United States at the behest of Lehigh Canal cofounder Erskine Hazard and had shown American ironmasters how to use anthracite coal to fuel blast furnaces, the race to acquire British technology was intense. As competition increased, Montour Iron was one of the leading players, inducing English and Welsh furnace operators, such as William Hancock and John Foley, to come to Pennsylvania to work for the firm. Many British subjects took up residency in America during those years, but what made Hancock and Foley so important was the knowledge they brought with them. While at Montour they were the first in America to roll the so-called "T" type railroad rail.

The actual "T" shape was created by John Stevens, Jr., of New Jersey. According to traditional accounts, Stevens was on a voyage to England to study railed transit, and to pass time while at sea he began carving pieces of wood into possible rail shapes. He created the "T" with a broad, firm

The Pennsylvania Iron Works near Harrisburg, Pennsylvania, in 1876

base. Turning it on its side caused some to call it the "H" rail. But the technology necessary to make Stevens's design did not yet exist in America. The first "T" or "H" rail was rolled in a mill at Merthyr Tydfil in South Wales. Americans tried to compete with this new product by making cast-iron rails, but railroad builders rejected the brittle and quickly deteriorating American product. When the Columbia & Philadelphia railroad let bids for rails, not one American ironmaster or his agent even attempted to compete with the more malleable iron produced in Great Britain.

The first "T" rail made in the United States was produced at Montour on October 8, 1845. One native of Danville, John E. Riley, described the process as "very complicated and tedious." The iron was first put through a set of rollers, from which it came out in heavy, flat bars about 3 inches wide and 3/4 inch thick. "These bars are then cut into three foot lengths and about fifteen or twenty are bundled together in what is known as a 'faggot' weighing nearly 400 pounds," writes Riley. The bundle was then placed into a furnace that was at white heat, later withdrawn, and crushed into a solid mass. It was then transferred to the rollers. Put through a series of rollers, the iron assumed the rail shape. "As it passes through, the bar is caught and supported by iron levers fastened to chains sus-

pended on pulleys from above. It first passes through the square groves (dies) of the rollers three or four times before it is run through the different groves that gradually bring it to the form of a T," reported Riley. The rails were then cut to a length of 18 feet. During the 1840s Montour's rails were shipped out by canal boat. When winter froze the canal, the rails were taken overland by sled to the nearest rail line at Pottsville.

It was the creation of the "T" rail that made Montour, later the Pennsylvania Iron Works, a huge success. Although Hancock and Foley left the company less than a year later, their technique led to the creation of a firm that made Danville a city. Known locally as the "Big Mill," Montour employed 1,800 men. The company town that was erected sprawled over 3,000 acres and housed 300 families. Its yearly payroll amounted to $1.1 million. Goods were sold at a company store, where all workers were required to shop. In 1872 the store employed 40 clerks and sold nearly $550,000 worth of goods.

In 1860, when the firm became the Pennsylvania Iron Works, it was owned by the five Grove brothers, two of whom, John Peter Grove and John Grove, were pioneers in the business. In 1839 they opened the Columbia Furnace, the first in Montour County.

According to a city directory of Danville printed in the 1870s, the ironworks was among the largest in the world exclusively owned and controlled by private, unincorporated capital. The principal owners in those years were Philadelphia capitalists Isaac Waterman and Thomas Beaver. Beaver was a community benefactor who built a free public library and a convent for the nuns of the local Catholic church.

At the peak of its operation the ironworks had six different units in operation. Its oldest furnaces were taken over from previous owners and dated from as early as 1840. Rolling Mill Number 1, where the first rail was rolled, was built in 1844. Stretching 345 feet by 166 feet, not including four encircling wings, it was then the largest in the world. It had 37 double and single puddling furnaces and 10 heating furnaces. Rolling Mill Number 2 measured 232 feet by 166 feet and had one wing. Built in 1853, it had 32 single puddling furnaces. The two mills together produced 2,500 tons of finished rails each month.

The panic of 1873 that led to the collapse of railroad building in the United States hit the Pennsylvania Iron Works hard. Within a few years steel rails began to displace iron rails in the returning rail market. The discovery of better quality ores in the upper Midwest dealt an even more severe blow to the Pennsylvania Iron Works. Despite the changing market and the firm's inability to compete either on the basis of price or quality, the Pennsylvania Iron Works lingered for many years, finally closing permanently in 1938.

References:

Arthur Toye Folke, *My Danville: Historical Sketches of his Hometown of Danville, Pennsylvania* (North Quincy, Mass.: Christopher Publishing, 1968);

Folke, *Picture Book for Proud Lovers of Danville, Montour County and Riverside, Pennsylvania* (North Quincy, Mass.: Christopher Publishing, 1976);

Albright Zimmerman, "Iron for American Railroads, 1830-1840," *Canal History and Technology Proceedings, Iron Issue,* 5 (March 22, 1986): 63-101.

Penokee & Gogebic Development Company

by Terry S. Reynolds

Michigan Technological University

The Penokee & Gogebic Development Company opened the first iron ore mine on the Gogebic Range of Michigan and Wisconsin, near what is now Bessemer, Michigan, in 1884. As early as 1848 surveyors had reported evidence of iron ore in the area of northern Wisconsin bordering the western extremity of Michigan's Upper Peninsula. A local hunter and trapper, Richard Langford, confirmed the presence of ore in the 1860s and had begun test pitting for ore on the Michigan side of the border near the site of Bessemer in 1868. Around 1873 Langford discovered commercially significant deposits in the same area. But the remoteness and ruggedness of the region discouraged development.

Rising iron prices in the early 1880s, however, encouraged exploitation of the area. Around 1883 Charles Colby, with financial backing from several other investors, including Pickands, Mather & Company, a newly formed, Cleveland-based iron ore and pig iron marketing agency, created the Penokee

& Gogebic Development Company to explore the site first uncovered by Langford. After exploratory mining had proven the commercial potential of the mine, called the Colby Mine, the Penokee & Gogebic Development Company subleased it for three years to Pickands, Mather. Pickands, Mather partner Jay Morse brought in Joseph Sellwood, a veteran mining captain, to begin commercial work at the location in 1884. That year the Colby shipped only 1,022 tons of iron ore, but it quickly became a major producer of high-quality iron ore. The Colby's success, moreover, set off a major mining boom on the Gogebic. In 1886, two years after the Colby Mine began shipping ore, the Gogebic produced more than 250,000 tons. By 1892 the Gogebic Range had become the nation's leading producer.

In subsequent years the Penokee & Gogebic Development Company was transformed into the Penokee & Gogebic Consolidated Mines and began

The south vein of the Colby mine on the Gogebic Range, October 1885 (courtesy of Marquette County Historical Society)

operating the Aurora and Tilden Mines, two potentially valuable properties, on the Gogebic Range. When the panic of 1893 hit the Lake Superior mining region, Charles Colby, the principal owner of the company, was unable to secure funds from bankers to carry on operations at the mines. To secure operating capital, Colby was compelled to sell controlling interest in Penokee & Gogebic Consolidated Mines to John D. Rockefeller.

In August 1893 the Penokee & Gogebic Consolidated Mines became a part of the Lake Superior Consolidated Mines, a company created in summer 1893 to combine the iron and steel holdings of John D. Rockefeller with those of the Merritt brothers, the pioneer entrepreneurs on the Mesabi Range and owners of the Duluth, Missabe & Northern Railroad. In late 1893 and early 1894, however, the Merritt brothers encountered severe difficulties in meeting their financial obligations, compelling them to sell their share of Consolidated stock to Rockefeller at depression prices. This placed the new company completely under Rockefeller control.

The Penokee & Gogebic Consolidated Mines played a major role in the 1894 lawsuit brought by Alfred Merritt against John D. Rockefeller in the aftermath of Rockefeller's takeover of Lake Superior Consolidated Mines. Merritt charged that he and

his brothers had been defrauded by Rockefeller in the arrangements creating the Lake Superior Consolidated Mines. Specifically, he claimed that Rockefeller had fraudulently inflated the values of the properties which he had contributed to the new company, particularly the value of the Spanish-American Iron Company and the Penokee & Gogebic Consolidated Mines. A jury made up largely of Minnesotans supported the charges in June 1895, awarding Alfred Merritt $940,000 in damages. The award was overturned on appeal and remanded back to the U.S. District Court for a new trial. In 1897, however, Rockefeller and the Merritts settled out of court for $500,000, the Merritts agreeing to sign a statement retracting all charges of fraud against Rockefeller.

In 1901, as part of the Lake Superior Consolidated Mines, Penokee & Gogebic Consolidated Mines passed into the United States Steel conglomerate.

References:

Henry Oliver Evans, *Iron Pioneer: Henry W. Oliver, 1840-1904* (New York: Dutton, 1942);

Walter Havighurst, *Vein of Iron: The Pickands Mather Story* (Cleveland & New York: World Publishing, 1958);

David A. Walker, *The Iron Frontier: The Exploration and Discovery of Minnesota's Three Iron Ranges* (Saint Paul: Minnesota Historical Society, 1979);

Fremont P. Wirth, *The Discovery and Exploitation of the Minnesota Iron Lands* (Cedar Rapids, Iowa: Torch Press, 1937).

Phoenix Iron Works

by Frank Whelan

Morning Call, Allentown, Pennyslvania

On May 3, 1783, the business which would become the Phoenix Iron Works began the manufacture of iron in Chester County, Pennsylvania. By 1790 the works consisted of rolling and slitting mills and a nail factory. The nail factory was the heart of the Phoenix Works during its first years of existence. Of course, the term factory was not applied as one would apply it today since an eighteenth-century nail factory still involved a great deal of handwork. Nails were in great demand after the Revolutionary War. In the early 1700s American colonists had hammered them by hand from a bar of hot iron. But in 1775 Jeremiah Wilkinson, an inventor from Cumberland, Rhode Island, developed a process for cutting nails from a sheet of cold iron. The improvement made production easier and faster.

During these years the iron-making facility was known as the French Creek Works. In 1813 the plant was given the name Phoenix Iron Works by a new partner in the venture, Lewis Wernwag, a builder and designer of wooden covered bridges. His chief claim to fame was the construction of the first bridge across the Schuylkill River at Philadelphia. This beautiful arched span in its day was one of the engineering wonders of America.

Why Wernwag chose the name Phoenix is not known. Perhaps the legend of the mythical bird rising from its own ashes was meant to be symbolic of a reborn American iron industry. The War of 1812 and the series of trade embargoes that preceded it had blocked the importation of foreign iron into America. This easing of foreign competition helped the fledgling American iron trade and aided the domestic manufacturers in developing a market for their products.

The postwar years, however, were not so prosperous for American iron makers, including Phoenix Iron. Following the return of the British to the American market, the panic of 1819 undermined the business climate. Finally, in 1821 new owners came to Phoenix Works, and they immediately began a program of expansion. In 1822 a new works was erected, including 54 nail-making machines and a merchant pig iron operation. By 1824 Phoenix Iron was being hailed as the wonder of American manufacturing. Described as the first and largest nail factory of its kind in the United States, Phoenix Iron possessed the capacity to produce about 40 tons of nails a week. In 1825, when the 100-horsepower steam engine was constructed to give power to the Phoenix Works, its designer, Mark Stackhouse, proposed using anthracite coal as a fuel. The experiment was a success and marked one of the first industrial uses of the fuel in the United States.

In 1827 the Phoenix Works was sold again. This time the purchasers were a partnership of Philadelphia capitalists—Firman Leaming, Benjamin Reeves, David Reeves, James Whitaker, and Joseph Whitaker. But it was the Reeveses who were to maintain controlling interest in Phoenix Iron well into the twentieth century. Phoenix Iron was one of many ventures into the iron trade by the Reeves family. In the 1840s, for example, under the name Reeves, Abbott, they built the massive Safe Harbor Iron Works in Lancaster County. The same year the new partners took over Phoenix Iron, a new rolling mill was built and the puddling of iron was introduced as part of the manufacturing process.

With the start of the railroad boom in the late 1830s Phoenix Iron entered a new phase. In 1837 the Philadelphia & Reading Railway was opened. In order to meet what the owners saw as a potential for growth in the railroad business, Phoenix Iron constructed a new blast furnace, one of the first of its kind, to be fueled by anthracite coal. Although the process of using anthracite coal as a furnace fuel would not be perfected until Welshman David Thomas's pioneering work for Lehigh Crane Iron Works in Catasuqua, Pennsylvania, in 1840, many ironmasters were already experimenting with the technology.

The 1840s were years of fantastic growth for Phoenix Iron. In 1841 six puddling furnaces were added. In 1842 a survey was done of the property. The works at that time consisted of a blast furnace with the capacity for making 1,500 tons of pig iron per year. There was also a refining furnace of equal capacity. Also included were a rolling mill capable of converting 3,000 tons of pig iron into bars per year and a nail factory with a capacity of making 32,000 kegs per year. The total number of employees was 147.

As demand for iron accelerated, Phoenix expanded in an attempt to keep up production. In 1845 two new blast furnaces, measuring 15 feet by 59 feet, were built. The following year another blast furnace of a similar size was constructed. That same year additional partners entered the firm, which was reorganized as Reeves Buck & Company. Under the new management a new rail mill was installed, and new puddling and reheating mills were added. In addition a small foundry, blacksmith shops, and pattern and machine shops were constructed.

In 1855 Reeves Buck & Company again took the name of Phoenix Iron Company. The same year the company began the manufacture of iron beams and a variety of structural shapes of iron, becoming the first mill in America to roll beams and shapes.

Much of the work was overseen by the plant's superintendent, John Griffen (sometimes spelled Griffin), who was hired in 1856 and the following year obtained a patent to roll large wrought iron beams. On January 22, 1861, the first 15-inch beams to be produced in the United States were rolled at Phoenix Iron. With the outbreak of the Civil War, Phoenix entered military production, creating a small cannon designed by and named for John Griffen. The "Griffen Gun" was the first spirally wrapped wrought iron gun to be manufactured in the United States and perhaps in the world. According to Phoenix company historian Robert Schaffner it was a Griffen Gun that fired the first artillery shot at the battle of Gettysburg. Over 75 pieces of ordnance made at Phoenix are still in place on the Gettysburg battlefield. Griffen left Phoenix in 1862, traveling north to Buffalo, New York, to erect an ironworks there. But Griffen was not to be gone for long, returning to Phoenix in 1868 and remaining superintendent until his death in 1884.

At the same time Griffen was headed to Buffalo, Samuel J. Reeves invented a new piece of structural iron called the Phoenix Column. It became so popular with builders that it was a lucrative export item, reaching areas as far flung as Japan, Russia, Australia, South America, and Africa. In addition to the Phoenix Column, on February 15, 1865, Phoenix rolled iron bars 8 inches square and 25 feet long, the largest squares to be rolled up to that time in the United States.

From 1862 to 1872 some of the most important bridges were manufactured from material made at Phoenix Iron–among them the Verrugas Viaduct in Peru and the International Bridge across the Niagara River near Buffalo. In 1871 a new rolling mill was erected at Phoenix with the most up-to-date equipment in the world. With the collapse of the iron market following the panic of 1873, the company began to experiment with going into steel production. The decision was made in 1886 to enter the steel market, and the company's first steel was produced in 1889.

The Phoenix Iron Company continued to be owned by the Reeves family until 1948. The company was acquired by the Barium Steel Corporation of Harrisburg, Pennsylvania, on August 9, 1949. That same year the name of the firm was changed to Phoenix Iron & Steel Company. Its primary product continued to be structural steel for bridges.

In 1960 Barium sold Phoenix Iron & Steel to Stanley Kirk, who in 1963 renamed the company Phoenix Steel. By the 1970s its days as a leader of the steel industry were over. Large parts of the old mill and the bridge division were sold for scrap. Today only a seamless tube mill acquired in 1956 remains intact on the site. In 1985, 202 years after it began, the Phoenix plant closed.

References:

J. Bennett Nolan, *Southeastern Pennsylvania: A History of the Counties of Berks, Bucks, Chester, Montgomery, Philadelphia and Schuylkill*, volumes 1-3 (Philadelphia: Lewis Publishing, 1943);

Telephone interview with Robert Schaffner, historian of Phoenix Iron Company, February 1988.

David Rittenhouse Porter

(October 31, 1788-August 6, 1867)

by Paul F. Paskoff

Louisiana State University

CAREER: Clerk, Pennsylvania surveyor general's office (1809-1813); clerk, manager, Barree Forge (1813-1814); partner, Sligo Iron Works (1814-1819); member, Pennsylvania State Assembly (1819-1823); clerk of court, Huntingdon County (1823-1827); clerk, Huntingdon County (1827-1835); member, Pennsylvania State Senate (1836-1838); governor, Pennsylvania (1839-1845); owner, Harrisburg furnace (1845-1867).

David Rittenhouse Porter was governor of Pennsylvania for two consecutive terms from January 1839 to January 1845, a period of profound economic and political turmoil. Before and after his service as governor, Porter maintained a close connection to the iron industry in his state as a producer and innovator and also showed an active interest in the health of American manufacturing.

Born on October 31, 1788, just outside Norristown in eastern Pennsylvania, Porter was the son of Andrew Porter, a Revolutionary War hero, and his second wife, Elizabeth. The Porters were a family of more than modest means, and David, the eleventh of his father's 13 children by two marriages, attended Norristown Academy, where he pursued the higher classical curriculum with the aim of going on to Princeton. The fire which forced Princeton to close for several months in 1802 abruptly altered his plans for further education.

In 1809, at the age of twenty-one, Porter moved with his father, the Pennsylvania surveyor general, to the capital in Harrisburg, where Porter was employed as his father's clerk. While in Harrisburg he began to study law with the intention of passing the bar and going into practice. Once again, however, Porter's educational plans were derailed, this time by illness. His father's death in 1813 removed the only compelling reason for remaining in Harrisburg, and Porter left that year to take a job

David Rittenhouse Porter

as clerk in an ironworks in the Juniata Valley of Huntingdon County.

The Juniata Valley, in the middle of Pennsylvania and across the Susquehanna River to the west of Harrisburg, had been the object of intense interest in iron mining and production even before the Revolutionary War. The long war for independence, however, had disrupted the ventures already under way to exploit the region's potential. The end of the war in 1783 and the ratification of the Constitution in 1789 stimulated renewed interest in iron making, and by the turn of the century, the Juniata region supported at least 20 furnaces and forges. The rapid growth of Pittsburgh during the 1790s and the first decade of the nineteenth century as the center of a burgeoning western trade in the Ohio River Valley provided a ready market for Juniata iron. Demand for this iron intensified as a result of the War of 1812, and when Porter arrived in 1813 at Barree Forge (also called Dorsey Forge) to assume his duties as clerk, he found business to be brisk.

Barree Forge had been built about 1795 by Greensburg P. Dorsey on a site along Spruce Creek near the town of Huntingdon. Within a short time after Porter's arrival the Dorseys promoted him to the position of manager of the forge. In 1814 Porter had saved sufficient funds to enter into a partnership with Edward Patton to purchase nearby Sligo Iron Works. Unfortunately, the investment was ill timed, having been formed on a thin capital base in the last good year, for many to come, to be enjoyed by domestic producers of pig and bar iron. The end of the war in early 1815, the subsequent dumping of British bar iron on the American market, and the panic of 1819 put the firm under severe stress, and it failed. The Sligo Iron Works failure in 1819 was only one of many among industrial, banking, and commercial ventures caused by the panic.

Porter seems to have handled his misfortune with a good deal of equanimity. The year that Sligo failed, Porter became a representative of Huntingdon County in the Pennsylvania State Assembly, continued to study law, and pursued a growing involvement in farming and livestock breeding. In 1820 he married Josephine McDermitt, daughter of the late William McDermitt, who himself had failed as an ironmaster along Spruce Creek. These years were something of a personal watershed for Porter. At slightly over thirty he had decided, perhaps with some goading by events, to make politics his chief pursuit. Reelected to the assembly in 1820 and again in 1822, Porter also began to assume an increasingly influential role as a county politician. By the early 1830s he was a formidable political power in the Juniata Valley, and in January 1836 he became a state senator. It was, however, as governor of Pennsylvania that Porter exercised his greatest influence upon state and national affairs.

As Pennsylvania's gubernatorial election of 1838 approached, feuding among the state's Democrats jeopardized their hopes of defeating the Whig incumbent, Joseph Ritner. Ritner had capitalized on the strident anti-Masonic sentiment of many voters who found in the Masonic Order a convenient scapegoat for their economic distress. Ritner had even been accused of having sanctioned an anti-Mason hysteria with the cynical aim of distracting the electorate from the ineptitude and graft which permeated Pennsylvania's state-funded internal improvements. The state's fiscal problems were aggravated by the financial chaos associated with the panic of 1837, which affected public and private enterprise throughout the country.

The economic difficulties besetting the state were not an insurmountable political problem for the Whigs, provided the Democrats failed to achieve organizational discipline and unity. For their part, the Democrats had to find a candidate whose nomination would command the loyalty and energy of all factions within the party. Porter was just such a candidate.

The election was hard fought, dirty, and close, with Porter edging out Ritner by approximately 1 percent of the vote. Because the election was so close, ethics took a backseat to strategy. The Democrats gleefully claimed in their newspapers that Ritner was an illiterate and a sexual pervert, calling him a "damned Dutch hog." Not to be outdone in the way of invective, the Whigs returned insult for insult and slander for slander. They accused Porter of fraud, perjury, and miscegenation, presenting in support of the last charge the affidavit of a black woman who swore that Porter had fathered her two illegitimate children. Porter, whose son, Horace, had been born just months earlier in April 1837, was disgusted when Whig newspapers printed the affidavit and gave it prominent play. The trading of lies and libels by the two parties was, in fact, the least unpleasant and dangerous aspect of the electoral contest.

The closeness of the count prompted Ritner's allies in the legislature to attempt to reverse the outcome by challenging the validity of thousands of Porter's votes. Their maneuver and the ensuing dispute in and out of the legislature's chambers were dubbed the "Buckshot War" and almost succeeded in overturning Porter's victory through intimidation and gunplay. There was even a reported plot to assassinate the state's canal commissioner, Thaddeus Stevens, then a Whig and Ritner loyalist and an aspiring United States senator. Despite the plots, real and imagined, Porter's victory was sustained.

There was a certain irony to the timing of Porter's election. He had been drawn into political life by his personal business failure during the panic of 1819. Twenty years later, in January 1839, he assumed the governorship just as the last rumblings of the panic of 1837 were fading and the first convulsions of the depression of 1839 to 1843 were being felt. Not surprisingly, he was preoccupied during his two consecutive terms in office with the state's economic problems.

One of the more pressing problems in 1839 and 1840, and one not unique to Pennsylvania, was the refusal by state-chartered banks to redeem their bank notes in specie, that is, in hard currency. Called "suspension," it was at one and the same time a reaction to and a cause of financial and commercial chaos. Holders of notes issued by banks that refused to honor them in cash suffered acutely while the defaulting banks, some of which were not necessarily without specie reserves, shored up their precarious positions. Politically and practically, this was an untenable situation. An allied problem was the growing demand within the state government and among the general population for repudiation of the part of the state debt represented by interest due on state bonds. Repudiation would have severely damaged Pennsylvania's domestic and foreign credit rating.

Porter tackled both problems, suspension and repudiation, in 1840 with sweeping legislation which forced state-chartered banks to resume redemption of their notes in specie on pain of losing their charters, and which provided for the payment in specie of the interest on state bonds. These legislative accomplishments were not unqualified victories, and, although Porter's insistence that state-chartered banks redeem their notes in specie was good popular politics, it proved to be an unworkable policy in practice. He found that while the banks certainly needed the support of the state, the converse was equally true as the deepening financial and commercial crisis compelled the government to turn to the state-chartered banks for the money needed to pay the interest on the state's bonds. Obviously, the banks could not simultaneously redeem their own notes in specie *and* support the integrity of Pennsylvania's state bonds. Recognizing this, Porter quietly relented in his pressure on the banks to abandon suspension. This accommodation between the state and the banks and Porter's widely applauded efforts to arrest the deterioration

of Pennsylvania's economy assured him of renomination and reelection in 1841, this time with a comfortable 55 percent share of the vote.

Although Porter's second term was marked by a gradual improvement in the state's economy, it was also marred by an intensification of the conflict between the state government's executive and legislative branches that had begun in the early months of his first term. In large part this struggle for control over the state government had to do with repeated attempts by the legislature to regain the powers taken from it and given to the state judiciary and the governor by Pennsylvania's 1838 constitution. But it was also due to Porter's advocacy of positions—including opposition to the rise of corporations—which antagonized various interests within the state. Despite his actions that garnered widespread support, such as using force to quell the nativist anti-Catholic riots of 1844 in Philadelphia and convincing a majority of the legislature to abolish imprisonment for debt, Porter's repeated clashes with the leadership of the legislature created a poisonous political atmosphere. At one point, in 1842, he had to fight off what ultimately proved to be an unsuccessful impeachment attempt.

There is reason to believe that Porter's tenure as governor would have been less convulsive had he enjoyed the wholehearted backing of his party. But that was impossible given his strong advocacy of a protective tariff on imported manufactured goods at a time when the Democratic party was staunchly opposed to one. When he left office in January 1845, Porter left behind a party still fractured along the lines of the mid 1830s and a somewhat ambivalent record of positions and accomplishments. He had been remarkably firm in defying a significant segment of his party by advocating such pro-business measures as a protective tariff and internal improvements, positions more commonly identified with Whigs than Democrats. At the same time, however, he had sought to inhibit the spread of the incorporated company just as that form of business organization was rapidly growing in Pennsylvania and most other states. Although he spoke out in support of reforming and augmenting the state's system of public education, he did nothing along those lines but speak, and then only seldom and not very effectively. At least in this respect Porter had much in common with the men who followed him as governor, especially his immediate successor, Francis R.

Shunk, a fellow Democrat who had served as Porter's superintendent of common schools.

The end of Porter's second term as governor marked the end of his career in politics and the resumption of his active interest in the iron industry. That year he completed the construction of his anthracite furnace in Harrisburg. One of only 10 such furnaces built that year and one of just 20 that had been constructed since the first had been erected in 1840, Porter's Harrisburg was the first anthracite coal furnace built outside the northeastern Pennsylvania counties of Lehigh, Luzerne, and Columbia. As such, it represented a bold initiative by Porter, who had seen the future of pig iron production earlier and more clearly than had most of his contemporaries in the industry.

Probably because the Harrisburg furnace was one of the first anthracite furnaces, it was not exceptional in an engineering sense. Its rated annual output capacity of 3,800 tons was slightly smaller than the average of 4,015 tons for anthracite furnaces built in the state that year. Still, within the state it was a significant and bold regional innovation and one which inspired other men to build the anthracite furnaces which, collectively, helped to transform the pig iron industry in the United States before the Civil War.

Porter's interests in the years after he left office were not confined to iron making. He maintained a warm friendship with the fast-rising Democrat James Buchanan and with Sam Houston, governor of Texas. While his relationship with Buchanan mixed friendship and politics, his association with Houston added the element of a common business interest. As fellow Democrats, Porter and Houston were allies in an effort to attract financial and congressional support for a southern route for the proposed transcontinental railroad. Their campaign was swept aside by the increasingly bitter political conflict in Congress between Democrats and Republicans during the last years before the outbreak of the Civil War.

During that war Porter remained true to his long-standing Unionist principles as one of the so-called "War Democrats." A source of pride for him was the fact that his son, Horace, served as a general in the Union army. By the end of the war David Porter was an old man. His economic influence as an innovative iron maker had long since waned as the industry had surged past him; he had begun to devote increasing amounts of time to his farm. He died in Harrisburg at the age of almost seventy-nine on August 6, 1867.

References:

W. C. Armour, *Lives of the Governors of Pennsylvania* (Philadelphia, 1872);

Arthur C. Bining, *Pennsylvania Iron Manufacture in the Eighteenth Century* (1938; reprinted, Harrisburg, Pa.: Pennsylvania Historical and Museum Commission, 1973);

Fawn M. Brodie, *Thaddeus Stevens, Scourge of the South* (New York: W. W. Norton, 1959);

Paul F. Paskoff, *Industrial Evolution: Organization, Structure and Growth of the Pennsylvania Iron Industry, 1750-1850* (Baltimore, Md.: Johns Hopkins University Press, 1983);

W. A. Porter, *Life of David Rittenhouse Porter* (N.p., n.d.);

Robert Sobel and John Raimo, eds., *Biographical Directory of the Governors of the United States, 1789-1978* (Westport, Conn.: Meckler Books, 1978);

James M. Swank, *History of the Manufacture of Iron in All Ages* (Philadelphia: American Iron and Steel Institute, 1892);

James Pyle Wickersham, *A History of Education in Pennsylvania* (Lancaster, Pa.: Inquirer Publishing, 1886).

Orrin W. Potter

(December 25, 1836-May 17, 1907)

by Alec Kirby

George Washington University

CAREER: Secretary and general manager (1864-1871), president, North Chicago Rolling Mill (1871-1889); cofounder and president, Illinois Steel Company (1889-1896).

Orrin W. Potter began his career in 1856 as a clerk for Eber B. Ward, the first of the American "iron kings." By the time of his death, Potter had become president of Ward's North Chicago Rolling Mill and, with Jay C. Morse, created the largest American steel conglomerate next to the Carnegie group.

Potter was born in Rochester, New York, on Christmas day, 1836. He received a basic education in public schools, although he was largely self-taught in higher mathematics and engineering. In 1856 he married Ellen Owen and moved to Wyandotte, Michigan, where Ward hired him as a clerk. Ward had a well-known ability to recognize talent and initiative in others, and Potter quickly moved up through the company. When Ward established a new firm, the North Chicago Rolling Mill Company, he transferred Potter to Illinois. In 1864 Potter became secretary and general manager of the new company and assumed its presidency seven years later. Under Potter's leadership the North Chicago Rolling Mill Company expanded quickly and came to control important iron- and steelworks in Chicago and Milwaukee. The firm thus had facilities including blast furnaces, open-hearth and Bessemer steel plants, mills for rail, beams, plates, and slabs, and could also produce spiegeleisen, which consisted of a triple compound of iron, manganese, and carbon.

Potter was a member of the distinct generation of American businessmen that followed a generation of industrial innovators who created the first large-scale enterprises and plants. This older generation included Eber Ward, who became a leader in the Great Lakes shipping industry before turning his attention to steel. The businessmen following Ward created huge corporations, not by starting from scratch, but by consolidating a number of large-scale plants and enterprises. This group included Andrew Carnegie and Orrin W. Potter.

On May 4, 1889, Potter, in association with Jay C. Morse, consummated a gigantic agreement: the merger of three major midwestern iron and steel corporations. Under this arrangement the North Chicago Rolling Mill became the basis of a merger with two other major corporations. The first of these was the Union Iron and Steel Company of Chicago. By 1889 this firm was under the direction of Morse and H. A. Gray and specialized in the creation of Bessemer steel. The second corporation involved in the merger was the Joliet Steel Company, run by Alexander J. Leith, with Horace Strong Smith serving as superintendent. As a result of the efforts of Potter and Smith, the steel and iron industries of the Chicago area followed the path of consolidation which had already been blazed in Pittsburgh by Andrew Carnegie and his associates.

The new Chicago consolidation was incorporated as the Illinois Steel Company. The new firm controlled five plants, including two in Chicago built in 1857 and 1880 by the North Chicago Rolling Mill Company and one in Milwaukee built in 1868, also by North Chicago. The other plants were the Joliet Works in Joliet, Illinois, constructed in 1870, and the Union Works in Chicago, built in 1863. Illinois Steel subsequently controlled the Chicago Lake Shore and Eastern Railroad Company. The combined plants encompassed 14 blast furnaces, an iron rolling mill, four Bessemer steel plants, and mills to produce almost a full line of finished steel products including rails, beams, merchant iron or steel nails, and iron rods. The new conglomeration was fully integrated because it also controlled its own coke ovens and enough ore and coal lands to provide it with sufficient raw materials for an estimated 25 years. All told, the Chicago firm had the largest steel-producing capability in the world.

Potter remained as president of Illinois Steel until his retirement in 1896. He died on May 17, 1907, from kidney disease.

References:
Herbert N. Cason, *The Romance of Steel* (New York: Barnes, 1907);
William T. Hogan, *Economic History of the Iron and Steel Industry in the United States*, volume 1 (Lexington, Mass.: Lexington Books, 1971);
Fritz Redlich, *History of American Business Leaders*, volume 1 (Ann Arbor, Mich.: Edwards Brothers, 1940).

Pottstown Iron Company

by Frank Whelan

Morning Call, *Allentown, Pennsylvania*

The Pottstown Iron Company was founded in Pottstown, Montgomery County, Pennsylvania, in 1846, but its roots go back much earlier. In 1717 Thomas Rutter constructed the first iron furnace in the region. His ironmaster was a Welshman named Thomas Potts, who in 1690 had settled in Germantown, just outside the nascent city of Philadelphia.

Potts and Rutter became friends, and, as time passed, Rutter turned more and more of the operation of the iron furnace over to Potts. Through marriage and by purchasing shares, Potts gradually took control of the operation.

The growth of Potts's holdings in the iron industry was swift. By the 1750s the Potts family held interests in more than a dozen forges. In 1752 Thomas Potts's son, John, purchased two tracts of land of 500 acres each which became the headquarters of the family's iron business. On the plantation, which was named Pottsgrove, Potts built a mansion and an iron forge. Soon after this purchase Potts also acquired the nearby Mount Joy forge, which is better known as Valley Forge. The mansion built by Potts at Valley Forge became General Washington's headquarters.

The Potts family made Pottsgrove plantation into one of the first centers of America's metalworking industry. The label of plantation was appropriate; the complex contained its own gristmill, store, offices, and shops. Farms, orchards, and gardens made the estate self-sufficient. The fuel used for smelting in the blast furnaces was charcoal made from the plantation's forests. One of the properties known as Warwick Furnace consumed 5,000 cords of wood annually and required 250 acres of woodland to supply its needs. In many ways these iron "plantations" were similar to their southern counterparts, and the ironmaster's home was equal to southern mansions. Well stocked with European furniture and domestic imported luxuries, Pottsgrove Mansion was the focal point of community social life.

The real heart of the plantation was the forge or furnace, which in those days was located next to a creek. The running water supplied the needed power to turn huge waterwheels, which ran the bellows and tilt hammers. The ore used was mined on the land from an open trench, sometimes 20 or more feet deep.

There was a profound socioeconomic gap among employees at Pottsgrove and employees at other ironworks. Because skilled metalworkers were in short supply, they could demand and receive high wages. Not as fortunate were the many unskilled laborers, many of which were indentured servants and were assigned to work for a master, who had paid for their passage from England. At the bottom of the hierarchy were the black slaves and "redemptioners." The last were of the lowest class in England and had sold themselves into virtual slavery in order to escape crime, debt, or grinding poverty. They and the blacks worked side by side at the most arduous jobs that the plantation had to offer.

The little world of Pottsgrove, which became Pottstown in 1815, was ready to take advantage of the changes sweeping the metals industry. By the be-

ginning of the nineteenth century, the semifeudal social world of the northern iron plantation was
gone. But the Pottses were still deeply involved in
the iron trade. In 1846 brothers Henry and David
Potts, the sixth generation of their family, built the
Pottstown Iron Works.

The works was located at Water Street, between Charlotte and Penn streets. In 1857 a local
man named Edward Bailey joined them, and the business became Potts & Bailey. By 1862 Bailey had
left the business and sold his interest back to the
Potts brothers. They changed the name to Potts
Brothers' Iron Works, a name the business held into
the twentieth century.

In 1880, at the height of its success, Potts Brothers' employed 670 men and was the largest employer in the community. By this time anthracite
coal fueled the blast furnaces. The county history
points out that the first iron truss bridge in this hemi-
sphere was built in the shops of Pottstown. Bridge
building was to become something of a specialty
for the little community. A visitor to the community in 1880 noted: "The eastern approaches to the
town are lined with iron works from the railroad to
the Schuylkill River. The western end is lighted by
the Warwick Furnace."

The prosperity of the metals industry lasted a
little longer in Pottstown than it did in other locations, and business at Potts Brothers' Iron Works continued strong through 1892. At that point the firm
was making boiler and tank plates and had a payroll of 190 men; annual production reached 10,000
tons of muck iron and 12,000 tons of plate iron.
But the firm was hit hard by the panic of 1893. In
1916 the Potts family sold its interest to the Nagle
Steel Company.

Although the Great Depression killed off most
of Pottstown's metal industry, the bridge-building
McClintic & Marshall Corporation survived. It was
taken over by Bethlehem Steel in the 1930s and fabricated the parts of both the George Washington and
Golden Gate bridges. But fate was less kind to
Potts Brothers' Iron Works. It stopped production altogether and by the 1950s was being used as a cold
storage warehouse.

References:

Paul Chancellor, Marjorie Potts, Wendell Wescott, and
    Paul Westcott, *A History of Pottstown* (Pottstown,
    Pa: Historical Society of Pottstown, 1953);
Michael J. Schwager and Jean Barth Toll, eds., "Montgom-
    ery County: The Second Hundred Years," *Montgom-
    ery County [Pa.] Federation of Historical Societies*, 1
    (1983): 528-549.

Puddling

by Paul F. Paskoff

Louisiana State University

Puddling was a process for making wrought iron from pig iron in a rolling mill's reverberatory furnace at substantially higher temperatures and with greater efficiency than could be obtained in a forge. The technique was developed and patented in England by Henry Cort in 1784 and was substantially improved in 1830 by Joseph Hall. It was the Hall process, sometimes called the "pig boiling" or "wet puddling" process, that came into widespread use in the United States after 1830.

The reverberatory furnace had two chambers that were partially separated by a short wall. On one side was a fireplace in which the fuel was burned, and on the other was the furnace's hearth which had a bowl-like depression in its floor in which boiling pig iron could form a "puddle." Next to the hearth was the flue through which exhaust gases were conducted to the furnace's stack. The furnace's three sections–fireplace, hearth, and flue–were contained under an overarching brick roof with a 10-degree interior gradient toward the flue to deflect the flames from the fireplace over and onto the pig iron in the hearth and, then, to conduct the gaseous waste to the flue.

Fuel was loaded into the fireplace and pig iron was put into the hearth. Once the fuel was ignited, its flames began to play on the iron which remained untouched by the fuel. This isolation of the iron from the fuel was one of the truly distinctive features of the reverberatory furnace and of the puddling process. Unlike a forge or conventional blast furnace in which fuel and iron were mixed, thereby allowing the fuel's impurities to react with the molten iron, the puddling furnace permitted the making of a purer, more controlled product. A related advantage of the reverberatory furnace was that it made possible the use of coke, a lower grade fuel, in terms of purity, than charcoal, the traditional fuel used in blast furnaces and forges. Because the fuel's purpose was to provide a source of heat, the

quality or purity of the form of carbon–whether charcoal, coal, or coke–was immaterial. The use of the less expensive coke in place of charcoal resulted in significant reductions in the cost of raw materials.

Another advantage of the reverberatory furnace was that its two-chambered architecture allowed the furnace man, or puddler, to work the iron in comparative safety and with considerably more ease and control than was the case at a forge. There forge men, especially hammer men, had to remove the heated iron from the forge fire, an arduous and dangerous task, transfer it to the trip-hammer for pounding and shaping and removing impurities, and then return it to the fire for further heating so that the entire process could be repeated until the pig iron had been worked up into wrought iron. By contrast, the puddler was able to "puddle," or stir, his iron while the reverberatory furnace's flames still played on it, keeping it malleable and, therefore, more readily worked. The elimination of hammering not only increased the productivity of the ironworkers but also significantly reduced necessary labor time, wages, and total costs.

In the course of the puddling process, as the pig iron was continuously stirred and heated, the boiling of the iron drove out most of the iron's impurities without the need for hammering. Although this was a desirable result, it was not achieved without cost. As the pig iron progressively lost most of its impurities, its melting point increased beyond the furnace's capacity to heat, thereby making the puddler's efforts to continue to work it more difficult. No longer molten, the purified iron was now in a semiplastic form. At this point the puddler rendered the iron into 60- to 100-pound balls consisting of an admixture of slag and very pure iron. Most of the slag was then removed from the balls by compressing them in a rotary squeezer. The squeezed iron was then passed through a set of

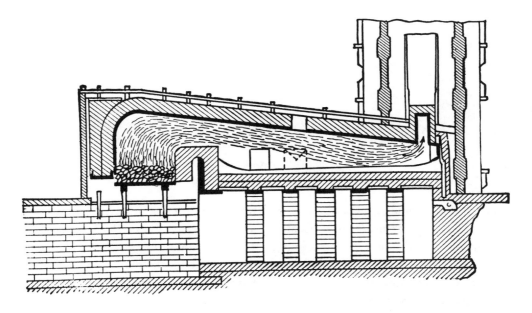

Diagram of a reverberatory furnace used in puddling pig iron into wrought iron

grooved metal rollers which shaped it into a "muck" bar, so called because of the strands of slag visible along its length. A number of these "muck" bars were then placed into one of the rolling mill's reheating furnaces to bond them to one another. These thicker bars were then withdrawn and run through a succession of grooved rollers of different shapes and sizes to form the rails, rods, and structural iron that were the products of the rolling mills.

The superiority of puddling and rolling over forge heating and hammering was overwhelming and was reflected in the rapidity with which the newer techniques spread throughout the industry. By the 1840s rolling and puddling had taken root in Pennsylvania, the premier iron-producing state, and together accounted for almost 80 percent of the state's wrought iron output in 1849. In 1856 the proportion of wrought iron made in Pennsylvania's rolling mills from puddling had increased to about 90 percent of the 520,000 tons of

wrought iron produced that year. This dramatic transformation of the iron industry was stimulated in large measure by the rapid growth of the railroad industry and its voracious appetite for rails and other railroad-related products, such as boiler iron. American rails came from the nation's rolling mills, and these installations relied almost exclusively on the iron made by puddling.

References:

William T. Hogan, S.J., *Economic History of the Iron and Steel Industry in the United States*, 5 volumes (Lexington, Mass.: Lexington Books, 1971);

Jeanne McHugh, *Alexander Holley and the Makers of Steel* (Baltimore: Johns Hopkins University Press, 1980);

Paul F. Paskoff, *Industrial Evolution: Organization, Structure and Growth of the Pennsylvania Iron Industry, 1750-1860* (Baltimore: Johns Hopkins University Press, 1983);

Peter Temin, *Iron and Steel in Nineteenth-Century America: An Economic Inquiry* (Cambridge, Mass.: M.I.T. Press, 1964).

Samuel J. Reeves

(March 4, 1818-December 15, 1878)

by Paul F. Paskoff

Louisiana State University

CAREER: Manager, Reeves family store in Philadelphia (1840-1846); member, Pennsylvania's Coal and Iron Association (1842); manager, Reeves, Buck & Company (1846-1855); vice-president (1855-1871); president, Phoenix Iron Company (1871-1878); vice-president (1864-1869), president, American Iron and Steel Association (1869-1878).

Samuel J. Reeves, an ironmaster, inventor, and industry leader, was born into an iron-making family at Bridgeton, New Jersey, on March 4, 1818. Both his father, David Reeves, and his uncle Benjamin Reeves had been iron producers since 1814 when the two retail merchants from Philadelphia had founded the Cumberland Nail & Iron Works at Bridgeton, the county seat of New Jersey's Cumberland County.

The decision to concentrate on making nails enabled the Reeves brothers to avoid much of the ruinous British competition in the American pig and bar iron markets which followed the end of the War of 1812. Moreover, Bridgeton's proximity to Philadelphia, Baltimore, and New York City ensured the Cumberland Nail & Iron Works of a growing and convenient market for its products. By 1826 a combination of locational advantage, prudent management, and an expanding economy had made the Reeveses' venture quite profitable. Under the name of Reeves, Whitaker & Company they extended their iron operations into Pennsylvania by purchasing a controlling interest in the Phoenix Iron Works on the Schuylkill River at Phoenixville in Chester County, about 25 miles northwest of Philadelphia.

While David and Benjamin Reeves expanded the business and increased the family's wealth, they made sure that Samuel received the finest education available in the Philadelphia area. He attended grammar school, then was enrolled in a private academy on the outskirts of Trenton, New Jersey, in preparation for Princeton University. When Reeves graduated from Princeton in 1837, he had had no formal experience or training which might have equipped him to take on the management of any part of the family's business. The Master of Arts degree which Princeton subsequently conferred on him in 1840 reflected only the university's customary policy of awarding the advanced degree to its graduates three years after the B.A. and not any accomplishment by Reeves. He was twenty-two, highly educated, and ignorant of business. He had, of course, listened to and observed his father, who saw to it that Samuel acquired both experience and training while managing the store which the family still maintained in Philadelphia. After he had served several years in that capacity, David Reeves decided that his son was ready for greater responsibility and in 1846 made him manager of the company's facilities in Phoenixville and Bridgeton. This marked the beginning of Samuel Reeves's direct involvement in the iron industry. That year was a watershed in Samuel Reeves's life for another reason: he married and began the family that eventually would include two sons and four daughters.

The firm founded by his father and uncle had by 1846 grown considerably from what it had been just 20 years earlier. In 1835 the company had rebuilt and expanded the then 45-year-old rolling mill in Phoenixville which it had purchased in 1826. When in 1846 Samuel assumed the management there and at Bridgeton, the firm was just completing the ambitious Phoenix Rail Mill, which was one of the largest rail mills in the state. The decision to build the rail mill was especially well timed because the railroads in Pennsylvania and in the adjacent states were poised for renewed growth following the retrenchment of the early 1840s. During the next 14 years the railroads within Pennsylvania more than tripled their trackage, from just over 730 miles in 1846 to more than 2,660 miles in 1860.

This dramatic growth meant a sharp increase in the railroad's demand for rails, a product which the Reeves were eager and able to supply from their newly expanded plants.

The firm's growth imposed new and greater demands on a manager's skill and wisdom and on the firm's capital stock. The need for additional capital compelled David Reeves in 1846 to offer Robert S. Buck a partnership in the company and smaller shares in it to other investors. The most readily visible evidence of the refinancing of the old firm was the change of its name to Reeves, Buck & Company. But the elder Reeves was determined to retain close family control over the direction of the new company, and it was primarily this consideration that led him to make his son the general manager. Despite his lack of managerial experience in iron production prior to 1846, Reeves was by no means uninformed about the conditions prevailing within the industry. In 1842 he had served as one of the members of a committee of the Pennsylvania Coal and Iron Association, which had conducted a survey of the state's iron industry to determine the levels of output of the various types of iron and the extent to which producers required tariff protection from foreign, that is British, competition.

Ostensibly, there should have been no need for the Association to undertake a special investigation of the health and performance of the iron industry because the United States census, conducted only two years before, had presumably covered the subject. The widely held opinion throughout the country, however, was that the census findings were highly unreliable and of doubtful value. It was the skepticism, in industry and government circles, of the worth of the 1840 census estimates of iron production that led to the formation of the Coal and Iron Association's survey committee. The committee's findings became the basis for the estimates of pig iron output used by U.S. Secretary of the Treasury Robert J. Walker in formulating his proposals for the tariff of 1846, subsequently known as the Walker Tariff.

That Reeves, as a member of the committee, was neither naive nor uninformed soon became evident to his colleagues and to those with whom he subsequently corresponded about the committee's findings and related matters. In an often quoted letter of November 1849 to William M. Meredith, Reeves expressed his contempt for both the statistical picture of the Pennsylvania iron industry pre-

sented by the 1840 census, and for those who had compiled it, saying that it was "no credit to those who had immediate charge of it." Reeves, already a protectionist where the iron industry was concerned, continued to monitor the industry's health and progress during the turbulent 1840s. In 1849, as an influential member of the Convention of Iron Masters which met in Philadelphia on December 20 of that year, he helped to design the most ambitious survey yet of Pennsylvania's iron industry. Reeves and his fellow iron producers hoped that the survey's results would provide compelling evidence to present to Congress in support of their demands for higher tariffs on imported iron. Far more reliable than either the U.S. Census of 1840 or that of 1850, the cogency and thoroughness of the Convention's report, which enjoyed widespread circulation when it was subsequently published in the *Journal of the Franklin Institute, Hunt's Merchants' Magazine*, and the *American Railroad Journal*, owed much to the clear mind and restrained hand of Samuel Reeves.

Reeves's managerial talents and informed understanding of the iron market enabled Reeves, Buck & Company to continue its growth and profitability. By 1850 the firm had the largest rolling mill in Pennsylvania and was, in key product lines, especially rails, a vertically integrated operation. Its mills at Phoenixville rolled iron far more efficiently, both in terms of fuel consumed and labor required, than any of its competitors. Reeves, Buck's production capacity and share of the market increased substantially with the formation in 1847 and 1848 of an allied venture, Reeves, Abbott & Company, at Safe Harbor, along the Susquehanna River in Lancaster County, Pennsylvania. The integrated Safe Harbor works included a large anthracite furnace of the most advanced design and one of the largest rolling mills in the country for making rails. In 1850 the vertically integrated mill complexes at Phoenixville and Safe Harbor accounted for about one-eighth of all production of rolled iron in the state, a proportion sustained throughout the next 10 years in the face of fierce domestic and foreign competition.

Just as the firm's rapid growth during the 1830s and early 1840s had prompted the Reeves family to seek additional capital and to restructure the firm, its expansion during the 10-year period from 1846 to 1855 required still more capital and greater administrative control. To attain these ends the firm was incorporated in 1855 as the Phoenix Iron Company (Reeves, Abbott & Company remained nominally independent), with David Reeves as president and Samuel Reeves as vice-president of the still family-dominated company. For all practical purposes, however, day-to-day control and management of the Phoenix Iron Company and the affiliated Safe Harbor plant of Reeves, Abbott & Company rested in the younger Reeves's hands. Under his guidance the mill at Safe Harbor achieved a significant advance in 1855 in the new and burgeoning field of structural iron when it began to produce 9-inch-deep I-beams. This was, after the fact, an obvious if not inevitable outgrowth of rail-making technology, but one not widely developed until Safe Harbor successfully introduced it. The I-beam success of 1855 was a straw in the wind for the Phoenix Iron Company and particularly for Samuel Reeves, whose interest began to focus increasingly on the fabrication and improvement of structural iron.

Reeves's concerns in 1855 were not confined to the affairs of Phoenix Iron Company, and he devoted considerable time to the condition of the iron industry in Pennsylvania and in the other producing states. He was instrumental in organizing the American Iron Association that year to articulate and advance a coherent, united political agenda for the industry in its perennial battle for a high protective tariff. This activity was a logical extension of Reeves's earlier efforts as a leading member of Pennsylvania's Coal and Iron Association and reflected the realization by Reeves and other iron producers that the industry's problems and objectives transcended the interests of individual states as much as they did those of individual firms. As a champion of this view Reeves helped to move the industry from its traditional parochialism to the more sophisticated level of discourse which characterized its subsequent lobbying efforts.

When the American Iron Association was reorganized as the American Iron and Steel Association in 1864, Reeves was made the organization's vice-president, and Eber B. Ward, the Illinois iron and steel magnate, became its president. This was the last full year of the Civil War and also one in which the industry enjoyed the voracious demand of a market that was substantially closed to foreign competition by high tariffs. The passage of the tariffs by a protectionist-minded Republican Congress was something of a personal victory for Reeves, who had lobbied for such legislation for almost 20 years. When the resignation of Ward in 1869 resulted in Reeves's election to the presidency of the Association, it was clear that the man and his new position were ideally suited to one another. As president, Reeves was, as ever, a forceful and articulate advocate of protectionism when he testified before congressional committees. But beginning with 1869, Reeves was, more than anyone else, the public and official voice of an industry. At fifty-one years of age, the prominence that came to Reeves as president seems to have left him unmoved. For him, it was the message and not the messenger that deserved attention.

Not long after his election as president of the American Iron and Steel Association, Reeves found himself in the potentially dangerous position of clashing with Andrew Carnegie, then already a man to be treated with circumspection. The dispute arose because of their expanding and converging interests in the field of structural iron, particularly railroad bridge building. Although both men had become involved in the business of building railroad bridges at about the same time—Carnegie in 1862 with his Keystone Bridge Company, and Reeves two years later—they had come to it from different directions. For Carnegie the construction of iron and, later, steel bridges led to greater involvement in the markets for iron and steel products which, ten years later, he set out to master as a producer. Reeves, of course, had traveled another route; his and his company's rolled iron interests and activities had enlarged to embrace not only rail making and the production of structural iron, but bridge building as well.

Both men knew a good deal by 1870 about structural iron as a product, including its manufacture and markets. Reeves had invented the "Phoenix wrought-iron column" for which he had received a patent in June 1862. The column, which was formed when four or six lengths of curved structural iron, each with flanged edges, were bolted together, found ready acceptance in the market and was widely used in the construction of bridges and commercial buildings. The invention of the "Phoenix Column" led Reeves and his father to consider how their firm might best profit from its exploitation. Less than two years later, in 1864, they launched the Phoenix Bridge Company for the purpose of building bridges for railroads and municipalities. Almost from the outset the Phoenix Bridge

Company became a leading buyer of the Phoenix Iron Company's structural iron. The largest part of the profits of each firm went, of course, to David and Samuel Reeves.

For his part, Andrew Carnegie wished to impose order on the fiercely competitive market for structural iron by forming a pool to fix prices. Such arrangements in other product lines had been at least temporarily successful in the past, and Carnegie had entertained fairly high hopes that negotiations which had begun in 1870 to establish the pool—on terms most favorable to his own interests, of course—would succeed. Although there were other parties involved in the talks, his chief counterpart was Samuel Reeves, whose interests in structural iron by then rivaled Carnegie's own in size. Carnegie's hopes for an agreement by year's end were disappointed when his competitors balked at his terms. Perhaps Carnegie's reputation for duplicity in such deals had put them off. In any case he made his disappointment plain to Reeves in a letter written at the end of December of 1870. The letter laid out cogently and succinctly the rationale which underlay all such proposed arrangements and also revealed Carnegie's propensity to inject into such disputes considerations of morality which his opponents could only have found to be ironic, at the very least.

Reeves in his capacity as vice-president of Phoenix Iron Company had conducted the negotiations with Carnegie. Although David Reeves retained the office and title of president of the firm, he was already a very old man and had, over the preceding decade, turned over much of his actual authority to his son. When the elder Reeves died in 1871 his son succeeded him as president, having been the firm's vice-president and chief tactician and strategist for 16 years. This long stint as corporate apprentice and journeyman executive served Reeves and Phoenix Iron Company especially well during the next several years when Reeves fought for the firm's survival in the wake of the panic of 1873.

The panic, which accentuated a general business contraction already underway, was a somewhat muddled affair, with respect to determining all of the major influences which helped to precipitate it and the major consequences which flowed from it. The proximate cause in this country—it was an international disaster—was the failure of the investment banking firm, Jay Cooke & Company, on September 18, 1873. Beyond that, however, was the flight of foreign, particularly British, capital from banks, commercial firms, and the stock exchange in New York City, the nation's chief money market. This exodus left the financial houses desperately overextended, many of them having engaged in lavish speculation in railroads and western lands. Preceding and immediately following the panic in September was an acute recession which lasted well into 1874. The recession quickly deteriorated into a deep depression which lasted through most of 1879 and, in a less virulent form, until 1896.

Phoenix Iron Company suffered, as did most businesses, from the crisis of 1873 and the year-long slide of prices which had led up to it. Prices of pig and bar iron and iron rails had fallen dramatically while costs, especially fixed costs occasioned by the extensive capital investment of the preceding three years, remained alarmingly and stubbornly high. During the last two years of the Civil War and in the immediate postwar period, Reeves had embarked upon an ambitious expansion and modernization of the Phoenix Iron Company's physical plant. By late 1875, however, it resembled to one observer "a sort of cemetery where millions of dollars are buried, awaiting a resurrection which can only come through a resuscitation in the business marts of the nation. . . . " The obituary was premature.

More than perhaps anything else, the demand for structural iron by builders and developers of urban rapid transport lines helped to sustain Phoenix Iron Company and its affiliates, including the newly formed Clarke, Reeves & Company, a bridge-building subsidiary, through the hard months and years following the panic of 1873. As early as 1869 Phoenix had been able to enter the emerging rapid transit market in New York City when it sold structural iron, including large numbers of the patented Phoenix Column, to the Greenwich Street Elevated Railway. This company had begun after Col. Charles T. Harvey had successfully tested a prototype above Manhattan's Greenwich Street in 1867 using the Phoenix column to support his elevated track. The satisfactory trials insured subsequent orders for Phoenix structural iron when construction of the full-fledged elevated line was under way. All of that, however, had preceded the panic, and, although Reeves's company was hardly moribund, Phoenix, by summer 1875, was hard pressed. Ever alert to opportunities to do business, the more so because of the depression, Reeves was favorably disposed to support tests on his company's property of a proposed elevated monorail system for New York City. The proposed system would, of course, use Phoenix structural iron.

Reeves's informed and energetic pursuit of orders for structural iron from municipal railway promoters and other developers helped him to guide Phoenix Iron Company through the depressed 1870s. In one year, 1878, the company sold 40,000 tons of its structural iron to the builders of New York City's largest elevated line. This order, remarkable enough for having been placed during a depression year was also noteworthy for having been the largest order for structural iron ever placed up to that time. As such, it represented not only a coup for Phoenix Iron Company but also a personal triumph of Samuel Reeves and a vindication of his confidence in the health of the firm and industry to which he had devoted most of his life. When he died only few months later at the age of sixty in Phoenixville on December 15, he was still president of the American Iron and Steel Association and of the company which his father and uncle had founded so many years before and for which he had done so much. Fittingly enough, his firm survived him, eventually to become Phoenix Steel.

References:

Robert W. Fogel, *Railroads and American Economic Growth: Essays in Econometric History* (Baltimore, Md.: Johns Hopkins University Press, 1964);

General Catalogue of Princeton University, 1746-1906 (Princeton: Princeton University, 1908);

William T. Hogan, S.J., *Economic History of the Iron and Steel Industry in the United States*, 5 volumes (Lexington, Mass.: Lexington Books, 1971);

Glenn Porter and Harold C. Livesay, *Merchants and Manufacturers: Studies in the Changing Structure of Nineteenth-Century Marketing* (Baltimore, Md.: Johns Hopkins University Press, 1971);

Fritz Redlich, *History of American Business Leaders: A Series of Studies*, 2 volumes (Ann Arbor, Mich.: Edwards Brothers, 1940);

Mark Reisenberg, "General Stone's Elevated Railroad: Portrait of an Inventor,"*Western Pennsylvania Historical Magazine*, 49 (July 1966): 185-195;

Charles E. Smith, "The Manufacture of Iron in Pennsylvania," *Hunt's Merchants' Magazine*, 25 (November 1851): 574-581 and tables following 656;

James M. Swank, *History of the Manufacture of Iron in All Ages* (Philadelphia: American Iron and Steel Association, 1892);

Peter Temin, *Iron and Steel in Nineteenth-Century America: An Economic Inquiry* (Cambridge, Mass.: M.I.T. Press, 1964);

John Frazier Wall, *Andrew Carnegie* (New York: Oxford University Press, 1970).

Reeves, Abbott & Company

by Frank Whelan

Morning Call, *Allentown, Pennsylvania*

Reeves, Abbott & Company was founded in Safe Harbor, Lancaster County, Pennsylvania, in 1848. It was more commonly known as the Safe Harbor Iron Works. Its founders were David and Samuel Reeves, Dr. Joseph Pancoast, and Charles and George Abbott, a group of Philadelphia iron men and financiers. The Reeves family had been deeply involved with the iron industry since the 1820s when they purchased the Phoenix Iron Works in Chester County.

The investors had several reasons for locating the plant at Safe Harbor. By using the region's extensive canal network, the firm could move to this location anthracite coal, which was needed to fuel the iron furnaces cheaply. Also, large iron deposits had been found in nearby Manor Township. Property for the ironworks was purchased along both banks of the Conestoga River. In order to ensure an outlet for its product and a necessary water supply, the new company found it necessary to purchase the franchise of the financially uncertain Conestoga Navigation Company.

Construction of the plant began in 1846 and was completed in 1848 at an initial cost somewhere in excess of $200,000. The blast furnace had a 45-foot-high stack and was 40 feet square at its base. The biggest building on the property, however, was the rolling mill for making railroad rails. It covered more than an acre of ground and was 165 feet long by 265 feet wide. Its roof consisted of 50,000 square feet of slate tiles. Inside was 1 single puddling furnace, 12 double puddling furnaces, 7 heating furnaces, 2 roller trains, and 16 boilers, each 40 feet long, so arranged over the top of the furnaces in the mill as to provide steam for operating the roller trains.

Historian John W. W. Loose writes in the *Journal of the Lancaster County Historical Society* that "This arrangement was an innovation which permitted double application of the heat." The rolling mill's primary product was the new "T" rail for use by the Pennsylvania Railroad. Running at full capacity, the mill produced, on average, 280 tons per week.

Although it would not be accurate to say Reeves, Abbott created Safe Harbor, it would be safe to say that the firm played an important role in the community's development. In order to house the large influx of workers, the company created its own village. Housing was uniform. About 70 two-story frame double houses were built in sizes varying from 24 feet by 28 feet to 32 feet by 32 feet. Water came from a spring at the corner of an aptly named Spring Street. It continues to flow there to this day. Into this little community came a torrent of workers, about 250 being employed at the ironworks in its heyday. Some came straight from their homes in Ireland. Others were from rural families and came from surrounding communities.

One of the people drawn to Safe Harbor was a young mechanic, John W. Fritz. In May 1849 the future innovator of the iron industry received a request from the Reeves, Abbott superintendent, John Griffin, to come and help with the installation of the new rail-making equipment. It was while he was waiting in nearby Lancaster to pick up his trunk that he ran into an old friend. Writing in his autobiography in 1912, Fritz recalled his friend warning him about Safe Harbor. He "tried to persuade me not to go there, saying it was the worst place in the whole state of Pennsylvania for fever and ague, and that no stranger escaped it. From the way he talked, the probabilities were that I would die with it. I told him my object in going there. He shook hands, smiled, and said good-by." Deaf to all of this, the strong young mechanic went to Safe Harbor and worked himself to exhaustion. He caught a severe case of ague which almost killed him. It was several months before Fritz recovered.

The 1850s were a period of boom and bust at Reeves, Abbott's Safe Harbor Iron Works. Demand for rails increased, and by 1854 the managers had finished an extensive expansion plan. But in 1855 the supply of capital suddenly dried up as creditors demanded payment from the cash-short concern. The owners begged for a grace period. At the same time the workers announced after a meeting of their own that they had confidence in the company and would be willing to work without pay until credit became easier. The creditors were impressed by this action and held off on foreclosure.

On May 5, 1855, Reeves, Abbott was incorporated by the state under the name, the Safe Harbor Iron Works. Just when things began to improve for the new company, another disaster hit. In February 1857 a flood on Conestoga River swamped the roller mill, causing a temporary suspension of work. In September the panic of 1857 hit. The ensuing collapse of railroad expansion stopped rail production. But once more the creditors held off.

During the Civil War, Safe Harbor took on new life as an arms maker. Its most famous weapon product was the Dahlgren gun, named after its designer, Adm. John A. Dahlgren. Its bulbous shape had given it the nickname, the "soda bottle" cannon.

The return of peace in 1865 required that Safe Harbor reconvert from weapons manufacture. But before the firm could complete its retooling, disaster struck Safe Harbor. On March 18, 1865, a massive flood washed out of the nearby Susquehanna River and overwhelmed the Conestoga. It wiped out the canal that had supplied iron ore to the mill, cutting the plant off from any transportation route. The Safe Harbor Iron Works closed down, and although the company declared its action was only temporary, the big blast furnace never went back into action again.

With the opening of the Columbia & Port Deposit Railroad in 1877, the puddling furnaces in the rail mill opened. In 1880 a rail spur was laid to the puddle rail mill. But the boom didn't last. In 1894 the Reeves family sold the mill to Adolph Segal of Philadelphia. For ten years he produced blue-tipped phosphorous matches there. The plant's final days were spent housing a machine shop and an air compressor for a small branch railroad. The building was finally razed in 1907.

References:

John W. Fritz, *The Autobiography of John Fritz* (New York: John Wiley, 1912);

John W. W. Loose, "Anthracite Iron Blast Furnaces in Lancaster County, 1840-1900," *Journal of the Lancaster County Historical Society*, 86, no. 3 (1982): 78-117;

Loose, "The Anthracite Iron Industry of Lancaster County: Rolling Mills, 1850-1900," *Journal of the Lancaster County Historical Society*, 86, no. 4 (1982): 129-144.

Riverside Iron Works

by John A. Heitmann

University of Dayton

Undoubtedly the most progressive of the nineteenth-century Wheeling, West Virginia, iron and steel manufacturers, the Riverside Iron Works illustrates the history of the role of technological innovation in maintaining profitable operations and in adjusting to changes in market conditions. The Riverside firm was a technological leader in the manufacture of cut nails and steel tubing during the 1870s and 1880s, and its significance and reputation transcended the local area of Wheeling.

The Riverside Iron Works had its origins in 1852, when Eliphalet C. Dewey of Cadiz, Ohio (later the hometown of Clark Gable), came to Wheeling, West Virginia, and erected a small plant which he called the Eagle Wire Mill. Consisting of five puddling and two heating furnaces, the Eagle works operated with limited success and in 1855 was closed. Three years later J. H. Pendleton, who had leased the works from Dewey, reopened the plant and made light gauges of bar iron and railroad spikes. This venture was profitable until 1859 when the works burned. Early in 1860 the mill was rebuilt and became the property of O. C. Dewey of Cadiz, Ohio, who formed a partnership with J. N. Vance and W. H. Russell of Wheeling. Initially capitalized at $15,000 and operating under the firm name of Dewey, Vance & Company, the firm was reorganized in 1866, new partners were added, and the decision was made to enter the nail business. Construction of a new plant consisting of a nail factory, heating furnaces, plate rolls, and shears progressed slowly during 1867. That same year a retired capitalist from New York City, William L. Hearne, became a partner in the company, and Hearne's energy and vision, along with that of his son Frank J. Hearne, eventually transformed this marginal firm into a dynamic operation.

The elder Hearne's influence in expanding operations was soon evident. In December 1867 construction of the mill was completed with 48 nail-making machines; another 42 machines were added by 1870. Concurrently, a number of important manufacturing improvements were incorporated at the Dewey, Vance & Company works during the early 1870s. To begin with, equipment was installed to enable the feeding of larger plate into the nail machines. Secondly, a new device for shoving plate into the nail machine was placed into production. Also, a modified mechanical arrangement was used to catch the larger number of manufactured cut nails. These refinements contributed to making a better-quality product with increased efficiency and at a lower cost. A blast furnace was erected in 1872, and continued expansion resulted in the increase of nail machines to 126 and the addition of railway facilities to handle both raw materials and finished product.

In January 1875 the firm of Dewey, Vance & Company was reorganized under the name of the Riverside Iron Works with Frank J. Hearne as general manager. By 1879 the firm employed 800 men, and, in addition to producing 300,000 kegs of nails per year, the firm made T-rails for narrow-gauge railroads. Although a relatively young man, Frank Hearne had previously served as chief engineer of the Hannibal & St. Joseph Railroad. Hearne's experience, coupled with a keen business acumen, led him to realize that the future lay in steel, not in iron, a belief anathema to other leaders of the Wheeling manufacturing elite. Hearne pushed forward with the construction of a Bessemer steel facility at the Riverside works which made its initial blow in 1884. The Riverside remained an important cut nail manufacturer in the 1880s, but with a capacity for steel the firm developed a lucrative trade in semi-finished steel such as steel slabs, bars, and tack plate. Steel was also locally rolled into rods for making wire and wire nails.

Because of Frank Hearne's ability to shift the activity of the Riverside works away from the manufacture of iron cut nails, the firm did not experience a serious downturn in business as did the other Wheeling nail manufacturers during the late 1880s. Indeed, Hearne was best known in the iron and steel industry for his innovation in the manufacture of welded steel pipe, a process that was developed at the Riverside Works in 1887 after the management had decided to erect and operate a tube works earlier that year. For some time the industry had developed a prejudice against the use of steel in making pipe. After a series of experiments in the manufacture of steel pipe, the company abandoned its forge and iron-making plant altogether.

The Riverside Works prospered during the 1890s and was an attractive target in the wave of consolidation that characterized American business during the latter part of the decade. In 1889 J. P. Morgan's interests purchased the firm and incorporated it into the National Tube Company (NTC). The Riverside plant became a key facility in this ven-

ture, and Frank Hearne became first vice-president in charge of operations for NTC. In 1901 Hearne resigned, however, and in 1903 assumed the presidency of the Colorado Fuel and Iron Company. Tragically, Hearne died suddenly that same year after ignoring the advice of physicians to slow his pace.

The Riverside Works, an important part of the Wheeling, West Virginia, economy, would remain a part of the National Tube Company until 1927, when it was sold to the Wheeling Steel Corporation.

References:

J. H. Newton, editor, *History of the Pan-Handle; Being Historical Collections of the Counties of Ohio, Brooke, Marshall and Hancock, West Virginia* (Wheeling, W.Va.: J. A. Caldwell, 1879);

Henry Dickerson Scott, *Iron & Steel in Wheeling* (Toledo: Caslon, 1929);

Earl Chapin May, *Principio to Wheeling, 1715-1945* (New York: Harper, 1945).

Roane Iron Company

by Brady Banta

Louisiana State University

While the Roane Iron Company never emerged as a giant in the nineteenth-century southern iron industry, it merits consideration for both the timing of its development and its technological achievements. What became the Roane Iron Company originated with the actions of John T. Wilder, a midwestern entrepreneur whose wide-ranging antebellum interests included the establishment of an iron foundry. During the Civil War Wilder remained alert and receptive to business opportunities. Commissioned as a lieutenant colonel, later rising by brevet to brigadier general, Wilder served with the Union Army in Eastern Tennessee. While on this assignment his keen business eye recognized the region's mineral-development potential, especially with regard to the hematite deposits in Roane County.

The possibility of profitably developing these resources intrigued Wilder, and after the war he decided to pursue this opportunity. Sensitive to the feelings of postwar Tennesseeans toward "carpetbag-

gers," Wilder, then a resident of Chattanooga, took as a business associate a former Confederate army captain, W. E. London. Together they scouted Roane County's iron-bearing properties. Based upon these explorations, Wilder and Hiram S. Chamberlain, another northerner then operating a small rolling mill in Knoxville, purchased 928 acres in Roane County. Building upon this foundation, Wilder and Chamberlain recruited five additional northern investors and organized the Roane Iron Company, which in December 1867 received a charter from the Tennessee legislature.

Despite having to contend with primitive transportation facilities, Wilder and Chamberlain chose to construct an iron-production facility near their recently purchased coal and iron deposits in Roane County. Also cognizant of advancing technology, they built coke ovens, the product of which fueled the South's first coke-fired iron furnaces. Put into operation in December 1868, these furnaces immedi-

The Roane Iron Company works in Rockwood, Tennessee (courtesy of Roane County Heritage Commission, Inc.)

ately demonstrated an advantage in effectiveness by producing roughly three times as much iron as could charcoal-fueled furnaces of similar capacity.

The location of these activities was the company-dominated town of Rockwood, Tennessee. The company grew quickly, and in 1870 it absorbed the Southwestern Iron Company of Chattanooga. Constructed and previously operated as a refashioner of damaged and worn rails, Roane used the Chattanooga facility to produce iron rails for the lucrative southern and midwestern markets. In the mid 1870s, however, steel rails swept into, and dominated, the market. While the Roane Iron Company attempted to take advantage of this transition, the high phosphorus content of its pig iron thwarted the efforts to produce steel rails with the existing technology. Unable to compete with northern producers of steel rails, the Roane Company lost a large share of its rail market.

The continuing attractiveness of the lucrative steel rail market prompted the Roane Iron Company to remodel the Chattanooga facility. As part of this modernization project, the company installed a Bessemer converter and in 1887 produced the South's first Bessemer steel. Unfortunately this development coincided with the collapse in the bull market in steel rails, and the company accumulated huge losses at the Chattanooga plant. With no imme-

diate prospects for escaping this economic morass, the Roane Iron Company sold the steel mill and concentrated on pig iron production at its Rockwood, Tennessee, facility.

The company never again enjoyed the prosperity experienced in the early 1870s. Prudent management enabled it to weather economic difficulties in the 1890s and the first quarter of the twentieth century, but during the 1920s a series of industrial accidents saddled the company with compensation claims filed by injured employees. While these judgments rocked the company's fiscal foundation, the dramatic fall in prices during the Great Depression spelled its doom. The company suspended operations in 1930, entered bankruptcy in 1932, and saw its remaining assets sold to the Reconstruction Finance Corporation in 1941.

Unpublished Document:

William H. Moore, "Rockwood: A Prototype of the New South," M.A. thesis, University of Tennessee, 1965.

References:

Victor S. Clark, *History of Manufactures in the United States, Volume II, 1860-1893* (New York: McGraw-Hill, 1929);

William H. Moore, "Preoccupied Paternalism: The Roane Iron Company in Her Company Town—Rockwood, Tennessee," *East Tennessee Historical Society Publications*, 39 (1967): 56-70.

Solomon White Roberts

(August 3, 1811-March 22, 1882)

by John W. Malsberger

Muhlenberg College

CAREER: Engineer, Lehigh Coal & Navigation Company (1827-1829); engineer, state of Pennsylvania (1829-1831); resident engineer and superintendent of transportation, Allegheny Portage Railroad (1831-1836); iron rail inspector, Philadelphia & Reading Railroad (1836-1838); chief engineer, Catawissa Railroad (1838-1841); president, Philadelphia, Germantown & Norristown Railroad (1842); president, Schuylkill Navigation Company (1843-1845); member, Pennsylvania House of Representatives (1847-1848); chief engineer, Ohio & Pennsylvania Railroad (1848-1856); chief engineer and general superintendent, North Pennsylvania Railroad (1856-1879).

Solomon White Roberts, a civil engineer who was connected for most of his career with railroad companies, played an important tangential role in the evolution of the nineteenth-century iron and steel industry. During a trip to Wales in 1836 to procure iron rails for railroads, White observed the successful efforts of George Crane and David Thomas in manufacturing iron with anthracite coal. Roberts's report led directly to the organization in 1838 of the Lehigh Crane Iron Company of Catasauqua, Pennsylvania, one of the first commercial producers of anthracite iron in the United States.

Born in Philadelphia on August 3, 1811, Solomon W. Roberts was the son of Charles and Hannah White Roberts. Roberts's father was a well-known resident of Philadelphia, having been a member of the Pennsylvania State Legislature, one of the original directors of the Franklin Fire Insurance Company, and manager of the Pennsylvania Company for Insurance on Lives. Because both of his parents were devout Quakers, Roberts was educated at the Friends' School regarded as one of the city's finest schools in the early nineteenth century. His formal education ended when he reached the

age of sixteen, and, like many young men of his generation, Roberts acquired occupational training through direct work experience.

In 1827 sixteen-year old Solomon Roberts left Philadelphia for Mauch Chunk, Pennsylvania, to go to work for his uncle, Josiah White, one of the principal founders of the Lehigh Coal & Navigation Company, which was building the Lehigh Canal to connect the anthracite coalfields of northeastern Pennsylvania with Philadelphia. Employed as an assistant to his uncle, Roberts directed the construction of the first railroad in Pennsylvania, a 9-mile route to carry coal from the mines at Summit Hill to the Lehigh River at Mauch Chunk. The satisfaction that he derived from this project was so great that Roberts determined at this point to pursue a career in the railroad industry.

After the railroad was opened for use in 1827, Roberts's practical education in engineering was furthered when he was assigned as a rodman to a group of engineers on the Lehigh Canal under the supervision of Canvass White, who formerly had been the chief engineer of the Erie Canal. Roberts displayed an unusual proclivity for civil engineering work and was quickly promoted to chief engineer for the construction of a section of the Lehigh Canal. By the time this section of the canal was completed in 1829, Roberts's practical education as an engineer was sufficient to enable him to strike out on his own.

Roberts's first job away from his family's Lehigh Coal & Navigation Company was as an engineer employed by the state of Pennsylvania to build a canal on the Conemaugh River, near Blairsville. When this project was completed, Roberts, though only twenty years old, became the principal assistant engineer for the Allegheny Portage Railroad, then being built to connect Johnstown with Hollidaysburg by traversing the Allegheny Mountains. The 36-mile portage railroad was regarded as one of the mechanical wonders of America because it used 11 levels, 10 inclined planes, and wire ropes attached to stationary engines to move the cars over the planes. In this project Roberts was also directly responsible for designing and supervising the construction of a viaduct over the Conemaugh River at Horseshoe Bend. This experience helped to establish Roberts's reputation as one of the leading civil engineers in the state. Roberts remained with the Portage Railroad Company after construction was completed, serving as resident engineer and superintendent of transportation until 1836.

A career change in 1836 led Roberts to the position from which he helped to alter the American iron and steel industry. That year he accepted a position as the iron rail inspector for the Philadelphia & Reading Railroad and was sent to Wales to procure iron rails for his new employer. Over the course of the two years he spent in Wales, Roberts developed a friendship with George Crane, who in 1837 had received a patent for the manufacture of iron through the hot blast smelting method using anthracite coal. Roberts reported Crane's method to the Franklin Institute in Philadelphia and to his uncle Josiah White who, along with other Americans, had been experimenting with anthracite iron production for more than a decade. Initially, White offered to organize a company with Roberts as its head if the Crane furnace proved successful. Roberts refused the offer, insisting that the railroad was his life's calling. Instead, Roberts encouraged his uncle to hire George Crane to build an anthracite iron furnace. When Crane refused, Roberts suggested David Thomas, Crane's chief ironmaster, for the position. The hiring of David Thomas in late 1838 led directly to the incorporation of the Lehigh Crane Iron Company, the first commercial manufacturer of anthracite iron in the Lehigh Valley of Pennsylvania and one of the first in America.

On his return to America in 1838 Roberts accepted a position as chief engineer of the Catawissa Railroad, a post he held for three years. In 1842 he became president of the Philadelphia, Germantown & Norristown Railroad, but resigned the following year to become president of the Schuylkill Navigation Company. Between 1843 and 1845 Roberts oversaw the enlargement of the Schuylkill Canal. Elected to the Pennsylvania House of Representatives in 1846, Roberts served one term during which he played an active role in securing the char-

ter for the Pennsylvania Railroad. When his term expired in 1848 Roberts became chief engineer of the newly chartered Ohio & Pennsylvania Railroad; Roberts guided its development, most notably in planning the railroad bridge across the Allegheny River at Pittsburgh to link his railroad with the Pennsylvania Railroad. He resigned from this position in 1856 and returned to his native Philadelphia as the chief engineer and general superintendent of the North Pennsylvania Railroad, which provided a direct rail link between the Quaker City and the burgeoning iron mills of the Lehigh Valley. Roberts remained in this position until his retirement from the business world in 1879.

Although Solomon W. Roberts was most directly associated with the American railroad industry in its formative era, he also clearly influenced the development of the nineteenth-century iron and steel industry. When Roberts entered the business world in the late 1820s there existed two substantial obstacles that prevented the expansion of the iron industry. One of those obstacles was the lack of a cheap and abundant supply of fuel. Prior to the late 1830s the two chief fuels used in American ironmaking were charcoal and coke, both of which were expensive to produce and, therefore, in relatively scarce supply. The introduction of the Crane furnace to America, in which Roberts played an important role, permitted anthracite coal, a cheaper, more abundant, and easily mined fuel, to be used in the manufacture of iron.

The second obstacle deterring the growth of the American iron industry was the lack of internal transportation. Prior to 1815 the American nation relied primarily on a system of roads for its internal transportation. Because many of these roads were little more than unpaved paths, the cost of transporting such heavy commodities as iron ore, coal, and pig iron was so high as to be prohibitive in most cases. Roberts, in his capacity as builder of canals and railroads, played an integral role in what historian George Rogers Taylor has termed the "transportation revolution," whereby dramatic improvements were made in the efficiency of America's internal transportation system between 1815 and 1860. In this sense, therefore, Roberts also contributed to the growth and development of the nineteenth-century iron and steel industry by providing an industrial infrastructure upon which it could rely.

Roberts married Anna Smith Rickey of Cincinnati, Ohio, in 1851 with whom he had three daughters and two sons. In 1865 he married Jane E. Shannon with whom he had a son and a daughter. He died on March 22, 1882, in Atlantic City, New Jersey.

Selected Publications:

An Account of the Portage Rail Road, Over the Allegheny Mountain, in Pennsylvania (Philadelphia: N. Kite, 1836);

Report to the Board of Managers of the Schuylkill Navigation Company, on the Improvement of the Schuylkill Navigation (Philadelphia: J. & W. Kite, 1845);

Obituary of the Late George Crane, Esq. (Philadelphia, 1846);

The Promotion of the Mechanic Arts in America (Philadelphia: J. C. Clark, 1846);

Ohio and Pennsylvania Rail-Road (Philadelphia: J. C. Clark, 1849);

The Destiny of Pittsburgh, and the Duty of Her Young Men (Pittsburgh: Johnston & Stockton, 1850);

Philadelphia and Wilkesbarre [sic] Short Line Railroad (Philadelphia: Crissy & Markely, 1859);

Memoir of Josiah White (Easton, Pa.: Bixler & Corwin, 1860);

Reminiscences of the First Railroad Over the Allegheny Mountain (Philadelphia, 1879).

References:

Craig Bartholomew, "Anthracite Iron Making and Industrial Growth in the Lehigh Valley," *Proceedings*, Lehigh County Historical Society, 32 (1978);

Biographical Encyclopedia of Pennsylvania of the 19th Century (Philadelphia: Galaxy, 1874);

James Moore Swank, *History of the Manufacture of Iron in All Ages*, second edition (Philadelphia: American Iron and Steel Associations, 1892).

St. Louis Ore & Steel Company

by Terry S. Reynolds

Michigan Technological University

The St. Louis Ore & Steel Company was the successor company to the Vulcan Steel Company, which in 1876 erected the first Bessemer converter west of the Mississippi River and the thirteenth in the United States. The history of St. Louis Ore & Steel illustrates the difficulties of establishing a steel industry in the West.

Although blast furnaces and forges had been built and iron products fabricated in Missouri as early as 1815 or 1816, the Missouri iron industry before 1850 was largely a rural enterprise serving a local market. An urban industry attempting to serve a regional market emerged in St. Louis only in the last half of the nineteenth century.

Iron fabrication in St. Louis began around 1850 when Harrison, Chouteau & Vallé built the Laclede Rolling Mill in north St. Louis. Other companies quickly duplicated this accomplishment, building the Missouri Rolling Mill in 1854 and the Pacific Rolling Mill in 1856. By 1880 St. Louis had seven rolling mills. In the meantime, St. Louis had also become a regional iron production center as well as an iron fabricator. The first St. Louis blast furnace was erected in 1863, but between 1863 and 1875 ten hot blast furnaces were built in or near the city, their ores coming largely from the Iron

Mountain and Pilot Knob mines, around 80 miles to the south.

One of the companies formed in this period of expansion was the Vulcan Iron Company. In 1869 Vulcan erected its first two blast furnaces, operating them with coke made from Illinois coal. In 1872 Vulcan erected the Vulcan Iron Works, a rolling mill located near its blast furnaces.

The Vulcan Iron Company had some initial success in selling its rails to a regional market and in 1875 created a subsidiary company, the Vulcan Steel Company, to erect adjacent to its rolling mill the first Bessemer steelworks west of the Mississippi. To build the new Vulcan steel plant the company brought in Alexander Lyman Holley, a prominent American metallurgical engineer. Holley's biographer, Jeanne McHugh, writes that the Vulcan plant was of special interest since it embodied the "latest improvements in the Bessemer practice and represented the high point in . . . the American Bessemer plant." Among the innovations in the plant design were the elimination of the English deep pit beneath the converters, the raising of the vessels to get working space under them on the ground floor, the substitution of cupolas for reverberatory furnaces, and the introduction of the intermediate ladle placed on scales to insure the accuracy of the charge being put into the Bessemer converters. The new works had an annual capacity of about 105,000 tons, and the first "blow" took place on September 1, 1876.

Although the Vulcan plant was equipped with the most advanced technology and was well located to supply the West, it soon encountered major problems. Attempts to use nearby Illinois coke as a fuel failed, as the pig iron produced was not of sufficient quality to use the Bessemer steel process. This forced Vulcan to import coal from the Pittsburgh area, which significantly raised production costs. Furthermore, the larger and better-financed Eastern and Chicago steel firms took actions to prevent the emergence of a strong regional competitor in their western markets. Soon after Vulcan began producing, the rail market declined, the company failed to meet its obligations, and the works closed down. In 1878 the Bessemer Steel Association, at the instigation of Andrew Carnegie, agreed to meet the $70,000 annual interest payment on Vulcan's $1 million mortgage, provided Vulcan would resume operations only when demand for rails equaled the production capacities of Association members, including Vulcan. This placed the decision on when to resume operations with the eastern- and Chicago-dominated Bessemer Association, not with Vulcan. And while Vulcan's equipment remained idle, these companies, and particularly Carnegie's, forged ahead, updating their equipment and adding additional production capacity.

Toward 1880 a revival of demand for steel and stability in steel prices stimulated a reactivation of the Vulcan works. Holley was again called in, this time to advise on overhauling the plant, and William Shinn, builder and manager of the Edgar Thomson Works, was hired to manage the revitalized plant.

Shortly after resuming production, the Vulcan works formally passed under the control of the St. Louis Ore & Steel Company, which was formally created by an 1882 merger of the Vulcan Steel Company and the Pilot Knob Company. The new company purchased the Grand Tower Mining Manufacturing Company, owner of the Grand Tower & Carbondale Railroad Company and the Mount Carbon Coal & Coke Company.

St. Louis Ore & Steel was still of modest size by eastern standards, but it had begun the process of vertical integration, owning a mine, a railroad,

and supplies of coal in addition to the Vulcan steelworks and Vulcan rolling mill. And it may have enjoyed a short period of prosperity. In September 1882, according to the testimony of E. A. Hitchcock, the company president, before the Tariff Commission, St. Louis Ore & Steel employed 3,000 men. By 1883 it had leased the Jupiter Iron Works, giving it control over blast furnaces with an annual capacity of 88,000 tons.

But Hitchcock also complained that depressed steel prices were making his company unprofitable. Even in good times, however, it is unlikely that St. Louis Ore & Steel could have been competitive over the long term. The growing scale and better geographical location of the eastern- and Chicago-area plants put St. Louis at a disadvantage. They could depend on iron ore supplies shipped cheaply in lake freighters from the Lake Superior ore district, while St. Louis's ore supplies at Iron Mountain and Pilot Knob proved much less abundant than initially expected. Continued attempts to use Illinois coal to produce Bessemer-quality pig iron failed, forcing coal carriage from Pittsburgh and putting St. Louis Ore & Steel at a further competitive disadvantage. If these problems were not enough, there were others to plague the company. The output of the Vulcan rail mill was limited by the problems involved in rolling steel in a mill built for rolling iron. The company also suffered from labor problems.

Exactly when St. Louis Ore & Steel disappeared as a production unit is not clear, but it probably occurred around 1892. Sister Mary Celeste Leger cites an 1892 letter indicating that the company had been sold by that date. The *Missouri State Gazetteer* for 1891 and 1892 still lists a St. Louis Ore & Steel Company, but the *St. Louis Directory* for 1900 does not. It is likely that the panic of 1893 insured the company's permanent demise.

Unpublished Document:

Sister Mary Celeste Leger, "A Study of the Public Career of Ethan Allen Hitchcock," Ph.D. dissertation, City University of New York, 1971, pp. 18-19.

References:

Arthur B. Cozzens, "The Iron Industry of Missouri," *Missouri Historical Review*, 35 (July 1944): 509-538; 36 (October 1941): 48-60;

Robert W. Hunt, "A History of the Bessemer Manufacture in America," *Transactions* of the American Institute of Mining Engineers, 5 (1876-1877): 214-215;

Jeanne McHugh, *Alexander Holley and the Makers of Steel* (Baltimore, Md.: Johns Hopkins University Press, 1980);

Report of Tariff Commission ... appointed May 15, 1882, U.S. Congress, 47th Cong. 2d. Session (House Misc. Doc. no. 6) (Washington: GPO, 1883): Pt. III, p. 1185;

J. L. Ringwalt, *Development of Transportation Systems in the United States* (Philadelphia: Privately printed, 1888);

James M. Swank, *History of the Manufacture of Iron in All Ages* (Philadelphia: Privately printed, 1884);

Joseph Frazier Wall, *Andrew Carnegie* (New York: Oxford University Press, 1970);

Kenneth Warren, *The American Steel Industry, 1850-1870: A Geographical Interpretation* (Oxford: Clarendon Press, 1973);

Year Book of the Commercial, Banking, and Manufacturing Interests of St. Louis, 1882-1883 (St. Louis: S. Ferd. Howe, [1883]).

George Whitfield Scranton

(May 23, 1811-March 24, 1861)

by John A. Heitmann

University of Dayton

CAREER: Teamster, Belvidere, New Jersey (1828-1829); clerk and partner, Judge Kinney's store (1829-1835); farmer (1835-1837); owner, iron furnace, Oxford, New Jersey (1837-1840); partner, Scrantons, Grant & Company (1840-1843); partner, Scrantons & Grant (1843-1846); partner, Scrantons & Platt (1846-1861); member, U.S. House of Representatives (1858-1861).

George Whitfield Scranton, a manufacturer, railroad executive, civic leader, and politician, was born in Madison, Connecticut, May 23, 1811, and died March 24, 1861. An extraordinary entrepreneur, Scranton possessed the vision and flexibility necessary to transform a struggling iron and coal company into an important supplier of T-rails for the rapidly growing railroad industry of the 1850s. His efforts greatly contributed to the emergence of Scranton, Pennsylvania, located in the northeastern section of that state, as a major industrial city on the eve of the Civil War. Yet his success was not merely the consequence of individual initiative, for the capital and expertise was drawn from a tightly knit kinship network that included his brother and several cousins.

George Scranton was the son of Theophilus and Elizabeth Warner Scranton and a direct descendant of John Scranton, who had emigrated from England in 1639 and had settled in the vicinity of Guilford, Connecticut. As a young man Scranton attended Lee's Academy in Madison, Connecticut, and at age seventeen received an employment offer from an uncle, Chapman Warner, to work as a team-

George Whitfield Scranton

ster in Belvidere, New Jersey. A few years later Scranton became the partner of a Judge Kinney in a local store, and like many contemporaries who would eventually distinguish themselves in business, he used his experience as an apprentice clerk as a stepping-stone to bigger things. During the late 1830s Scranton and his brother Selden purchased the lease and stock of an Oxford, New Jersey, iron furnace, which had been formerly owned by Henry Jordan & Company. It was in this venture that the Scrantons learned the fundamentals of the iron trade, an industry that was undergoing rapid technological transition as anthracite coal was increasingly supplanting timber in certain locales as a fuel in the manufacture of pig iron.

The movement of the Scrantons from New Jersey to northeastern Pennsylvania had its origins in developments during the late 1830s. While much of the speculative land fever of the 1830s had been quelched with the panic of 1837, a visionary mineralogist, explorer, and promoter, William Henry, the father-in-law of Selden Scranton, had placed his financial hopes on an area in the Lackawanna Valley known then as Slocum's Hollow. Henry had been familiar with this area since the mid 1830s when he had been involved in a scheme to construct a railroad between the Lackawanna Valley and New Jersey; he had conducted preliminary mineral surveys that indicated rich reserves of limestone, iron ore, and coal—all the substances necessary for the manufacture of iron.

In March 1840 Henry and a partner took an option on over 500 acres of land at Slocum's Hollow for $8,000 but ran into difficulties in raising the money necessary to purchase the tract. In desperation Henry turned to his son-in-law, Selden Scranton, and George Scranton for assistance. The Scrantons quickly recognized the site's potential in terms of natural resources and realized that the Lackawanna Valley would be an ideal place to expand their operations. In August 1840 a group that included George and Selden Scranton, Sanford Grant, and Henry arrived at Slocum's Hollow with the purpose of erecting an iron furnace. What they came upon was a small decaying village consisting of three small houses and an old stone mill surrounded by forest. The group was encouraged, however, by their discovery of formations of usable coal deposits, "huge boulders of iron ore showing in the ravines," and limestone cliffs bordering Nay Aug Creek. A partnership was formalized, with the new firm being named Scrantons, Grant & Company. Its partners included Selden, George, and Joseph Scranton (a cousin of George's working in the mercantile business in Augusta, Georgia), Sanford Grant, local Pennsylvania banker Philip Mattes, and the mercurial William Henry.

Throughout the fall and winter of 1840 and 1841 work continued on the construction of an iron furnace at Slocum's Hollow, now renamed by Henry as Harrison in honor of the late Whig president William Henry Harrison. During 1841 two experiments aimed at making iron using anthracite were attempted, and in both cases the runs ended in disaster. The mass of reacted materials had undergone a process known as cementation, and the impurities had not been separated out with the slag. On both occasions it took weeks to remove the solid mass from the furnace with hammers and hardened steel chisels, and new linings had to be installed. Changes were most certainly necessary; the cause of these initial problems were ascribed to the use of a poor grade of limestone and to iron ore that had not been properly pulverized. Accordingly, a new source of limestone was discovered in Columbia County, and iron ores destined for the furnace were more carefully prepared and blended. To supervise the furnace operations a recent emigrant from Wales, John F. Davis, reputed to be familiar with English anthracite iron-making techniques, was hired. In early January a blast, originating from a water-powered bellows, was put on the furnace, and shortly thereafter a large tonnage of iron was made. Despite Henry's rather thorough geological survey, it was soon concluded that the iron-making operation needed a high-quality limestone that was obtained only with great difficulty, brought to the site in wagons from a deposit over 50 miles away. Thus, an operation that was initially thought to be extremely profitable was, after start-up, viewed in a more realistic light. Other unforeseen complexities of the business would soon quickly erode the remaining optimism of the small group of initial investors.

As a result of these challenges George Scranton took on more and more of the responsibilities for the venture, while William Henry was gradually eased out of the company's affairs. While the new firm succeeded in making pig iron, its owners soon discovered that high transportation costs and a sluggish market for the product were eating up the anticipated profits. In response to this dilemma, George Scranton decided to shift production from pig iron to nails, a decision that would require far more capital.

One source of capital for this undertaking came from within the Scranton family, particularly from Joseph and Erastus Scranton, cousins in the mercantile trade. But capital outside the family was also necessary, and thus Scranton called on a New York City mercantile firm, Howland & Company, with which he previously had dealt. After sending an agent to the Lackawanna Valley to investigate matters, Howland agreed to invest $20,000 in the nail-making venture.

By May 1843 Howland & Company and Scrantons, Grant & Company agreed to terms, and the New Yorkers were made partners in a firm reorganized as Scrantons & Grant. Now possessing a capitalization of $86,000, the directors designated George W. Scranton, Selden T. Scranton, and Sanford Grant as active partners. Despite this infusion of relatively large amounts of outside capital, the Scranton family remained in control of the organization, owning 51.6 percent of the firm's shares. As a result of this arrangement George Scranton possessed the means to purchase the equipment necessary to erect a large-capacity, highly efficient nail-making plant.

In early May 1843 George Scranton began supervising the installation of the newly purchased equipment on lots adjacent to the iron furnace. In one building a 110-by-114-foot double-cylinder rolling mill was erected, powered by a 90-horsepower waterwheel. This device took the pig iron produced from five nearby puddling furnaces and shaped it into the flat plate feed material for the nail-making machines. In a second shop 20 nail machines and a spike machine were readied for operation.

As operations commenced in early 1844 an air of optimism surrounded Scranton and his colleagues, but this mood quickly evaporated when it was discovered that the nails produced were of extremely poor quality. Despite the best efforts of George Scranton to demonstrate the strength of his nails by hammering them into logs before the eyes of skeptical visitors, in reality his product was brittle and frequently broke upon impact. In the terminology of the trade, the iron produced using Lackawanna Valley materials was "cold-short," possessing coarse fibers. Thus the company, already in dire straits financially, had made a disastrous error by sinking a large amount of capital into an effort that could only be viewed as a dismal failure. Tons of these "cold-short" brittle and frail nails had to be sold to merchants at enormous discounts just to move the product off the company's premises.

Despite this serious setback George Scranton's cool head prevailed, and his creative flexibility once more focused upon overcoming a formidable manu-

facturing challenge. This time Scranton decided to harness his productive capabilities by making T-rails for the burgeoning railroad industry, a product in very short supply during the mid 1840s. As fate would have it, the very properties that made Scranton's iron unsuitable for nail making—the material's coarse fibers—made it almost ideal for fabricating railroad rails.

While pioneer railroads had their origins during the late 1820s with the chartering of the Baltimore & Ohio, the industry developed rather slowly until the mid 1840s, when a frenzy of construction activity began to occur. Concurrently, British railroads were in the process of a dramatic burst of expansion, and thus most of the T-rails made in Britain went to sustain that nation's own transportation boom. By 1846 T-rail exports to the United States had largely ceased because of British internal demand; as a result of the relative immaturity of its iron industry, American railroad companies had no alternative domestic sources.

Scranton, recognizing a market when he saw one and on the rebound from the expensive reverse in the nail-making operation, chose to concentrate on making iron track. Immediately north of Scranton's operations the Erie Railroad had received a charter from New York State and a $3 million appropriation for the purchase of right-of-way lands and equipment. But the state assembly had called for a deadline in the completion of a line from Port Jervis to Binghamton, and thus lengthy delays in the procurement of British rails would be catastrophic to the railway firm.

Confronted with a shortage of necessary rails and the time stipulations of the charter, William E. Dodge, a major figure at the Erie and a stockholder in Scrantons, Grant, made an arrangement with George Scranton to manufacture and deliver T-rails to several sites along the surveyed line. To assist Scranton in converting his rolling operations from nails to rails Dodge convinced the Erie's directors to advance $100,000 to the Lackawanna Valley iron manufacturer. In responding to the challenge Scranton succeeded not only in quickly transforming his plant into the first T-rail rolling firm in the United States but also in meeting the deadlines to the Erie for delivery, thus establishing a solid reputation in the business as well as finding an important niche within the iron and steel industry.

Scranton, possessing renewed confidence from this victory, began to set his sights on other objectives, and these plans would require fresh injections of outside capital. During the financial crisis of the mid 1840s Sanford Grant's position within the firm had weakened, and ultimately he had been replaced by another in-law of the Scrantons, J. C. Platt. Reflective of these changes the iron and coal firm was reorganized in 1846 as Scrantons & Platt, and the latest family partner assumed Grant's responsibilities dealing with mercantile functions in Pennsylvania. But family support could take George Scranton only so far, especially since he now dreamed of constructing a network of railroads extending both north and east with the city of Scrantonia (formerly Harrison and by the early 1850s renamed Scranton) as a hub.

Scranton quickly realized the inherent advantages of establishing railroad links to the Erie line directly north, thus tapping into the rich areas of the Midwest, and to the Delaware River and New Jersey, thus providing access to New York and Philadelphia. The railroads not only would serve as markets for his iron products and coal but also convey these goods to other potential industrial and consumer purchasers.

Scranton's designs were initiated in 1849 when he gained the rights to a previously chartered but unconstructed rail line known as the Liggett's Gap Road. Investment capital was secured and work on the road was begun shortly after the company's organizational meeting in January 1850. Construction was placed in the hands of Peter Jones, an engineer of considerable experience who had just completed the Cayuga & Susquehanna Railroad, which ran from Owego, New York, on the shores of Lake Ontario, to Ithaca. By fall 1851 the line was in place and in operation, thus connecting Scranton with Great Bend, New York, and near the east-west Erie route. Meanwhile, a second venture, with an eastern terminus on the Delaware River, was under way, chartered as the Cobb's Gap Railroad. By 1856 this route was in use under the name of the Delaware, Lackawanna & Western Railroad. Concurrently Scranton obtained a lease on the Ca-

yuga & Susquehanna, and thus Owego, New York, became the first lake port for coal destined for Canada and the West.

With these additional successes in the 1850s Scranton began to direct his energies into politics, an activity he had shunned earlier after he had worked hard but unsuccessfully to elect Henry Clay in 1844. As did many American manufacturers of the 1850s, Scranton became increasingly protectionist, and on this stance he was elected to Congress in 1858 and 1860.

But either the strains of his earlier career as a manufacturer and entrepreneur or later efforts as a politician eventually caught up with him, and a heart condition weakened him on the eve of the Civil War. As Scranton's physician B. H. Throop would later state, Scranton's "machinery of life was worn out." He died on March 24, 1861, leaving the business that he did so much to create and the city that grew around it to play a major role in settling the conflict between the North and South.

References:

Robert J. Casey and W.A.S. Douglas, *The Lackawanna Story* (New York: McGraw-Hill, 1951);

Burton W. Folsom, Jr., *Urban Capitalists: Entrepreneurs and City Growth in Pennsylvania's Lackawanna and Lehigh Regions* (Baltimore, Md: Johns Hopkins University Press, 1981);

Frederick L. Hitchcock, *History of Scranton and Its Peoples*, 2 volumes (New York: Lewis Historical Publishing, 1914);

W. David Lewis, "The Early History of the Lackawanna Iron and Coal Company: A Study in Technological Adaptation," *Pennsylvania Magazine of History and Biography*, 96 (1972): 424-468;

Nicholas E. Petola, *Scranton, Once Upon a Time* (Scranton, Pa., n.d.);

Dwight D. Stoddard, *Prominent Men of Scranton and Vicinity* (Scranton, Pa.: Tribune Publishing, 1906);

Benjamin H. Throop, *A Half Century in Scranton* (Scranton, Pa.: The Scranton Republican, 1895).

Archives:

The Scranton family papers are at the University of Delaware, Newark, Delaware.

Joseph H. Scranton

(June 28, 1813-June 6, 1872)

by John A. Heitmann

University of Dayton

CAREER: Merchant (1830-1847); manager, Scrantons & Platt Company (1847-1853); general manager (1853-1858), president, Lackawanna Iron & Coal Company (1858-1872); commissioner, Union Pacific Railroad (1867-1872).

Joseph H. Scranton, a merchant and manufacturer, was born in Madison, Connecticut, on June 28, 1813, and died in Baden Baden, Germany, June 6, 1872. An important investor in his cousin George W. Scranton's iron-making activities in northeast Pennsylvania during the early 1840s, Scranton relocated to the area in 1847 and became the general manager of the Lackawanna Iron & Coal Company in 1853. Under his judicious leadership, the firm emerged on the eve of the Civil War as a leader in the manufacture of T-rails for the rapidly expanding railroad industry.

Scranton's early education consisted of a combination of theoretical knowledge taught to him in the local tuition schools in Madison and the practical experience gained during vacation periods from assisting the family business in the construction of wharves, breakwaters, lighthouses, and other similar projects. Like many nineteenth-century contemporaries who later attained success in business ventures, Scranton began his career serving as an apprentice clerk to a New Haven merchant. After learning fundamental business skills in his clerkship, Scranton worked in his relative Daniel Hand's mercantile firm located in Augusta, Georgia. Within a decade the young man took control of the operation and in the process amassed a substantial fortune. And it was his surplus capital that became increasingly tied to cousin George Scranton's gamble during the 1840s of transforming the Lackawanna Valley into a center of the antebellum iron trade.

Beginning in the late 1830s the Scranton family became involved in an iron-making scheme initi-

Joseph H. Scranton

ated by the mercurial yet visionary mineralogist and promoter William Henry. Henry's preliminary surveys of a tract known as Slocum's Hollow in 1839 appeared promising, since there existed every indication that the crucial materials necessary for the manufacture of pig iron–limestone, iron ore, and anthracite coal–were present in abundance on the site. Henry's efforts to purchase the property and to erect a furnace were dependent on his ability to raise large sums of capital, and after the premature death of a partner he turned in desperation to his son-in-law Selden T. Scranton and Selden's brother George.

During the 1840s George Scranton took the lead in developing the technological know-how of manufacturing iron from anthracite coal at Slocum's Hollow and later made crucial decisions related to what products would be made. In 1843 an abortive attempt was made to convert pig iron to nails, but the physical properties of the local iron ores resulted in the production of a brittle, poor-quality final product that frequently shattered upon impact. However, George Scranton responded to the setback with a second market maneuver, this time shifting production equipment to T-rail manufacture, a strategy which ultimately proved successful. It was during this time of start-up, initial failure, and product changes that the capital provided by Joseph Scranton enabled the Scranton family to maintain control of the fledgling yet increasingly costly and complex enterprise.

With the production of T-rails the struggling firm, first known as Scrantons & Grant Company and in 1846 reorganized as Scrantons & Platt Company, achieved a level of stability. In 1847 Joseph Scranton moved from Augusta to Slocum's Hollow, already renamed as Scranton, to assist in the management of the rapidly developing business. Family cohesiveness and control was further strengthened during this time when Scranton married a sister of one of the venture's partners, J. C. Platt, a New England merchant who had married Katherine Scranton in 1844. With a reorganization in 1853 changing the firm's name to the Lackawanna Iron & Coal Company, Scranton became the firm's general manager, quickly distinguishing himself. In the opinion of a contemporary, Benjamin Throop, Joseph Scranton "brought an executive ability and practical management. Keen, shrewd, far seeing, yet just, his mind comprehended a wide-sweeping knowledge for all the conditions necessary for the success of every undertaking he had in hand."

A civic-minded entrepreneur, Scranton became involved in numerous business activities in the Lackawanna Valley and beyond. In 1867 Scranton was appointed to be a commissioner of the newly organized Union Pacific Railroad. In addition to his position with the Lackawanna Iron & Coal Company, Scranton served as president of the Lackawanna & Bloomsburg Railroad, the First National Bank of Scranton, the Scranton Gas & Water Company, and as a director of the Delaware, Lacka-

wanna & Western Railroad, the Sussex Railroad of New Jersey, Mt. Hope Mineral Railroad Company, Franklin Iron Company, Scranton Trust Company, Scranton Savings Bank, Dickson Manufacturing Company, Moosic Powder Company, and Oxford Iron Company.

Joseph's managerial control over the Lackawanna Iron & Coal Company gradually passed to his son, William Walker Scranton, who directed the firm's transition into the steel business. William Walker Scranton, born in 1844, was educated at Phillip's Academy in Andover, Massachusetts, and Yale University, from which he graduated in 1865. After serving in several apprentice positions within the family company he was appointed superintendent of a new mill in 1867 and became general production superintendent in 1871. In 1874 William Walker Scranton traveled to Europe to study the latest steel-making processes and returned later that year to supervise the installation of a Bessemer converter that ultimately doubled the existing plant's output. Involved in the resolution of several labor disruptions during the 1870s, William Scranton organized the Scranton Steel Company in 1881, a firm that rolled 120-foot steel rails from ingot. And following a career pattern like his father, William Scranton became a civic leader, boosting and promoting the city of Scranton and the Lackawanna Valley.

Joseph Scranton's successful career illustrates the value not only of energy, business acumen, and individual initiative but also that of kinship ties. His efforts, along with those of other family members and a small group of outside and local investors, did much to contribute to the rise of Scranton as an important nineteenth-century industrial city.

References:

Burton W. Folsom, Jr., *Urban Capitalists: Entrepreneurs and City Growth in Pennsylvania's Lackawanna and Lehigh Regions, 1800-1920* (Baltimore: Johns Hopkins University Press, 1981);

Frederick Hitchcock, *History of Scranton*, 2 volumes (New York: Lewis Historical Publishing, 1914);

Dwight J. Stoddard, *Prominent Men of Scranton and Vicinity* (Scranton, Pa.: Tribune Publishing, 1906);

Benjamin H. Throop, *A Half Century in Scranton* (Scranton, Pa.: The Scranton Republican, 1895).

Archives:

The Scranton family papers are located at the University of Delaware, Newark, Delaware.

Joseph Sellwood

(December 5, 1846-February 24, 1914)

by Terry S. Reynolds

Michigan Technological University

CAREER: Miner and mine contractor, Michigan copper and iron ranges (1865-1884); superintendent, Colby Mine (1884-?); president, Brotherton Mining Company (1885?-1898?); superintendent, Chandler Mine (1888?-1892); vice-president, Duluth & Iron Range Railroad (1892-1898); vice-president, Minnesota Iron Company (1892-1898); general superintendent of mines, American Steel & Wire Company (1898-1901/1902?); president, Masaba Steamship Company, (?-1914); president, City National Bank of Duluth (?-1914).

Joseph Sellwood, one of the leading figures in opening up the Gogebic iron range in Michigan and Wisconsin and the Vermilion and Mesabi iron ranges in Minnesota, was born on December 5, 1846, in Cornwall, England, to Richard and Elizabeth Carter Sellwood, a mining family. At nine years of age he began work as a miner's helper in a Cornish tin mine. He acquired some public school education but at thirteen began mining full time. Around 1865 he immigrated to America, finding employment as a miner at the Mount Hope Mine in New Jersey.

In the 1860s the expanding copper and iron mines of Michigan's Upper Peninsula attracted large numbers of Cornish miners. Because of their long mining tradition and knowledge of English, Cornishmen were generally awarded most of the supervisory positions in the Michigan mines. Within a few months of his arrival in America, Sellwood joined the Cornish migration to upper Michigan. Between 1865 and 1870 he worked in Michigan's "Copper Country," beginning at the Ogemaw Mine in Ontonagon County.

In August 1870 he accepted a position in the New York Mine at Ishpeming, Michigan, on the Marquette iron range. Although Sellwood began his iron mining career as an ordinary miner, he soon began to take contracts for mining ore from exist-

ing mines. A big man who drove himself hard, often working 18-hour days, Sellwood generally delivered more than was expected. As a result, within a few years he was handling contracts for the entire output of mines such as the New York and Cleveland mines of the Cleveland Iron Mining Company.

In 1884 Mather, Morse & Company sent Sellwood to the Gogebic Range along Michigan's far western border with Wisconsin. There Sellwood opened the first mine on that range, the Colby, near Bessemer, Michigan. Sellwood was given a 25-percent interest in the property. In 1885 Sellwood opened the Brotherton Mine at nearby Wakefield for Pickands, Mather & Company. In 1885 or 1886 Sellwood became president of the Brotherton Mining Company, which secured ownership of the Brotherton from Pickands, Mather. Shortly following this acquisition, Sellwood also leased the adjacent Sunday Lake mine and returned it to profitability. At the Brotherton Sellwood popularized the "caving system" of mining, one of the major methods used in iron ore mining in the Lake Superior region. He was also among the first to use steam-powered shovels to load ore cars.

With an established record as a mine superintendent and operator, Sellwood was sent in 1886 to the newly opened Vermilion range in Minnesota to secure properties for the Chicago & Minnesota Ore Company. For a time he divided his time between the Gogebic and Vermilion ranges but in 1888 moved to Duluth. He opened the Chandler Mine on the Vermilion Range and operated it from 1887 or 1888 to 1892.

In 1892 Sellwood retired from the active operation of mines but remained active in mining. Through the remainder of the decade he explored properties on the Mesabi and elsewhere for the Minnesota Iron Company and served as a vice-president of that company and its subsidiary, the Duluth & Iron Range Railroad. He resigned these positions in

1898 to become general superintendent of the iron mining properties of the American Steel & Wire Company in the Lake Superior region, continuing in this position until the company was absorbed by U.S. Steel in 1901.

After 1901 he inspected and purchased properties on several Lake Superior iron ranges, especially the Gogebic and Mesabi, for himself and for a number of the independent steel firms, including Wheeling Steel & Iron and Central Iron & Steel. For a time he also supervised the mines of International Harvester Company.

Through his connections with the iron ore mining industry, Sellwood also became interested in lake shipping. He was one of the owners of the *V. H. Ketchum*, a large lake freighter, and at the time of his death was president of the Masaba Steamship Company. A large lake freighter, the *Joseph Sellwood*, launched in 1906, was named in his honor. In his last years Sellwood's interests shifted from mining to banking. He had large interests in

and served as an officer of several banks but most notably served as president of the City National Bank of Duluth in the years preceding his death in Duluth on February 24, 1914.

During his active mining career Sellwood was widely regarded as one of the best mining men in the nation and at his death was one of the wealthiest and most influential men living in the Michigan-Minnesota iron mining regions.

References:

"Captain Joseph Sellwood," in *History of Minnesota*, volume 3 (West Palm Beach: Lewis Historical Publishing, 1969), pp. 377-378;

Henry A. Castle, *Minnesota: Its Story and Biography*, volume 3 (Chicago & New York: Lewis Publishing, 1915), pp. 1478-1480;

"Joseph Sellwood," *Proceedings of the Lake Superior Mining Institute*, 19 (1914): 292-293;

"Joseph Sellwood," in *Encyclopedia of Biography of Minnesota*, volume 1, edited by Charles E. Flandrau (Chicago: Century Publishing, 1900), pp. 353-354.

Peter Shoenberger

(October 16, 1782-June 18, 1854)

by John A. Heitmann

University of Dayton

CAREER: Owner, assorted iron forges and furnaces (1815-1854); co-owner, Cambria Iron Works (1844-1854).

Peter Shoenberger was born in Hannover, Germany, on October 16, 1782, and died in Marietta, Pennsylvania, on June 18, 1854. Shoenberger was the foremost ironmaster of Pennsylvania during the first half of the nineteenth century; his accomplishments were recognized by Andrew Carnegie, who remarked long after Shoenberger's death: "I have always considered him [Shoenberger] my predecessor. He was to the iron industry what I later became to the steel.... If he were living today with his wonderful brain and organizing genius, [he] would almost surpass the Pierpont Morgans & Rockefellers of the present time."

Peter Shoenberger, along with his father, George, mother, Frances, uncle Peter, and other members of the family, arrived in Philadelphia on

the square-rigged sailing ship *Adolph* on August 27, 1785. Soon the family moved west, first to Lancaster County and then to Huntington County near Altoona, where beginning in 1796 George and uncle Peter Shoenberger became active in a number of iron-making enterprises along with their partner Samuel Fahnestock. In 1804 George Shoenberger and Fahnestock established the Juniata Forge on the Little Juniata River, and four years later George independently set up the Huntington Furnace in Franklin Township.

With the death of his father in 1815 Peter Shoenberger took over the family iron business. Several sources indicate that he was a graduate of the Jefferson Medical College in Philadelphia and had started to practice medicine in Pittsburgh until his health failed. Apparently the outdoor life at the iron furnaces agreed with his physical constitution, and his business acumen resulted in the building of an iron empire during the next 35 years. Because of

Peter Shoenberger

the quality of his product he found a ready market at the Harpers Ferry Armory, where his iron was used to make gun barrels and trigger mechanisms.

Shoenberger had inherited both the Huntington Furnace and Juniata Forge from his deceased father and later acquired Marietta Furnace near Philadelphia. Always prospecting for new ores, Shoenberger discovered rich deposits in the Morrison's Cove area of south central Pennsylvania during the 1820s, and in quick succession between 1828 and 1832 he erected in McKee's Gap the Upper Maria, Middle Maria, and Lower Maria furnaces, named in honor of one of his daughters. Subsequently Shoenberger set up Martha Furnace and Forge on the Roaring Spring stream at McKee's Gap, Rebecca Furnace on Clover Creek near Fredericksburg, Pennsylvania, and Sarah Furnace at Sprout, Pennsylvania, naming all of these ironmaking operations in honor of either his daughters or his wife. Other facilities established by Shoenberger and located in the central Pennsylvania region included Elizabeth Furnace, built in 1827 near Woodbury and later moved to Bloomfield

where large deposits of ore were found, and the Allegheny Forge near Duncansville. In addition to these furnaces and forges Shoenberger set up the first rolling mill in Pittsburgh in 1824, the Juniata Iron Works, which was later absorbed by the American Steel & Wire Company.

At the peak of Shoenberger's career he became half-owner, with George S. King, of the Cambria Iron Works located in Johnstown, Pennsylvania. Shoenberger had entered a partnership in 1844 with King in the Cambria furnace, an ironmaking firm that sold its pig iron in Pittsburgh. Cambria continued to expand its operations during the boom times brought on by the tariff of 1842 and by the mid 1840s consisted of four furnaces. With the repeal of the protective tariff in 1846 the business returned only marginal profits, and Shoenberger proposed that a foundry be erected to fabricate large iron kettles for the rapidly expanding sugar trade located in south Louisiana. However, King suggested that investment should be made in a rolling mill to take advantage of the demand for rails for the Pennsylvania Railroad that was completed in 1852 and which passed through Johnstown. After securing the support of Boston financiers the Cambria Iron Company and Cambria Iron Works were established with an initial capitalization of $1 million and with Shoenberger as president. Consisting of a rolling mill, four hot-blast coke furnaces, and other facilities, Cambria was one of the first American firms to supply the rails necessary for America's spectacular revolution in railroad transportation that began in earnest at mid century.

Shoenberger's wide-ranging enterprises required large amounts of timber, limestone, and iron ore, and as a result of these demands for raw materials the "iron king" owned more than 100,000 acres of land in central Pennsylvania. A small army of workers were employed to prepare charcoal fuel, quarry limestone, and dig iron ore from open pits. Shoenberger owned hundreds of horses and mules to convey his products and materials, and his employees—often farmers from the nearby area who worked whenever they were able to free themselves from agrarian chores—purchased goods from stores owned by Shoenberger and located throughout the iron-manufacturing district.

Despite the fact that Shoenberger's influence was felt most directly in central and western Pennsylvania, he also played an important role in establishing the Wheeling, West Virginia, area as an iron-making district. In 1832 Shoenberger and partner David Agnew set up an ironworks in Wheeling, the result of Shoenberger's realization that Wheeling's location on the Ohio River made it a gateway for products marketed in the rapidly growing West. In the opinion of Shoenberger, the Wheeling site was ideal for the manufacture of finished products like nails and iron sheets. Shoenberger's "Top Mill" consisted of puddling furnaces, a sheet mill, and a nail factory. Pig iron brought down from Pittsburgh was converted into pasty balls of wrought iron in puddling furnaces, and then forged under the tilt hammer into crude slabs suitable for rolling into plate. The plate was subsequently sheared into strips which after heating were fed into nail machines where knives would cut the final product. The "Top Mill" was the first of many such establishments located in Wheeling during the nineteenth century, and by the early 1870s this small town located on the Ohio River would be known as the "Cut Nail Capital of the World."

Peter Shoenberger has attracted little attention on the part of historians. Forging an industrial empire in antebellum America, his contributions have been overlooked by those fascinated with the lives of men like Andrew Carnegie and others of the Gilded Age. Yet men like Shoenberger paved the way in terms of organization and production technology. The generation that followed and led America into the age of steel and industrial maturity owed an enormous debt to ironmasters like Peter Shoenberger.

References:
Frank C. Harper, *Pittsburgh: Forge of the Universe* (New York: Comet Press, 1957);

Calvin W. Hetrick, *The Iron King* (Martinsburg, Pa.: Morrisons Cove Herald, 1961);

Henry Dickerson Scott, *Iron & Steel in Wheeling* (Toledo: Caslon, 1929);

James M. Swank, *Cambria County Pioneers* (Philadelphia, 1910);

Swank, *History of the Manufacture of Iron in All Ages*, second edition (Philadelphia: American Iron and Steel Association, 1892);

Swank, *Introduction to a History of Ironmaking and Coal Mining in Pennsylvania* (Philadelphia, 1878).

William Henry Singer

(October 2, 1835-September 4, 1909)

by Bruce E. Seely

Michigan Technological University

CAREER: Clerk, Wallingford & Company (c. 1853-?); apprentice, G. & J. H. Shoenberger ironworks (dates unknown); partner (1858-1900), president, Singer, Nimick & Company (c. 1881-1900); president, Pittsburgh Bessemer Steel Company (1879-1883); director, Carnegie Steel Company (1883-1901); director, Crucible Steel Company of America (1900-1909).

During the nineteenth century, western Pennsylvania provided a stage for a long-running drama; the story concerned the growth in the American iron and steel trade. Usually the men who mastered the Bessemer process and produced railroad rails in vast quantities were cast in the leading roles. The expansion of the American steel industry was not, however, simply a tale of Andrew Carnegie and the Bessemer process. Other steel men and other processes were as important, and other branches of the trade also were centered in Pittsburgh. William Henry Singer was one of the leading steel men in the cast-crucible-steel trade. He joined his brother's business at an exciting time, just as it began production of high-quality cast-crucible steel. By the 1880s he was the dominant partner and the firm was a leader in its field. Singer also recognized, however, that the crucible-steel industry faced technological threats that limited its prospects. With greater success than most other crucible-steel makers, Singer found and made opportunities to move in new directions. The most important of these chances was the formation of the Pittsburgh Bessemer Steel Company, which provided Singer with a connection to Andrew Carnegie.

William Henry Singer was born in Pittsburgh on October 2, 1835, just as the iron industry was

emerging as the key to the city's economic prosperity. In the fashion of the time, he received good instruction in the public and private schools of the city and then began a business career as a clerk in the commercial house of Wallingford & Company. His older brother, John F. Singer, may already have been a partner in this establishment. Singer seems not to have stayed long at Wallingford, however, for he soon joined one of the city's oldest ironworks, G. & J. H. Shoenberger. The Shoenbergers had established one of the first blast furnaces near Pittsburgh in the 1810s and had built the Juniata Iron Works with a rolling mill and nail works in the 1820s. The Shoenberger name was long connected with iron-making operations in Pittsburgh and Allegheny County, and in 1835 the company launched one of the earliest attempts to manufacture cast-crucible steel in the country.

Singer served the equivalent of an apprenticeship at the Shoenberger ironworks until about

1858, when he became a partner in an enterprise of his brother's. That firm, founded in 1848, is usually assumed to have been called Singer, Nimick & Company, although some evidence suggests that it carried the name Singer, Hartman & Company until 1859. It has been established that William Kennedy Nimick owned a successful commercial house, Nimick & Company, which dealt in pig iron, and went on to become a leading figure in the Pittsburgh financial community. Thus, there were good reasons for John F. Singer to join forces with Nimick in 1848, because Singer had decided to manufacture blister and spring steel. Other Americans had entered this business earlier, although British domination made the manufacture of steel a very risky undertaking. John Singer's operation survived, but a measure of the problems it faced was the name given the steel plant: the Sheffield Works, a clear reference to the company's real competitors. By 1853 Singer, Nimick & Company was successfully making the lower grade steels, and the partners decided to begin producing cast-crucible steel, the material that commanded the highest prices. Their first cast steel was intended for saws and agricultural implements, uses that did not demand the finest quality product, but the company clearly hoped to make finer steel. Although claims of success appeared, the quality problems that damaged the reputation of every early American company also plagued John Singer. Unlike other steel pioneers, however, Singer, Nimick & Company seems to have struggled to make high-quality cast steel throughout the 1850s.

This was the situation when William Singer joined the company in 1858. Importantly, his arrival coincided with favorable developments at other Pittsburgh firms in the crucible-steel industry. In the early 1860s first Hussey, Wells & Company and then Park Brothers & Company finally mastered the production of high-quality cast-crucible steel, aided in part by the opportunities and demands of the Civil War. After 1863 Singer, Nimick & Company also produced top-grade cast steels, using the traditional Huntsman process. Singer's role in this success is not clear, although one source identified Singer's other brother, George, who was treasurer for over 40 years, as the dominant partner in 1860.

In any event the firm grew quickly into one of the largest and most respected steel producers in Pittsburgh. By the mid 1870s the facilities included

12 puddling furnaces, 8 converting furnaces, 17 heating furnaces, 30 coke-fired melting holes, 4 of the new 24-pot Siemens regenerative melting furnaces, 12 roll trains, and 11 steam hammers; the Sheffield Works had an annual capacity of 12,000 tons. By 1881 the firm, now a joint stock company with William Singer as president, was the largest specialty steel maker in the city, with a production capacity of 23,000 tons a year and an impressive product line. Two particular specialties were carriage springs and axles; the works produced 40 to 50 tons of springs a month, and another 100 tons of axles. Other products included homogeneous steel plates for boilers and fireboxes, saw steel (including circular plates up to 74 inches in diameter), sheet and plate of all gauges, round machinery steel, and steel for railroad use. An entire department was devoted to making steel for agricultural implements, including rakes, hoes, picks, and hammers, while other steels went into axes, skates, cutlery, augers, coal and granite wedges, and magnets. Special attention was also given, as at most crucible steel firms, to producing the best cast steels for edge tools.

William Singer's role is clearer in this period of growth than in earlier years. In the 1870s he was responsible for several important technical developments, continuing a company tradition of innovation first manifested in the early adoption of the Siemens gas-fired melting furnace. Singer's first invention was a machine for rolling a bevel on plow coulters and harrow disks. He then developed "liquid compression" saw steel that limited the splitting of saw teeth, and introduced a soft center steel plow that was less likely to break. Another indicator of Singer's technical interest was his early membership (1873) in, and support for, the American Institute of Mining Engineers. Singer made sure that the newest steel-making technology was introduced into the works in the 1880s, including a 5-ton open-hearth furnace and a pair of Lauth patent 3-high rolling mills. Singer, Nimick & Company was the only U.S. patent licensee for this design, and the entire rolling department was equally advanced. In the mid 1880s the firm boasted that "We can roll a plate of steel 78 inches wide, 6 inches thick, and 12 feet long. The daily capacity of this train of rolls, 24 hours, on plates of that size, would be about fifty tons." The company also produced cold-rolled sheet and strip steel for watch springs, corsets, shoe shanks, keys, and other hardware.

Both these products and the presence of an open-hearth furnace indicated that Singer, Nimick & Company was, by the 1880s, more than a cast-crucible-steel maker. Changing technology, especially the open-hearth process, was forcing a large number of crucible-steel firms out of business because it produced nearly equal quality at lower cost. It was no accident that a Pittsburgh cast-crucible-steel maker built the first American open hearth in 1873. By the mid 1880s Singer, Nimick & Company used open-hearth steel for boilerplates, locomotive fireboxes, smokestacks, machinery steels, and carriage and wagon axles, products previously fabricated only with crucible steel. Adopting the new open-hearth process prevented a loss of markets and demonstrated the value of staying abreast of technical innovations.

A more unexpected surprise was the competition of Bessemer steel. Along with several other firms in the city, Singer, Nimick & Company learned with some surprise that it was possible to make high-quality Bessemer steel. Thomas Carnegie apparently conducted experiments at the Edgar Thomson Works to demonstrate carriage springs, plows, and railroad axles made from Bessemer steel. When excess Bessemer ingots became available from the Edgar Thomson Works in 1877, Singer's firm and others bought them, remaining quiet, of course, about this substitution, for crucible-steel makers marketed their steel by stressing quality.

In 1879, however, the rail mill at the Edgar Thomson Works was remodeled, absorbing the excess capacity that for two years had been sold to the crucible-steel makers. This loss explained why William Singer led other crucible-steel makers in Pittsburgh to build their own Bessemer plant. The Pittsburgh Bessemer Steel Company was organized in 1879, with Singer as president. Joining Singer in the firm, and each investing $50,000, was William G. Park and C. G. Hussey, the other large producers in Pittsburgh. Reuben Miller and William Clark invested $40,000 each, and Andrew Kloman put in $20,000. Kloman had been Carnegie's partner until the panic of 1873 embarrassed him financially, and Kloman entered the company determined to get revenge on Carnegie. Kloman guided construction of the new plant at Homestead and designed equipment built by MacIntosh, Hemphill & Company. Kloman also built his own rail mill on adjoining property to compete directly with Carnegie's Edgar Thomson Works.

Kloman died before the mill was complete, but his legacy was evident when the first steel was poured in March 1881 and the first rails rolled in August. Construction had taken only 15 months, an amazing achievement. Labor relations, however, proved far more troublesome. The problem was tension between the skilled workers of the Amalgamated Association of Iron and Steel Workers at the Homestead Works and the openly antiunion plant manager, William Clark. Inflammatory statements and physical confrontations produced a series of real and threatened strikes. Although the partners made several compromises with the union, further union demands, a lockout, and Clark's insistence on a yellow-dog contract poisoned the atmosphere. Moreover, these problems took place during a declining market for steel. By 1883 the owners wanted out, and their request that Carnegie purchase the works was accepted. Carnegie absorbed the Homestead Works for nearly cost in October 1883, gaining a truly modern facility. Within 18 months the rail mill had been converted to structural shapes, while the converters supplied steel to Carnegie's Union Iron Mills and Keystone Bridge Company.

Andrew Carnegie was not the only person to benefit from Homestead's problems, for William Singer also found opportunity amidst adversity. Of the original investors in the Pittsburgh Bessemer Steel Company, he alone chose to accept payment in Carnegie Brothers Company stock and become a director of the company. Few business decisions have paid off so handsomely, as Singer entered the inner circle of the American steel industry.

Singer's contribution to Carnegie's iron and steel ventures is difficult to judge for several reasons. First, it would be hard for anyone to shine in the glare of Carnegie, Frick, and the other truly dynamic figures in Carnegie's steel empire. Moreover, Singer's activity is obscured by a lack of information. He was, of course, involved in many of the fascinating conflicts within Carnegie's domain, including the acquisition of the Pittsburgh, Shenango & Lake Erie Railroad, Frick's legal battle over the Iron Clad Agreement, and the creation of the United States Steel Corporation in 1901. But his influence in these events is unclear. Henry Oliver Evans's biography of Henry W. Oliver presented Singer as "a valuable member of the Carnegie Companies," but James Bridge noted that most of the forceful men around Carnegie were eventually forced out, since Carnegie tolerated no real rivals. Those who stayed, argued Bridge, were almost unimportant, and he dismissed Singer with the sarcastic remark that he "conscientiously attended the Board meetings, and his ambition was more than satisfied with the prerogative of making the motion for dividends."

Bridge's comment refers to Singer's position on one of the consistently disruptive debates within Carnegie Steel: with the exception of Carnegie, Frick, and Schwab, the shareholders opposed Carnegie's insistence on pouring money back into the firm. The smaller stockholders wanted higher dividends. Regardless of dividend policy, however, Singer benefited handsomely from his ties to Carnegie. In April 1899 Singer's personal account in the Carnegie Steel Company totaled almost $359,000. In addition Singer held $5 million in Carnegie Steel Company, Ltd., stock and over $773,000 in shares in Henry Clay Frick Company. After the Frick suit over the Iron Clad Agreement was settled in 1900, Singer owned 2,830 shares of stocks and $2,886,000 in bonds in the Carnegie Steel Company. By 1901 Singer was the seventh largest stockholder in the Carnegie companies, and his $50,000 investment in the Pittsburgh Bessemer Steel Company was worth about $8 million when U.S. Steel was formed.

Although Bridge's description of Singer was too harsh, he was not a dynamic force in the company. Singer was, rather, in the position of almost all the junior partners of the Carnegie Steel Company—bit players surrounded by giants. This picture emerges in Joseph Wall's biography of Carnegie, for Wall shows Singer and the other board members taking their lead from Carnegie, Frick, and Schwab. On some matters all agreed. Singer, for example, spoke for all members of the board during a discussion in early 1900 regarding renewed attempts by the Amalgamated Association of Iron and Steel Workers to organize Homestead. "I would not let them get a foothold under any circumstances—not under any conditions—stop it at once. I would mean not only Homestead but the rest of the Works. We have gone through that condition of affairs, and it cost us a good deal of money to get our works back again; but it was money well spent. . . ."

On other subjects, however, Carnegie's wishes prevailed, even though the smaller shareholders might express initial objections. This was the case on July 16, 1900, when the board debated building new mills for producing finished steel products.

Singer and several others were cautious, with Singer in favor of delaying a decision until September. A motion to that effect lost, however, whereupon Singer voted for Schwab's motion to allocate $1.4 million for rod, wire, and nail mills at the Duquesne Works.

In short, Singer never opposed Carnegie on any substantive issue. This statement, however, underestimates Singer's business acumen. He turned the Pittsburgh Bessemer Steel Company's problems to advantage. He also joined the plan developed largely by William and David Park of the Black Diamond Steel Works to create a crucible-steel trust. In the late 1890s the two brothers wished to see their assets in liquid form, and they organized the Crucible Steel Company of America in 1900. The new company brought together the 13 largest firms in that industry, including Singer, Nimick and Company, and controlled specialty steel making in this country. Singer, Nimick & Company had slipped from being the largest crucible-steel maker in the city by this time; its capacity had fallen to 16,000 tons annually. Nonetheless, William Singer became a member of the original board of directors of the Crucible Steel Company of America on July 21, 1900, serving until his death in 1909.

Clearly, then, Singer was a very successful businessman although many details of his career are unavailable. Not surprisingly, Singer the man is as hard to define as his role in Carnegie Steel. A portrait shows him with the muttonchop sideburns and mustache of this era, hair parted in the middle—in short, the picture of a respectable business leader of the late nineteenth century. He belonged to three prestigious country clubs and was a vestryman at the Trinity Episcopal Church. Biographers called him intensely public spirited, but unlike several pioneering crucible-steel makers, Singer showed little evidence of charitable or civic activity. *The National Cyclopedia of American Biography* says little about his personality, noting only that "His mind was vigorous and direct, moving according to strictly logical formulas. His speech was incisive, yet preserved from curtness by the kindly play of an unfailing humor, and an unfailing charity." However much allowance one makes for good humor, it seems clear that the steel business was the core of William Singer's life. His outside affiliations were limited, including only the American Institute of Mining Engineers and the Engineers' Society of Western Penn-

sylvania. Singer was married and the father of four children; his only son, William, Jr., became a landscape painter of some repute.

The Encyclopedia of Pennsylvania Biography suggests that William Henry Singer was the most important figure in the Pittsburgh steel industry because his career spanned almost 50 years. Certainly, he was widely recognized in his hometown, although it is difficult in the 1980s to appreciate Singer's position because he walked, quite literally, in Andrew Carnegie's shadow. Yet for 40 years Singer was a principal leader of one of the largest and most successful non-Bessemer steel companies in Pittsburgh. He was an innovator who appreciated the value of the new technical capabilities of his age. To be sure, Singer apparently lacked the desire of Carnegie and Frick to win total dominance of their industries, but no other steel maker, crucible or otherwise, shared that aggressiveness. Singer did, however, make the most of his opportunities. Singer, Nimick & Company, begun as a family partnership, grew and prospered under his direction. Singer also turned what seemed to be the Pittsburgh Bessemer Steel Company's disaster at the Homestead Works into his opportunity; he built his fortune from that point. That choice cast him into Carnegie's arena, and in comparison he may seem diminished. Out of Carnegie's shadow, however, Singer's accomplishments seem more substantive.

References:

James Howard Bridge, *The Inside History of the Carnegie Steel Company: A Romance of Millions* (New York: Aldine Book Company, 1903);

Crucible Steel Company of America, *Fifty Years of Fine Steelmaking* (Pittsburgh, 1951);

Henry Oliver Evans, *Iron Pioneer: Henry W. Oliver: 1840-1904* (New York: Dutton, 1942);

J. S. Jeans, *Steel: Its History, Manufacture, Properties, and Uses* (New York: E. & F. Spon, 1880);

John W. Jordan, ed., *The Encyclopedia of Pennsylvania Biography* (New York: Lewis Historical Publishing, 1921);

Pennsylvania Historical Review: Cities of Pittsburgh and Allegheny, Leading Merchants and Manufacturers (New York: Historical Publishing, 1886);

George H. Thurston, *Pittsburgh As It Is* (Pittsburgh, 1857);

Thurston, *Pittsburgh's Progress, Industries, and Resources* (Pittsburgh, 1886);

Joseph Frazier Wall, *Andrew Carnegie* (New York: Oxford University Press, 1970).

Horace Strong Smith

(December 28, 1826-October 17, 1899)

by Alec Kirby

George Washington University

CAREER: Master mechanic, Chicago & Alton Railroad (1865-1875); manager (1875-1879), general superintendent (1879-1889), director and vice-president, Joliet Steel (1889-1894).

Horace Strong Smith is widely credited as being the leading spirit behind the Joliet Steel Company. Born in Dunstable, New Hampshire, on December 28, 1826, to Benjamin and Anna M. Smith, he received only a common school education. Trained as a mechanic, Smith moved to Illinois in 1865 to take a position as master mechanic for the Chicago & Alton Railroad.

In 1875 Smith was appointed by the receiver of the financially troubled Joliet Iron & Steel Company to manage the firm. By 1879, when the company was reorganized as the Joliet Steel Company, Smith had placed the works in top condition. He also introduced several innovations, including auto-matic nail and billet rollers and a technique for rolling rails directly from a Bessemer converter. In 1887 Joliet steel was the largest producer of steel rails in the world, that year producing over 200,000 tons.

In 1889 Smith participated in the negotiations which merged Joliet Steel, Union Steel, and North Chicago Rolling Mill into the Illinois Steel Company. At the time Illinois Steel was the largest steel manufacturer in the world, outstripping even the Carnegie conglomerate. Smith served as a director of the Joliet subsidiary and as vice-president in charge of operations. He retired in 1894 and died in Chicago, Illinois, on October 17, 1899, the year after Illinois Steel was merged into Federal Steel.

Reference:

Fritz Redlich, *History of American Business Leaders*, volume 1 (Ann Arbor, Mich.: Edwards Brothers, 1940).

Spang, Chalfant & Company

by Paul F. Paskoff

Louisiana State University

Spang, Chalfant & Company, located in the Pittsburgh, Pennsylvania, suburb of Etna, was one of the largest producers of tubular iron and steel in the United States by 1930. The firm began on a very modest scale in 1828 as a family-owned concern under the name of Henry S. Spang & Son. In that year the elder Spang bought the Etna Iron Works and began to make cut nails and wrought-iron tubing at the works' rolling mill from blooms produced in the Juniata Valley of Pennsylvania.

The growth of the tubular-iron market encouraged the Spangs to expand their operations, and in 1845 they reorganized the firm as Spang & Company. By 1850 it was one of the largest of the 23 rolling mills in western Pennsylvania. The retirements of Henry Spang and another of the company's senior partners in 1858 and the admission of John Weakley Chalfant as a full partner prompted a change of name for the firm to Spang, Chalfant & Company.

Almost from its inception in 1828 the firm had enjoyed the distinction of having been the first rolling mill west of the Allegheny Mountains to have made iron pipe. This product and other types of boiler iron were the company's mainstays during the 1840s and 1850s as demand from the rapidly expanding railroad industry stimulated the growth of advanced segments of the iron industry. Spang, Chalfant continued its innovative course after the Civil War when, in the late 1870s, its rolling mills, under the management of George Alexander Chalfant, John's younger brother, began to produce iron pipe using natural gas instead of coal as a fuel. Although the substitution of gas for coal had first been made in late 1874 at the Siberian Rolling Mill of the firm Rogers & Burchfield of Armstrong County, Spang, Chalfant's use of gas for puddling iron was the first such application in Allegheny County and encouraged other firms to adopt the practice.

The ready adoption during the late 1870s and early 1880s of natural gas as a fuel was hardly coincidental. By 1882 output of natural gas in the United States for industrial and household consumption had grown to 3.4 billion cubic feet, from which level it increased more than 20-fold over the next three years. This enormous increase in production, which just barely kept pace with demand, meant a rapidly growing market for tubing. At about the same time, the number of municipal water companies began to increase dramatically, a development which also increased demand for iron pipe. The intense interest on the part of the rolling mills in using natural gas in the production of iron tubes and pipe for that market is quite understandable.

By the late 1880s the growth of Spang, Chalfant's business had made the firm one of the largest manufacturers of tubular iron in the United States. This growth, along with the continuous challenge to keep pace with the rapid rate of technological change within the iron and steel industry, made the firm's 1899 incorporation a necessary and logical step. In that same year the company's management, led by George A. Chalfant, who took over the company after John W. Chalfant's death in 1898, entered into negotiations with the National Tube Company concerning the latter firm's proposed purchase of Spang, Chalfant. National Tube, by far the largest steel pipe and tube producer in the United States, had already absorbed by merger and buy-out several other competitors and included Spang, Chalfant on a list of acquisition targets. The negotiations to effect the takeover broke down, however, and Spang, Chalfant continued its independent existence.

In 1902 Henry Chalfant became the firm's president. Chalfant, the son of John W. Chalfant, was a Harvard graduate and had served as Spang, Chalfant's treasurer since the firm's incorporation in 1899. Only thirty-five years old at the time,

The works of Spang, Chalfant & Company in Pittsburgh (courtesy of USX Corporation)

Henry Chalfant was aggressive and innovative as he continued the expansion that his father had initiated. Under the younger Chalfant's leadership Spang, Chalfant began to produce welded steel tubing and pipe, products already turned out by other, larger firms and for which a rapidly growing market already existed, primarily because of the oil industry's construction of pipelines.

Largely as a result of its movement into the steel pipe and tubing market, a product line which required a larger scale of operations, Spang, Chalfant enlarged its capital base from just $900,000 in 1902 to $7 million in 1922 and to $17.5 million in 1928. The year 1928 was a momentous one in the company's history on several counts. Of most immediate importance to the firm was the death of Henry Chalfant in August and, with his passing, the loss of a knowledgeable and accessible chief. But Chalfant had worked almost to the end on a successful program to modernize and expand the company's operations, by both internal growth and acquisition. He had moved Spang, Chalfant into the fabrication of seamless steel tubing, then one of the most advanced products of the industry and one for which demand outstripped supply. The large increase in the firm's capitalization in 1928 was used to finance this expansion and to enable Spang,

Chalfant to buy the Standard Seamless Tube Company and that firm's already operating seamless steel pipe mills in suburban Pittsburgh. This takeover made the expanded Spang, Chalfant & Company the nation's third largest tubular steel maker on the eve of the Great Depression.

References:

William T. Hogan, S. J., *Economic History of the Iron and Steel Industry in the United States*, volume 1 (Lexington, Mass.: Lexington Books, 1971);

Stefan Lorant, *Pittsburgh: The Story of an American City* (Garden City, N.Y.: Doubleday, 1964);

Louis McLane, *Documents Relative to the Manufactures of the United States*, 4 volumes (1833; New York: Burt Franklin, 1969);

Neal Potter and Francis T. Christy, Jr., *Trends in Natural Resource Commodities: Statistics of Prices, Output, Consumption, Foreign Trade, and Employment in the United States, 1870-1957* (Baltimore, Md.: Johns Hopkins University Press, 1962);

Fritz Redlich, *History of American Business Leaders: A Series of Studies*, 2 volumes (Ann Arbor, Mich.: Edwards Brothers, 1940);

Charles E. Smith, "The Manufacture of Iron in Pennsylvania," *Hunt's Merchants' Magazine*, 25 (November 1851): 574-581 and tables following 656;

James M. Swank, *History of the Manufacture of Iron in All Ages* (Philadelphia: American Iron and Steel Association, 1892).

Powell Stackhouse

(July 16, 1840-February 4, 1927)

by Stephen H. Cutcliffe

Lehigh University

CAREER: Various positions, Cambria Iron Company (1855-1861); superintendent, Johnstown Manufacturing Company (1866-1868); assistant superintendent, Cambria Iron Company (1868-1874); general agent, Republic Iron Company (1874-1876); superintendent of real estate, woolen mill, and brickyard (1876-1878), acting general manager (1878-1880), comptroller (1880-1884), vice-president (1884-1892), president, Cambria Iron Company (1892-1927); president, Cambria Steel Company (1898-1910).

Powell Stackhouse was an important and well-respected leader in the late-nineteenth- and early-twentieth-century iron and steel industry; he was connected with the Cambria Iron and Cambria Steel Companies and their affiliate subsidiaries throughout his lengthy 70-year business career.

Powell Stackhouse was born on July 16, 1840, the son of Joseph Dilworth and Sarah Phipps Shaw Stackhouse. His two grandfathers, Powell Stackhouse (after whom he was named) and Alexander Shaw, were among the founders of the Franklin Institute in Philadelphia. Following his mother's death in 1851 Stackhouse lived with his paternal grandfather and attended public school until the age of fifteen, when he was hired by the Cambria Iron Company of Johnstown, Pennsylvania. Daniel Morrell, his uncle by marriage, was general manager of the Cambria Works. Although briefly interrupted by a period of active military duty during the Civil War, Stackhouse remained associated with the Cambria company or its corporate affiliates until his death in 1927.

Stackhouse married four times, and these marriages, beyond their sheer number, reflect the importance of family connections and the close-knit nature of the nineteenth-century iron- and steel-making community. His first wife, Lucinda Elizabeth Roberts of Johnstown, whom he married

Powell Stackhouse

on December 28, 1863, died in 1866, leaving one child, Daniel Morrell Stackhouse. The child presumably was named after Daniel J. Morrell, the general manager of Cambria Iron Company, who had married Stackhouse's aunt, Susan Lower Stackhouse. On August 24, 1868, he married Genevieve Royer Swank, also of Johnstown, who was presumably the sister or, at the most distant, a cousin of James M. Swank of Johnstown, the statistician and writer who served as secretary and later vice-president and general manager of the American Iron and Steel Association. James Swank was also a close friend of Morrell, who served a term as president of the industry group. Although Stackhouse's third and fourth marriages to Anna Elizabeth Wheeler (1876) and Lucinda M. Buchanan (1879) do not suggest the same closeness regarding the makeup of the iron- and steel-making elite, it is instructive to note the importance of the familial ties in the formation and social structure of the late-nineteenth-century business community.

Stackhouse's military career, although brief, was notable. Rising from a volunteer corporal and participating in the battles of Antietam, Fredericksburg, and Chancellorsville, he later recruited and commanded his own company of Pennsylvania Volunteers. During this command he was promoted for bravery to major and selected with his regiment under Maj. John Stanton to accept the final surrender of Gen. Robert E. Lee at Appomattox.

Stackhouse, who eventually succeeded to the presidency of Cambria, did not leave behind him an extensive historical legacy that permits a detailed assessment of his contributions. However, his steady progression through the ranks does suggest the important role he played in bringing the company into the twentieth century. At the close of the war, Stackhouse returned to the Cambria Iron Company and in 1866 became superintendent of its subsidiary, the Johnstown Manufacturing Company. In 1868 he became assistant superintendent of the Cambria Iron Company itself. During the period from 1874 to 1876 he served as general agent for the Republic Iron Company in Marquette, Michigan, also a Cambria subsidiary, returning in the latter year to Cambria as superintendent of the company's real estate department, woolen mill, and brickyard. He held this position until 1878, when he was appointed assistant and acting general manager. Two years later he was promoted to comptroller. In 1884 he was appointed vice-president, becoming president in 1892. When the Cambria Steel Company was organized in 1898, Stackhouse became president of the new company as well. He remained president of Cambria Steel until 1910 but retained his presidency of Cambria Iron until the time of his death. During his long career Stackhouse also served as president of a number of other Cambria affiliates, including the Mahoning Ore & Steel Company, the Penn Iron Mining Company, the Republic Iron Company, and the Manufacturers' Water Company.

Stackhouse's career with the Cambria Iron and Steel Companies was both long and notable, overseeing one of the foremost iron- and steel-manufacturing operations in late-nineteenth-century America. His contributions to the industry were also reflected through his position as a director and vice-president of the American Iron and Steel Institute from its inception until 1918 and as an honorary member of the American Iron and Steel Association.

References:

John Fritz, *The Autobiography of John Fritz* (New York: John Wiley & Sons, 1912);

William Hogan, *Economic History of the Iron and Steel Industry in the United States* (Lexington, Mass.: Lexington Books, 1971);

Robert W. Hunt, "Evolution of the American Rolling Mill," *Transactions* of the American Society of Mechanical Engineers, 13 (1892): 45-69;

Hunt, "History of the Bessemer Manufacture in America," *Transactions* of the American Institute of Mining Engineers, 5 (1877): 201-15;

Jeanne McHugh, *Alexander Holley and the Makers of Steel* (Baltimore: Johns Hopkins University Press, 1980);

Elting E. Morison, *From Know How to Nowhere* (New York: Basic Books, 1974);

Morison, *Men, Machines, and Modern Times* (Cambridge, Mass.: M.I.T. Press, 1966);

Fritz Redlich, *History of American Business Leaders: A Series of Studies*, 2 volumes (Ann Arbor, Mich.: Edwards Brothers, 1940);

James M. Swank, *History of the Manufacture of Iron in All Ages* (Philadelphia: American Iron and Steel Association, 1892);

Swank, *Introduction to a History of Ironmaking and Coal Mining in Pennsylvania* (Philadelphia: J. M. Swank, 1878);

Peter Temin, *Iron and Steel in Nineteenth-Century America: An Economic Inquiry* (Cambridge, Mass.: M.I.T. Press, 1964).

Amasa Stone, Jr.

(April 27, 1818-May 11, 1883)

by Marc Harris

Ohio State University

CAREER: Carpenter and builder (1837-1842); partner, Boody, Stone & Company (1842-1847); superintendent, New Haven, Hartford & Springfield Railroad (1845-1846); construction contractor, Cleveland, Columbus & Cincinnati Railroad, Cleveland, Painesville & Ashtabula Railroad, Chicago & Milwaukee Railroad (1848-1855); superintendent (1850-1854), director, Cleveland, Columbus & Cincinnati Railroad (1852-1854); director (1852-1854), president, Cleveland, Painesville & Ashtabula Railroad (1857-1869); managing director, Lake Shore & Michigan Southern Railroad (1869-1875).

Born into a farming family in central Massachusetts, Amasa Stone, Jr., was in many ways an exemplary self-taught entrepreneur of the nineteenth century. Trained as a carpenter, with no significant formal or technical education but endowed with drive and intelligence, he progressed from building bridges to building and managing railroads and finally amassing large iron and steel and financial holdings. Along with his brother Andros B. Stone, whose career developed closely with his, Stone acquired skills as he needed them and capitalized on his family and business connections to become a principal figure in the midwestern railroad, banking, and iron and steel industries and an important industrial leader in the emerging center of Cleveland.

The Stone family traced its descent from Puritan forebears who had immigrated to New England before 1636. The family lived on a farm in Charlton, a country town southwest of Worcester, Massachusetts, where Amasa, Jr. was born on April 27, 1818, and Andros, the youngest, was born in 1824. Both remained on the family farm and attended the district school when it was in session, and both left the farm as youths. When Stone was seventeen, he apprenticed himself to an older brother, a carpenter and joiner, to learn building.

Amasa Stone, Jr.

After two years with his brother, he supplemented his common-school education with a term at an academy in Worcester. The curriculum included the Greek and Latin classics, along with some modern geology; possibly he also learned the trigonometry he would later need in order to design and build bridges. Except for the apprenticeship with his brother, however, none of Amasa Stone's early education was primarily technical in nature, and it was acquired unsystematically. Andros left the family farm at fifteen, four years after Amasa, and also apprenticed himself to the older brother to learn carpentry.

Their early business success, like that of many of their contemporaries, owed much to timing and to family connections. When the Stones began their working careers, southern New England had been feeling the effects of the transportation revolution for at least a generation. The area opened to an expanding flow of trade, credit, and economic activity; more specialized skills were needed, and the more complex economy could in turn support a wider range of specialized occupations. Moreover, New England rebounded in the early 1840s from a brief depression, restoring high demand for builders in a growing economy. Amasa Stone began his building career in a modest way, at first contracting along with two brothers to build a church. In 1839, when he was twenty-one, his brother-in-law William Howe, an established builder, hired him as a foreman to oversee construction of several buildings in the town of Warren, Massachusetts. Later, Howe won a contract with the Western Railroad Company to bridge the Connecticut River at Springfield using his newly patented wood-and-iron truss design, and he hired Amasa Stone to supervise much of the work. From then until 1842 Stone oversaw several other Western Railroad projects for Howe, concentrating on bridges and depots. Howe, obviously delighted to employ his capable brothers-in-law, hired Andros to supervise construction of several bridges in 1842. Both Amasa and Andros Stone were closely associated with Howe's design for several years.

That design, Howe, and his brothers-in-law stood at the center of a fundamental change in American civil engineering that solved a major problem in railway construction by turning bridge building from an empirical seat-of-the-pants process into a systematic technology. Before the 1840s American bridge builders, mainly carpenters, combined wooden arches and trusses according to design handbooks and their own experience; the rustic covered bridge was one result. Such bridges, designed in ignorance of Continental practice, sufficed for local road traffic and turnpikes, but they were totally unsuited to railroads. Railways carried unprecedented, but calculable, loads and demanded large numbers of bridges because their principal constraints, curvature and grade, virtually dictated that designers would have to bridge small irregularities that wagons and foot traffic handled with ease.

Two new developments answered these difficulties and became central to extension of the transportation revolution: the use of iron structural elements and of simple trigonometric design princi-

ples. Howe pioneered and developed the first innovation in 1840, using iron rods as vertical tension members, in effect bolting the horizontal and diagonal wooden members together. Competitors also turned to iron, but Howe's design was probably the most popular wood-and-iron construction in use before the Civil War. Trigonometric design principles were introduced independently between 1847 and 1851 by at least three men. Perfectly suited to truss structures, which were little more than articulated triangles, this technique of analysis enabled men like the Stones, with no engineering training, to lay out the design of a truss bridge, calculate the loads on each member, and precut timbers at a lumber pit for later assembly. A carpenter's knowledge of wood remained vital for the selection of appropriate timbers. Thus the vastly increased demand of the late 1840s for inexpensive, reliable bridges could be met, and the technology remained predominant until just before the Civil War, when builders turned increasingly to all-iron structures.

For the Stone brothers, bridge building proved to be the path to their real careers in the railroad and iron industries and in finance. Amasa, the older, committed himself to it in 1842, when he and Azariah Boody formed a company to buy the New England patent rights to Howe's truss design. The company built several railroad lines and bridges and secured Stone's reputation as a designer and construction manager. In 1844 or 1845 he solved a technical problem in the Howe design which exposed bridges to damage from shrinking timber and had threatened to make Boody, Stone's $40,000 investment worthless. His solution saved his business interests and the Howe design's future, but that exploit, combined with other technical successes, may have convinced Stone that he knew more than he actually did about bridge materials and design and thus indirectly led to a major disaster.

Andros, better known as A. B., worked as a construction superintendent for the firm until he turned twenty-one in 1845, when he began to collaborate with his brother on projects in Maine. In 1847 Boody and Amasa Stone agreed to split the patent rights and accordingly dissolved their partnership. Stone retained the southern New England rights, while A. B. Stone and Boody took the remainder. At this point the brothers' careers diverged for several years before they collaborated once again in the iron industry in Cleveland. Amasa Stone enjoyed great success, serving briefly as superintendent of the New Haven, Hartford & Springfield Railroad. He took up residence in the growing city of Springfield, and his income was large enough that he could invest in and act as a director of a bank and a cotton mill in the city.

His expanding reputation and business interests apparently took him on travels to several states, and in 1848 he formed a partnership with Frederick Harbach and Stillman Witt, who had contracted to build the Cleveland, Columbus & Cincinnati Railroad. At the time it was the largest single railroad construction contract undertaken by one firm. The "Three-C," also known as the "Bee Line," had originally been chartered in 1836 during a national burst of railroading enthusiasm, but its prospects were dashed by the panic of 1837 and the ensuing depression. When revived in 1845, the project had the backing of several of Cleveland's most influential business and political leaders, including Alfred Kelley, who had spearheaded construction of the Ohio Canal. The Three-C was planned to be a key link in the railroad system that connected the Northeast with the midwestern and central states.

Stone arrived in Cleveland just as the city's economy was turning to industry, and he had an important hand in its growth. Though Moses Cleaveland had chosen the site as the capital of "New Connecticut"–the Western Reserve's original designation–it remained a modest lake port of no special importance until 1825. But in that year and the years following, the Ohio Canal terminus made it a major shipping and wholesaling center and northern Ohio's main entrepôt. Its economic life revolved around transshipping agricultural produce from a large hinterland to eastern markets and acting as a port of entry for goods imported from eastern centers. Cleveland's business leaders, like those of other commercial cities, aggressively developed railways during the 1840s in large part to prevent rivals in other cities from taking control of their marketing region. Additionally, in the decade before the Civil War, Cleveland businessmen began to augment their commercial pursuits by turning to manufacturing. These efforts were made possible by exploitation of new coal mines near Youngstown after 1845; Cleveland lacked waterpower and needed an inexpensive fuel if it was to take up manufacturing. In the early 1850s an iron industry became feasible when Charles Whittlesey, John Lang

Cassels, and others located and opened iron ore ranges in the upper Great Lakes and secured them for Cleveland businessmen. Thereafter, primary iron production was added to an existing small milling industry, and Cleveland's commercial and industrial leaders were to find Stone a valuable and willing partner in these enterprises.

Originally he was to have confined his role in the Three-C to that of financier and labor contractor, but in 1850 he moved permanently to Cleveland to supervise the work himself. His part in the railroad's construction also led him indirectly into ownership and management. Like many builders of other roads, he had approached the Three-C contract as a speculation; his firm did part of the work in return for stock rather than payments. This arrangement made him a major stockholder, and after he finished building the road he accepted its superintendency as well.

A similar strategy allowed him to extend his holdings eastward and westward into what has been called an "empire." Before he completed work on the Three-C, he and his partners had also contracted to build the vital link between Cleveland and Erie, Pennsylvania, that would tie the Three-C with the New York Central system. Stone accepted the superintendency of that line–the Cleveland, Painesville & Ashtabula (CP&A)–when construction was finished, and was also its president from 1857 to 1869. In the latter year the CP&A was merged into a new line, the Lake Shore & Michigan Southern, which linked Erie with Toledo; Stone was the managing director of the new road. Because of the technical difficulties involved in building a line close to the lakeshore, he took greatest pride and interest in the CP&A. His last railroad construction contract was to build the Chicago & Milwaukee line in 1855, and he was its president and a director at various times. He was to have other business interests in Chicago as well. Although he resigned active superintendency of the roads in 1854, his empire continued to grow. He sat on the boards of directors of several railroads in addition to the Three-C, the Lake Shore, and the Chicago & Milwaukee. He was also a director of several banks in northern Ohio, founder of a textile mill, and a heavy investor in iron production.

In these investments he resumed his collaboration with his brother. A. B. Stone had closed out his New England business interests in 1852 and moved to Chicago, where he and his partners continued to build Howe bridges for the many new railroads then being constructed in Illinois. At the same time he also began constructing rolling stock for which iron fittings and equipment were very important. Amasa Stone's own shops were also building cars for a number of railways, and the brothers' mutual interest in the iron industry was strong. In 1860 A. B. Stone moved and they invested in the Cleveland firm of Chisholm & Jones, renaming it Stone, Chisholm & Jones. The former company, formed in 1857 to reroll worn iron rails, had invested in a blast furnace and needed more capital to expand its iron-producing and -rolling capacity. In 1864 the firm was recapitalized and incorporated as the Cleveland Rolling Mill Company, with the Stone brothers as major stockholders; Amasa Stone and Henry Chisholm together owned nearly half the firm's shares. Under usual conditions Stone's railroad companies would be a major source of business for the mill. A. B. Stone served as the company's president until 1878, when he moved to New York as vice-president to look after its eastern interests. Under the day-to-day leadership of Henry Chisholm, the firm was responsible for many technical innovations in the use of iron and Bessemer steel.

The Stone brothers developed an extremely complex network of business interests which were integrated through interrelated ownership involving many associates, including the Chisholm family, Stillman Witt, and Jephtha Wade. Of the Stones, A. B. was the more active in iron and steel, and Amasa prevailed in financial matters. As such, however, Amasa Stone was the linchpin in the Stone-Chisholm industrial empire. Together with the Chisholms, who were technically more knowledgeable, the Stone brothers invested in 1863 in a branch plant in Chicago, the Union Rolling Mill Company, of which A. B. was the nominal president until Henry Chisholm's son William succeeded him. In 1871 they invested along with the Chisholms in Union Steel Screw Company to exploit William Chisholm's development of a technique of making screws from Bessemer steel. A. B. Stone also had major holdings in the Kansas Rolling Mill Company, and, like Amasa, in railroads and other heavy users of iron products. Several Stone associates, including Fayette Brown, operated or controlled ore beds in the Lake Superior region which were thus accessible to the producing companies. One associate, Samuel Mather, married a daughter of Amasa Stone. These interlocking holdings controlled a vast commerce in ore, shipping, iron and steel production, and rail transportation. Its value, though difficult to quantify, was immense.

By modern standards, it must be said, some of Stone's business practices were questionable, although perhaps no different from those common during the era. The use of family members and business associates to form a network of interlocking corporate relations is one such practice. Again, as managing director of the Lake Shore, Stone agreed to give rebates to Cleveland petroleum shippers in order to counter the price advantages of Pennsylvania refiners. Although he did not discriminate among Cleveland shippers at that time, he and Stillman Witt made large investments in Standard Oil just prior to Rockefeller's organization of the South Improvement Company in 1872. South Improvement was used by Standard Oil in an attempt to gain control of the oil industry by controlling freight rates, a plan which depended on cooperation by the railroads. The Lake Shore agreed to the scheme of discriminatory rates and rebates, and Stone was widely believed to be a director of South Improvement.

In addition to his correction of the Howe design's shortcomings, Amasa Stone was responsible for some innovations in the use of truss architecture and in railroads. He is credited with the first long-span swing bridge and with the creative use of truss designs to roof large floor spaces. He built the former lakefront Union Terminal in Cleveland, a large pillarless space, to his own design. He also has been credited with innovations in railroad car design, including an 8-wheeled gravel car. But Stone's faith in his untutored powers of design was probably responsible for the second of two major tragedies in his life. The first, in 1865, was the death of his only son, Adelbert, who drowned in the Connecticut River while a student at Yale College. The other, a notorious disaster, occurred in December 1876 near Ashtabula. When the CP&A was built, Stone used his authority as president of the road to insist that the tracks be carried over the 75-foot-deep Ashtabula River gorge on an all-metal Howe truss bridge. In doing so he defied the advice of his engineers, who warned of safety problems. They were proved right when a train pitched off the bridge in a violent blizzard, killing 85 people. The Ohio General Assembly's investigation blamed

Stone for using an experimental design, and awareness of his responsibility for the collapse caused a decline in his health. He was plagued by nightmares for the rest of his life; active in the Presbyterian church, he must have regarded this tragedy as a severe trial of conscience and faith. His death was by suicide.

Central to Stone's life as an entrepreneur was a combination of fierce independence and, later, a powerful sense of responsibility. The tendency toward domination which led him to overrule his engineers was a fixed facet of his personality; he was said to be unable to work as anyone's subordinate. Lacking personal pretensions, he impressed others as energetic, physically robust, highly self-confident, and overbearing in business. Later in his career, however, the strong sense of responsibility predominated. His reaction to the bridge disaster can be attributed partly to it, and by then he preferred to style himself a philanthropist. In 1874 he began a systematic program of benefaction to aid homeless children and aged women. During the same time he financed Western Reserve College's relocation from Hudson to Cleveland and granted it a large endowment in his son's memory. He could also be prevailed upon to assume responsibilities he preferred to avoid, as when Commodore Vanderbilt convinced him to resume direction of the Lake Shore after he had already retired, and when President Lincoln inveigled him into consulting on problems of wartime military transportation.

In addition to his personal qualities, Stone's success also owed much to good fortune. He was trained in a craft for which great demand erupted just as he reached adulthood, and his family connections, particularly with his brother A. B. and brother-in-law Howe, enabled him to exploit that craft to reach a position of prominence and great wealth. Subsequent family connections enlarged his business and political reach. He was also fortunate in his choice of business associates and in those who chose to associate with him. All these factors together led him to become a central figure in one of the major steel, transportation, and financial empires of the Old Northwest.

References:

George M. Danko, "The Evolution of the Simple Truss Bridge 1790-1850: From Empiricism to Scientific Construction," Ph.D. dissertation, University of Pennsylvania, 1979;

John Hay, "Amasa Stone," *Magazine of Western History*, 3 (1885-1886): 108-112;

William Howe, *Historical Collection of Ohio*, centennial edition (Cincinnati, 1902);

Maurice Joblin, *Cleveland, Past and Present* (Cleveland: World, 1869);

Crisfield Johnson, compiler, *History of Cuyahoga County, Ohio* (Cleveland: World, 1879);

James Harrison Kennedy, *A History of the City of Cleveland: Its Settlement, Rise, and Progress, 1796-1896* (Cleveland: World, 1896);

William Granson Rose, *Cleveland: The Making of a City* (Cleveland & New York: World, 1950);

David B. Steinman and Sara R. Watson, *Bridges and Their Builders* (New York: Putnam's, 1941);

Ida Tarbell, *The History of the Standard Oil Company*, 2 volumes (New York: McClure, Phillips, 1904).

Archives:

Material relating to Amasa Stone, Jr., is located in the Samuel Mather family papers at the Western Reserve Historical Society, Cleveland, Ohio.

George L. Stuntz

(December 11, 1820-October 24, 1902)

by Terry S. Reynolds

Michigan Technological University

CAREER: Surveyor, prospector, and merchant (1842-1902); consultant, Northern Pacific Railroad (1869-1873).

George L. Stuntz, the surveyor-prospector who initiated the chain of events that led to the opening of the first of Minnesota's iron ranges, was born in Erie County, Pennsylvania, on December 11, 1820. After a common-school education Stuntz continued his studies in 1840 at the Grand River Institute in Ohio, where he took a 2-year course in mathematics, chemistry, engineering, and surveying. He then taught school for several years in Illinois and worked as an axman on a survey of Keokuk County, Iowa. He next moved to Wisconsin, where in 1847 he was elected surveyor and later sheriff of Grant County. In the late 1840s he secured a position as a deputy U.S. surveyor. For the next 50 years Stuntz worked off and on as a surveyor on both government and private contracts.

In 1852 George Sargent, surveyor general of the Iowa, Wisconsin, and Minnesota District, sent Stuntz to run land lines and subdivide townships in northwestern Wisconsin and northeastern Minnesota. In 1853 Stuntz settled in what is now Duluth, Minnesota; he was the first settler of the community. In the mid 1850s he established a trading post and a sawmill there, operating them until 1858.

In 1865 Stuntz was one of the first to seek gold when it was reported on Vermilion Lake, some 80 miles north of Duluth. He failed to find gold but while prospecting collected samples of iron ore. In 1868 Stuntz received a contract to construct a state road to the Vermilion Lake area, and in 1869 he completed it with a federal appropriation. As a result of his surveying and road-building work Stuntz, by the late 1860s, had become an expert on northeastern Minnesota's wilderness. Thus when Jay Cooke began work on his Northern Pacific Railroad in 1869, he retained Stuntz as a consultant on

George L. Stuntz

land, timber, and mineral values of the country the Northern Pacific was to traverse.

Between 1870 and 1873 Stuntz worked on railway and land surveying for Cooke. In the same period he became acquainted with George C. Stone, a Duluth businessman. Stone's business ventures were destroyed by the panic of 1873, but through Stuntz and others Stone had become convinced that large iron ore deposits lay in the interior of northeast Minnesota. With ore samples provided by Stuntz, Stone set out to reestablish his fortune by interesting eastern capital in Minnesota iron. In 1875 Stone secured the support of Charlemagne Tower, a

Pennsylvania capitalist. With Tower's financial backing Stone retained Stuntz to guide Albert Chester and others on exploratory expeditions to the Mesabi and Vermilion iron ranges in 1875 and 1880-1881. The first operating Minnesota iron mine, the Soudan, began production near Vermilion Lake on the Vermilion Range in July 1884 at a spot which Stuntz had first noted in 1865 and examined in more detail in 1875.

Between June 1880 and October 1882, retained by Stone and Tower for iron exploration but officially operating as a government surveyor, Stuntz explored and laid out several townships south and east of Vermilion Lake. In 1882 Stuntz also conducted surveys for the railroad (the Duluth & Iron Range) which was to connect Tower's Soudan Mine to Lake Superior shipping. After 1882 Stuntz's active role in iron exploration and promotion ended. His later years were devoted to studying aboriginal artifacts in Minnesota. He died at the Red Cross Hospital in West Duluth, October 24, 1902, a poor man.

One of the first to suspect the presence of large iron ore deposits on the Vermilion and Mesabi ranges, Stuntz played an important role as a prospector, explorer, and promoter in opening Minnesota's iron ranges. His name is perpetuated in Stuntz Township near Hibbing and Stuntz Bay on Lake Vermilion.

Unpublished Document:

Burleigh K. Rapp, "The Life of George L. Stuntz," 101-page MS, 1958, in Minnesota Historical Society, St. Paul, Minnesota.

References:

"George L. Stuntz and Stuntz Township in Minnesota," *Skillings' Mining Review*, 50 (June 24, 1961): 1, 4, 26;

Walter Van Brunt, *Duluth and St. Louis County, Minnesota: Their Story and People*, 3 volumes (Chicago & New York: American Historical Society, 1921);

David A. Walker, *Iron Frontier: The Discovery and Early Development of Minnesota's Three Ranges* (St. Paul: Minnesota Historical Society, 1979);

Dwight E. Woodbridge and John S. Pardee, editors, *History of Duluth and St. Louis County*, volume 1 (Chicago: C. F. Cooper, 1910).

James Moore Swank

(July 12, 1832-June 21, 1914)

by Brady Banta

Louisiana State University

CAREER: Editor, *Cambria Gazette,* renamed as the Johnstown *Tribune* (1852-1869); clerk, House Committee on Manufacturers (1869-1871); clerk, Department of Agriculture (1871-1873); secretary (1873-1885), vice-president and general manager (1885-1912), American Iron and Steel Association.

Born in Westmoreland County, Pennsylvania, on July 12, 1832, James Moore Swank grew up in Johnstown, Pennsylvania. His parents, George and Nancy Swank, oversaw his primary and secondary education. One year at Jefferson College in Canonsburg, Pennsylvania, completed his academic career, after which he taught school, clerked in his father's store, and briefly studied law.

The real break in Swank's career advancement came in 1852 when he assumed the editorship of the local Whig newspaper. Finding journalism challenging and satisfying, Swank resurrected the *Cambria Gazette,* a weekly Whig newspaper that had suspended publication. Under his leadership the paper evolved into the *Johnstown Tribune,* a successful daily publication.

Swank's association with the *Tribune,* intermittent until he emerged as its sole owner in 1864, lasted until December 1869. At that point Swank sold the paper to his brother and moved to the nation's capital to become clerk of the House Committee on Manufacturers, a position he obtained through friendship with Pennsylvania representative Daniel J. Morrell, chairman of the committee. Unfortunately, Morrell lost his bid for reelection in 1870, and the end of his term in 1871 terminated Swank's service to the House Committee on Manufacturers.

James Moore Swank

Seeking other opportunities in government service, Swank obtained an appointment as a fourth-class clerk in the Department of Agriculture. Assigned to the division of statistics, Swank advanced within six months to the position of chief clerk. His principal accomplishment was the writing of *The Department of Agriculture: Its History and Objects* (1872), the first history of that government department.

More importantly, while with the Department of Agriculture Swank developed the statistical skills that, along with his practical journalistic experience, underlay his career as secretary of the American Iron and Steel Association. On January 1, 1873, Swank inherited the secretaryship of an organization in disarray. He found that the association's management had neglected the gathering of trade statistics, and Swank immediately undertook a project to remedy this situation. In November 1873 his

first annual report to the association included a compilation of blast furnaces and rolling mills in the United States along with their production statistics. This report evolved into the *Directory of the Iron and Steel Works of the United States*, 17 editions of which appeared intermittently during Swank's tenure as secretary.

The initial volumes of this series established Swank's international reputation as a statistician and as an expert on the growth of the American iron and steel industry. This recognition led to a request by the Pennsylvania Board of Centennial Commissioners that Swank prepare a historical account of the state's iron and steel industries. Subsequently, in 1878 Swank published *Introduction to a History of Ironmaking and Coal Mining in Pennsylvania*. The appearance of this volume prompted Gen. Francis A. Walker, superintendent of the 1880 census, to engage Swank to gather statistical data on the iron and steel industry and to provide historical sketches of iron making in colonial America and the industry's development in each state and territory. Swank's pursuit of this project resulted in the publication in 1884 of the *History of the Manufacture of Iron in All Ages*. Having continued to compile data, in 1892 Swank published a second edition of this work. These volumes provide an authoritative account of the iron and steel industry and are still valuable sources of information on the industry.

Swank's publications furnished a legacy to the iron and steel industry. The esteem in which he was held by colleagues, however, stemmed from his indefatigable advocacy of tariff protection for iron and steel. Swank regularly appeared before congressional committees to oppose tariff-for-revenue legislation that would have lowered the protective rates on iron and steel. Additionally, during the 1880s Swank orchestrated the distribution in the South and West of millions of protariff tracts. Swank and other key protectionists designed these educational campaigns to counter what they considered to be antitariff editorial policies found in the newspapers of these regions. As secretary of the American Iron and Steel Association, Swank furnished key leadership in all of these efforts. His contemporaries acknowledged Swank's unrivaled contribution and accorded him considerable credit for the fact that the iron and steel industry did not lose a tariff battle from 1873 to 1909.

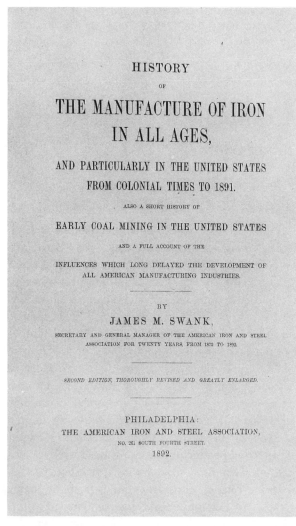

HISTORY

OF

THE MANUFACTURE OF IRON IN ALL AGES,

AND PARTICULARLY IN THE UNITED STATES FROM COLONIAL TIMES TO 1891.

ALSO A SHORT HISTORY OF

EARLY COAL MINING IN THE UNITED STATES

AND A FULL ACCOUNT OF THE

INFLUENCES WHICH LONG DELAYED THE DEVELOPMENT OF ALL AMERICAN MANUFACTURING INDUSTRIES.

BY

JAMES M. SWANK,

SECRETARY AND GENERAL MANAGER OF THE AMERICAN IRON AND STEEL ASSOCIATION FOR TWENTY YEARS, FROM 1872 TO 1892.

SECOND EDITION, THOROUGHLY REVISED AND GREATLY ENLARGED.

PHILADELPHIA:
THE AMERICAN IRON AND STEEL ASSOCIATION,
NO. 261 SOUTH FOURTH STREET.
1892.

Title page of Swank's most important work on the iron and steel industry

Selected Publications:

Early Iron Enterprises In Cambria, Somerset, Westmoreland, and Indiana Counties (Pittsburgh: W. S. Haven, 1857);

Essays and Sketches (Washington, D.C.: H. Polkinhorn, 1863);

The Department of Agriculture: Its History and Objects (Washington, D.C.: GPO, 1872);

The American Iron Trade in 1876 (Philadelphia: American Iron and Steel Association, 1876);

Country Days (Philadelphia: Porter & Coates, 1876);

The Industrial Policies of Great Britain and the United States (Philadelphia: American Iron and Steel Association, 1876);

Alexander Chesterfield Mullin (Philadelphia, 1878);

Introduction to a History of Ironmaking and Coal Mining in Pennsylvania (Philadelphia: Privately printed, 1878);

Linton-Lacock, 1831-1881 (Philadelphia, 1881);

Preliminary Report Upon the Iron and Steel Industries of the United States in the Census Year 1880 (Philadelphia: American Iron and Steel Association, 1881);

Footprints of the English Lion (Philadelphia: American Iron and Steel Association, 1882);

The Tariff on Iron and Steel Justified by Its Results (Philadelphia, 1882);

History of the Manufacture of Iron in All Ages (Philadelphia: Privately printed, 1884; revised, Philadelphia: American Iron and Steel Association, 1892);

A Bird's Eye View of the Production and Characteristics of Iron Ores in the United States (Philadelphia: American Iron and Steel Association, 1885);

The American Iron Industry From Its Beginning in 1619 to 1886 (Washington, D.C., 1886);

Twenty-one Years of Progress in the Manufacture of Iron and Steel in the United States (Philadelphia, 1886);

Papers Relating to Iron and Steel and Iron Ores, 1885 to 1889 (Philadelphia: American Iron and Steel Association, 1887);

The Iron and Steel Industries of the United States in 1887 and 1888 (Philadelphia: American Iron and Steel Association, 1888);

Our Bessemer Industry (Philadelphia, 1888);

Iron and Steel Industries of Pennsylvania (Philadelphia: American Iron and Steel Association, 1891);

The American Iron Trade Built Up By Protective Duties Since 1860 (Philadelphia: American Iron and Steel Association, 1895);

Our Tinplate Industry Built Up By the McKinley Tariff (Philadelphia: American Iron and Steel Association, 1896);

Present Conditions of the Iron and Steel Industries of the United States (Washington, D.C.: GPO, 1896);

The American Iron Trade in 1897 and Immediately Preceding Years and the Foreign Iron Trade in 1897 and Immediately Preceding Years (Washington, D.C., 1897);

Iron and Steel and Allied Industries in all Countries (Washington, D.C., 1897);

Notes and Comments on Industrial, Economic, Political and Historical Subjects (Philadelphia: American Iron and Steel Association, 1897);

American and Foreign Iron Trades in 1899 (Washington, D.C.: GPO, 1900);

Iron and Steel at the Close of the Nineteenth Century (Washington, D.C., 1901);

Iron and Steel, Iron Ore, and Coal Statistics for the United States, Great Britain, Germany, France, and Belgium (Philadelphia: American Iron and Steel Association, 1902);

Progressive Pennsylvania (Philadelphia: Lippincott, 1908);

Cambria County Pioneers (Philadelphia: Allen, Lane & Scott, 1910);

Our Iron Industry and the Protective Policy (Philadelphia: American Iron and Steel Association, 1911).

References:

"Death of James M. Swank," *Iron Age*, 93 (June 25, 1914): 1582-1582;

John Bruce McPherson, "James Moore Swank: Protectionist," *Bulletin of the National Association of Wool Manufacturers*, 43 (September 1913): 260-273;

A. T. Volwiler, "Tariff Strategy and Propaganda in the United States, 1887-1888," *American Historical Review*, 36 (October 1930): 76-96.

Tennessee Coal, Iron & Railroad Company

by Brady Banta

Louisiana State University

In 1907 the United States Steel Corporation absorbed the Tennessee Coal, Iron & Railroad Company, completing the transfer to northern capitalists of what had developed into the South's largest coal, iron, and steel producer. Tennessee Coal, Iron & Railroad's roots date back to 1852 when entrepreneurs in Nashville, Tennessee, organized the Sewanee Mining Company. Nashville endured chronic coal shortages, as river-borne supplies from western Pennsylvania were seldom plentiful and always expensive. Recognizing an opportunity for profit, the Nashvillians attracted investment capital from New York financiers, purchased over 20,000 coal-bearing acres on the Cumberland Plateau, opened two mines, and constructed and equipped a rail spur.

The local entrepreneurs underestimated, however, the cost of these undertakings. Moreover, title litigation affected more than 25 percent of their land, including one of the mine sites. With the Sewanee Mining Company facing bankruptcy, in 1859 the New York investors forced the liquidation of its assets, a portion of which they obtained and reorganized as the Tennessee Coal & Railroad Company. This effort to restore profitability collapsed as the Civil War erupted. Finding themselves unable to conduct business in Tennessee after secession, the New Yorkers suspended operation and invested the company's funds in United States government securities.

The company's facilities were not, however, inactive during the war. In 1859 local creditors, represented by Arthur St. Clair Colyar, had instituted state court proceedings to gain control of the Sewanee Mining Company. Although victorious, previously decided federal court action favoring the New York investors preempted any attempt to settle their claim. But Tennessee's secession in 1861 terminated federal court jurisdiction in the state, and

Colyar once again advanced the consolidated local claims, emerging as the president of a resurrected and reactivated Sewanee Mining Company. Upon the war's end, therefore, the company and its property had two claimants. In subsequent negotiations the New York interests relinquished their claims, and Colyar became the head of a reorganized Tennessee Coal & Railroad Company.

Over the next decade the Tennessee Coal & Railroad Company became the state's leading coal mining enterprise. To attract industrial consumers Colyar kept prices low. While this policy facilitated industrial development, it limited sales receipts without affecting production- and expansion-related obligations. Moreover, Colyar's low-price policy exacerbated a financial crisis stemming from the panic of 1873. To free himself and his company from this dilemma, in 1876 he sold his Tennessee Company stock to Thomas O'Connor, William Morrow, and William Cherry.

Thomas O'Connor emerged as the dominant new owner, and the five years under his direction were a period of transition. Colyar had developed the company into Tennessee's leading coke producer. In an effort to stimulate coke sales, O'Connor led the company into pig iron production by participating in the construction of a 30-ton blast furnace near Cowan, Tennessee. As the nation recovered from the panic of 1873, coal mining and iron production were favorite investments among capitalists and speculators. Active in both endeavors the Tennessee Coal & Railroad Company was a doubly attractive investment opportunity.

Never having lost interest in the Tennessee Company, Colyar orchestrated an investment strategy that culminated with the emergence of John Hamilton Inman as the majority stockholder in a reorganized corporation, the Tennessee Coal, Iron & Railroad Company. A finance capitalist and ruth-

The Tennessee Coal, Iron & Railroad works in Birmingham, Alabama (from the Photographic Collections of Birmingham Public Library, Birmingham, Alabama)

less speculator with little interest in operational details, Inman delegated management authority to Nat Baxter, Jr., and Alfred Montgomery Shook, as president and general manager, respectively. Freed of daily responsibilities Inman directed his efforts toward expanding pig iron production. Unfortunately, this step coincided with a declining demand for iron products that lasted for several years. The Tennessee Company soon found itself in a precarious situation, as it had borrowed extensively to finance this expansion. Desperate to raise capital, the owners turned to the public, listing the company's stock on the New York Stock Exchange. As they had hoped, stock prices rose. What they could not anticipate, however, was that an investor, William Duncan, through purchases and arranging proxies, had obtained voting rights to a majority of the stock and was able to take control of the company.

Duncan retained Baxter as president, but his dismissal of Shook as general manager opened the way for major changes in the company's direction. Having lost his position with the Tennessee Company, Shook cultivated a business relationship with Thomas Hillman, a disenchanted minority stockholder in the Pratt Coal & Iron Company of Birming-

ham, Alabama. Hillman's Alice Furnace Company had been absorbed by Pratt Coal & Iron, one of the South's largest coal and iron companies, during an expansion program directed by its energetic owner, Enoch Ensley. Although this transaction had made Hillman the second largest holder of Pratt stock, he had no effective voice in its management. Hillman wanted more authority, and he and Shook devised a plan to achieve that goal through consolidation of Pratt Coal & Iron with the Tennessee Company. Shook, with Hillman's blessing, approached Baxter and Inman with a complicated stock boosting and option scheme through which they could dominate Pratt Coal & Iron and merge it into the Tennessee Company. A speculator of long standing, Inman reacted enthusiastically, and by fall 1886 a once again Inman-directed Tennessee Company, with Enoch Ensley as its new president, had absorbed Pratt Coal & Iron and become the South's largest steel and iron producer.

The Pratt transactions brought the Tennessee Company into Alabama and into competition with new rivals, the largest of which was the De-Bardeleben Coal & Iron Company. Organized in 1886 by Henry Fairchild DeBardeleben to pursue mining and industrial opportunities in northern Alabama, DeBardeleben Coal & Iron owned vast coal and iron acreages and operated seven blast furnaces, four of which were concentrated at Bessemer, Alabama. The Tennessee-DeBardeleben competition coincided, however, with a significant national decline in basic iron prices. Seeking to avoid mutual destruction, in 1892 the Tennessee Company's management negotiated the purchase of the DeBardeleben Company. This transaction nearly doubled the size of the Tennessee Company and made it the nation's third largest pig iron producer.

The company's management anticipated a prosperous future, but the panic of 1893 and the ensuing depression dramatically reduced the demand for pig iron. Renewed prosperity returned when the company began to produce steel; in 1896 experiments with an open-hearth furnace produced acceptable steel from Tennessee Company pig iron. Believing steel to be their economic salvation, the directors began liquidating the company's railroad holdings to fund the new operation. This decision ignited a period of unparalleled activity as the Tennessee Company became a diversified organization producing coal, coke, iron, and steel.

While steel production received considerable attention, a quieter but ultimately more significant change began in the early 1890s as northern capitalists acquired large blocks of the company's stock. John Inman's death in 1896 brought control of the company to New York financiers. Believing that the current management was not adequately conducting the business, the New York owners hired Don H. Bacon, former president of the Minnesota Iron Company, to run the company. But Bacon was unable to mold this southern establishment into his type of operation. His controversial tenure with the Tennessee Company concluded in 1906 when new financiers, headed by the flamboyant John W. Gates, assumed control of the company's board.

Committed to expanding steel production, the Gates group significantly increased the company's indebtedness. Unfortunately, this expansion occurred as fundamental deficiencies threatened to undermine the American economy. Caught in the middle of this dislocation was the New York brokerage firm of Moore & Schley, the solvency of which rested on extensive loans secured with stocks and bonds of the Tennessee Coal, Iron & Railroad Company. In fall 1907, as the so-called panic of 1907 approached a crisis, these loans were maturing. Unfortunately, the vast block of Tennessee Company obligations held by Moore & Schley could not be sold at prices sufficient to cover these loans.

The vulnerability of Moore & Schley reached crisis proportions over the first weekend of November. On Friday morning, November 1, J. Pierpont Morgan learned that the brokerage firm would likely fail on Monday unless it could be refinanced before the New York Stock Exchange opened at 10:00 A.M. Hoping to blunt the growing panic by saving the brokerage firm, Morgan offered to loan it $5 million, but the directors of Moore & Schley rejected his proposal as insufficient. Morgan next urged the United States Steel Corporation to purchase the Tennessee Company, thus substituting its stocks and bonds for the unmarketable Tennessee Company securities held by Moore & Schley.

Confronted on Friday evening with Morgan's plan to save Moore & Schley, Elbert Gary, the chairman of the United States Steel Corporation, rejected the proposal. Gary appreciated the financial dilemma, but he also understood that the Tennessee Coal, Iron & Railroad Company had been only marginally profitable, and he seriously doubted its future prospects. Morgan overcame Gary's reluctance by stressing the gravity of the immediate problems. But Gary still had a serious reservation: Would the federal government view United States Steel as a monopoly and consider its absorption of a competitor, even if defended as an altruistic attempt to prevent a financial crisis, a violation of the Sherman Act?

Unwilling to act until put at ease, Gary hurriedly journeyed to the nation's capital to place his concerns before President Theodore Roosevelt. At their early morning meeting on Monday, November 4, Gary described a scenario in which United States Steel would purchase Tennessee Coal, Iron & Railroad Company in order to save an unnamed major financial institution. Understandably worried about the worsening economic situation, but also careful to avoid making a statement that might be construed as binding on the government, Roosevelt expressed no opposition to the purchase. Thus reassured, Gary completed the transaction. The United States Steel Corporation became the owner of valuable southern iron ore and coal deposits, and the 55-year life of the Tennessee Coal, Iron & Railroad Company as an independent coal, iron, and steel producer came to an end.

Unpublished Document:

Justin Fuller, "History of the Tennessee Coal, Iron, and Railroad Company, 1852-1907," Ph.D. dissertation, University of North Carolina, 1966.

References:

Ethel Marie Armes, *The Story of Coal and Iron in Ala-*

bama (Birmingham, Ala.: Birmingham Chamber of Commerce, 1910);

William Henry Harbaugh, *Power and Responsibility: The Life and Times of Theodore Roosevelt* (New York: Farrar, Straus & Cudahy, 1961).

David Thomas

(November 3, 1794-June 20, 1882)

by John W. Malsberger

Muhlenberg College

CAREER: Ironworker, Neath Abbey Iron Works (1812-1817); general superintendent of blast furnaces and mines, Yniscedwyn Iron Works (1817-1839); general superintendent, Lehigh Crane Iron Company (1839-1855); superintendent of construction, Thomas Iron Works (1854-1855); founder and president, Catasauqua Manufacturing Company (1863-1879).

Although he was not the first person in America to manufacture iron using anthracite coal as a fuel, David Thomas perfected the hot-blast method which produced commercial quantities of iron using anthracite coal. Because of his innovation Thomas is generally regarded as the "father of the American anthracite iron industry." In addition to his technological contributions, Thomas also played a major role in superintending or founding some of the most important anthracite iron companies in mid-nineteenth-century America, including the Lehigh Crane Iron Company, the Thomas Iron Works, and the Catasauqua Manufacturing Company. Through both his technological and managerial talents, David Thomas contributed substantially to the development of the American iron and steel industry in the nineteenth century.

Born November 3, 1794, in Tyllwyd, of the Cadoxtan parish in Glamorganshire, Wales, David Thomas was one of four children of David and Jane Thomas. His father was a farmer of moderate means but was a respected member of his community who held the parish positions of church warden and overseer of the poor. Both of Thomas's parents were devout members of the "Independents" Religious Community at Maesyrhaf Chapel and consequently instilled in their children a strong

David Thomas

religious and moral training that emphasized the need of the individual to care for the less fortunate members of society. This training and the example set by his father as overseer of the poor apparently had so profound an impact on Thomas that as an adult he regularly devoted a considerable portion of his energies to active community service.

Because he was their only son, Thomas's parents were determined to give him the best education they could afford. When he was nine, for instance, they enrolled him in the most advanced school in their parish at a cost of 21 shillings per quarter. Throughout his school years Thomas demonstrated both a great capacity for and a great appreciation of learning, and the experience apparently awakened in him a powerful ambition to rise above his agrarian roots. After completing his formal education, Thomas worked for several years on his family's farm. But in 1812 the seventeen-year-old Welshman accepted employment in the fitting shops and at the blast furnaces of the Neath Abbey Iron Works, located near his home. After five years in this position, Thomas's mastery of the iron-making process was so complete that he was hired as the general superintendent of the blast furnaces and mines at the Yniscedwyn Iron Works in the Swansea Valley; he remained in the position for the next 22 years. The years he spent at Yniscedwyn proved to be some of the most fruitful of his life, for it was during this period that Thomas perfected the method for manufacturing anthracite iron.

Like so many of the technological breakthroughs that altered the nineteenth-century iron industry, Thomas's experiments with anthracite coal arose out of practical concerns. Although the Yniscedwyn Works were located on the only deposit of anthracite coal in Great Britain, when Thomas was hired the company's blast furnaces were still fueled by coke that had to be transported 15 miles. In an effort to reduce the cost of producing iron by making use of the abundant coal deposits, Thomas began experimenting in 1820 by adding a small proportion of anthracite coal to the coke in the furnaces. Thomas continued the experiments over the next five years by varying the proportions of coal and coke. These early efforts succeeded in producing commercial-grade iron but failed to overcome objections by the ironworkers, who blamed the coal for every problem encountered. Between 1825 and 1830 Thomas used different amounts of coal to fuel a small blast furnace he had built especially for his experiments, but, although this method produced even higher grade iron, it was ultimately abandoned as unprofitable.

Thomas's experiments with anthracite coal finally succeeded in the late 1830s when he adapted a hot-blast furnace, that had been developed several years earlier by a Scottish inventor, James B. Neilson. Neilson's efforts had begun in 1828 when he developed a method that heated the blast before it was introduced into the furnace. By 1834 Neilson had patented his hot-blast furnace and it had been adopted by iron makers throughout Scotland and Wales who continued to use charcoal and coke as fuel in the new stove. When David Thomas learned of the new stove in 1836 he sensed immediately that the furnaces's high temperatures would permit the burning of anthracite coal. With the encouragement of George Crane, owner of Yniscedwyn Works, Thomas traveled to the Clyde Iron Works in Glasgow to observe the furnace's operations. Convinced that the new stove offered the solution to his earlier problems with anthracite coal, Thomas purchased a license from Neilson to construct a hot-blast furnace at the Yniscedwyn Works, and hired one of Neilson's mechanics to supervise construction. When the new furnace was blown in on February 5, 1837, it proved an immediate success, consistently producing 34 to 36 tons of pig iron per week, using only anthracite coal as fuel. Although David Thomas had been the individual who had experimented with anthracite coal and had recognized the importance of Neilson's furnace, George Crane, Yniscedwyn's owner, received from the British government a patent for the hot-blast smelting method for manufacturing iron.

Thomas's success in producing anthracite iron was of particular interest to the Lehigh Coal & Navigation Company, an American firm which had built the Lehigh Canal in the 1820s to connect its coalfields in northeastern Pennsylvania with the Philadelphia market. The company had begun to experiment with anthracite iron production in the early 1830s in order to diversify its product line. When the principal stockholders in Lehigh Coal & Navigation, Josiah White and Erskine Hazard, learned of Thomas's hot-blast smelting process from White's nephew, Solomon W. Roberts, a civil engineer who was in Wales to buy iron rails for Pennsylvania railroads, they sought immediately to bring this new furnace to America. In July 1838 White, Hazard, and several other principals organized the Lehigh Crane Iron Company, named in honor of George Crane of the Yniscedwyn Works. In the fall of that year Hazard traveled to Wales in an effort to persuade Crane to come to Pennsylvania to construct a furnace. When Crane demurred Hazard made a similar proposal to David Thomas, whose initial reluctance to leave his native Wales was overcome by Lehigh Crane's lucrative offer. Thus, the driving ambition that earlier had impelled Thomas to leave his family's farm for the iron industry was again apparently a factor in his decision to come to America.

The agreement he signed with Lehigh Crane Iron on December 31, 1838, pledged Thomas to construct an anthracite iron furnace and to serve as the furnace manager for at least five years. In return, the iron company was to pay Thomas's moving expenses, provide him with a house and coal to heat it, and pay him an annual salary of £200 until the first blast furnace was put into operation. Thereafter, Thomas's salary was to increase £50 for each additional blast furnace he erected. Prior to setting sail for America, Thomas had a blowing engine built in England for his new American blast furnace. Unfortunately, the hatch on his ship, the clipper *Roscius,* was too small to accommodate the cylinders of the engine, and thus, after arriving in America in June of 1839, Thomas and his new employer had to search for a foundry capable of casting cylinders with a bore of 5 feet. Because few of the foundries then operating in America had ever built machinery of this great size, Thomas's efforts were unsuccessful until he persuaded Merrick & Towne of the Southwark foundry of Philadelphia to retool its shop to produce the necessary cylinders. As a result of the delay, ground was not broken for Thomas's first blast furnace at Lehigh Crane Iron until August 1839, and work on it was not completed until July 3, 1840. The first cast of iron, amounting to about 4 tons, was made the following day, July 4, 1840, and marked the first blast furnace in America capable of producing commercial quantities of iron using only anthracite coal as fuel.

Built near present-day Catasauqua, Pennsylvania, Thomas's first American furnace was situated on the banks of the Lehigh Canal in order to utilize waterpower. It closely resembled the charcoal furnaces then prevalent in the iron industry except for the addition of the hot-blast stove. Placed next to the furnace, the stove consisted of four chambers fired by coal, each with 12 arched pipes of 5-inch diameter connected together so that as cold air was forced through the pipes it was heated to approximately 600 degrees Fahrenheit before entering the furnace. Changes of 8 feet in the water level of one of the canal's locks provided the power to drive the furnace's two blowing cylinders, each of which were 5 feet in diameter and had a 6-foot stroke. The waterpower was translated into mechanical power by means of a breast wheel 12 feet in diameter and 24 feet long, geared by segments on its circumference to a spur-wheel attached to the blowing cylinders.

The furnace built by David Thomas for Lehigh Crane Iron remained in continuous blast until August 1842, except for a 5-month period between January and May 1841, when its fires were extinguished by a flood on the Lehigh River. In this period Thomas's furnace produced 4,396 tons of pig iron, with 52 tons the largest weekly output. Encouraged by this success David Thomas erected four additional blast furnaces for Lehigh Crane Iron in the 1840s. Each of these additional furnaces drew heavily on the technological abilities of the Welsh ironmaster because it was recognized that the canal's waterpower was insufficient for more than two furnaces. Although furnace #2, which was put into operation in November 1842, was powered by water turbines, furnace #3, built in 1845 and 1846, as well as furnaces #4 and #5, constructed in 1849 and 1850, were powered by steam engines. Thus, when Thomas resigned his position as general superintendent of Lehigh Crane Iron in 1855 he could look back at his first 16 years in America with great satisfaction. Not only had he succeeded in developing Lehigh Crane Iron into one of the leading iron producers in the United States but also by perfecting the method for anthracite iron production, David Thomas had contributed substantially to the development of the American iron industry.

For the next 25 years, Thomas's ambition led him into other important ventures within the iron industry, although he continued to maintain a close connection with Lehigh Crane Iron. In 1855, for instance, Thomas oversaw the construction of the furnaces for the Thomas Iron Works in Hokendauqua, Pennsylvania, which was organized by a group of local investors and named in honor of the Welsh ironmaster. When the furnaces were put into production in late 1855, they were the largest and most productive anthracite iron furnaces in America. Two years later Thomas helped to organize and thereafter assumed the presidency of the Catasauqua & Fogelsville Railroad, which was built as a joint effort by Lehigh Crane Iron and the Thomas Iron Works to connect their production facilities with iron ore beds in rural Lehigh County. In addition, David Thomas founded the Lehigh Fire Brick Works in 1868 to manufacture bricks used to line the blast furnaces of the local iron mills.

The most important venture in which Thomas became involved in the years following his resignation from Lehigh Crane Iron was the founding and managing of one of the leading rolling mills in the Lehigh Valley region. In 1863 Thomas, together with several local investors, established a firm, subsequently named the Catasauqua Manufacturing Company, to produce armor plate for naval vessels during the Civil War. Although the war ended before Thomas's rolling mill was able to produce any armor plate, the Catasauqua Manufacturing Company, with Thomas as its president, expanded rapidly to become one of the region's major producers of flue- and boilerplate, as well as bar and skelp iron. By the time David Thomas retired from the presidency in 1879, his Catasauqua Manufacturing Company operated rolling mills in Ferndale, Pennsylvania (now Fullerton), and Catasauqua, Pennsylvania, and had a combined annual output of 36,000 tons of iron and a work force of 600.

Despite the numerous business and industrial ventures with which he was involved, Thomas, influenced by his parents' teachings and example, gave freely of himself to improve the life of his adopted American community. When, for instance, the borough of Catasauqua was incorporated in 1853 Thomas was chosen as its first burgess and held the position for many years. He also actively promoted the construction of homes and helped found the first Presbyterian Church in Catasauqua so that the local ironworkers would have decent places both to live and to worship. Indeed, so strong was Thomas's concern for the welfare of his town that its residents affectionately referred to him as "Father Thomas."

The Welsh ironmaster's concern for the community welfare was not, however, limited to his immediate town. He served as a trustee of Lafayette College in Easton, Pennsylvania, and as a trustee of St. Luke's Hospital in Bethlehem, Pennsylvania. Additionally, he ran for Congress as a Republican candidate in 1866 but failed to win election, mainly because he refused to campaign actively for himself.

Thomas married Elizabeth Hopkins in 1817, and they had five children: Samuel, John, David, Jr., Jane, and Gwenllian. His three sons were all prominently connected with the American iron industry.

References:

Craig Bartholomew, "Anthracite Iron Making and Industrial Growth in the Lehigh Valley," *Proceedings,* Lehigh County Historical Society, 32 (1978): 129-183;

James F. Lambert and Henry J. Reinhard, *A History of Catasauqua in Lehigh County Pennsylvania* (Allentown, Pa.: Searle & Dressler, 1914);

Alfred Matthews and Austin N. Hungerford, *History of the Counties of Lehigh and Carbon* (Philadelphia: Everts & Richards, 1884);

Ed Roberts, "The Late David Thomas, Catasauqua, Pa., The Father of the Anthracite Iron Trade," *The Cambrian,* 4 (April 1885): 104-108; 5 (May 1885): 133-135;

Samuel Thomas, "Reminiscences of the Early Anthracite-Iron Industry," *Transactions of the American Institute of Mining Engineers,* 29 (1899): 901-928.

Joseph Thatcher Torrence

(March 15, 1843-October 31, 1896)

by Alec Kirby

George Washington University

CAREER: Chief salesman, Reis, Brown & Berger (1865-1870); engineer, Joliet Steel Company (1870-1874); cofounder, Joseph H. Brown Iron & Steel Company (c. 1872); founder, Chicago & Calumet Terminal Railroad Company (1886).

Joseph Thatcher Torrence, an iron manufacturer, industrial innovator, U.S. army veteran, and businessman, was born on March 15, 1843, in Mercer County, Pennsylvania, the son of James and Rebecca Torrence. For reasons that are unclear, he was taken away from his parents when he was nine years old and reared by relatives in Sharpsburg, Pennsylvania. At Sharpsburg he worked for three years as a blast furnace operator. By the outbreak of the Civil War he had learned the blacksmith trade before becoming an assistant foreman in a blast furnace in Ohio.

In August 1862 Torrence volunteered for military service but was soon wounded. He retired from the Union army to take a position as chief salesman for the iron manufacturing firm of Reis, Brown & Berger in New Castle, Pennsylvania. He held this position for five years before becoming an engineer for the Joliet Steel Company in 1870. Four years later he became a consulting engineer for the Green Bay & Bangor Furnace Company. During this period he was one of the organizers of the Joseph H. Brown Iron & Steel Company, which later was merged with the Calumet Iron & Steel Company. He also was a founder of the Chicago & Calumet Terminal Railroad Company in 1886, which constructed railroad lines around the city of Chicago.

In addition to his business activities during this period Torrence was commissioned in 1874 as a colonel of the Illinois National Guard, becoming a brigadier general two years later. He became military director of Chicago during the railroad strikes of 1877. Through his military activities Torrence be-

Joseph Thatcher Torrence

came acquainted with Gen. John A. Logan and later actively supported the 1884 Blaine-Logan campaign for the presidency. Through his political connections Torrence was able to help secure a congressional appropriation for the improvement of the Calumet River in Illinois.

Torrence married Elizabeth Norton, daughter of Judge Jesse O. Norton of Chicago, on September 11, 1872. They had one daughter. Elizabeth Norton died in 1877. Torrence, who remained in Chicago for the rest of his life, had a major and lasting impact on the area's development. He died on October 31, 1896.

Charlemagne Tower

(April 18, 1809-July 24, 1889)

by Terry S. Reynolds

Michigan Technological University

CAREER: Lawyer, private practice (1836-1871); director (1873-1879), chairman, finance committee, Northern Pacific Railroad Company (1874-1879); owner, Duluth & Iron Range Railroad (1852-1887); president, Minnesota Iron Company (1882-1887).

Charlemagne Tower, the financier responsible for the opening of Minnesota's first iron range, was born in Paris, Oneida County, New York, on April 18, 1809, the oldest child of Reuben and Deborah Taylor Pearce Tower. His father was a prosperous farmer who engaged in the livestock business and owned a small distillery.

After a period in the local common schools, Tower attended the Oxford, Clinton, and Utica academies. Beginning at age fourteen he taught school in Oneida County for two years, and in 1825 he was assistant teacher in the Utica Academy. After preparatory tutoring he entered Harvard in 1827, graduating in 1830.

Shortly after his graduation from Harvard, Tower began to study law at Albany. His father's death in 1832 forced him to abandon his studies temporarily and return to Waterville, New York, in order to carry on his father's businesses. In 1835 Tower resumed his legal studies in New York City and was admitted to the bar in October 1836. He again returned to Waterville in 1837 to assist in running the family businesses.

Although he practiced law on a small scale from 1836, the failure of the family enterprises in 1843 pushed him more heavily in that direction. In 1844 he was retained by Alfred Munson, a wealthy Utica manufacturer, to investigate and clear the titles to his land holdings. Munson had potentially valuable holdings (the Munson-Williams estate) in Schuylkill County, Pennsylvania, but these lands were tied up by a morass of conflicting legal claims. Munson liked Tower and his work and in 1848 of-

Charlemagne Tower

fered him half interest in his Pennsylvania coal claims for his services. In return, Tower was to prove Munson's title and to acquire additional lands. To accomplish these tasks Tower moved to Pennsylvania in 1848. There he worked for 25 years on title problems and coal-land purchases, acquiring in the process a high reputation as an expert on title questions and as a hardworking, patient, and skilled lawyer of considerable personal integrity. Tower's legal career was interrupted briefly by the Civil War. He served as captain of a company of three-month volunteers in 1861 and as U.S. provost marshal for the Tenth Congressional District of Pennsylvania from April 1863 to May 1864.

By 1867 Tower had largely secured Munson's title to the disputed coal lands, working as a partner to Alfred Munson's son, Samuel, after the former's death in 1854. In 1871 Tower and Samuel Munson sold the lands, Tower netting around $3 million for his share. After the sale Tower devoted the remainder of his career to management of his investments and to service on the boards of several of the companies in which he invested.

In the early 1870s Jay Cooke persuaded Tower to invest in his Northern Pacific Railroad. Cooke's plans collapsed in the panic of 1873, but Tower remained convinced of the Northern Pacific's ultimate success. He thus accepted appointment as a director of the railroad in 1873 and served as chairman of its finance committee during its reorganization from 1874 to 1879.

The greatest and most successful work of Tower's career, however, was the opening of the first of the Minnesota iron ranges. As early as the 1860s there were reports of iron in the Vermilion and Mesabi regions of northeast Minnesota. Because of remoteness, climate, and low iron prices these reports were ignored. In the mid 1870s, however, Duluth businessman George C. Stone, who had been devastated by the panic of 1873, attempted to recoup his fortune by recruiting capitalists to invest in Minnesota iron development. In 1875 Stone called on Tower. Tower, along with Samuel Munson, eventually agreed to share exploration expenses and hired Stone to be general manager of the enterprise. Stone, in turn, employed George Stuntz, a Duluth prospector and surveyor who had provided him with ore samples, to guide an expedition. Under the direction of Albert Chester, a Hamilton College geologist, the expedition collected samples and dug test holes in summer 1875 on both the far eastern end of the Mesabi Range and on the nearby Vermilion Range. The samples taken from the eastern Mesabi were disappointing (the high-grade Mesabi ores were farther to the west), but on the Vermilion the samples yielded a high-grade iron ore.

Tower and Munson initially went no further. Ore markets were too depressed and development expenses too high. By 1880, however, the iron market had improved, and in February 1880 Stone began secretly acquiring lands in the area of Vermilion Lake for Tower and Munson, who simultaneously financed additional explorations in the area by Chester, Stuntz, and Richard Lee (Tower's son-in-law). Munson died in 1881, but Tower, by now convinced of the potential of the region, bought his partner's share and continued exploration. In late 1881 or early 1882 Tower acquired a new partner, Edward Breitung. Together they formed the Minnesota Iron Company on December 1, 1882.

Developing the Minnesota properties acquired proved more difficult than anticipated. Stone, whose talents were in the area of promotion and lobbying, secured passage in the Minnesota legislature of bills which exempted mine properties from taxation until production and which transferred a state land grant to the railroad (the Duluth & Iron Range Railroad) on which Tower planned to carry Vermilion ores to Lake Superior. By somewhat dubious means he also acquired large landholdings for the Minnesota Iron Company in the Vermilion Lake area. But these achievements were not enough. Neither railroad nor mining investments in the wilds of Minnesota were financially attractive, and Tower's junior partners in the venture—Edward Breitung and, later, Samuel Ely—proved incapable of meeting their financial obligations. Thus the burden of development expenses fell largely on Tower and his reserves began to dwindle.

Under the direction of Tower's son (Charlemagne Tower, Jr.) and his son-in-law (Richard Lee), the railroad from Lake Superior to Vermilion Lake was completed on July 31, 1884; the first load of iron ore from Tower's Soudan Mine was shipped shortly after. Despite this success Tower's financial situation was, by then, desperate. To make matters worse the market for iron ore collapsed in 1884. In late 1884 Tower desperately struggled to keep his enterprise afloat, borrowing from close friends and former associates.

In 1885 and 1886 the ore market began to revive, and ore sales soon rescued Tower from a very precarious situation. Under the management of his son the Minnesota Iron Company flourished and became one of the nation's largest ore producers. But success brought imitators. In late 1886 Henry H. Porter of Chicago organized a syndicate—which included Marshall Field, Cyrus H. McCormick, and John D. and William Rockefeller—to tap into the mineral wealth Tower had uncovered on the Vermilion

Range. After acquiring iron lands near Tower's claim, Porter threatened to construct a rival railroad, undercut Minnesota Iron Company ore prices, and attack Tower's somewhat suspect land claims on the Vermilion if Tower did not sell the Duluth & Iron Range Railroad to the syndicate.

Tower reluctantly agreed to sell. He was seventy-eight years old, his partner Edward Breitung had just died, and he had no desire to fight alone. In May 1887, after long negotiations, Tower sold the Minnesota Iron Company and the Duluth & Iron Range Railroad for $6.4 million in what the *Philadelphia Inquirer* called "one of the largest financial transactions" ever completed in Philadelphia.

Tower did not long survive the sale of his Minnesota properties. He died shortly after, on July 24, 1889, at his country residence in Waterville, New York. Although Tower's agents missed the iron deposits of the Mesabi, his success on the Vermilion provided American iron furnaces with new deposits of high-grade hard ores and stimulated further exploration in northeast Minnesota that led to the discovery of Mesabi's iron wealth a few years later.

Publications:

Our Book, by Tower and G. W. Dillingham (New York: G. W. Dillingham/London: S. Low, Son, 1889);

Tower Genealogy: An Account of the Descendants of John Tower of Hingham, Massachusetts (Cambridge, Mass.: J. Wilson, 1891).

References:

Hal Bridges, *Iron Millionaire: Life of Charlemagne Tower* (Philadelphia: University of Pennsylvania Press, 1952);

Herbert N. Casson, *The Romance of Steel* (New York: A. S. Barnes, 1907);

Theodore Christianson, *Minnesota* (Chicago & New York: American Historical Society, 1935);

C. S. R. Hildeburn, *The Charlemagne Tower Collection of American Colonial Laws* (Philadelphia: Historical Society of Pennsylvania, 1890);

Grace Lee Nute, "Charlemagne Tower: Developer of the Vermilion Iron Range," *Gopher Historian*, 6 (April 1952): 6-7, 16;

David A. Walker, *Iron Frontier: The Discovery and Early Development of Minnesota's Three Ranges* (St. Paul: Minnesota Historical Society, 1979).

Archives:

The Charlemagne Tower papers are located in the libraries of Columbia University, New York, New York.

Charlemagne Tower, Jr.

(April 17, 1848-February 24, 1923)

by Terry S. Reynolds

Michigan Technological University

CAREER: Lawyer, private practice (1878-1882); director and treasurer (1882-1887), managing director, Minnesota Iron Company (1886-1887); director (1882-1887), president, Duluth & Iron Range Railroad Company (1883-1887); vicepresident (1887-1890), president, Finance Company of Philadelphia (1890-1891); minister to Austria-Hungary (1897-1899); ambassador to Russia (1899-1902); ambassador to Germany (1902-1908).

Charlemagne Tower, Jr., president of the railroad which first shipped Minnesota iron ore, was born on April 17, 1848, in Philadelphia, Pennsylvania. He was the eldest child and only son of Charlemagne and Amelia Malvina Boitte Tower. Tower spent his youth in Orwigsburg and Pottsville, Pennsylvania, where his father practiced law and, with a partner, accumulated coal lands.

Tower began his education in the common schools of Pottsville, but at age ten he was sent to York, Pennsylvania, to board and study with A. C. Heffelfinger. Later he was sent to military school at New Haven, Connecticut, and to Phillips Exeter Academy in New Hampshire to prepare for admission to Harvard. He entered Harvard in 1868 and graduated in 1872. At the encouragement of his father he spent the next four years in Europe informally studying history, languages, and culture.

On his return to America in 1876 he elected to follow his father's vocation, entering the office of William Henry Rawle of Philadelphia as a law student and attending law lectures at the University of Pennsylvania. He was admitted to the bar in September 1878. Tower made his home in Philadelphia practicing law until 1882, when his father's investments

Charlemagne Tower, Jr.

in Minnesota iron properties brought him new responsibilities.

In 1882 and 1883 he worked closely with his father in the business arrangements which created the Minnesota Iron Company and its subsidiary, the Duluth & Iron Range Railroad Company, becoming a director in both enterprises. When the first president of the railroad resigned, Tower was given the post and moved to Duluth.

Although his father maintained close control over his Minnesota iron enterprises through detailed correspondence, Tower was himself an able and hardworking administrator. His greatest strengths were not in original thought or personal dynamism, but in respect for and efficient use of the specialized training of his associates and in the systematic and orderly manner in which he oversaw affairs. He played a major role in the planning,

construction, and operation of the Duluth & Iron Range Railroad, successfully pushing the contractor to complete it on time. As a result the first shipment of Minnesota iron ore departed on schedule in summer 1884. When a depressed ore market and his father's strained finances threatened the Minnesota Iron Company, he took a leading role in successfully negotiating several loans.

After the Duluth & Iron Range Railroad was completed in mid 1884, Tower's administrative responsibilities were extended to cover the management of the Minnesota Iron Company's Soudan Mine. In 1886 the bylaws of that firm were amended to recognize this, and he was named managing director of the, by then, increasingly profitable operations.

Following his father's decision to sell the Minnesota iron properties and railroad in 1887, Tower

341

Charlemagne Tower, Jr., (right) with Richard H. Lee during construction of the Duluth & Iron Range Railroad during the late 1870s

returned to Philadelphia, where he became vice-president of the Finance Company of Philadelphia, becoming its president in 1890. He remained active in business, especially coal mining and financing, until 1891.

The estate Tower inherited after his father's death in 1889 made him a wealthy man. In 1891 he took advantage of this, relinquishing many of his offices to devote himself to history. He published *The Marquis de La Fayette in the American Revolution* in 1895. He also served as vice-president of the University of Pennsylvania, Department of Archaeology and Paleontology, as well as vice-president of the Historical Society of Pennsylvania.

Like his father, Tower was a staunch Republican. This, along with his wealth, scholarly reputation, and European education, made Tower a strong candidate for a diplomatic post under McKinley's administration. His diplomatic career, which began in 1897, spanned 11 years. He served as minister to Austria-Hungary from June 1, 1897, to early 1899. In January 1899 he was appointed ambassador to Russia, a post he held to November 1902. In 1902, as American relations with Germany weakened due to German threats against Venezuela, Tower was sent to Berlin. Among his first tasks as ambassador to Germany was to protest German actions to force Venezuela to pay German claims. Tower was influential in negotiating the submission of the Germano-Venezuelan dispute to the Hague Tribunal. Tower resigned from his diplomatic post in 1908 and returned to Philadelphia. There he continued to pursue his private interests until his death on February 24, 1923.

Although he spent only four years in important positions in the iron industry, Charlemagne Tower, Jr.'s management of the Duluth & Iron Range Railroad and the Soudan Iron Mine between 1882 and 1887 were critical to the success of the venture which first opened Minnesota's vast iron resources.

Selected Publications:

The Marquis de La Fayette in the American Revolution, 2 volumes (Philadelphia: Lippincott, 1895);

Essays Political and Historical (Philadelphia and London: Lippincott, 1914).

References:

Hal Bridges, *Iron Millionaire: Life of Charlemagne Tower* (Philadelphia: University of Pennsylvania Press, 1952);

David A. Walker, *Iron Frontier: The Discovery and Early Development of Minnesota's Three Ranges* (St. Paul: Minnesota Historical Society Press, 1979).

Archives:

The Charlemagne Tower papers are located in the libraries of Columbia University, New York, New York.

Edward Y. Townsend

(October 4, 1824-November 5, 1891)

by Stephen H. Cutcliffe

Lehigh University

CAREER: Merchant, Wood, Abbott & Company and Wood, Bacon & Company (1842-1855); partner and vice-president, Wood, Morrell & Company (1855-1861); vice-president (1862-1873), president, Cambria Iron Company (1873-1891).

Edward Y. Townsend became involved with the Cambria Iron Works in 1855 shortly after its founding and remained active in the firm first as its vice-president and then as president until his death in 1891. He was responsible for placing the company on a solid financial footing and helping to turn it into one of the leading iron- and steel-producing firms of the day.

Townsend was born in West Chester, Pennsylvania, on October 4, 1824, to John and Sybilla Price Townsend. Townsend remained in school until he was eighteen, when he entered the wholesale dry goods firm of Wood, Abbott & Company in Philadelphia. Eventually he became a partner in the firm, by then reorganized as Wood, Bacon & Company. In 1850 Townsend married a daughter of Henry Troth of Philadelphia and fathered two sons, Henry and John. Henry Townsend became president of the Logan Iron & Steel Company, while John rose to be the vice-president of Cambria Iron Company.

Townsend remained in the mercantile field until 1855 when the firm of Wood, Morrell & Company was formed by a group of creditors of the Cambria Iron Company to lease the works in an attempt to make them profitable. He entered into partnership with Richard D. Wood, the senior member of Wood, Bacon & Company, Wood's brother Charles Wood, Daniel J. Morrell, and three others, each of whom put up $30,000. Townsend, Morrell, and Charles Wood became the active managers of the business. Following Cambria Iron's reorganization in 1862 Wood became president, Townsend vice-president, and Morrell the general manager.

Edward Y. Townsend

In his capacity as a Cambria partner and vice-president, Townsend obviously held positions of importance. Yet he left few historical footprints and as a result has remained less well known to historians than others like John Fritz, whose central role in the development of the three-high rolling mill has been so well documented. However, Townsend's support for Fritz's plan was, in the latter's mind at least, critical. Fritz's solution to the problem of rolling high-quality iron rails was the introduction of a three-high set of rolls rather than the traditional two. Although the solution seemed obvious to Fritz, Cambria's management vacillated, unable to make up its mind. One Sunday morning during this period of indecision, Townsend visited the mill where he found Fritz engaged in equipment repairs. Fritz described how the two men sat "on a pile of discarded rails with evidences of failure on every side . . . talking over the history of the past, the difficulties of the present, and the uncertainties of the future. . . ." Apparently Townsend became convinced of Fritz's position, for upon rising to leave, he turned and said, "Fritz, go ahead and build the mill as you want it," and in response to Fritz's query as to whether Townsend was saying this officially, the latter replied, "I will make it official." Although reserved by nature, once Townsend had determined upon a course of action, he carried it out forcefully. In Fritz's words, "to no other person so deservedly belongs the credit, not only of the introduction of the three-high-roll train but also of the wonderful prosperity that came to the Cambria Company. . . ."

In 1873 Townsend succeeded Wood as president upon the latter's death. At that time the company's capital stock was $2 million, and it was burdened by a large floating debt. Townsend focused his attention on reducing this debt, eventually eliminating it altogether, to solidify the company's fiscal condition. During his presidency Cambria slowly evolved into a steel-producing firm, reaching a capacity of approximately 1,000 finished tons daily with a capitalization of $15 million. The financial stability, which was largely a result of Townsend's skill and conservative management, enabled the company to withstand the destructive Johnstown flood of 1889 despite the great loss of property suffered. However, the flood and the personal losses it inflicted were apparently a great shock to Townsend and affected his health adversely, eventually contributing to his death on November 5, 1891.

Townsend's historical importance is due largely to his managerial role at Cambria, which was so crucial to the firm's success, but he also retained an interest and played an active supporting role in promoting the industry as a whole and became one of the original members of the American Iron and Steel Association, where he served as an officer.

References:

John Fritz, *The Autobiography of John Fritz* (New York: John Wiley & Sons, 1912);

William Hogan, *Economic History of the Iron and Steel Industry in the United States* (Lexington, Mass.: Heath, 1971);

Robert W. Hunt, "Evolution of the American Rolling Mill," *Transactions of the American Society of Mechanical Engineers*, 13 (1892): 45-69;

Hunt, "History of the Bessemer Manufacture in America," *Transactions* of the American Institute of Mining Engineers, 5 (1877): 201-215;

Jeanne McHugh, *Alexander Holley and the Makers of Steel* (Baltimore: Johns Hopkins University Press, 1980);

Elting E. Morison, *From Know How to Nowhere* (New York: Basic Books, 1974);

Morison, *Men, Machines, and Modern Times* (Cambridge, Mass.: M.I.T. Press, 1966);

Fritz Redlich, *History of American Business Leaders: A Series of Studies*, 2 volumes (Ann Arbor, Mich.: Edwards Brothers, 1940);

James M. Swank, *History of the Manufacture of Iron in All Ages* (Philadelphia: American Iron and Steel Association, 1892);

Swank, *Introduction to a History of Ironmaking and Coal Mining in Pennsylvania* (Philadelphia: J. M. Swank, 1878);

Peter Temin, *Iron and Steel in Nineteenth-Century America: An Economic Inquiry* (Cambridge, Mass.: M.I.T. Press, 1964).

Tredegar Iron Works

by Brady Banta

Louisiana State University

Chartered by the Virginia legislature in 1837, the Tredegar Iron Works of Richmond, Virginia, grew from a small foundry, forge, and rolling mill into an industrial complex furnishing armor, ordnance, munitions, and numerous other iron products for the Confederate war effort. The company emerged intact from the Southern defeat, but it never regained the industrial importance that it had achieved during the Civil War. At the helm throughout much of this transformation was Joseph Reid Anderson, and in large measure the Tredegar's struggle to achieve and maintain preeminence involved overcoming, or at least neutralizing, a seemingly endless series of difficulties.

Not the least of these problems was a shortage of native, white labor. Anderson responded by making greater use of slaves and by importing northern and foreign workers. Slaves traditionally occupied subordinate, unskilled positions, but in 1848 Anderson initiated their use as puddlers—those individuals who conducted the process of transforming pig iron into wrought iron. This action precipitated a strike which Anderson swiftly crushed; afterwards and up to the Civil War, an average of 77 slaves were working at the ironworks at any one time.

Retained principally as a cost-cutting measure, slave laborers did not figure prominently in the expansion of the ironworks. Indeed, during the 1850s the Tredegar's labor force increased from 250 to 800, but the number of slaves declined from 100 to 80. Concurrently, the Tredegar's management increasingly relied on northern and foreign sources of labor; whether arriving directly from Europe or lured to Richmond from the North, the labor force became predominately Irish and German.

While the company's labor force became increasingly nonsouthern, so did its purchases of the basic material of production, pig iron. Unfortunately, the local pig iron industry had not matched its northern competitor's pace in adopting technological advancements, especially anthracite-fueled iron furnaces. Virginia's charcoal-fueled foundries generated acceptable iron, but their production costs significantly exceeded those of the anthracite furnaces. During the 1850s this cost-of-production differential contributed to the loss of roughly half of Virginia's pig iron capacity.

Although preferring local pig iron, Joseph Anderson reluctantly turned to Pennsylvania suppliers for this basic ingredient of production. By adding significant transportation costs to already high labor expenses, however, this reliance on Pennsylvania iron hampered the Tredegar's ability to compete in national markets. Anderson responded by recruiting southern markets, frequently alluding to the growing sectional tensions when encouraging southerners to patronize local industries. This campaign was effective. By the late 1850s the principal markets for Tredegar iron products were southern railroads and bridge construction companies.

But the cultivation of southern markets almost spelled doom for the Tredegar Iron Works. As political controversies swirled during spring and summer 1860, Anderson prepared the ironworks for the fall production season. But orders for iron products failed to materialize. Moreover, numerous southern customers, especially railroads, defaulted on notes used in previous years to purchase Tredegar iron. Soon unable to meet the company's financial obligations, the Tredegar's owners resorted to short-term loans. While this expedient relieved the immediate crisis, the threat of collapse loomed on the horizon.

Ironically, rapidly escalating sectional tensions generated a demand for Tredegar products that rescued the company from bankruptcy. Abraham Lincoln's election as president of the United States prompted inquiries from southern states regarding the purchase of ordnance and munitions. What

The Tredegar Iron Works during the Civil War

began as a trickle became a torrent as throughout spring 1861 orders arrived so rapidly and in such magnitude that the Tredegar's management doubled the capacity of the ordnance division and employed additional shifts to meet delivery schedules.

The Tredegar Iron Works rapidly emerged as the principal ordnance supplier for the Confederate States of America. This distinction brought with it, however, the arduous and frequently bewildering task of maintaining satisfactory production levels. The first difficulty to arise was a critical labor shortage. During the antebellum years the Tredegar had relied heavily on northern and foreign employees, but as war erupted many of them quit and headed north. While this did not cripple production, a brisk and continuing competition for experienced ironworkers developed in the Richmond area. The intensity of this rivalry for skilled ironworkers found expression in wage inflation. In spring 1861 the Tredegar Iron Works paid its skilled employees in a range from $2.50 to $3.00 per day. Four years later these wages had risen to $9.00 to $10.00 per day, plus monthly rations equivalent to those issued by the army to its soldiers.

Exacerbating this situation was the fact that the Tredegar's traditional and much-preferred source of gun iron, the Cloverdale furnace in Botetourt County, closed as the war began because its owner won election to the Virginia House of Delegates. This sent Joseph Anderson scrambling to obtain pig iron. Eventually he negotiated exclusive

purchase contracts with six Virginia iron furnaces, but throughout 1862 they consistently failed to fulfill their obligations. While labor and material shortages partially explain these deficiencies, profiteering also led producers to sell pig iron promised to the Tredegar in other markets. Unfortunately, the latter was especially true of the management team that Anderson and the Tredegar management helped to install at the reopened Cloverdale furnace.

This inability to secure Cloverdale iron was critical as Tredegar-produced artillery fashioned from alternative iron supplies proved unreliable. These quality deficiencies and an inability to satisfy government delivery schedules convinced a reluctant Joseph Anderson that his company would have to produce its own pig iron. Anderson had consistently maintained that expansion into blast furnace operation was beyond the Tredegar's financial capabilities. To overcome his reservations, the government signed an ordnance contract with the Tredegar that included a $300,000 interest-free loan with which to underwrite pig iron production. For his part Anderson agreed that Tredegar-produced pig iron would be used to satisfy government ordnance and armor contracts.

Regrettably, the acquisition of blast furnaces did not satisfy the Tredegar's pig iron appetite, but it did aggravate labor and supply problems. Provisioning the enlarged work force—approximately 2,000 employees and their dependents—proved difficult. The ravages of war and continuing military demands for food and fodder in Virginia drove Tredegar agents beyond the state's boundaries in their quest for supplies. Initially sources in East Tennessee satisfied their needs, but as the war progressed Tredegar agents increasingly purchased provisions in central and southern Georgia. While successful, these efforts consumed managerial talent and fiscal resources that were desperately needed in the productive aspects of the Tredegar's business.

Expansion into pig iron production also aggravated already difficult staffing problems. Upon attempting to reopen idle blast furnaces, Tredegar officials frequently found that their managers, artisans, and employees were in the army. Efforts to have them detailed to the furnaces or released from military service were infrequently successful. Moreover, attempts to hire foreign replacements generally proved fruitless. Given these conditions, the Tredegar management used convict labor when available and turned increasingly to slave labor. Indeed, by 1864 slaves comprised well over half of the company's total labor force. But with this increasing reliance on slave labor came another problem. The rapidly escalating costs of obtaining and maintaining the slaves drained the Tredegar's financial resources and contributed to a steadily decreasing margin between the company's costs of production and its profits.

Despite these labor and material problems, the Tredegar delivered 214 pieces of ordnance to the confederate governments in 1861 and a wartime high of 351 in 1862. While this figure declined thereafter, in just over four years the Tredegar delivered 1,099 pieces of artillery to the Confederacy. While this pales in comparison to the more than 8,000 pieces received by the federal government from northern works, it is more enlightening to recognize that only one northern manufacturer, R. P. Parrott, produced more ordnance (1,557 pieces) than the Tredegar Works during the Civil War.

While ordnance work was always predomi-

nant, the Tredegar Iron Works made other significant and interesting contributions to the Confederate war effort. The most fascinating of these was the Tredegar's role in refurbishing and outfitting the U.S.S. *Merrimack* as a Confederate ironclad. Having received a contract to produce and oversee the installation of the *Merrimack*'s armor plate, the company needed rolled iron to fashion into iron plate. The solution to this supply dilemma lay in stripping, with the cooperation of the Confederate war department, rails from the railroad lines of northern Virginia. The actual refashioning of rails into plate progressed slowly as the naval officer overseeing the project changed the specifications from a triple layer of 1-inch plates to a double layer of 2-inch plates. Regardless, by fall 1861 the Tredegar had the initial consignment of plates ready for delivery. Shortages of flatcars caused extensive shipping delays, but by February 1862 the Tredegar had delivered to the Norfolk shipyard the 725 tons of iron plates needed to outfit the ironclad.

As a producer of armor, artillery, munitions, and rails the Tredegar Iron Works rapidly emerged as, and remained, a vital cog in the Confederate industrial complex. Largely as a result of the unstinting efforts of Joseph Anderson, the company survived the transition from war to peace. But it could not retain its position of preeminence in the southern economy. Indeed, after the panic of 1873 the company's standing in the iron industry steadily deteriorated. In capsule the Tredegar Iron Works had a mercurial history, rapidly rising and declining. But the company peaked during the Civil War when it was a crucial element in sustaining the Confederate army in the field.

References:
Kathleen Bruce, *Virginia Iron Manufacture in the Slave Era* (New York: Century, 1931);

Charles B. Dew, *Ironmaker to the Confederacy: Joseph R. Anderson and the Tredegar Iron Works* (New Haven & London: Yale University Press, 1966).

Union Steel Screw Company

by Marc Harris

Ohio State University

The Union Steel Screw Company of Cleveland was known primarily for pioneering the use of Bessemer steel to make screws and other rod and wire products. The firm had its origin in the related Cleveland Rolling Mill Company, whose part-owners and principal operators, Henry and William Chisholm, had established in 1868 the second Bessemer steel converter installation in the west. In common with other early steel-rail producers, the Chisholms faced a serious scrap problem. In rail rolling, as much as 20 percent of the valuable steel went to scrap, and contemporary practices severely limited what could be done with it.

The Chisholms, however, were not bound by steel-making tradition. In 1870 Henry Chisholm was able to roll Bessemer rail scrap in rod and wire mills designed for iron stock, and William Chisholm, by all accounts the more inventive of the two brothers, also set to work developing new products to manufacture and market. He had been producing such goods as spikes, bolts, and horseshoes from malleable iron, and the following year he demonstrated that screws could be made from Bessemer steel.

Union Steel Screw was established in 1871 in order to manufacture and market the new product and was one of several extensions of the family's interests. The Chisholm brothers and their partners consistently aimed to integrate their operations vertically and to expand their product line, usually through affiliated enterprises. They had invested in blast furnaces very shortly after forming their original partnership and afterward installed a pair of Bessemer converters. In 1863 the partners set up a Chicago plant, which was reorganized in 1871 as the Union Rolling Mill Company. Union Steel Screw specialized in screw and nail production, primarily under William Chisholm's direction. Both new investments were primarily financed by profits

from operation of the Cleveland Rolling Mill Company, the center of their operations.

Many prominent Cleveland industrialists were associated with the Chisholm brothers in Union Steel Screw, including their partners in the Cleveland Rolling Mill Company, Amasa Stone, Jr., and his brother Andros B. Stone. Others who served as company officers included Stillman Witt, a contractor and partner of Amasa Stone, Jephtha Wade of Western Union, and Fayette Brown, who owned some Lake Superior ore mine properties and represented others. Union Steel Screw was calculated to be a sizable enterprise from the beginning, with a capitalization of $1 million (double the original capitalization of the parent company in 1864), and in 1887 the firm declared its daily production to be "12,000 gross of screws of all sizes and about three tons of nails and tacks."

As did its parent company, Union Steel Screw welcomed innovation. In 1879 the firm developed a high-carbon, relatively high-sulfur steel stock suitable for machine screws. It has been claimed that this happened by accident; if so, it only demonstrates that the firm's operators remained open to innovations, whatever their source.

The Chisholm companies coordinated their operations fairly closely. Union Steel Screw depended on Cleveland Rolling Mill for its steel and rod stock and concentrated primarily on fabricating screws, nails, and tacks. Another family firm, William Chisholm & Sons, also relied on Cleveland Rolling Mill for the steel it used to manufacture shovels and other digging equipment, presumably from plate stock. Cleveland Rolling Mill, the center of the brothers' industrial empire, relied on holdings, to which they had access through their partners, in coal and iron ore and in the Great Lakes shipping that carried the ore down from its Lake Superior beds. Union Rolling Mill in Chicago was directed by a son of Henry Chisholm and relied on access to the same holdings. This type of organization essentially mirrored Andrew Carnegie's accomplishment of integrating as many aspects of iron and steel production as possible within Carnegie Steel.

But Union Steel Screw, like its parent company, was unable to retain its independence during the great merger wave at the turn of the twentieth century. In 1908 it was absorbed into the National Screw & Tack trust.

References:

Stephen L. Goodale, compiler, *Chronology of Iron and Steel* (Pittsburgh: Pittsburgh Iron & Steel Foundries, 1920);

Walter Havighurst, *Vein of Iron: The Pickands Mather Story* (Cleveland & New York: World Publishing, 1958);

The Industries of Cleveland (Cleveland, 1888);

William Ganson Rose, *Cleveland: The Making of a City* (Cleveland & New York: World Publishing, 1950);

"William Chisholm," *Magazine of Western History*, 3 (1885-1886): 174-177.

Eber Brock Ward

(December 25, 1811-January 2, 1875)

by Alec Kirby

George Washington University

CAREER: Shipbuilder, manufacturer, investor (1830-1872); owner, Eureka Iron Company (1855-1875); coowner, Kelly Pneumatic Process Company (1863-1875); president, American Iron and Steel Association (1864-1869); chairman, Chicago Rolling Mill Company (1872-1875).

Eber Brock Ward, a manufacturer, shipbuilder, industrial innovator, and investor, was born in Canada on December 25, 1811. His parents, who had moved to Canada, returned to the United States, when Ward was nine years old, to establish a store in Kentucky. Incredibly, the Wards began their journey in the dead of winter, traveling through the snow in a canvas-covered sleigh. In a rural area of Pennsylvania, Ward's mother contracted pleurisy and died three days later. She was buried near a roadside oak tree.

His father chose to abandon his Kentucky ambitions after the loss of his wife, and the family settled in New Salem (now Conneaut), Ohio, which was the home of Ward's uncle, Sam Ward. Sam Ward soon moved from Ohio to a location near Detroit, where he was joined by his brother's family after four years. By then Sam Ward had invested in several sailing vessels, and having no children of his own, he developed a close relationship with his nephew. As Ward grew older, he was made a cabin boy in one of his uncle's ships.

In this position Ward demonstrated the initiative and ambition that marked his mature personality. Within a few years Ward mastered the craft of sailing, became captain of his uncle's ship, and acquired the first of a fleet of vessels he would eventually purchase or build. By the mid nineteenth century Ward was known as the "steamship king" of the Great Lakes, owning a fleet of 30 ships. Concurrently, he expanded his financial holdings. Ward quickly came to own or control interests in coal, copper, salt, iron, ore, silver, timber, newspapers, railroads, and a plate glass factory. Ward was also involved in building the Sault St. Marie Canal, which made accessible the iron ore deposits of Lake Superior. It was Ward's schooner, the *Columbia*, that brought the first shipment of iron ore through the canal on August 17, 1855.

While Ward amassed a fortune through diverse interests, seminal events in the development of the steel industry were proceeding on both sides of the Atlantic. At this time the cost of producing steel made the metal about as practical for industrial use as silver. The railroads used iron rails, which wore out in less than two years. Clearly, there was a need–and a market–for high-quality, inexpensive steel. A major step toward meeting this demand was taken at the Suwanne Iron Works near Eddyville, Kentucky, in 1846, where William Kelly discovered a pneumatic process for making malleable iron from pig iron. In this process air was blown through

Eber Brock Ward

molten pig iron contained in a vessel, thereby oxidizing and removing impurities and leaving a fluid product. Kelly did not apply for a patent until 1857, when his application was accepted. In 1855, however, a British patent had been granted to Henry Bessemer, who had independently developed a pneumatic process for the making of steel. The pneumatic process, which both in America and in Britain came to be known as the Bessemer process, seemed to offer the possibility of inexpensive steel.

Meanwhile, Ward was beginning to invest in the iron ore industry. In 1855 he was a founder of the Eureka Iron Company in Wyandotte, Michigan, where he constructed a furnace and, later, a rolling mill. Soon after, Ward and his associates took advantage of the location of six important railroads in Chicago and founded a mill in that city for rerolling iron rails. By the end of the Civil War Ward had added a plant in Milwaukee to his Chicago and Wyandotte holdings and had earned a reputation as the first of the "iron kings."

An important adviser to Ward during the creation of the Wyandotte furnace was Zoheth S. Durfee, who was an expert in the technology of iron production. It was apparently Durfee who first informed Ward of Kelly's patent, although this may have been done by Daniel Morrell, who was to become another longtime Ward associate. Ward was initially very cautious about investing in steel, despite the fact that he had rarely hesitated to plunge into industries when he had only a superficial knowledge of the technology involved. Nevertheless, Ward soon decided to enter into a business relationship with Durfee, and in 1861 the two men obtained the rights to the pneumatic process from Kelly.

Durfee then traveled to England to study the Bessemer process and to obtain American rights for its use. Demonstrating characteristic impatience, Ward, immediately after Durfee's departure, asked his cousin William E. Durfee to construct an experimental plant at Wyandotte, utilizing the pneumatic process. In doing so Ward was anticipating the acquisition of the Bessemer patent, because the proposed plant drew directly on Bessemer's (as well as Kelly's) ideas. Unfortunately, Zoheth Durfee was not able to secure the American rights to the Bessemer patent, despite his reportedly vigorous efforts. Although it is unclear why he failed, one explanation is that Durfee may have offended Bessemer by his initial skepticism of the pneumatic process. Another explanation is that Bessemer did not want to have his work associated with the as yet untried Kelly process. Whatever the reason, Durfee returned home empty-handed.

Ward's decision to enter into an alliance with the Durfee cousins and to proceed vigorously with an experimental plant is a telling example of Ward's sense of timing and business acumen. The steel industry took on vital significance as the Civil War dragged on, and the period after 1865 continued to be a time of considerable growth in both the iron and Bessemer steel industries. This was largely stimulated by the demands of the railroad industry, as an average of over 3,000 additional miles of track were laid each year between 1865 and 1880. The output of the iron industry nearly doubled, and the price exploded from $18.60 per ton to as high as $73.60 per ton. Yet Bessemer steel, as Ward correctly perceived, was to dominate the future. By 1880 the production of the new Bessemer steel would exceed the total iron production of 1860.

Much of this demand was due to the dissatisfaction of railroad companies with iron rails. Representatives of British steel companies were soliciting business from American railroad firms for Bessemer steel, and the emerging U.S. steel industry was anxious not to be shut out of the market.

Investment in the steel industry required expertise, which the Durfee cousins–among Ward's rapidly expanding list of associates–supplied. Ward himself was not a master of the details of the steel-making process and certainly had little understanding of the intricacies of the Bessemer process. Yet in his anxiety to make financial success of his steel venture, he was determined to surround himself with experienced individuals. He hired agents to look for skilled workers on Ellis Island and was known to pay the travel expenses of families relocating from England to Detroit.

In May 1863 Ward and Zoheth Durfee joined with Daniel J. Morrell, William M. Lyon, and James Park, Jr., to organize the Kelly Pneumatic Process Company, utilizing the Kelly patent at the experimental plant under construction in Wyandotte. Kelly was to receive a percentage of the profits of the company. The Kelly patent, however, was inadequate for the entire production process. Essentially, the Kelly patent controlled only the first of three steps, which collectively came to be known as the Bessemer process.

The first step involved the use of air as fuel, a technique discovered by Kelly. Ward and his associates owned the patent rights for this process. The second step involved removing all the carbon from the molten iron, as in the production of wrought iron, and then adding some back to give the steel its proper consistency. This was a process discovered by Robert F. Mushet. In 1864 Ward acquired this patent by admitting Mushet and his associates into the Kelly Process Company. The third step, involving the use of a tilting converter and casting ladle, was created by Bessemer, who owned the patent rights.

The Bessemer process, as commonly understood, required all three steps, yet Ward and his associates controlled only the first two. Ward had authorized the creation of his experimental Wyandotte plant on the assumption that he would obtain the patent rights controlled by Bessemer. Failing in this, he nevertheless finished the construction of the plant. On September 6, 1864, using the charcoal pig iron from Lake Superior, the first Bessemer steel produced in the United States was created at the Wyandotte plant. This date was to mark a new epoch in the steel industry, making available for the first time high-quality, inexpensive steel. Those on hand to witness the event were aware of its significance. For one of the apparently few times in his life, Ward was visibly nervous, anxiously shifting silver dollars from one hand to another.

Yet Ward had succeeded in producing steel before his company had succeeded in obtaining the patent rights which would legally entitle him to do so. Quickly, Ward found himself in litigation. His opponent was Alexander L. Holley, who had purchased the Bessemer rights. Neither Ward nor Holley, as matters stood, could produce Bessemer steel without infringing on the legal rights of the other. Meanwhile the steel industry began its post-Civil War expansion. This expansion was steady before it exploded into a sustained boom in 1870. From that year to 1880 total railroad mileage increased 75 percent, from 52,922 miles to 93,267 miles. This increase occurred even through the depression following the panic of 1873 and helped to sustain the steel boom. The expansion put pressure on the American steel industry to resolve its legal disputes in order to make further investment legally safe.

In 1866 a compromise was reached that consolidated all of the American patents. Under this arrangement the titles to the Kelly, Mushet, and Bessemer patents were vested in John F. Winslow and John A. Griswold (both associates of Holley) and Daniel Morrell (an associate of Ward). Morrell held a 30-percent interest in trust for the Kelly Process Company.

This arrangement was not a profitable one for Ward, and why he chose to accept it remains a mystery. The Kelly patent (which his company owned completely) did not expire until 1878, and it was doubtful that American courts would uphold the Bessemer patent held by Holley. The expanding ownership of the Kelly Process Company suggests that the firm was experiencing financial difficulty, but there is little evidence to document this thesis. There is one intriguing possibility. During the 1860s Ward had become deeply involved in the Spiritualism movement then sweeping America. Frequently Ward was accused by his associates of seeking the advice of a medium before completing important business transactions. Some of Ward's contemporaries believed that Ward's eldest son Frederick committed suicide because his father believed a Spiritualist's claim that the young man was not in fact Ward's son. Perhaps Ward sold his interests cheaply because of the advice of a medium. Whatever his motive, it appears to be the only disastrous business decision he ever made.

As was common for men such as Ward, business was not big enough to contain all his interests. He was concerned also with the larger political and economic affairs of the nation, especially in relation to the iron and steel industry. Ward was opinionated, and he did not hesitate to voice and promote his ideas, although he apparently never had any ambitions for public office. In 1864 the American Iron Association, which had been formed in Philadelphia in 1855, was reorganized into the American Iron and Steel Association, a change in nomenclature which reflected the technological changes of the industry. Ward, perhaps because of his leading contribution to this change, was selected as the first president of the new organization. The purpose of the association was to "procure regularly the statistics of the trade both at home and abroad; to provide for the mutual interchange of information and experience; to collect and preserve all works relating to iron and steel . . . and generally to take all proper measures for advancing the interests of the trade in all its branches."

Ward was a devout believer in a protectionist trade policy in order to protect American steel against British imports. He expressed his ideas in a series of strongly worded political tracts in the 1860s. In August 1860, for example, he published a seven-page essay entitled *Reasons Why the North-West Should Have a Protective Tariff and Why the Republican Party is the Safest Party to Trust With the Government*. The work was unabashed political propaganda "written for the especial benefit and for the information of the WORKING PEOPLE and FARMERS of MICHIGAN." Although the tract was a frank appeal for the Republican party, Ward emphasized that it was "published entirely independent of any of the political organizations of the country."

Ward's declaration of political independence was a reflection of his self-reliant nature. When President Grant offered to appoint him secretary of the treasury, Ward refused on the grounds that he was too preoccupied with his business concerns. Ward was simply interested in promoting his political and economic convictions, to which he was more at-

tached than to any political party. He declared bluntly, for example, that "party is of no account—principles everything. Vote [for the Republican party] while they adhere to their present principles."

Ward has been described as a man of extremes, one who could be alternately self-controlled and passionate, shrewd and credulous, intolerant and open-minded. He was able to delegate authority well, and he was a good judge of character. Many of his associates, especially the Durfee cousins, were known for their ability.

Physically Ward had never enjoyed robust health. In his preoccupation with business and political activities, he generally ignored his poor health, even after a severe heart attack in 1869. He recovered after a year, but did not change his work habits. He collapsed on a Detroit street and died on January 2, 1875, at the age of sixty-three. At the time of his death his estate was valued at over $5.3 million, making him one of the wealthiest men in the Midwest. He was survived by his wife and seven children.

Selected Publications:

Reasons Why the North-West Should Have a Protective Tariff and Why the Republican Party is the Safest Party to Trust With the Government (Detroit, 1860);

Protection vs. Free Trade, National Wealth vs. National Poverty (Detroit, 1865);

The Farmer and the Manufacturer (Detroit: Daily Post, 1868);

Eber Brock Ward (Detroit, 1875);

The Absorbing Questions of the Day (N.p., n.d.).

References:

Herbert M. Boylston, *Iron and Steel* (New York: John Wiley & Sons, 1928);

James H. Bridge, *The Inside History of the Carnegie Steel Company* (New York: Aldine Book Co., 1903);

Herbert N. Casson, *The Romance of Steel: The Story of a Thousand Millionaires* (New York: A. S. Barnes & Co., 1907);

William T. Hogan, *Economic History of the Iron and Steel Industry in the United States* (Lexington, Mass.: Lexington Books, 1971);

Jeanne McHugh, *Alexander Holley and the Makers of Steel* (Baltimore: John Hopkins University Press, 1980);

Peter Temin, *Iron and Steel in Nineteenth-Century America: An Economic Inquiry* (Cambridge, Mass.: MIT Press, 1964).

James Ward

(November 25, 1813-July 24, 1864)

by Marc Harris

Ohio State University

CAREER: Rolling-mill operator (1832-1841); proprietor, James Ward & Company (1842-1864).

Many nineteenth-century American ironmasters were British born, and many were Pittsburgh men. James Ward was both, and a pivotal figure in developing the Mahoning Valley iron industry. He operated the first rolling mill west of Pittsburgh, was the first to demonstrate that iron made from raw Mahoning Valley bituminous coal could be worked as well as any other, and helped to lay the basis for future expansion of an endangered industry.

Ward was born in Staffordshire in England's industrial Midlands, on November 25, 1813. His father, an ironworker, took the family to Pittsburgh in 1817 when Ward was four years old. In Pittsburgh he was educated in the common schools up to the age of thirteen. At that point he began to work at his father's forge producing wrought-iron nails and during the next six years learned the practical aspects of iron working. At nineteen, in 1832, Ward decided to study the emerging discipline of rolling-mill engineering. For the next nine years he studied and worked in and around Pittsburgh, specializing in steam-driven rolling mills.

In 1841 Ward struck out for the upper Ohio Valley with his brother William, who was an experienced iron roller, and Thomas Russell, a stock heater. They first set up shop in Lisbon, in Columbiana county on the Ohio River, but the next year they moved northward to a settlement known as Heaton's Furnace, or Niles, near the confluence of the Mahoning River and Mosquito Creek, where good quality steam coal was beginning to be exploited. Niles had been little more than a hamlet until 1839, when a crosscut canal linked the Mahoning Valley to the main line of the Ohio Canal near Akron. With an eye on the Great Lakes steamship market, local landowners began to

Bust of James Ward

exploit coal deposits on their properties, and it was this coal that lured Ward to Niles. The coal could power his mill, and an extant local iron industry could supply him with pig iron.

Ward was one of the first of the Pittsburgh men to move into the Western Reserve iron industry. Since the beginning of settlement there, furnace men had set up stacks at scattered locations. These men were from New England or New York or Westmoreland, Pennsylvania. The Mahoning Valley was an active iron-producing region because of its deposits of bog and kidney ore and of limestone. Although the valley was part of the Reserve and hence heavily affected by New England influences, it also fell within Pittsburgh's orbit because of its geography. The Mahoning Valley is located near the Pennsylvania border at Sharon, and the river flows into the Beaver, a tributary of the Ohio River. It is the only part of the Reserve with direct access to the Ohio River watershed, and most of the area's early iron production had been sold to Pittsburgh mills. The firm of James Ward & Company was the first to roll iron in what would later become one of the nation's major iron-producing and iron-working regions.

But in 1841 the Mahoning Valley iron industry was primitive and seemed almost on the verge of extinction; none of the early furnaces was able to operate profitably for any length of time. Like almost all their contemporaries, local ironmasters used charcoal fuel and ore taken from surface deposits in the vicinity. From the iron they made some castings but mainly produced hammered blooms and bar stock for the Pittsburgh market. Theirs was an industry like those in many other places at the time: dependent on local materials, small scale, and loosely organized. By the time Ward & Company began operations in 1842, local ore was already becoming scarce and charcoal difficult and expensive to obtain, although limestone remained plentiful.

However doubtful its future, the local industry did produce a usable grade of pig iron which Ward & Company could successfully work in its mills. The company began operations with three puddling furnaces and a muck bar mill, using pig iron produced by other concerns. This was the first true puddling operation in the Mahoning Valley and probably the first west of Pittsburgh. Puddling iron and rolling muck bars, a technique of producing wrought iron imported by British and Welsh ironworkers, was necessary before the age of Bessemer steel in order to make the iron ductile enough for further rolling. From this intermediate stage the company produced several products, including horseshoe iron, sheet and bar stock, and a mixture of steel scrap with iron. The latter could be welded easily and had good wear characteristics, which recommended it for use on wagon wheels; it was called "Dandy Tire."

In order to assure himself a supply of iron, Ward began leasing at least one of the existing area furnaces, the Maria furnace, built by James Heaton, who was part of a large and somewhat eccentric family of local entrepreneurs and ironmasters. Heaton was responsible for renaming the settlement at Heaton's Furnace in honor of one of his Whig party idols, the newspaper editor Hezekiah Niles. Ward used the Maria furnace until 1854, when he leased another furnace in Youngstown. In 1859 the company built its own blast furnace across Mosquito Creek from the main works.

In 1845 Ward had an important hand in two developments that broke the Mahoning Valley iron industry out of its charcoal and bog ore phase and put it on the footing on which it was to remain for the next generation. The first such development resulted from a miner's discovery. Underlying one of the coal mines near the Ward works, the Brier Hill mine, was a layer of what was widely assumed to be slate. A Welsh miner named John Lewis, however, recognized it for the black-band iron ore it was, because it resembled ore he had mined in Scotland before he left for the United States. Ward & Company charged the Maria furnace with this ore. When combined with other ores it produced a fine-grained casting iron similar to Scotch iron and therefore known as "Brier Hill Scotch." This was the first use of black-band ore in the United States. The discovery, made possible only because of Lewis's personal knowledge and experience and Ward's willingness to experiment, revitalized the local industry and enabled it to grow through the Civil War years. This ore, mixed with others, carried the industry through the early 1870s.

Also in 1845 and 1846, Wilkes, Wilkeson & Company built a blast furnace at Lowellville, a few miles from Niles, designed to use only local Brier Hill bituminous coal for fuel. The quality of iron produced from raw coal remained to be demonstrated; Ward & Company rolled it successfully and thus showed that local raw coal, known as black coal, could be used as a furnace charge. Valley coal fueled the local iron industry for the next generation. Together with the black-band ore, Valley coal not only made the industry independent of its traditional and fast-disappearing resources, but provided the basis for further growth.

From 1841 until after the Civil War, Ward & Company and its successors were largely responsible for the growth of Niles into a respectable town. The firm developed into one of the major industrial installations in the Mahoning Valley, and at its height under James Ward it employed about 100 workers. Ward was a major civic leader in Niles until his untimely death, and he was one of the few men in the town whose organizational ability could have made Niles rather than Youngstown the industrial capital of the valley. During and after the Civil War other manufacturers installed plants in Niles, but the more advanced and aggressive organizations were already becoming established in Youngstown.

Ward, whose training was as much lore as science, remained an iron man all his days and missed an early chance to move into steel. During the 1850s William Kelly approached him more than once about a new pneumatic process he had developed to decarbonize iron, virtually the same process Bessemer developed at about the same time. As Joseph Butler, Ward's protégé and later a prominent steel man in his own right, recalled later, Kelly presented the process as a means of producing wrought iron; Ward, like many other ironmasters, failed to grasp the principle involved and privately declared that Kelly was crazy. It was left to Daniel J. Morrell to midwife Kelly's efforts, and no steel was produced in the Mahoning until the 1890s.

During the Civil War the Ward works at first fell on hard times, then saw its business greatly expand. Operating his plant at high capacity worried Ward, and he habitually made nightly rounds of the grounds to see that things were running smoothly. One summer evening in 1864, visiting the Elizabeth blast furnace, he was shot and killed by a carousing drinker.

For a short time Ward's brother William ran the company, and his son James, Jr., inherited the works in 1866. Neither had the business acumen of the elder Ward. After the Civil War James Ward, Jr., invested in several new furnaces and mills. One of these was intended to reproduce a highly polished and expensive stove iron known as Russia iron, but the attempt failed, and the expansive scale of new investments made the company vulnerable to an economic contraction in 1873. In 1874 the original business failed; the consequences for Niles were disastrous because of the large amount of Ward & Company scrip in circulation among townspeople and local merchants. When the successor company failed in the next decade, its property was taken over by the Mahoning Valley Iron Company, later part of Republic Steel.

References:

Joseph G. Butler, *Autographed Portraits* (Youngstown, Ohio: Butler Art Institute, 1927);

Butler, *Fifty Years of Iron and Steel*, seventh edition (Cleveland: Penton, 1923);

Butler, *History of Youngstown and the Mahoning Valley* (Chicago & New York: American Historical Society, 1921);

Kenneth Warren, *The American Steel Industry 1850-1970: A Geographical Interpretation* (New York: Oxford University Press, 1973);

H. Z. Williams, *History of Trumbull and Mahoning Counties* (Cleveland, 1882).

Ichabod Washburn

(August 11, 1798-December 30, 1868)

by Alec Kirby

George Washington University

CAREER: Partner, Washburn & Goddard (1822-1835); partner, I & C Washburn (1842-1849); co-founder and president, Washburn & Moen Manufacturing Company (1850-1868); member, Massachusetts State Senate (1860-1861).

Ichabod Washburn, an inventor, manufacturer, and philanthropist, was born in Kingston, Massachusetts, on August 11, 1798, son of Charles and Sylvia (Bradford) Washburn. Washburn's mother was a fifth generation descendant of William Bradford, the famous early settler of Plymouth. His father, Charles Washburn, a sea captain also from a prominent New England family, died of yellow fever when Washburn was two months old.

The death of Charles Washburn led to severe financial difficulties for the family. At the age of nine Washburn was sent to live as an apprentice with a chaise and harness maker in Duxbury, Massachusetts, where he remained for two years. He then returned home to work in a cotton factory, where he assisted in the operation of a handloom. While working in the factory Washburn became interested in machinery and decided upon a career as a machinist. In 1814 he moved to Leicester, Massachusetts, and served a 4-year apprenticeship with a blacksmith. During this period he managed to further his education at a local academy.

At age twenty Washburn completed his apprenticeship and began to work as a journeyman in Millbury, Massachusetts. He held this position for only two months before his stepfather offered him a position in a grocery store in Portland, Maine. After a brief stint in the grocery business, Washburn returned to Millbury and opened his own plow manufacturing firm. He soon abandoned this effort to take a job in an armory operated by Col. Asa Waters, and then was hired by William Hovey, a machinery manufacturer in Worcester,

Ichabod Washburn

Massachusetts. Working for Waters and Hovey allowed Washburn to obtain a detailed knowledge of the forging and finishing of machinery. In 1821 he put his knowledge to use, entering into a partnership with W. H. Howard to manufacture lead pipe and machinery used in the production of woolen goods. The following year he purchased Howard's interest.

During this period the demand for machinery to use in the wool industry was increasing dramatically. To develop the capacity to meet the demand for both this machinery as well as for lead pipe,

Washburn entered into a partnership with Benjamin Goddard in 1822, creating the firm of Washburn & Goddard. The company quickly expanded and soon had 30 employees. In the operation of the firm Washburn benefited from his experience as an apprentice to a blacksmith, during which time he became familiar with the manufacturing of cards for cotton and wool machinery. While the firm of Washburn & Goddard was manufacturing lead pipe, which was drawn out from iron rods, it occurred to Washburn that wire might also be drawn out in a similar manner. Washburn devised a machine for this purpose and by 1830 had so perfected it that Washburn & Goddard purchased a waterpower plant in Northville, a suburb of Worcester, for the production of wire. The early technology, based on a pair of self-acting pincers drawing out about a foot of wire at a time, was crude, and production could not exceed 50 pounds per day. Washburn gradually increased output by refining the machinery. By 1833 he developed the drawing block, by which 2,500 pounds of wire could be drawn out in a day.

With this increased productive capacity, Washburn believed that the firm needed to move to a location where more power could be generated. Goddard opposed this expansion, and the partnership was sold on January 30, 1835. Washburn moved to Grove Mill, Massachusetts, where he established a new wire-production facility, taking his twin brother, Charles, into a partnership in 1842, under the name of I & C Washburn. The firm was dissolved in 1849. In 1850 Washburn established a new company with his son-in-law, Philip L. Moen, who served under Washburn as vice-president.

After the establishment of this new firm Washburn began experiments in the manufacture of steel wire for piano strings and developed a procedure for wire production that allowed the firm to break into a market which had been monopolized by a British company. Meanwhile, the commercial demand for crinoline wire and steel wire for sewing machine needles increased. Reorganizing to meet the demand, the partnership of Washburn & Moen became the I. Washburn & Moen Wire Works, and a new plant was constructed in South Worcester. A further expansion in 1868 led to the adoption of the final name of Washburn's firm, the Washburn & Moen Manufacturing Company, which grew to become one of the largest of its kind in the United States.

While expanding his business career, Washburn gained a widespread reputation as a philanthropist, focusing his efforts largely on behalf of education. He made large donations to Bangor Theological Seminary, Wheaton College, Berea College, and to Lincoln College in Kansas, which was renamed Washburn College after Washburn's death. He also served for 32 years as a deacon of the Union Congregational Church in Worcester and sat in the Senate of Massachusetts in 1860 and 1861.

Washburn was married twice–to Ann Brown of Worcester in 1823, and, after her death, to Elizabeth Cheever of Hallowell, Maine, in 1859. In February 1868 Washburn suffered a stroke, and he died on December 30 of that year.

Reference:

J. D. Van Slyck, *Representatives of New England* (Boston: Van Slyck & Company, 1879).

Washburn & Moen Manufacturing Company
by Alec Kirby

George Washington University

The Washburn & Moen Manufacturing Company grew out of a partnership established in 1850 by Ichabod Washburn and his son-in-law, Philip Louis Moen. Specializing in the production of wire, by 1895 the firm was the largest company of its kind in the United States, with an annual output of 100,000 tons of wire of various types.

Ichabod Washburn, cofounder and first president of the company, had been active in the iron industry for 28 years before beginning his partnership with Philip Moen. In 1830 Washburn created an innovative method for drawing wire out of iron rods. Both the method and the machine it used were crude and required considerable improvement, but Washburn nevertheless established a water-powered plant in Northville, Massachusetts, a suburb of Worcester, to produce iron wire. By 1833 Washburn had invented a new system, the "drawing block," which dramatically increased production. Recognizing the potential of his innovations, Washburn hoped to move his plant operations to a location where more power could be generated. But his partner, Benjamin Goddard, opposed the expansion, and on January 30, 1835, the partnership was dissolved.

Washburn then established a new wire-production facility in Grove Mill, Massachusetts, in a partnership with his twin brother, Charles. This partnership lasted until 1849. At this point Washburn joined forces with his son-in-law, Philip Moen, who purchased a half interest in I. Washburn & Company. Moen became vice-president while Washburn continued to serve as president. Despite his administrative duties, Washburn continued his experiments, at this time focusing on the manufacturing of steel wire for piano strings. A new process developed by Washburn allowed the partnership to become a force in the piano string market, which had for years been monopolized by the British. Meanwhile, the commercial demand for crinoline wire and steel wire for sewing machine needles expanded dramatically. To attract capital I. Washburn & Company reorganized in 1850 into a corporation, becoming the I. Washburn & Moen Wire Works, and a new plant was constructed in South Worcester. In 1868 the company again reorganized, merging with the Quinsigamond Iron & Wire Works, a corporation of which Moen was president. The new Washburn & Moen Manufacturing Company was incorporated with a capital of $1 million.

Washburn was appointed president of the new firm, with Moen again serving as vice-president. On December 30, 1868, Washburn died, and Moen assumed the presidency and, in 1875, became treasurer as well. Under Moen's leadership the firm continued to grow, expanding its operations in 1890 to include a rolling mill and wire-producing plant in Waukegan, Illinois, the first Washburn & Moen facility outside New England. At the turn of the century Washburn & Moen remained a major producer of wire products.

Reference:
J. D. Van Slyck, *Representatives of New England* (Boston: Van Slyck & Co., 1879).

Samuel Thomas Wellman

(February 5, 1847-July 11, 1919)

by David B. Sicilia

The Winthrop Group, Inc.

CAREER: Draftsman and engineer, Nashua Iron Company (1865-1867); consulting engineer, Siemens Company and independently (1867-1870); chief engineer and assistant superintendent, Nashua Iron Company (1870-1873); chief engineer and superintendent, Otis Iron & Steel Company (1873-1888); president, Wellman Steel Company (1890-1895); president and later chairman of the board, Wellman-Seaver Engineering Company and its successor companies (1896-1900).

Samuel Thomas Wellman was born in Wareham, Massachusetts, on February 5, 1847, the first child of Samuel Knowlton Wellman, whose ancestors emigrated from Wales to Lynn, Massachusetts, in 1625, and Mary Love Besse, a descendant from early settlers of Farmington, Maine. Samuel K. Wellman was employed at a local ironworks when his son was born. The Wellmans moved to Nashua, New Hampshire, where Samuel K. Wellman took a job as a heater for the Nashua Iron Company in 1850. He would remain there until 1876, rising to superintendent and guiding the company as it grew into one of the region's largest and most prosperous iron and steel manufactories, its plant covering 15 acres and employing as many as 400 men.

Samuel T. Wellman attended public schools and took an interest in his father's vocation. After working at the ironworks and for a local machine shop, he completed a course of study in engineering at Norwich University, Vermont, during 1862 and 1863. He then enlisted in the Union army, where he served as a corporal in Company F of the First New Hampshire Heavy Artillery.

Following his discharge in 1865 Wellman returned to the Nashua Iron Company as a draftsman and engineer. A turning point came in 1866. One of the company's products at that time was steel locomotive tires, which it rolled from British

Samuel Thomas Wellman

hammered steel blooms. To improve this process Nashua Iron had ordered a Siemens regenerative gas furnace, the second to be licensed in the United States (a rolling mill in Troy, New York, had been first). Wellman was directed by his father to construct the furnace from plans supplied by the Siemens Company.

The Siemens furnace had been developed by William C. and Frederick Siemens of Germany during the 1850s and was known as the regenerative furnace because it utilized its own exhaust gases to preheat its air supply. Hot gases from combustion were channeled through a checkerboard refractory brickworks in one chamber, while intake air was drawn through a second chamber, constructed and preheated in the same way. This process was alternated continuously between the two chambers, resulting in unprecedented high temperatures as well as fuel reductions of as much as 80 percent. When gas was introduced as a fuel in 1861, the thermal efficiency of the Siemens furnace became still more dramatic. In 1862 several European firms experimented with the Siemens design, and a few unauthorized units were constructed in the United States to manufacture copper, glass, and other materials. The most successful modifications of the Siemens furnace design were made at the Sirieul Works near Paris by Pierre and Emile Martin, after which the technology was known as the Siemens-Martin process.

As Wellman later recalled the Nashua incident: "The drawings were sent over and turned over to me, who had never seen a drawing of that kind before. . . . I had just finished the furnace and had a drying-out fire in it when a big black-whiskered Englishman walked into the office and announced that he was the Siemens engineer who had been sent from England to build this particular furnace." Impressed with the thoroughness and precision of Wellman's work, the Siemens engineer, J. T. Potts, offered the young engineer a position as his assistant in carrying out Siemens installations at other locales, an offer that was eagerly accepted.

Wellman's first assignment with Potts was to supervise the start-up and operation of the first crucible steel furnace in the United States at Anderson, Cook & Company of Pittsburgh. As hoped, the furnace permitted a tremendous savings of fuel; whereas the old furnaces had consumed three tons of prime coke (costing $2-$10 per ton) to melt a ton of steel, the Siemens furnace performed the same task with a half ton of nut coal (about $1 per ton).

After a few months Wellman was reassigned to Cooper, Hewitt & Company of Trenton, New Jersey. Cooper, Hewitt had attempted unsuccessfully to manufacture Bessemer steel in the 1850s and since 1866 had produced steel-headed rails using an adaptation of the Mushet-Heath process (in which wrought iron and special carbonized irons were fused in a furnace). Less expensive than imports and capable of being rerolled, these iron and steel rails found a ready market, prompting Abram S. Hewitt to seek more efficient methods of production while serving as commissioner for the United States at the Paris Exposition of 1867.

Hewitt became interested in the possibility of using a Siemens regenerative furnace to produce steel by heating it externally in a shallow vessel or open-hearth. By increasing the heat beyond the melting temperature of wrought iron, the open-hearth furnace might eliminate the need to manipulate the metal to remove carbon. Hewitt obtained the American rights to the process and sent the firm's engineer, Frederick J. Slade (brother-in-law of American Bessemer steel pioneer Alexander L. Holley), to France to study the technology. Siemens's American agents drew up the plans for a 5-ton unit, and Wellman arrived in time to assist with the start-up and operation of the furnace in 1868.

The installation at Cooper, Hewitt & Company was plagued with difficulties. Wellman was able to solve problems with the gas-producing apparatus, but other troubles were more persistent. The bath was comprised of franklinite pig iron, puddled bars, and scrap steel. In the absence of ferromanganese, franklinite had to be used for recarbonization, resulting in a steel that had to be rolled at high temperatures and was brittle. In addition there was much to be learned about working the molten metal at such unprecedented high temperatures. The men had to work within an uncomfortably narrow margin between the casting temperature of the steel and the melting temperature of the furnace itself, and keeping the tap on the crucible from chilling and hardening was a tricky task. After a year or two of sporadic operations, Cooper, Hewitt abandoned the project and returned exclusively to the Mushet-Heath method.

Wellman, however, remained confident about the open-hearth process and soon applied the knowledge he had gained in Trenton toward a second attempt in a new setting. Ralph Crooker of the Bay State Iron Works in South Boston, an iron man with more than half a century of experience, was also interested in applying the open-hearth process to the manufacture of steel-headed rails. After an unsuccessful attempt to secure plans for a furnace from Siemens's agents, Crooker sought help from Wellman, who by this time had been working independently from the Siemens Company. Wellman accepted. Working from plans supplied by French engineers, he completed a 5-ton unit at Bay State in 1870.

It was soon apparent that the Boston works had become an American technological milestone—the first commercially successful open-hearth steel plant in the nation. Wellman had improved the design of the furnace in several ways, making it "a very powerful one for its size, sharp and quick in its working." He also deepened the bath and developed an ingenious turntable apparatus to carry the ingot molds. More importantly, Wellman solved the ferromanganese problem by installing several auxiliary units to produce the substance, a first in the United States. It was an expensive process, but only small qualities of ferromanganese were consumed, and the quality of the product was improved significantly. (When the market later shifted in favor of all-steel rails because of their superior strength and stiffness, Bay State successfully used its open-hearth furnace to produce flange plates for boilers and fireboxes.)

By this time Wellman was busy with several Siemens open-hearth installations, such as for Singer, Nimick & Company of Pittsburgh and the Chrome Steel Company of Brooklyn. Although his own accounts of the chronology of these early installations differ, it is clear that in 1870 he returned to the Nashua Iron Company, where he was instructed to build and operate a new steel department comprised of a 5-ton open-hearth furnace and a plate and bar mill. Construction began in 1871 and was completed the following year.

Again, Wellman made important modifications. He improved casting with an elaborate ladle arrangement and eliminated the preheating furnaces. Wellman continued to display an almost obsessive dedication to the work, later recalling that when no steel had been produced after the first week, "I went home, that Saturday night, sick in mind and body. I do not remember the sermon that Sunday; but before Monday I found out the trouble ... [and] from that time on, we had no serious trouble in making any steel we wanted." The Nashua Iron & Steel Company (renamed in 1872) made open-hearth steel for railroad and marine forgings, boiler and firebox plates, and, later, locomotive and car wheels. Wellman remained at the Nashua company for three years, serving as chief engineer and filling the position of superintendent once occupied by his father.

During summer 1873 Wellman began work on an open-hearth furnace for the Otis Iron & Steel Company of Cleveland. The company recently had been organized for $300,000 by Charles Augustus Otis, who had sold his iron-forging business after the Civil War but now hoped to profit from the new open-hearth technology. On Wellman's recommendation, Alexander Holley was appointed consulting engineer for the project, which began operations in October 1874. Holley's design included important improvements in materials handling, among them a raised charging platform and a crane-suspended ladle.

Wellman stayed at the Otis works as chief engineer and superintendent for 13 years, a period that for him proved exceptionally creative and productive. During this time he was granted many of the nearly 100 patents he would receive during his career for the invention of iron and steel machinery and other equipment. These included the Wellman hydraulic crane in 1878 and the Wellman electric open-hearth charging machine in 1888. The charging machine, used for feeding white-hot steel into open-hearth furnaces, solved one of the greatest economic problems with the open-hearth method, the high cost of labor for furnace charging.

Moreover, while at Otis, Wellman installed the first successful basic open-hearth apparatus in America. In basic steel production the acid lining of the refining vessel was replaced with a basic lining (an oxide of a metal), which allowed phosphorus in iron to combine with the lining and be discharged with the slag. This innovation was one of the key factors leading to the dominance of open-hearth over Bessemer steel in the United States, where high-phosphorus ores were common.

Wellman observed the basic open-hearth method (developed by Percy Gilchrist) while in Europe during the 1885. He arranged for a shipment of Styrian magnetite to the United States and installed a lining of calcined magnetite on a silica bottom. The Otis works produced the first basic open-hearth steel in the United States in 1886, and for a time, Wellman explained, the process "was kept a great secret." Still, not entirely pleased with its basic steel, the Otis company returned to the acid process after producing only 1,000 ingots of the metal, and basic steel was not manufactured continuously until the Homestead works took up production in 1888 after important patent disputes were settled.

Wellman left the Otis company on the first day of 1889 and for a short time served as a consulting engineer at the Illinois Steel Company. In 1890 he joined with his half brother, Charles, in forming the Wellman Steel Company, which took over the works of the Chester Rolling Mills of Chester, Pennsylvania. Wellman served as president of the new company until its failure in 1895.

Undaunted, the Wellman brothers and John W. Seaver formed in 1896 a consulting engineering firm specializing in iron and steel manufacture, the Wellman-Seaver Engineering Company. This company was succeeded by the Wellman-Seaver-Morgan Engineering Company in 1902, which in turn was incorporated in Ohio the following year as the Wellman-Seaver-Morgan Company to acquire the property and business of Webster, Camp & Lane of Akron. Wellman served as president and chairman of these enterprises during various years, building them into leading manufacturers of mining machinery and coal and ore handling equipment. He retired in 1900 and died 19 years later on July 11, 1919, in Stratton, Maine, while on his way to a hunting camp.

The great steel magnate Charles Schwab once characterized Samuel T. Wellman as "the man who did more than any other living person in the development of steel." Indeed, Wellman contributed to the development of the American steel industry in several important ways. By understanding the open-hearth process better than any other American, he was able to make critical design changes and incremental improvements that made the new technology commercially viable. In this way Wellman played a role similar to that of Alexander Holley, who successfully introduced the Bessemer process into the United States. In addition, by repeating this process at several key steel manufactories, Wellman served as the focal point for the diffusion of open-hearth steel making in America. For example, more than 1,000 steel makers learned the open-hearth method at the Otis works during Wellman's tenure.

Moreover, Wellman made significant contributions not only to the open-hearth process itself (through his pivotal early modifications as well as later innovations such as the rolling open-hearth furnace in 1895) but also to critical auxiliary functions in iron and steel manufacture. Among his key innovations in materials handling was an electromagnet for handling scrap steel and pig iron (patented in 1895). Both the charging machine he developed while at Otis and the Wellman electromagnet became standard equipment in large open-hearth steel plants throughout the United States, Great Britain, France, Japan, Spain, Egypt, Germany, and elsewhere. Wellman estimated that these innovations saved hundreds of millions of dollars in operating costs per decade during his later life, noting that "every open-hearth plant of any size in the world today is equipped with these inventions, and they are considered as much a necessary part of the equipment as the furnace itself."

Thanks to the efforts of Wellman and other open-hearth pioneers, the open-hearth method surged in popularity after 1880 and eventually overtook Bessemer as the leading method of steel making in the United States. Whereas the Bessemer process relied on the burning of carbon and silica within the iron ores, open-hearth was more suited to the remelting of scrap iron and steel and could rely on the phosphoric ores common to North America. Although slower than the dramatic blasts of the Bessemer, the open-hearth process could be controlled and sampled more easily, and many experts also believed it yielded higher quality products.

Between 1869 and 1900, as demand for specialty steel products and the availability of scrap increased markedly, total output of open-hearth steel surged from 893 to 3.4 million tons, while the average size of open-hearth furnaces grew from 5 to 50 tons. Although Bessemer steel production stood at 6.7 million tons at the turn of the century, it was surpassed by open-hearth output in 1908, and by 1937, nine-tenths of all American steel was produced using the open-hearth process.

Wellman married Julia Almina Ballard of Stoneham, Massachusetts, on September 3, 1868; the couple had two daughters and three sons, the latter playing key roles in related family businesses: W. S. Wellman as president of Wellman Products Company; and M. C. and F. S. Wellman at the Wellman Bronze Company. A Republican and a dedicated Congregationalist (active in the Euclid Avenue Congregational Church of Cleveland), Wellman was a member of the American Society of Civil Engineers, the American Institute of Mining Engineers, the British Institute of Mechanical Engineers, the British Iron and Steel Institute, and he served terms as vice-president and president of the Cleveland Engineering Society. He also joined the American Society of Mechanical Engineers in 1881, where he served as manager from 1885 to 1888, as vice-president from 1896 to 1898, and as president from 1900 to 1901.

Selected Publications:

"Machinery for the Charging of Heating- and Melting-Furnaces," *Transactions* of the American Institute of Mining Engineering, 19 (1891): 313-317;

"The Early History of Open Hearth Steel Manufacture in the United States," *Transactions* of the American Society of Mechanical Engineers, 23 (1902);

George Westinghouse (New York, 1914).

References:

Joseph G. Butler, Jr., *Fifty Years of Iron and Steel* (Cleveland: Penton Press, 1923);

Victor S. Clark, *History of Manufactures in the United States, 1860-1914* (Washington: McGraw-Hill, 1928);

Jeanne McHugh, *Alexander Holley and the Makers of Steel* (Baltimore: John Hopkins University Press, 1980);

Edward E. Parker, ed., *History of the City of Nashua, N. H.* (Nashua, N.H.: Telegraph Publishing, 1897);

Fritz Redlich, *History of American Business Leaders: A Series of Studies*, volume 1 (Ann Arbor: Edwards Brothers, 1940);

Peter Temin, *Iron and Steel in Nineteenth-Century America: An Economic Inquiry* (Cambridge, Mass.: M.I.T. Press, 1964);

Wellman Engineering Company, *Wellman Engineering Co: A Pictoral Record of Engineering Achievements* (Cleveland: Wellman Engineering Company, 1944);

Wellman-Seaver-Morgan Company, *The Open Hearth: Its Relation to the Steel Industry, Its Origin and Operation* (Cleveland: Wellman-Seaver-Morgan Company, 1920);

Victor Windett, *The Open Hearth: Its Relation to the Steel Industry; Its Design and Operation* (New York: U.P.C. Book Company, 1920).

Joseph Wharton

(March 3, 1826-January 11, 1909)

by John W. Malsberger

Muhlenberg College

CAREER: Clerk, Waln & Leaming (1845-1847); partner, Wharton Brothers (1847-1851); manager, Lehigh Zinc Company (1853-1863); director, Saucon Iron Company (1857-1861); director, Bethlehem Iron Company (1861-1899); partner, Wharton & Fleitmann (1864-1866); owner, nickel-mining operations (1866-1909); director, Bethlehem Steel Corporation (1899-1909).

A noted metallurgist, businessman, financier, and philanthropist of nineteenth-century Pennsylvania, Joseph Wharton used his skills to develop a wide range of business and industrial interests. The first American to succeed in producing commercial quantities of metallic zinc and nickel, Wharton also played a prominent role in the iron and steel industry. As the largest single stockholder in the Bethlehem Iron Company and as one of its directors, Wharton was instrumental in diversifying and in expanding the firm to become one of the leading American iron and steel manufacturers. Wharton also became a manufacturer of iron in his own right, constructing blast furnaces in northern New Jersey and purchasing coal mines, ore beds, and coke ovens to become one of the largest individual ironmasters in the United States. A strong believer in education and the work ethic, Wharton founded the Wharton School of Finance at the University of Pennsylvania

Joseph Wharton

to educate people to become productive members of society. Thus, as scientist, businessman, and philanthropist, Joseph Wharton made signal contributions to the advancement of the American industrial economy in the nineteenth century.

The fifth child of William and Deborah Fisher Wharton, Joseph Wharton was born into a life of privilege on March 3, 1826. The Whartons were one of the oldest and wealthiest families in nineteenth-century Philadelphia. The progenitor of the family, Thomas Wharton, had arrived in Philadelphia about 1685; when he died in 1718, Wharton was one of the wealthiest men in the city. Because of the family's great wealth, William Wharton, Joseph's father, was able to devote his adult life to philanthropy. Deborah Fisher Wharton, Joseph's mother, was a descendant of John Fisher, who came to America with William Penn in 1682 on the first voyage of the ship *Welcome*.

Unlike most young men in early-nineteenth-century America, Joseph Wharton received a good deal of formal education. Trained initially at the Friends' School, a Quaker establishment regarded as one of the finest schools in Philadelphia, Wharton also prepared for college at a select school in the city run by Frederick Augustus Eustis. Wharton's goal was to enroll at Harvard University, but ill health forced him to abandon the idea. Instead, in 1842 at age sixteen Wharton went to work on the Chester County farm belonging to Joseph S. Walton, a family friend, hoping that the experience would restore his health. Wharton spent the next three years of his life learning the rudiments of agriculture while his family arranged to continue his formal education. Ever the child of privilege, Wharton spent the growing season on the farm, but for three months each winter he returned to Philadelphia in order to study chemistry in the laboratory of Martin H. Boyé during the day and French and German in the evening. The training he received, especially in chemistry, figured prominently in determining the path of his career.

The three years of farm life served their recuperative purpose, allowing Wharton in 1845 to direct his educational efforts to an entirely new tack. Like many young men of his generation, Wharton was attracted by the powerful allure of the commercial spirit that dominated Jacksonian America. At age nineteen he entered the business world by accepting employment at the Philadelphia dry goods store of Waln & Leaming. Wharton's position was largely educational, however, for he worked two years without compensation, seeking mainly to learn bookkeeping and general business procedures. Eventually he became so adept as a bookkeeper that he was put in charge of more than 800 ledger accounts. After two years in the dry goods business Wharton's education was largely complete, and he was ready to strike out on his own. As a member of one of Philadelphia's wealthiest families, he had had the luxury of developing well-rounded training in both science and commerce, an opportunity that few of his era enjoyed. Wharton demonstrated quickly, however, that he was not a dilettante, but rather a talented metallurgist and shrewd businessman.

The first venture that combined Wharton's scientific and entrepreneurial abilities came in 1847 when he formed a partnership with his eldest brother, Rodman, to establish a plant to manufacture white lead, then the chief ingredient used in paint. After several years of successful operations the Wharton brothers sold their business to the Philadelphia firm of John T. Lewis & Brothers. The event which most directly influenced Wharton's business career, however, occurred in 1853 while he was making a horseback trip through eastern Pennsylvania. During the trip Wharton stopped at Friedensville, near Bethlehem, to inspect a zinc mine operated by an unincorporated group of Philadelphia capitalists which at the time was supplying ore to two local entrepreneurs for the manufacture of white paint. Wharton convinced the capitalists that his metallurgical training, together with his experience in the manufacture of paint, equipped him to manage their firm most efficiently. He was hired later in the year at an annual salary of $3,000. In 1855, largely through Wharton's efforts, the firm was incorporated as the Pennsylvania & Lehigh Zinc Company (later the Lehigh Zinc Company) for the mining of ore and the manufacture of white paint. Wharton's tenure as manager was interrupted, however, by the panic of 1857, which forced the company into receivership. Seeing an opportunity, Wharton persuaded Lehigh Zinc's directors to lease the entire firm to him, and within several months Wharton had reorganized the company, restored it to a profitable basis, and resumed his former position as manager. When he left the reorganized Lehigh Zinc in 1863 to pursue new interests, Wharton's gamble had netted him a profit of more than $30,000.

Fresh from his triumph at Lehigh Zinc, Wharton embarked on a new venture that tested his scientific training, the manufacture of metallic zinc or spelter from the ore mined at Friedensville. Despite the failure of several earlier efforts to produce this form of zinc in America, Wharton was convinced that his background in chemistry would enable him to succeed. After convincing the directors of Lehigh Zinc to grant him exclusive authorization to experiment with the process, Wharton in 1859 hired the services of Louis De Gée of De Gée, Gernant & Company of Liege, Belgium, to superintend construction of the works. Over the next five years Wharton imported 60 zinc workers from Belgium and at his own expense built 16 furnaces capable of using anthracite, rather than bituminous, coal to reduce the silicate of zinc to metal. By the time the terms of his initial agreement with the zinc company expired in 1863, Wharton's efforts had succeeded beyond his wildest expectations, producing more than 9 million pounds of spelter. Following his great success, however, the Lehigh Zinc Company refused to renew their agreement with Wharton, believing that they could run the spelter manufacturing business more profitably by themselves. Thus, at age thirty-seven Wharton was cut loose by his employer, free to direct his talents to new challenges.

Wharton turned his abilities next to the manufacture of a malleable form of nickel, a metal which up to that time was not in wide use commercially because of its brittleness. Learning that the U.S. government was in search of a larger and more dependable supply of nickel for its mint in Philadelphia, Wharton in 1864 formed a partnership with Dr. Theodore Fleitmann to meet this need. For the next two years Wharton & Fleitmann successfully operated both a nickel manufacturing plant in Camden, New Jersey, and the Gap Nickel Mine in Lancaster County, Pennsylvania, making Wharton the first American to manufacture commercial-grade nickel. When their manufacturing plant was destroyed by fire in 1866, Wharton bought out his partner and expanded the business to become the largest producer of nickel in America. His nickel was renowned for its purity and uniformity, and for many years Wharton produced one-sixth of the world's output of nickel. Wharton made huge profits with this business, moreover, when, at the conclusion of the Franco-Prussian War in the early 1870s, he secured a contract to supply nickel to the Prussian government for coinage. His achievements in advancing the manufacture of nickel were recognized at the Paris Exposition of 1878 where he was awarded a gold medal.

Wharton's connection to the iron and steel industry dates to the years he spent in South Bethlehem in the late 1850s and early 1860s managing the manufacture of white paint and spelter for the Lehigh Zinc Company. At this time Bethlehem, as well as the entire Lehigh Valley, was in the midst of a great iron-making boom. Only several years before Wharton assumed the management of the zinc company, David Thomas, the Welsh ironmaster, had built in nearby Catasauqua one of the first blast furnaces in America capable of producing commercial quantities of iron using anthracite coal. The success of Thomas's furnace, together with the abundance of anthracite coal in the region and the transportation network of railroads and canals that linked it with the markets of Philadelphia, had by the late 1850s produced a spate of iron furnaces, rolling mills, and foundries throughout the Lehigh Valley. Perhaps the most notable of these iron companies was the Bethlehem Iron Company (later Bethlehem Steel Corporation), which was constructed in the early 1860s near Wharton's spelter manufactory.

Because of his metallurgical experience and shrewd business sense, Wharton recognized through this exposure in the Lehigh Valley the great potential of the iron and steel industry in America. Accordingly, he began in the 1860s to acquire numerous holdings within the industry. Wharton, for instance, constructed his own blast furnaces in northern New Jersey, which at their peak produced 1,000 tons of pig iron daily, and also purchased the Andover Iron Company of Phillipsburg, New Jersey. To assure that his furnaces would have constant access to the necessary raw materials, Wharton built his own coke ovens and purchased over 5,000 acres of iron ore land throughout the Northeast, as well as 7,500 acres of coal deposits in Indiana County, Pennsylvania, and 24,000 acres of coalfields in West Virginia and northern New York. These holdings made Wharton one of the largest individual ironmasters in late-nineteenth-century America.

In addition to developing his own iron manufacturing facilities, Wharton took a strong financial interest in the Bethlehem Iron Company. One of the original investors in the Saucon Iron Company, the forerunner of Bethlehem Iron, Wharton was so impressed by the mechanical skills of John Fritz, the young engineer hired to superintend the Bethlehem firm, that he expanded his holdings steadily as the company grew. Ultimately, Wharton became the largest single share owner in Bethlehem Iron as well as one of its directors, enabling him to play an important role in shaping the company's development. Wharton's most important contribution to the growth of Bethlehem Iron came in the early 1880s when the U.S. government decided to modernize its navy. Because of his metallurgical experience, Wharton was convinced by the arguments of the company's general superintendent and chief engineer, John Fritz, that the future of the American navy lay in steel armor-plated vessels with guns of large caliber. Thus, he believed Bethlehem Iron had a rare opportunity to expand into this relatively new field. In 1886 Wharton persuaded his more conservative fellow directors, who only several years earlier had vetoed the development of a structural steel plant, to authorize the construction of a mill to manufacture armor plate and heavy castings and forgings, one of the first to be built in America. Later that year when the government asked for proposals on armor plate, Bethlehem Iron was the only company in the position to bid on all the contracts. When it received $4.5 million in government contracts Wharton traveled to England and France to study their methods of armor plate and gun manufacture and was one of the individuals most responsible for introducing these techniques at Bethlehem Iron. In 1901, when a syndicate headed by Charles Schwab sought to purchase the Bethlehem firm, Wharton was empowered by the board of directors to conduct and complete the transaction. Thus Wharton's metallurgical knowledge and encouragement contributed substantially to the development of Bethlehem Iron from a small manufacturer of iron rails into the principal manufacturer of steel and nickel-steel armor plate and gun forgings, all of which were essential to the construction of the modern United States Navy and to American shipbuilding.

In addition to his iron, steel, and nickel holdings, Wharton also had a wide range of business interests. He was the owner of copper mines around Lake Superior which provided the ore for his stamping mill, known initially as the Delaware Mining Company and later as the New Jersey Mining Company. Wharton also developed and was the principal owner of the Menhaden Fisheries which operated a fleet of fishing boats and processing plants all along the Atlantic Coast.

Like his father before him, Wharton in his later life devoted a considerable portion of his time to philanthropy. A firm believer in the work ethic, Wharton became a strong opponent of proposals for government welfare made by reformers of the late nineteenth century in their efforts to adapt society to the impact of industrialization. Believing that all Americans could become useful members of society through education, Wharton donated $500,000 to the University of Pennsylvania in May 1881 to create the school of finance which bears his name. Similarly, Wharton was one of the founders of Swarthmore College and served as chairman of its board of managers from 1883 until his death. Together with New York businessman Samuel Willetts, Wharton established a scientific laboratory at Swarthmore and personally endowed a chair in history and political economy.

Politically Wharton was a lifelong Republican, serving as a presidential elector for William McKinley in 1896 and agreeing, like most within the iron and steel industry, with the party's advocacy of high protective tariffs. Believing that high tariffs were essential for the protection and stimulation of home industry, Wharton helped to organize the Industrial League of Philadelphia in the late 1860s to promote this cause. When this organization merged with the American Iron and Steel Association in 1875, Wharton was elected its first vice-president. The Philadelphia metallurgist was also honored for his contributions to American life by election to the Academy of Natural Sciences and the American Philosophical Society.

Selected Publications:

Project for Reorganizing the Small Coinage of the United States of America (Philadelphia: King & Baird, 1864);

Speculations Upon a Possible Method of Estimating the Distance of Certain Variably Colored Stars (New Haven, Conn., 1865);

The National Banks, Greenbacks, Resumption (Philadelphia, 1868);

Observations Upon Autumnal Foliage (New Haven, Conn., 1869);

International Industrial Competition (Philadelphia: H. C. Baird, 1870);

National Self-Protection (Philadelphia: American Iron and Steel Association, 1875);

Memorandum Concerning Small Money: With Illustrations and Descriptions of Existing Nickel Alloy Coins (Philadelphia: Collins, 1876);

The American Ironmaster (Pittsburgh, 1879);

The Industrial League to its Constituents, by Wharton, Henry C. Lea, and Wharton Barker (Philadelphia, 1879);

Brief Statement of the Action in Behalf of the Eaton Bill for a Tariff Commission (Philadelphia, 1881);

The Duty of Nickel (Philadelphia, 1883);

Is a College Education Advantageous to a Business Man? (Philadelphia, 1890);

The Poteophone (Philadelphia: Allen, Lane & Scott, 1890);

The Creed in the Discipline (Philadelphia, 1892);

Dust From the Krakatoa Explosion of 1883 (N.p., 1894);

The Water Supply of Philadelphia and Camden (Philadelphia, 1895);

Mexico (Philadelphia: Lippincott, 1902);

Poseidon in America (N.p., 1905);

The Last Will and Codicils of Joseph Wharton, Deceased (Philadelphia: Lippincott, 1909);

Speeches and Poems by Joseph Wharton, Collected by Joanna Wharton Lippincott (Philadelphia: Lippincott, 1926).

References:

Frederic A. Godcharles, *Pennsylvania: Political, Governmental and Civil* (New York: American Historical Publishing, 1933);

The Manufactories and Manufacturers of Pennsylvania of the 19th Century (Philadelphia: Galaxy Publishing, 1875);

W. Ross Yates, *Joseph Wharton: Quaker Industrial Pioneer* (Bethlehem, Pa.: Lehigh University Press, 1987).

John Wilkeson

(unknown-April 4, 1894)

by Marc Harris

Ohio State University

CAREER: Partner and operator, Arcole Furnace (1830-1845?); partner and operator, Wilkes, Wilkeson & Company (1845-1851?); owner, Arcole Iron Company (1852-1894).

John Wilkeson moved to northeastern Ohio with several of his brothers to help his Buffalo-based family enterprises exploit local iron deposits; in the process, he developed one of the largest western charcoal iron furnaces then in operation. Later in his career Wilkeson built a new furnace which was the first in the West to run continuously on raw bituminous or block coal. In so doing he helped to lay the groundwork for continued operation and expansion of the Mahoning Valley, Ohio, iron industry.

John Wilkeson was of the second generation of Wilkesons to participate in the industrial development of the Great Lakes region. His father, Judge Samuel Wilkeson, played a pivotal role in the development of Buffalo, New York, after the War of 1812 and started the family businesses which John and his brothers ran and coordinated in the eastern Lakes region.

The elder Wilkeson was of a Scotch-Irish family which had moved to Delaware in 1760 and was staunchly Presbyterian and strongly anti-British in

John Wilkeson (courtesy of Buffalo and Erie County Historical Society)

feeling. Born in Carlisle, Pennsylvania, in 1781, while his father served out a Revolutionary enlistment, Samuel Wilkeson grew up on the family farm in western Pennsylvania. Like his father, Judge Wilkeson also staked out a pioneer farm, this one in southeastern Ohio. But after what seemed a particularly interminable day clearing trees from his new farmstead, he decided to throw over the whole enterprise and take up a different career. Ambitious and impatient, he no longer wanted to wait the long years it would take to see his farm labors succeed. Like so many men on the new nation's frontiers, Samuel Wilkeson turned to commerce to better his fortunes and speed his economic ascent.

He enjoyed reasonable success as a frontier merchant but did not attain great wealth. Wilkeson traveled between Pittsburgh and Buffalo, trading manufactured goods for salt from the Onondaga region. After a time he added a small fleet of boats on Lake Erie to his enterprises, but in 1812 the lakes trade ceased with the outbreak of war. Wilkeson served in a militia unit assigned to defend Buffalo from invading British troops. They failed, and the victors burned Buffalo to the ground as the defenders retreated to safety. Samuel Wilkeson, however, recognized a good commercial site when he saw one and decided to return to Buffalo before the war ended and help rebuild and develop the town.

A large, forceful man, Wilkeson quickly made his presence felt as a public official. Early justices of the peace were as much constable as judge, and Wilkeson's bulk aided him in his tasks. His greatest success, however, came as the constructor of Buffalo's harbor during summer 1820. He and others originally contracted for an experienced builder to construct the harbor, but the builder reneged. Wilkeson took over the project himself in order to save the state loan that had been procured to finance the project. Two years later he took the lead in persuading DeWitt Clinton to terminate the Erie Canal at Buffalo rather than neighboring rival Black Rock and thus helped to assure the city's commercial future.

Having secured the canal terminus, Wilkeson's own business grew along with the town. He had begun with a general store, soon added a small fleet of lake boats, and then concentrated on developing industry in Buffalo. Iron was a particular interest, especially cast iron products; pots, kettles, and other cast goods found a ready market on the expanding frontiers of the time. He established the first iron foundry in Buffalo, producing steam engines and stoves. Probably through his knowledge of Lake Erie commerce, Judge Wilkeson learned of good bog ore deposits near Madison in Geauga County, Ohio, owned and worked by the Erie Furnace Company. Erie began working the deposits in 1826, and four years later Wilkeson added the operation to his business interests. He renamed it the Arcole Furnace. John Wilkeson, Judge Samuel's oldest son, moved west from Buffalo in 1830 to represent his father's interest in the Arcole Furnace and to merchandise its products and supervise the company store. Arcole produced simple castings for sale in the lower Lakes region and supplied the pig iron that fed Judge Wilkeson's Arcole foundry in Buffalo.

The Wilkeson family's widespread business interests resembled many others of the time. In the pre–Civil War United States it was common for families to provide both capital resources and commercial contacts. Family members who moved to developing areas around the Great Lakes established banks, stores, and other enterprises, drawing on family capital and business contacts in older areas; western settlers served as forwarders, commission agents, and salesmen for family businesses. Such practices allowed capital to move where it could secure a return and created reliable commercial networks at a time when most businesses operated as partnerships, and personal contact and knowledge of character were the bases of business relationships.

In its heyday the Arcole company was a very large operation; it was said to be the largest single industrial installation in Ohio in 1835 when it was reorganized and capitalized at $100,000. The firm produced between 1,000 tons and 1,500 tons of iron annually, employing 200 men to chop wood for charcoal, another 150 to wash the ore, and 200 molders. During its operating years the company employed as many as 2,000 men, although many of them must have been farmers working part-time. The company store's annual billing reached $150,000 before the onset of the panic of 1837.

For a short time the Arcole Furnace made Madison and its nearby port prosperous towns. The company's payroll dominated the local economy, and its notes passed as currency. The Erie Furnace Company had built its stacks near a small creek with a good harbor; the Arcole company shipped in limestone from mines on Kelley's Island, near Sandusky, through the port. This source of supply demonstrates how fragmented the Western Reserve iron industry was before 1839. Furnaces in the Mahoning Valley, about 50 miles away, relied on plentiful local supplies of good limestone, but land transportation apparently drove the cost of local limestone too high to compete with supplies obtained more cheaply from farther away by water. In 1839, when the newly completed Pennsylvania & Ohio Canal linked the Mahoning Valley with the Ohio Canal's main line, cheap water transport began to knit the region's industry together by allowing Cleveland ironmakers to use Mahoning Valley coal and the Valley to use Great Lakes ore supplies.

Perhaps because coal became cheaply available just as charcoal was running out, John Wilkeson began experimenting in 1840 with raw coal as a furnace fuel. He did not adopt it, however, and the Geauga County iron industry continued to rely on charcoal and local bog ore deposits. By 1850 those resources ran out. In the late 1840s—probably because of his father's death in 1848—Wilkeson sold the Arcole Furnace to the one company left in the county, the Geauga Furnace, which had better access to the lake and to outside supplies; Arcole's port had begun silting up, and the creek itself was too small to handle substantial ore shipments. Then, too, Wilkeson's own experiments with coal eventually helped to make the Arcole Furnace unable to compete with emerging firms.

Wilkeson's interests had shifted to the Mahoning Valley even before he sold the Arcole works. In 1845 he and his brother-in-law Frederick Wilkes began, as Wilkes, Wilkeson & Company, to build a new blast furnace in Lowellville. They received financial backing from Judge Wilkeson and from David Tod, a lawyer and politician who owned land and mines at Brier Hill, Ohio. The furnace was designed to use raw coal as the only fuel; Lowellville was known to have reserves of limestone and was also near Mount Nebo and Brier Hill coal mines. Both mines, along with others, had yielded only a small output until the canal's opening made wider markets available.

The question of who first blew in a furnace on raw coal has led to some controversy. Wilkeson's 1840 trials were probably not the only ones, because few furnaces of the time used coke in their search for alternatives to charcoal. Before builders completed the Wilkes, Wilkeson furnace in

1846, Himrod & Vincent of Mercer County, Pennsylvania, successfully used uncoked coal in their Clay furnace, though they did not use the process exclusively. This stack was overseen by a former Arcole employee of Wilkeson's who knew about the 1840 trials. Joseph Butler, a steel man who researched the question near the turn of the century, declared that David Tod, the mine owner, deserved the real credit because of his stake in Wilkeson's Lowellville furnace. The stack was blown in 1846 and produced up to 20 tons of iron a day at peak levels, using block coal. The iron was demonstrated to be workable by James Ward, who owned a rolling mill in nearby Niles.

Along with the nearly contemporaneous discovery of black-band iron ore in Tod's mine (in which Ward also had a hand), Wilkeson's furnace inaugurated a new era in midwestern iron production. The Mahoning Valley industry took on new life and expanded even though its charcoal supplies and bog ores had almost played out. The coal itself, a clean-burning variety unsuited to coking, was thought to be inexhaustible but began to run short after the end of the Civil War; Mahoning furnaces began to use Connellsville coke and Lake Superior ore in increasing proportions after 1867, following the more familiar modern pattern.

John Wilkeson operated the firm's blast furnace for several years. To assure adequate supplies of coal, the company bought the Mount Nebo mine and worked it for both coal and black-band ore until deep water forced its abandonment. Wilkeson himself gave up control of the furnace in about 1852. The new owner improved its construction and operated it until 1864; after two later sales it was operated by Ohio Iron & Steel Company, a supplier to Youngstown Sheet & Tube.

Little is known of John Wilkeson's subsequent career. He lived in Buffalo from 1852 until his death and participated in the Buffalo Historical Society's affairs. It is likely that he had returned there about the time he sold the Lowell furnace in order to take up his interest in his father's businesses.

References:

Joseph G. Butler, Jr., *History of Youngstown and the Mahoning Valley* (Chicago and New York, 1921);
The Commercial Advertiser Directory for the City of Buffalo (Buffalo: Jewett, Thomas, 1852);

Catharine C. DeWitt, "Bog Iron in Lake County," *Historical Society Quarterly* (Lake County), 13 (August 1971): 1-7;

History of Trumbull and Mahoning Counties (Cleveland, 1882);

James M. Swank, *History of the Manufacture of Iron in All Ages*, second edition (Philadelphia: American Iron and Steel Institute, 1891);

Samuel Wilkeson, Jr., "Biographical Sketch," *Publications of the Buffalo Historical Society*, 5 (1902): 135-146.

Contributors

Brady Banta–*Louisiana State University*
Allida K. Black–*George Washington University*
Stephen H. Cutcliffe–*Lehigh University*
Ernest B. Fricke–*Indiana University of Pennsylvania*
Northcoate Hamilton–*Columbia, S.C.*
Marc Harris–*Ohio State University*
John A. Heitmann–*University of Dayton*
John N. Ingham–*University of Toronto*
Alec Kirby–*George Washington University*
Stuart W. Leslie–*Johns Hopkins University*
John W. Malsberger–*Muhlenberg College*
Paul F. Paskoff–*Louisiana State University*
Terry S. Reynolds–*Michigan Technological University*
Larry Schweikart–*University of Dayton*
Bruce E. Seely–*Michigan Technological University*
David B. Sicilia–*Winthrop Group, Inc.*
Julian Skaggs–*Widener University*
Frank Whelan–Morning Call, *Allentown, Pa.*

Index

The following index includes names of people, corporations, organizations, laws, and technologies. It also includes key terms such as *anthracite coal, labor,* etc.

A page number in *italic* indicates the first page of an entry devoted to the subject. *Illus.* indicates a picture of the subject.